Management Practices in High-Tech Environments

Dariusz Jemielniak
Kozminski Business School, Poland

Jerzy Kociatkiewicz
University of Essex, UK

Information Science REFERENCE

INFORMATION SCIENCE REFERENCE

Hershey · New York

Acquisitions Editor:	Kristin Klinger
Development Editor:	Kristin Roth
Senior Managing Editor:	Jennifer Neidig
Managing Editor:	Jamie Snavely
Assistant Managing Editor:	Carole Coulson
Copy Editor:	Larissa Vinci
Typesetter:	Sean Woznicki
Cover Design:	Lisa Tosheff
Printed at:	Yurchak Printing Inc.

Published in the United States of America by
Information Science Reference (an imprint of IGI Global)
701 E. Chocolate Avenue, Suite 200
Hershey PA 17033
Tel: 717-533-8845
Fax: 717-533-8661
E-mail: cust@igi-global.com
Web site: http://www.igi-global.com

and in the United Kingdom by
Information Science Reference (an imprint of IGI Global)
3 Henrietta Street
Covent Garden
London WC2E 8LU
Tel: 44 20 7240 0856
Fax: 44 20 7379 0609
Web site: http://www.eurospanbookstore.com

Library of Congress Cataloging-in-Publication Data

Management practices in high-tech environments / Dariusz Jemielniak and Jerzy Kociatkiewicz, editors.

p. cm.

Summary: "This book leads to emergence of new, insufficiently analyzed and described organizational phenomena. Thoroughly studying this from international comparative cross-cultural perspective, Management Practices in High-Tech Environments presents cutting-edge research on management practices in American, European, Asian and Middle-Eastern high-tech companies, with particular focus on fieldwork-driven, but reflective, contributions"--Provided by publisher.

ISBN-13: 978-1-59904-564-1 (hbk.)

ISBN-13: 978-1-59904-566-5 (e-book)

1. High technology industries--Management. 2. Information technology--Management. 3. Knowledge workers--Management. I. Jemielniak, Dariusz. II. Kociatkiewicz, Jerzy.

HD62.37.M353 2008

658--dc22

2007042679

British Cataloguing in Publication Data
A Cataloguing in Publication record for this book is available from the British Library.

Table of Contents

Table of Contents

This chapter, based upon extensive interviews, describes the ways in which employees of high-tech corporations make sense of their plight in light of the processes of globalization, and the ways in which they attempt to accommodate, resist, or subvert the often detrimental changes to their work, particularly the threats of outsourcing and loss of benefits due to cheap competition from abroad. The study highlights the impact of dominant ideology on the options available to and considered by the employees.

Australian Cooperative Research Centers are temporary hybrid industry-research organizations bringing together academic, government, and industry personnel. This chapter shows how expectations of similarity, prevalent among all participants in these organizations, heighten communication difficulties and frustration arising from considerable differences in occupational cultures. Problems are further exacerbated by the centrality of discourse and argumentation for knowledge creation in the examined organizations. Overt discussion of cultural norms is proposed as a way of alleviating some of the experienced frustration.

This chapter, based on a study of researchers and designers working on developing a product using the human-centered design (HCD) approach, examines the relations between interaction with end-users and resultant design decisions. The analysis highlights difficulties involved in gathering information about end-users, but also in making sense of the received data, as well as the ethical qualities of the design process.

A study of IT specialists in Poland and their construction of their own professional roles and reality, this chapter examines how hierarchical control exercised by top management influences processes of role-creation, while taking into account other factors such as established community values and ongoing peer relations. It examines the strategies IT specialists use to maintain and enhance their professional and organizational status in different settings and circumstances, as well as the resultant variations in professional role definition.

proaches to the issue, and offers an explanation for the lack of cooperation in creating knowledge found in many studies. The text examines notions of trust and distrust, as well as issues of dealing with cultural differences within organizations.

An ethnographic study of Indian information technology workers whose local experiences become intermeshed with the transnational setting of a global software organization, this chapter argues that the workplaces described therein can be viewed as both model of and models for globalization processes, as well as milieus deeply imbued with personal, social and cultural relations and processes. By delineating the various forms of culture in a GSO, the study highlights the dialectical relationship between the local and global. Further, it explicitly demonstrates the ways by which GSOs and their workers constantly construct meaning and coherence in a volatile and international business context.

This study examines the consequences of post-Tayloristic managerial practices in manual work, comparing them against the high-tech industry environment. By analyzing the latter's knowledge-intensive character, it highlights the available possibilities for employees to rebel and oppose the traditional hierarchy. The chapter also discusses the notion of knowledge as the common denominator and barter unit among software engineers, concluding with remarks on the influence of this fact on employees' interactions with management.

This chapter examines the issue of power in information systems evaluation, applying the Foucauldian perspective, and focusing on ethical issues in IS. The research, based on an empirical study from Colombia, reveals the interplay of institutional and local, contingent level of organization. The analysis of the proposed model gives hints on how potential new practices of information systems evaluation could develop.

Preface

INTRODUCING MANAGEMENT TO HIGH-TECH

The Idea

High-tech corporations rely heavily on the normative/ideological control over their employees. While in manual labor the strict reign over body movements in Taylor-like manner suffices to achieve satisfactory results (a person is but a substitute to a machine), in the case of knowledge-work, it rarely does. The companies supplement the control of bodies with the control of minds. Paradoxically, in this sense, "blue collar" workers may enjoy more freedom of thought and more intact integrity than the "white collar" workers. Where behavior cannot be bracketed and prescribed by procedure or direct supervision, the self-definition and devotion of the worker play crucial role.

Although many excellent books have been published dealing with high-tech organizations, notably by Kunda (1992), Hochschild (1997), and Perlow (1997), these tended to shy from examining the full scope of international variation and similarity between practices in American, European, and Asian high-tech companies, while books adopting the global perspective, such as Castells (2000), were light on offering hands-on, fieldwork-driven but reflective insights. Yet others like Khosrowpour (2001) limited themselves to a particular subset of the high-tech field.

Out of these considerations came the idea to collect contributions from organization scholars all over the world studying high-tech workplaces, presenting a diverse view of management concerns and opportunities peculiar to high-tech organizations in the increasingly globalized world. We wished to examine not only the glamorous centers of high technology, but also organizations from regions often passed over in most discussion of the cutting edge such as Eastern Europe or, to a lesser extent, India.

High-Tech Management

High-tech organizational settings are of course characterized by the presence (and dominant position) of cutting edge equipment (the exact specifications thereof shifting to the tune of rapid technological change), but also by a shift from manual labor to professional and technical work, by increasing geographical dispersion within organizations, and by a move away from normative or bureaucratic forms of control (Bell, 1976; Zuboff, 1984). They are the closest current organizations come to the ideal of the postindustrial (or postmodern) society, characterized by its advocates as increasingly relying on knowledge and information rather than any physical (or, in the case of employees—behavioral) assets. This process is well exemplified by Burrell (1997), who points to the case of The Official Airline Guide being worth more than many of the described airlines, or the American TV Guide, in 1989 worth 3 billion dollars, significantly more than many of the TV stations covered in its publications, etc. As the systems of acquiring, storing and cataloguing the information, as well as of knowledge management rise to importance, the idea that those who project and program them gain power seems quite reasonable. Widely recognized theorists such as Drucker (1993) or Stewart (1997) make exactly this point: in their view the postindustrial revolution relies on the significance of information, which replaces capital and material goods as the most important mean of production. The "organization of the future" relies on the intellectual capital of the specialists. IT experts, rooting their power in true knowledge, replace in this model the traditional managers. The organizational structures become flexible, the career no longer is vertical, etc. High-tech specialists become empowered, and endowed with authority by the organization and its employees.

However, critics of the rosy picture of high-tech workplace relations, irrespective of the theory, point to dominant practices: in many, if not most high-tech organizations IT (information technology) specialists are permanently overworked (Perlow, 1997) and often conflicted with the management (Jemielniak 2007). While in case of many knowledge-intensive jobs the argument of gaining power through specialist knowledge many be valid (Alvesson, 2000), the situation of employees in high-tech environments may be just different. In opposition to other knowledge-intensive occupations (e.g. architects, see Larson, 1995), holders of posts in IT business rarely undergo the process of professionalization. Professions theory (e.g. Carr-Saunders, 1928/66; Alvesson, 1993) describes the successful liberation from management in case of many occupational groups long before the informational revolution and software engineers (along with other IT specialists) do not show much resemblance to those (for example, they do not unionize, they do not standardize education, they do not limit the entrance to the occupation, etc.). It also needs to be noted that high-tech organizations include not only IT specialists and engineers, but also masses of lower status (though not necessarily lower qualified, as Raganathan and Kuruvilla point out in their chapter) employees, including the paradigmatic job of the early twenty-first century IT sector, the call center staff.

Still, on the theoretical level, innovation management (Tidd et al., 2005) and learning organization (Senge, 1990) concepts strongly emphasize the highest role of human/intellectual capital of the company and the crucial function of knowledge in modern society. The popularity of this discourse in management literature consulting and official organizational language leads to an interesting paradox: on one hand, knowledge-workers are perceived as the most valued members of an organization, but on the other hand, they are being manipulated and "engineered", commonly driven to burn-out, and deprived of family life. Such a discrepancy between the official managerial language and the actual practice is by no means new, but in the case of high-tech companies, it is particularly striking. Moreover, in the case of high-tech employees, it is concurrent with a very specifically-developed occupational culture. Manager-worker conflict is taken to a different level.

At the same time, this is hardly a brand new development. High-tech environments, or settings perceived as high-tech, can be said to have existed at least since the industrial revolution. Even within the relatively new field of information technology, the specific issues involved in human resources management in IT businesses were being studied already in the 1970s. Philip Kraft (1977) described the motivational and cultural differences between the programmers and managers, and the ethnographical descriptions by Tracy Kidder (1981) clearly showed that high-tech environments are very unique. Yet, since the very beginning of the human resources management in IT the engineers have been treated as other white collar workers. Thus, their treatment has been following the traditional bureaucratic approach. However, two different views of programmers have been present simultaneously. According to the first one, they belong to the class of specialists, they form the crème de la crème, the true elite of professionals (Knell, 2000; Barley and Kunda, 2004). On the other hand, some recent studies found them effectively to be nothing more than a new category of shop-floor workers or crafted technicians

(Zabusky and Barley, 1996; Whalley and Barley, 1997). The already mentioned recent global proliferation of (and media attention to) call centers witnesses the rise in cultural prominence of these low-paid, low-status high-tech jobs.

The studies collected in the present book look into both high and low status high-tech jobs, organizations in developed and developing countries, examining the many encountered phenomena from managers', employees' and academic perspectives. Based on thorough studies of actual practices pervading high-tech organizations they, taken together, offer a comprehensive look into the issues of contemporary management as well as pointers towards the shape of the still emerging trends.

Structure of the Book

The book is divided into four parts, examining different aspects of work in high-tech environments. The division is by no means rigid, and we recommend the reader interested in only a single theme to take a look at different sections as well, as they all, though with different emphasis, give insights into high technology management practices.

We start out with the study of different high-tech workplaces, and issues deriving from the particularity of these environments. Vidar Hepsø takes us to the oil fields off the coast of Norway, and discusses the boundary-spanning aspects of the engineering-work, all the more remarkable as it is documented within the setting usually described as dominated by formal procedures as well as strict division of labor. Yet, as Hepsø demonstrates, the engineers' work is dependent on the ability to move between different contexts, and to translate information and action between them.

Ester Barinaga also deals with the issues of translation (and also examines Scandinavia): the process of translating the discourse of information society from a general idea to a realized project of a high-tech neighborhood, striving to provide general access to information services for all its inhabitants. She follows the process of translation from Swedish parliament into suburban streets, examining the ways in which social and ethnic inequalities, embedded in society and conveniently ignored by the emancipatory discourse of information society, hinder, subvert, and distort the implementation process.

Next, we travel to Seattle area in the United States, and the effects the processes of globalization have on the high-tech workers. Jasmine Folz, based upon extensive interviews, describes the ways in which employees make sense of their plight, and the ways in which they attempt to accommodate, resist, or subvert the often detrimental globalization-driven changes to their work, particularly the threats of outsourcing and loss of benefits due to cheap competition from abroad. She highlights the impact of dominant ideology on the options available to and considered by the employees.

Kate Hayes and Anneke Fitzgerald present Australian Cooperative Research Centres, temporary hybrid industry-research organizations bringing together academic, government, and industry personnel. They show how expectations of similarity, prevalent among all participants, heighten communication difficulties and frustration arising from considerable differences in occupational cultures. Problems are further exacerbated by the centrality of discourse and argumentation for knowledge creation in the examined organizations. Authors propose overt discussion of cultural norms as a way of alleviating some of the experienced frustration.

In the next section, we move the focus from the workplace, towards the main actor of high-tech environments, the knowledge worker. Marc Steen, studying researchers and designers working on a product via the human-centred design (HCD) approach, examines the relations between interaction with end-users and resultant design decisions. He analyzes the difficulties involved in gathering information about end-users, but also in making sense of the received data, as well as the ethical qualities of the design process.

Agnieszka Postuła's focus is on IT specialists in Poland and their construction of their own professional roles and reality. She is particularly interested in how hierarchical control exercised by top management influences this process, but also looks at other factors such as established community values and ongoing peer relations. She examines the strategies IT specialists use to maintain and enhance their professional and organizational status in different settings and circumstances, as well as the resultant variations in professional role definition.

The next chapter, by Aruna Ranganathan and Sarosh Kuruvilla, explores the problem of high turnover observed in the high-tech business process outsourcing sector in India. Many of the issues can be traced to the situation in which relatively well-educated employees perform jobs requiring low skills and offering low pay. The authors examine different strategies used by employers to deal with high turnover, ranging from providing instrumental incentives to promote employee retention to creating organizational culture designed specifically to engage the company's employees. Still, none of these strategies prove overwhelmingly successful, which the authors argue is due to their half-hearted implementation as well as the apparent clashes between promoted organizational culture and actual workplace conditions experienced by the employees.

Pauline Gleadle continues the exploration of professional status, taking us again to the United States, to a recent historical study of high-tech engineers in one leading US firm, Techco. Her chapter describes the engineers' position as fluctuating between being seen as highly privileged knowledge workers and as 'caught in the middle' between management and labor with no shelter from the harshest dynamics of global capitalism. The study follows the introduction of a range of new measures by company's senior management, leading to the disruption of existing organizational timings and threatening both the work and the self identity of engineers. She examines the impact of interplay between managerial decisions and outside technological change on the organizational status of the engineers.

In the next chapter, Dominika Latusek looks at the developing relation of trust between solution providers and their clients in the IT industry in Poland. She examines the question of how trustworthiness is socially constructed in the interaction between the suppliers and their potential customers, and how success or failure of this construction process affects their business relation. The importance of the process is further emphasized by the commonly expressed lack of trust which serves as the framework for the interaction. She sees trust as a tranquilizer that suspends the feeling of vulnerability and enables action regardless of uncertainty involved in the situation.

At this point, we move into the next part, and shift our focus from a particular organizational actor to the issues of power in high-tech organizations, and to workplace relations structuring the behavior of different groups within a high-tech setting. Knowledge management is at the heart of the matter, and Marie-Josée Legault proposes a sociological framework for the study of knowledge creation and diffusion in organizations, based on a framework which addresses many of the failures of commonly accepted approaches to the issue, and offers an explanation for the lack of cooperation in creating knowledge found in many studies. Her study touches upon the notions of trust and distrust discussed in the previous chapter, as well as upon cultural differences within organization as presented by Hayes and Fitzgerald.

Marisa D'Mello moves us back to discussion of globalization and back to India, with an ethnographic study of Indian Information Technology workers whose local experiences become intermeshed with the transnational setting of a global software organization.She argues that the workplaces she describes can be viewed as both model of and models for globalization processes, as well as milieus deeply imbued with personal, social and cultural relations and processes. By delineating the various forms of culture in a GSO, the study highlights the dialectical relationship between the local and global. Further, it explicitly demonstrates the ways by which GSOs and their workers constantly construct meaning and coherence in a volatile and international business context.

Eric Piñeiro and Peter Case amplify on the consequences of post-Tayloristic managerial practices in manual work, and they compare them against the high-tech industry environment. By analysing its knowledge-intensive character, they discuss the various possibilities for employees to rebel and oppose the traditional hierarchy. Later on they ponder on knowledge being the common denominator and barter unit among software engineers, to conclude with remarks on the influence of this fact on interactions with management.

In a somewhat similarly directed endeavour, José-Rodrigo Córdoba and Wendy Robson bring insight on the notion of power in information systems evaluation. By applying the Foucauldian perspective, they focus on ethical issues in IS. The research, based on an empirical study from Colombia, reveals the interplay of institutional and local, contingent level of organization. The analysis of the proposed model gives hints on how potential new practices of information systems evaluation could develop.

Rajneesh Chowdhurry and Alan Nobbs continue the practical focus in their chapter on Strategic Assumption Surfacing and Testing (SAST) methodology. They delve into the subject of healthcare information systems, using the example of National Health Service (NHS) in England. Their research bases on a study of healthcare professional workers, under the process of SAST introduction. The involvement of all actors and stakeholders is analyzed, and pinpointed as crucial in the long-term success of organizational change.

In the next chapter Ben Passmore brings us to Czech Republic, and describes the transformation the high-tech Czech corporations have undergone. By linking the global changes in the industry (the transfer in the value chain from manufacturing to knowledge work) with the post-Soviet political and economic specifics, he analyses the managerial attempts at participative planning. The ethnographical account of an engineering company from Brno brings conclusions of a more general nature on trust creation, control, and workplace culture.

The next chapter follows smoothly on the issue of participation, and opens the last part of this book, covering the issues of self-management practices. Elisabeth Kelan brings up the issue of flexible career planning. The Swiss example from information communication technology is presented as typical for modern self-entrepreneurial career model. A comparison of two companies is offered, to conclude with an analysis of responsibility-based and network-based career paths.

James J. Keenan goes beyond regional focus, by examining two knowledge-work groups operating on a global scale and across culturally divergent settings. His main interest is the social construction of power and dialogue within the team. He discusses the self-management theory as compared to the actual practice as seen in a qualitative study. The conclusions reveal that communication (and, in particular, informal interaction) is essential in self-managing team construction.

Staying within the same, constructivist framework, Maria Aggestam demystifies the modern institutions of innovative industry. She examines the classical opposition of bureaucratic rationality and authoritarian approach one the one hand, and the brave new innovative, flat organizations on the other. The study focuses on the implications of the possibilities contemporary technology offers to entrepreneurs. The author is skeptical about the popular beliefs on virtual organization and discusses them in comparison to her findings.

Maria-Liisa Trux takes us to Finland, to analyze the identity construction in data security workplaces in a Finnish company, as well as its Silicon Valley subsidiary. She criticizes main-stream normative management theory, by pointing to the complex and ambiguous nature of organizational culture. In an ethnographical study, she focuses on the tension between the employees and managers and describes the manipulative practices in high-tech environment, as well as forms of participant resistance.

The concluding chapter by Chris Russell summarizes many of the contemporary discussions on control versus innovation in knowledge businesses. The research is based on a case-study of mobile information system development. The story is developed in details, and the author reveals consecutive stages of development of a particular socio-technical artifact, starting out with its being desired and demanded by engineers, and continuing through the taming of the innovation process to its final exploitation by the vendor.

Despite the wide-ranging and comprehensive nature of the various chapters, we by no means believe we have managed to exhaust the topic within the covers of this book, but rather to sketch out the field that demands much further study, for which we tried to have provided a foundation. We hope this book will serve as an inspiration for more research, confirmations and critiques of the theses presented therein, as yet another chapter in the eternal academic debate, and as a case for more humane and participative management practices.

REFERENCES

Alvesson, M. (1993). Organizations as rhetoric: Knowledge-intensive firms and the struggle with ambiguity. *Journal of Management Studies, 30*(6), 997-1016.

Alvesson, M. (2000). Social identity in knowledge-intensive companies, *Journal of Management Studies, 37*(8), 1101-1123.

Barley, S. & Kunda, G. (2004). *Gurus, hired guns, and warm bodies: Itinerant experts in a knowledge economy*. Princeton: Princeton University Press.

Bell, D. (1976) *The coming of post-industrial society: A venture in social forecasting*. New York: McGraw-Hill.

Burrell, G. (1997). *Pandemonium: Towards a retro-organization theory.* London: SAGE.

Carr-Saunders, A. M. (1928/66) Professionalization. In H.M. Vollmer & D.L. Mills (Eds.), *Professionalization.* Englewood Cliffs: Prentice Hall.

Castells, M. (2000). *The rise of the network society, second edition.* Oxford: Blackwell

Hochschild, A.R. (1997). *The time bind: When work becomes home and home becomes work.* New York: Metropolitan Books.

Jemielniak, D. (2007). Managers as lazy, stupid careerists? Contestation and stereotypes among software engineers, *Journal of Organizational Change Management, 20* (4), 491-508.

Kidder, T. (1981). *The soul of a new machine*. New York: Avon Books.

Khosrow-Pour, M. (2001). *Managing information technology.* Hershey: Idea Group Inc.

Knell, J. (2000). *Most wanted: The quiet birth of the free worker*. London: The Industrial Society.

Kraft, P. (1977). *Programmers and managers.* The routinization of Computer Programming in the United States, New York: Springer Verlag.

Larson, M.S. (1995). *Behind the postmodern facade: Architectural change in late twentieth-century America.* Berkeley: University of California Press.

Perlow, L.A. (1997). *Finding time. How corporations, individuals, and families can benefit from new work practices.* Ithaca-London: ILR Press.

Senge, P.M. (1990). *The fifth discipline: The art and practice of the learning organization.* New York: Doubleday.

Stewart, T. A. (1997). *Intellectual capital: The new wealth of organizations*. London: Nicholas Brealey Publishing.

Tidd, J., Bessant, J., & Pavitt, K. (2005). *Managing innovation: Integrating technological, market, and organizational change.* Hoboken: John Wiley & Sons.

Whalley, P. & Barley, S.R. (1997). Technical work in the division of labor: stalking the wily anomally. In S.R. Barley & J.E. Orr (Eds.), *Between craft and science* (pp. 20-52). Ithaca: Cornell University Press.

Zabusky, S.E. and Barley, S.R. (1996). Redefining success: Ethnographic observations on the careers of technicians. In P. Osterman (Ed.), *White Collar Careers* (pp. 185-214). Oxford: Oxford University Press.

Zuboff, S. (1984). *In the Age of the smart machine: The future of work and power.* Oxford: Heinemann.

Dariusz Jemielniak, *PhD – assistant professor of management at Kozminski Business School (Poland). He was a visiting researcher at Cornell University (2004-2005), Harvard University (2007), University of California Berkeley (2008). He also co-edited of Handbook of Research on Knowledge-Intensive Organizations (2009). He was a recipient of scholarships and awards from Fulbright Foundation, Foundation for Polish Science, Collegium Invisibile, Kosciuszko Foundation. He published in journals such as Journal of Organizational Change Management, Knowledge Transfer, The International Journal of Knowledge, Culture and Change Management, or The International Journal of Technology, Knowledge and Society. His research focuses on knowledge-intensive workplace and professions, which he analyzes by the use of qualitative methods.*

Jerzy Kociatkiewicz, *PhD - lecturer at the School of Accounting, Finance, and Management at the University of Essex in Colchester, UK, having previously worked in Sweden and Poland. Most of his research focuses on narrativity, space, and technology in organizations. His publications include articles in journals such as Human Resource Development International; Knowledge Transfer; Qualitative Sociology; and Studies in Cultures, Organizations, and Societies. He has also published numerous book chapters, and co-edited, with Dariusz Jemielniak, Handbook of Research on Knowledge-Intensive Organizations (2009).*

Acknowledgment

The editors would like to sincerely thank Deborah Yahnke from IGI Global for her infinite patience and constant professional assistance with the editing process.

We are also indebted to Joanna Jemielniak, Krzysztof Klincewicz, Monika Kostera, Stella Kounelakis, Damian Makowski, Patrik Persson, Józef Tkaczuk, Kevyn Yong, and other reviewers who preferred to stay anonymous. We would like to thank all the authors involved for the journey together, and for their ability to endure our idiosyncrasies, last-minute announcements, and changing demands.

Dariusz Jemielniak's work on this project was possible thanks to generous support from the Minda de Gunzburg Center for European Studies at Harvard University, Kozminski Business School, and Polish Ministry of Science, for which he is truly grateful.

Jerzy Kociatkiewicz would like to thank Monika, as always, for enduring his writing habits, as well as all the colleagues from Växjö Universitet and the University of Essex for help and inspiration.

Dariusz Jemielniak and Jerzy Kociatkiewicz, Editors

Section I
The High-Tech Workplace

Chapter I
"Boundary-Spanning" Practices and Paradoxes Related to Trust Among People and Machines in a High-Tech Oil and Gas Environment

Vidar Hepsø
Statoil Research and Technology, Norway

ABSTRACT

This chapter follows subsea engineering coordinators (SEC) at Statoil, a major Norwegian oil company, and their collaboration with subsea engineering/operational support personnel and external vendors. This is a high-tech business that tends to be described by formal procedures and a strict division of labor, or in other words, strict hierarchy and market coordination mechanisms. Still, engineers in this setting perform substantial informal boundary work to be able to do their work efficiently. Their self-definition and devotion is realized through boundary-spanning interaction with various material resources and through extensive management of trust. The consequence of this knowledge intensive operational practice is that the engineers have to live continuously with paradoxes. In the light of the situation of these engineers, we address some of the dynamics of collaboration and control that such professionals must cope with in today's high-tech environments.

BOUNDARY-SPANNING AND TRUST IN INTERORGANIZATIONAL COORDINATION AND COLLABORATION

Over the years, boundary-spanning has attracted substantial attention in organization studies (i.e., Aldrich & Herker, 1977; Caldwell & O'Reilly, 1982; Friedman & Polodny, 1982). Furthermore, the increasing focus on knowledge management in organizations has given the concept renewed popularity. While much focus has been on how to build competences within a particular community of practice (Wenger, 2002), more recent work has focused on the competence and ability of agents to span multiple boundaries in practice (Carlile, 2002, 2004; Cross & Parker 2004; Levina & Vaast, 2005; Orlikowski, 2002; Pawlowski & Robey 2004). Orlikowski (2002) demonstrates how organizational competence in spanning boundaries is embedded in the everyday practices of its members. Levina et al. (2005) focus on a new joint field where such everyday practices are produced or enacted. This chapter builds on this thinking but focuses on how the management of trust is important in boundary-spanning in such knowledge intensive-organizations. It also considers how strategies for trust building, maintenance, trade-offs, and living with paradoxes are vital to understand the practices of boundary-spanning in such settings. If we are to understand how high-tech formal organizations actually work and how these mechanisms fill the space between formal organizational requirements and day-to-day practices, we must address these informal mechanisms. How can we describe the work boundary-spanners do in high-tech environments? Cross et al. (2004) define them as key individuals who facilitate the sharing of expertise by linking two or more groups of people separated by location, hierarchy, or function. Levina et al. (2005) argue that much attention in research on boundary-spanners has been on identifying and classifying the role such boundary-spanners should have, but that such roles often come into conflict and that it is more interesting to look at how the ideal roles function in practice. They make a distinction between nominated boundary-spanners and boundary-spanners-in-practice (Levina et al., 2005, p. 339-342). The first category of boundary-spanners is defined as agents assigned by the empowered agents, such as managers, in a field to perform certain roles in spanning boundaries of diverse fields. Boundary-spanners-in-practice are agents who, with or without nomination, engage in spanning (navigating and negotiating the meaning and terms of the relationships) boundaries of diverse fields. The present chapter describes boundary-spanners-in-practice.

Boundary-spanners are formally and informally integrated in organizations that have institutionalized certain ways to coordinate activities. Groups develop coordination mechanisms to manage interdependence among individuals, groups, organizational units, and activities when confronted by behavioral uncertainty. They are the ways to represent information, guide appropriate types of behaviour, and sketch routines for coordinating actions (Adler, 2001; McEvily, 2003; Powell, 1990; Williamson, 1993). Coordination mechanisms are ways of working with the problem of interdependence and uncertainty, and represent a logic in which work can be coordinated and information handled.

Recent research (Adler, 2001; Kramer & Cook, 2004; Lane, 1998; McEvily, 2003) shows that the coming of more knowledge-intensive corporations increase the importance of trust as a coordination mechanism and that trust becomes an increasingly attractive mechanism to economic agents. It is therefore relevant to the study of management practices in high-tech environments. If trust exists in the relationships it means that much of the work a boundary-spanner invests in monitoring and controlling others becomes of less importance (McEvily, 2003, p. 92-93) and this reduces the transaction costs associated with boundary-spanning. The large increase in the number and variety of exchange relations and the increased complexity and uncertainty of the business environment cannot be handled without the presence of interpersonal and interorganizational trust (Lane, 1998).

While trust as an organizing principle or mechanism is an important prerequisite for collaboration, and in understanding boundary-spanning, it can never be applied alone in organizational and interorganizational relations. At the same time, there hardly exist market and hierarchy coordination mechanisms that work properly without trust (Callon, 1999). We call this setting where market, hierarchy, and community coordination mechanisms co-exist an ecology (Nardi & O'Day, 1999).

However, there is an element of power lurking behind trust and boundary-spanning work. As Hardy, Phillips, and Lawrence (1998, p. 64-87) argue it is important to describe, first how trust is generated, and second differentiate between trust-based relationships and relationships where power creates façades of trust behind a rhetoric of collaboration. This façade and rhetoric can be used to promote vested interests through the manipulation and capitulation of weaker partners.

The main research questions are how is trust institutionalized in human and non-human resources in the high-tech operation and maintenance of subsea systems? What is the role of boundary-spanners in this process? Are there paradoxes that engineers have to handle in this high-tech environment? The structure of the paper is as follows. It starts by addressing the particular methodology used in the case. Then the subsea operation in Statoil case is introduced. The focus is enquiring about the maintenance of trust among engineers in this knowledge-intensive setting, the human and non-human elements of trust developed through boundary-spanning work. It describes how subsea engineering coordinators (SEC), as boundary-spanners, are able to sustain trustful relationships across internal and external borders to re-establish the technical condition of subsea systems. It presents the trust element as the glue, or the foundation for a flexible structure of communication and enrolment realized through boundary work. Finally, there are four paradoxes and the dynamics of collaboration and control that such professionals must cope with in such high-tech environments.

METHODOLOGY

Statoil is a major Norwegian oil company with 24 000 employees. The author is a researcher at Statoil R&D and has worked with organizational development enabled by new information and communication technologies for over 15 years. He has a PhD in anthropology, but petroleum engineering has been the field of study in this period. The particular project reported in this chapter was set up to sketch future subsea operation and maintenance scenarios in Statoil. A substantial part of this project was to describe the present socio-technical practices within the subsea environment. The author spent considerable time with SECs in various parts of the company to understand their work practices in 2002 and 2003, to learn the necessary ropes of the ecology, develop personal relations with the insiders, and earn trust. The work is based on participant observation, daily diaries/fieldwork notes and over 30 interviews with personnel such as SECs and managers in various parts of the ecology. Statoil internal electronic archives of e-mail communication and Intranet have been used extensively as sources. Several workshops have been held with subsea personnel both in Statoil and among vendors where the case and the findings have been discussed. The case has been part of a larger action research project that will not be addressed in this chapter.

THE CASE: STATOIL SUBSEA OIL AND GAS OPERATIONS

Subsea operation and maintenance cover all activities and interfaces necessary for the execution of subsea work using manned or unmanned intervention, including essential activities in connection with the subsea systems, start up, operation, maintenance, shutdown and removal of subsea pipelines and subsea systems. In 2003, Statoil operates over 200 subsea wells, which makes Statoil the second largest subsea operator in the world. A subsea production system is typically connected to a topside installation as satellites, producing hydrocarbons from nearby reservoir pockets (see Figure 1). Subsea production systems in Statoil are systems that consist of subsea completed wells, subsea production trees/well-heads, seabed tie-ins to flowline systems and control facilities to operate the wells. Such systems are unique when it comes to remoteness in installation, service, and operation. There are valves/sleeves that can be remotely operated on the subsea wells to control the amount of produced flow of oil and gas (see Figure 1).

The Participants in the Setting

Let us now have a closer look at the organizational actors in the setting. The **business assets** are responsible for the operation and maintenance of subsea systems and are customers in this ecology. They have the resources, and plan the work to be undertaken in the subsea domain. Assets have their operational and technical tasks undertaken by subsea engineer coordinators (SECs). These are the technical experts for the subsea equipment, they have the overall responsibility for following up the technical condition of the subsea systems and plan the intervention on non-functioning subsea systems. SECs deliver the availability of subsea systems, known as uptime. Uptime means that the systems have the desired performance in relation to oil and gas production and the technical condition to meet targeted health, environment and safety standards. SECs work onshore but have frequent communication with offshore personnel, and they spend considerable time offshore on vessels during subsea intervention campaigns. The hands-on and day-to-day operation of subsea equipment is undertaken by the central control room (CCR) operators at the offshore installations. These CCR operators follow-up the performance of subsea systems when monitoring the oil and gas production in real-time. Additional Statoil units plan vessel operations and provide engineering support. Particular tools and equipment used during subsea intervention are placed at **Statoil bases** on the West coast of Norway, and the chartered vessels must get hold of the correct equipment before they can start the subsea intervention work offshore.

In addition to Statoil's internal organizational units, a number of **subsea system suppliers** do substantial work. Two important mechanisms are important in this arrangement. The first, the subsea pool is a pool of equipment shared between the assets used during intervention on subsea systems. The second, or the Statoil-FMC , is a framework contract initiated for technical services and provision of subsea equipment. Subsea suppliers like FMC have workshops close to the West coast bases where they repair, overhaul, and prepare subsea equipment for installation or replacement. Much of this work is part of the SSA-framework contract. **Shipping companies** provide vessels that are hired by short- or term-long term contracts to execute the intervention work. Once the vessel has been mobilized and is intervening subsea, the actual ROV or subsea robot operation, installation, or repair of the subsea equipment is undertaken by ROV operators under supervision of Statoil and the supplier (i.e., FMC) equipment expert onboard (see Figure 1).

The total operational support for Statoil subsea installations is an intricate organization of Statoil internal and external actors. When asked how this networked support organization works, one

Figure 1. Top: The Gullfaks field with subsea infrastructure. Middle: The intervention and supply vessel Viking Poseidon with subsea equipment. Bottom left: Part of a subsea system stored at the base. Bottom right: A ROV is intervening on a subsea system. Pictures and illustrations courtesy of Statoil (Øyvind Hagen and Arnfinn Olsen) or private.

of the vendors argue:

It is a pretty open system and collaboration across the companies where we share work practice irrespective of company allegiance: we are there to do a job.

The hierarchy and market mechanisms are made strong by formalized contracts between Statoil and the vendors. In normal operations, mechanisms like the SSA and the tool pool contract tend to be background resources, but are brought forward when disagreements have to be handled. Let us now have a look at how these formal roles are enacted in practice.

THE MAINTENANCE OF TRUST, SECS' STRATEGIES AND BOUNDARY-SPANNING WORK TO DELIVER AVAILABILITY OF SUBSEA SYSTEMS

The development of trust in the subsea setting is the consequence of a long process that I will not be able to address in detail in this chapter. It starts with a few people in the mid 1980s that developed and operated Statoil's first subsea installations, and the organization has grown in many small steps since then. Most of the SECs have a background from the Statoil internal subsea engineering support unit (UVPS) and have worked with subsea concept, project development, and operations. There are a number of key individuals with subsea domain expertise and long experience that link the engineers together. The special collaboration with FMC was founded in a contract but it has enabled both Statoil and FMC to fill complementary roles and develop the foundation for a long-lasting relationship. The tight collaboration with FMC is the consequence of a joint development of several generations of subsea equipment over the last 15 years. In this situation, Statoil personnel have had close contact with FMC, often sitting in their office facilities during the development of new equipment. Close interaction has also developed during building, testing, and deployment of the subsea equipment. Throughout the years, a substantial number of personnel have switched employment from Statoil to FMC, from FMC to Statoil.

In order to acquire into how trust mechanisms are maintained in this setting, the chapter addresses this work through the perspectives and programs of the subsea engineering coordinators (SECs) who are the obligatory passing point for establishing or re-establishing availability of subsea systems. The author acknowledges that by granting the voice of the SECs primary access the chapter is not able to portray other voices like the various vendors in the same way.

The SEC of the asset is typically an MSc engineer with valuable experience from various technical disciplines. Until recently no female engineer has filled this position in Statoil. This work is knowledge-intensive in the sense that he or she knows the subsea systems, the constraints, and possibilities on the receiving topside installations. He or she has typically more than 7-10 years of operational experience. A SEC has a large network of contacts in the offshore and onshore organization of the asset and among the internal and external suppliers, he or she nurtures contact with the SECs of the other assets for peer assistance and colleague communication. His or her experience normally consists of a general understanding of petroleum and process engineering (i.e., the hydrocarbon-flow from the reservoir to the export of the produced gas and fluids). These are the elements in which the subsea systems are integrated. The SEC will also have detailed knowledge related to the construction and elements of the subsea systems. It is because of this experience the coordinator is able to diagnose the malfunctioning subsea equipment often in collaboration with control room operators and subsea suppliers like FMC. These engineering skills must also be combined with project management, since in addition to diagnosis, the coordinator also evaluates the costs related to replacement of subsea modules. He or she writes the repair procedures for the replacement of the equipment and must "walk the talk" in following-up the procedures he or she has written. In doing so, he or she tends to get direct feedback on what works and what does not. Much of the world of subsea production evolves around the work of the subsea engineering coordinator (SEC) and the resources he or she mobilizes when the subsea production systems develop hick-ups or must be closed down due to malfunctions.

In what follows I will present an example of the work undertaken to mobilize the necessary human and non-human resources to replace a malfunctioning subsea component and re-establish the technical condition of the subsea system. Let us start with a SEC's own description of the communication intensity of this process reported in a mail (Box 1).

As the mail from the SEC indicates, subsea intervention work is an activity that involves numerous human elements both within and outside the Statoil organization. In addition, non-human elements of subsea installations, rigs, vessels, control rooms, choke valves, plugs, control cables, and oil and gas test separators are elements that link the human resources together.

Let us now follow such an operation. An operator in the central control room of an offshore installation receives an alarm from the safety and automation control system (SAS) that monitors the state of critical components and systems on the topside and subsea installation. An alarm in the offshore safety and automation system can typically indicate a sudden drop of pressure/temperature on one of the subsea wells or a subsea choke-valve that does not work. A SEC tells:

My experience is that most control room operators use "rule of the thumb" methods to see if something can be done with the alarm. He or she can use the manuals for the systems as well. If the operator is not able to solve the problem, he/she notifies me, via mail or phone. If he/she does this, I help diagnosing the malfunction where we try to isolate parts of the system at the time. Tricky, but we are able to solve some malfunctions this way

Box 1.

> From: xxxxx
> To: xxxxx
> Subject:.........
> Hi, I write this while I am on Normand Mjølne (*my comment: an intervention vessel*). We have a campaign on Gullfaks and Statfjord where the biggest work packages are:
> - Change isoplugs on Gudveig and inspect a choke valve on Gullfaks South while this valve is operated from Gullfaks A. These operations are coordination intensive vis-à-vis the installation.
> - Change shunt plugs/iso-plugs on template E, F, and M on the Statfjord North subsea satellites. This activity is coordinated with Statfjord C.
> - Measure the loads of cuttings at Statfjord B and inspect pipelines at Statfjord B before Safe Britannia arrives. This requires coordination with all three Statfjord platforms.
> - Adjust choke valves and measure sand production on several wells on templates M and E on the Statfjord East and North subsea satellites. Coordination with the operational manager in the CCR and the subsurface representative on Statfjord C plus the rig Borgland Dolphin is necessary. This operation is dependent on an available test separator on Statfjord C and that the Borgland Dolphin rig moves from the M template to the Norne field.
>
> Fluenta, FMC Service, Oceaneering, UVPS, and Statfjord operational support personnel are onboard the Norman Mjølne together with me. On Gullfaks A, FMC Service and Gullfaks operational support personnel are represented. Present at Statfjord C are Fluenta, FMC Service and Statfjord subsurface reps. Collaboration between Statfjord and Gullfaks have been intensive both in the planning and execution of this operation."
>
> Best regards
> xxxx

by pushing buttons in given sequences using past experience and by following manuals. Contamination in the service flowlines tend to accumulate in the hydraulic actuators on the subsea systems. When it comes to more complex problems we use operative procedures to find malfunctions in for instance a subsea control system (SCM).

In this process, the SEC is unable to see the same data interface as the CCR operator. He will have to picture the interfaces through which the operator is working.

Some operators want to have my help, others do not, and it is very important how I approach them. I tend to know many of them and how to handle them. If they perceive me as a 'besser wisser,' I will have serious problems helping them. This can be a problem because many of them do not have the same detailed understanding of the subsea facility as I have. I must be humble and over time I have earned their confidence and trust.

The SEC is keen on presenting himself as a friendly helper and under-communicates his competence in the domain. A control room operator phrased the same relation in the following manner:

The SEC is helping us, there are wonders he can do to find the malfunctions. I have most of my experience from traditional non-subsea platform wells and must admit that I can have problems grasping the full complexity of the subsea systems. When I present the information located in the operator station and discuss the details with him, we have a good dialogue. He suggests things, I click on the buttons and we evaluate the reaction via the SAS system. One by one the potential failure modes are discarded. In most cases the two of us are able to find a solution to the problem we are facing.

If we go back to the identified malfunction, an example of such a diagnosis and the proposed action related to this situation is as follows: "Subsea electronic module A on well X subsea control module suffers from low insulation resistance (short circuit). To rectify the situation the faulty SCM has to be pulled to surface for repair and replaced with a fresh subsea control module." If the problem is not to be solved by the subsea engineering coordinator in collaboration with the CCR, he mobilizes a larger network:

I can take contact with some SEC colleagues in other assets, which I have a lot of contact with and discuss the problem. One of them is sitting nearby. I can contact the subsea engineering service organization or approach suppliers like FMC directly because of the SSA framework contract. We often need the supplier of the equipment to handle the specific details of the particular subsea system. The shipping company and marine contractor for ROV-services do an excellent job but I do not know them that much. They are usually contracted for 2-3 years but have little to contribute in this part of the process. We plan and diagnose and they execute.

Once the diagnosis is set up and there is a need for an intervention job, the SEC creates a scope of work (SOW) text document if intervention activities are needed. An intervention is planned when the redundancy of the subsea components is running so low that the regularity of the subsea systems are threatened. There is normally a high degree of technical redundancy in the subsea systems. If one component breaks down there are redundant components that prolong the life of the malfunctioning unit. When the redundancy is gone a number of new resources must be mobilized if the SEC is to get the job done. Hearings, mail, telephone, or meetings are used where different personnel contribute in this planning process and the development of a scope of work; other SECs, platform managers, CCR production managers, operators, and Statoil subsea support organizations.

This scope of work (SOW) document is a boundary object (Star & Griesemer, 1989) that contains information related to HSE: important telephone numbers, contact persons, lines of authority, and communication. It addresses the malfunction and the problem to be solved; why it should be done in a particular manner, the conclusions of the diagnosis, asset field information with maps over templates and pipelines, the work process sequence during intervention and repair; preparations before replacement, the replacement process in detail, the close-down sequence of the well, the module or equipment to be replaced and requirements/needs for tools and documentation. Each SOW tends to have a "cut-and-paste" character since much information can be re-used from past interventions. Still, it is a boundary object of a heterogeneous character in the sense that the various actors that participate in the subsea intervention will be able to take out the elements they need to undertake their part of the work and it functions as a least common denominator. A SEC argues on the SOW:

You must not forget that many of these operations are repeated because it's not the first SCM repair procedure I write. I focus on what I know will be important based on things that have not worked in the past. These procedures cannot be too detailed; the people doing these jobs are experts in their fields. I draw the various elements together into a unity that will work for all participants.

There are two ways to organize such interventions. Campaign, the first, means several interventions/job packages are taken at the same time. Second is a single trip/ad-hoc assignment. If the malfunction is critical in relation to the diagnosis undertaken or assessments related to HSE; substantial leaks in hydraulic fluids or hydrocarbons, lost production or water interjection or wells that do not produce, single intervention trips are planned. A rough estimate is taken when these big problems occur to see if a single trip (one job package) covers the expenses (i.e., costs and consequence costs). Intervention is then of utmost importance and must be planned and undertaken quickly. In most assets the SECs do not have the budget for these activities themselves and appeal to asset managers to have the budget released and argue why this should be prioritized before other activities. A list of the most urgent subsea interventions legitimated via cost and risk assessments are made by the SECs in collaboration with the logistic planner. A cost sharing and coordination system has been developed, and this coordination mechanism is important for the smooth execution of subsea operations and maintenance. During campaign mobilization, demobilization and transition costs are shared among the participating parties. This cost structure or principle for sharing and reporting costs was itself a consequence of a compromise between the requirements of different actors in the setting. On the one hand the SECs need to follow up intervention costs on their systems, on the other hand the other participants also need to have a flexible structure that could show their part of the intervention costs, split, and aggregation.

Other institutionalized hierarchical and market-related coordination mechanisms exist as resources for the SECs. As with the previous cost-structure they are the consequence of a process not described in this chapter. The subsea intervention tools used are part of the subsea tool pool, formalized via the "detail pool agreement" (DPA). Assets that are members of the tool pool will receive the scope of tools needed in the near future interventions to maintain stable production. For the SEC, the subsea pool and the DPA are resources he or she can use to mobilize the intervention and which he or she needs to get the job done. Fixed prices exist for FMC activities via the SSA contract like replacement and repair of SCMs, mobilization and demobilization costs for vessels. These coordination mechanisms are existing resources that can be used or enrolled by the SEC as resources (Callon, 1999). General intervention priorities are developed based on the following criteria: are there malfunctions we know of based on symptoms and feedback from CCRs, on what systems have we used up the redundancy? Other important criteria are shared in the group of SECs:

I am willing to admit that I look after my systems and want my asset's needs to be placed up in front. However, we share some basic rules that in the long run will help us all even though it in the short run will not pay off for my asset, when I have an urgent situation coming up and need an intervention vessel. These are that financial and safety criteria are the basis for which subsea work packages are to be prioritized. We also have some additional criteria. We all want to minimize transit time and to give each asset a fair share of 'prime time' in the summer season. Even though the amount of jobs for each asset may differ, the campaigns must be divided into time slots (sub campaigns). Each asset has been given a time slot in the prime time of the year, as well as time slots in the seasons with possible waiting on weather (WOW) time during autumn and winter seasons. We report our business asset's needs via the 'vessel need' LOTUS NOTES-database. A logistic planner proposes a schedule or slot for each asset in an overall scheduling plan. Because of the good informal communication between me, the SECs in different assets and the logistic planner this plan is often improvised in order to meet emerging problems that have arisen since the last intervention needs were reported.

The SECs and the logistic planner agree in most cases on the priorities and are able to improvise when needed without appealing to management. Changes in the plan occur quite frequently because many potential contingencies can arise: waiting on weather, unforeseen problems during interventions, and delay of equipment just to mention a few. If the work package can wait it is placed ahead at some point in time on the main plan.

It often happens that something critical has happened in another asset. I know where the Viking Poseidon is, if something critical occurs on one of my subsea templates I try to reschedule the campaign by contacting the logistic planner and the respective SEC that is onboard at the time. We often get a deal. I get the vessel to do the critical work even though it is within the time-slot of another asset by persuasion. There is some reciprocity here since I owe him a future favour. I am obliged to help him when the redundancy is lost on one of his subsea systems. I know him well, and he knows that he can trust my word. Still, these actions are of course logged and the decisions taken are documented.

The SOW is overall and considerable articulation work is needed to fill in the details and makes the SOW plan rather robust in use. A SEC reports on the scope of work:

The operation should of course in principle be undertaken according to the scope of work. Adjustments must be made and improvisations undertaken to deal with contingencies. The descriptions in the SOW are on a high level because we acknowledge that the participants have core competence and are experts in their fields. I focus my details on the interfaces and overlapping work tasks.

This network of humans and non-human resources must also be mobilized by the SEC if he or she is to restore the technical condition of his or her subsea systems. In this process he or she must develop trustful relations with numerous resources. One SEC describes these relationships and roles:

It is the platform manager on the installation that has the overall responsibility for the intervened subsea installations so the CCR must be involved since the intervention will influence the flow of hydro carbons on the installation. The people on the platform must be confident that we can handle the situation. We typically involve them via radio or telephone. Most of them know me from the past. I typically monitor and supervise the operations from the vessel to ensure that interventions are performed according to approved procedures and Statoil's safety standards. In addition I follow up the progression of the intervention activities according to the main plan and make sure that the necessary resources are available. I can change priorities in the work tasks. This is seldom done without communicating with the SECs involved, but if we change priorities they are logged via mail to the people that are involved.

A typical example of this operation is related to the change of a subsea control module (SCM) on a subsea template. See the bucket bottom-right in Figure 1. It is too complicated to address the nitty-gritty details of a subsea control module (SCM) replacement in this chapter, but it requires a delicate collaboration between the SEC, the installation's CCR, subsea vendor, shipping vessel crew, and ROV operators. Statoil is leading the operation playing on the other participants' resources. The SEC is always in charge of running intervention activities on his or her asset's well on behalf of the platform. The SEC checks operational logs from the various participants, coordinates the operation and reports the status to the CCR. The CCR is responsible for logging operations (i.e., subsea system valve status, position open-shut on the installation). ROV operator logs all his or her operations. Suppliers like FMC always run the intervention-equipment. The vessel crew make sure that the vessel is on the right dynamic positioning (DP) position vis-à-vis the seabed subsea systems and for placing the equipment on the right place on the cargo deck. All participants have continuous communication with the SEC during operations and the others report to him or her after execution of the intervention activity.

To summarize this section, we have described the boundary-spanning activities involved in mobilizing human and non-human resources for subsea interventions. There are numerous heterogeneous elements that work together in this division of labor. The subsea organization of Statoil represents the company's peak competence and has grown in many small steps over the last 15-20 years. In 2003, it was still almost invisible in the company hierarchy and was fragmented across several organizational units. If we look at the organization chart of Statoil it will be difficult to find activities and actors associated with subsea operation and maintenance, and how these activities and roles are aligned to perform the work presented in this chapter. The boundary-spanner (SEC) has no formal management position; he or she is just a skillful and experienced engineer that must mobilize budget resources. It is his or her practice that enables him or her to fill this boundary-spanning and integrating function. The key boundary-spanners in the subsea ecology are the SECs. They are obligatory passage points for much of the activities. SECs have a great need to know both the offshore and the onshore organization of the assets that are participating in subsea interventions. The logistic planner that schedules the interventions is another obligatory passage point for the supplier resources. These two in particular also know the internal and external service providers and are able to link a variety of sub-organizations in an ever-changing network in which suppliers, shipping companies and others contribute to the overall Statoil enterprise based upon their core competencies. The subsea organization is an ecology where boundary-spanning activities are important to tie the niches of the ecology together.

The chapter has already started to describe some paradoxes related to control, authority and identity management evolving in the boundary-spanning activities of SECs in such ecologies. Let us address these paradoxes in more detail.

TRUST AND THE MANAGEMENT OF PARADOXES AS A BOUNDARY-SPANNING ACTIVITY IN HIGH-TECH ENVIRONMENTS

Up to now we have addressed the boundary-spanning work of the SECs as a maintenance process, a mobilization of human and material resources in an ecology to re-establish the technical condition of the subsea systems and have described all the minor but vital details that must be aligned after the discovery of the malfunctioning subsea control module to the subsea wells are functional again. The mobilization of this interorganizational high-tech environment reveals a number of paradoxes that are relevant to our research questions on boundary-spanning and trust. In this knowledge information-based mode of production or operation it is important to share sensitive information and core competence. At the same time the intensified competition of today's businesses makes it more difficult to preserve and develop trust within and between organizations. This forms the back curtain of the challenge that SECs as knowledge workers must handle in trust-building and boundary-spanning activities. These are paradoxes in the sense that they cannot be overcome without hampering the delicate dynamics and enabling properties that characterize this interorganizational collaboration.

Our first paradox is that trust and power are enmeshed in each other but that power relationships tend to be downplayed and a more egalitarian attitude develops between the parties. Hardy et al. (1998, p. 69-70) argue that trust can be conceptualized as a communicative, sense-making process that bridges disparate groups. This approach emphasizes the *shared* meanings that partners use to signal trust and trustworthiness to each other: "Trust is therefore an intersubjective social "reality" that cannot exist, regardless of the good intentions of partners, *unless* the symbols used to signal trustworthiness have meaning for all parties." Hardy et al. argue (1998, p. 75) that the existence of sincerity and mutual goals will not generate trust, unless partners can communicate with each other. In our case meaning is demonstrated in action during a subsea SCM intervention campaign. Power is also important in the relationship and is vital to understand action. The generation of trust is a complex, dynamic, and continuous process. Trust depends upon signalling trustworthiness in ways that create meaning for others. Sincerity and mutually compatible goals will not suffice if their meaning cannot be demonstrated in action. The SEC boundary-spanner enrols the necessary human and non-human elements in order to mobilize the ecology to sustain the availability of his subsea systems. He gets the work done, maintains present power structures and minimizes the risks associated with the job. In most respects he is the obligatory passage point for the budget and scope of work.

Let us discuss this boundary-spanning work in different parts of the ecology in relation to Hardy's four types of trust (Hardy et al., 1998). Some elements are more important if we are to understand the boundary-spanning work and the maintenance of trust among niches in this ecology. These elements are: potential participation in the community of SECs, core competence related to subsea engineering, company boundaries and identity/allegiance. Seen from the perspective of the SECs there is the most trust in the relationship among SECs. Trust is reduced when moving across company borders and when the technical domain knowledge of subsea engineering becomes more peripheral. In this sense trustworthiness is proportional with the subsea engineering competence and consequently a reduction in trust occurs when we move from subsea engineering competence to marine and shipping operations. The relationships between SECs can be described as a kind of **spontaneous trust.** Such trust situations develop where trustful relationships emerge naturally in the absence of deliberate attempts or intent to create them. SECs have responsibility for their separate systems, but they have a "community of practice" (CoP) relationship among themselves, with high degrees of shared meaning evolved through training, experience and practice (Wenger, 2002). The SECs have reached their positions via learning the tricks and ropes of their discipline in very similar ways. However, these activities can hardly be defined as boundary-spanning activities since they are within the CoP of SECs. **Generated trust** develops when cooperation is created via management of meaning and it resembles the relationship the SECs have in relation to Statoil internal vendors like the logistic planner function and to a certain extent FMC. The latter do not share the same symbols and discourse as that of the niche of SECs, but shared meaning is constructed via the SECs enrolment process and boundary-spanning (i.e., the SOW that is written and through ongoing communication through action during subsea intervention work). Hardy et al. also describes a third form called **manipulation.** This is the first façade of trust where cooperation is power-based but cooperation is achieved through management of meaning. On the surface it is similar to generated trust. The dominant partner uses symbolic power to reduce risk, increase predictability, and calculative benefits with the consequence that synergy related to collaboration is reduced. Since trust is a risky investment, it should not be surprising if SECs prefer to use power to achieve the necessary coordination in intervention campaigns vis-à-vis the other participants. Even though it may not bring about the creative synergy it increases the likelihood of predictable behaviour (Hardy et al., 1998). The consequence is that counterpart alternatives are reduced, therefore also the synergy in the interorganizational collaboration. This type of trust resembles much of the relationship with the subsea vendor FMC, but is a complex issue and I will return to this shortly. **Capitulation** is the final type or the second façade of trust. This type is also power-based through dependency and socialization and resembles the trust relationship toward shipping companies and marine operation contractors. These have the skills and competence that can be most easily contracted to other companies. These two niches act as a tool that works on behalf of the dominant partner Statoil. The risk to the dominant partner is low, but the synergy is also low. The weaker party is to a larger extent socialized to accept unquestioningly, its limited room for manoeuvre and does not interfere with the domain of subsea engineering.

Hardy et al. has given us some ideal types of trust that are important to understand boundary-spanning work, *but in real life these ideal types of trust have shortages when it comes to catching the dynamics of this interorganizational ecology.* If we look at the relationship between Statoil and FMC, the main issue is domain knowledge and skills. Statoil does not possess the core-competence to have full manoeuvrability. It is company policy that non-core competence work tasks should be performed by external vendors. Statoil has the critical competence and skills within the technical disciplines of subsea operation and maintenance to be a demanding customer. Statoil's subsea organization does not possess the domain knowledge related to execute marine and ROV operations. They are therefore dependent upon collaborating with external niches that have domain knowledge of Statoil systems and practices but have core competence in areas where Statoil personnel do not have the skills and expertise. The external supplier needs time to develop this competence, and this does not make it easy to switch vendors. Let us take FMC the main vendor of subsea systems as an example. Statoil has until the turn of the century mainly had an installed base of FMC subsea equipment and must rely on support from FMC personnel that knows these systems pretty well. This reduces Statoil's ability to use the power that is latent in the relationship. One is not always satisfied with the products and services of the vendor but is dependent upon having a long-term relationship with them.

The SEC must master the paradox of earning the trust of the other participants in his or her boundary-spanning activities to get the intervention job done while at the same time he or she leaves no doubt that he or she is the obligatory passage point and that the other participants involved in the subsea intervention campaign work on his or her behalf and according to his or her scope of work. This paradox is related to having control while at the same time not having control. Even though the SEC decreases the hierarchic control Statoil still has the operator ship and the responsibilities that come along with subsea operations. The paradox is that a certain level of stability, harmony, values, decisions, planning, procedures and goals must be in place, while at the same time the SEC as boundary-spanner must accept that instability, disagreement, risks, conflict, diverging goals and values exist in the subsea ecology at the same time. These are the requirements he must balance to get the potential out of this loose organization.

The second paradox is that this form of trust and boundary-spanning is both universalistic and particularistic. Boundary-spanning is a typical example of communicative activities that build on both universalistic and particularistic forms of trust. Adler (2001) argues that where market and hierarchy mechanisms are high on universalism and low on particularism, trust is high on particularism and low on universalism. Adler (2001, p. 227) also maintains that trust comparable to what we present in this chapter is a modern normative and reflective trust of the kind of values that characterize modernity (universalism, disinterestedness, and organized scepticism). It is inclusive and open but values integrity and competence more highly than a traditional kind of trust that values loyalty. A major point on paradoxes of trust in relation to universalism and particularism is that the SEC is able to handle this paradox in his boundary-spanning work and switch between the two forms of trust in the enrolment process of the SCM replacement. He develops a particularistic trust in his relation to his SEC colleagues. At the same time the work with the SOW, the planning and execution of the subsea intervention campaign rely upon universalistic mechanisms founded on market and hierarchy coordination mechanisms. The relationship with the logistic planner and the control room operator in particular is a delicate mixture and balance of the two forms of trust.

The third paradox is that boundary-spanning work that builds trust is non-material but is dependent upon material elements. We have shown that trustful relationships are vital for handling complexity and uncertainty. At the same time it is fragile and weak and therefore not sustainable without the help of material and non-human resources, or actor networks with nodes like contracts, telephones, tool pool, e-mail, SOW, and others across internal and external borders (Latour, 1991). The SEC manages to take advantage of the situation in Statoil, in a setting where the subsea organization of Statoil is an ecology of interdependent organizations. He mobilizes a temporary network of independent companies and Statoil subunits through his boundary-spanning work. The SEC is the main architect that creates the access and the foundation for a temporary structure, or loose coalition of operational and administrative services that come together for a specific business purpose in his asset; repair subsea equipment. Major parts of this enrolled assemblage disassemble when the purpose or scope of work has been met (i.e., subsea repair executed). This assemblage of human and non-human resources can exist temporally regardless of his enrolment on a long lasting base as they do in Statoil but the boundary-spanner brings them together in a practical arrangement to restore the technical condition of the subsea system. Through his boundary-spanning activities the SEC is able to align and move the assembly or collective of human and non-human resources in the direction he wants (repair subsea systems).The consequence is also that trust is sustained. There is an element of multiplexity related to trust here. McEvily (2003) argues that the tie sustaining trust becomes thicker because there are additional dimensions and relational contents. In addition to exchanging information and advice, friendship is also developed.

Paradox four is that boundary-spanners like the SEC handle dual and shared loyalties. The employees of every partner in the ecology must identify themselves with the job they are doing for Statoil but also identify themselves with their own company. They work on behalf of Statoil and must comply with the values and the business processes of Statoil. At the same time they have their own corporate culture, a culture that can be very different from that of Statoil. The life-world of a shipping company is different from that of an oil company. If we look at the work associated with the replacement of the subsea control module, the identity and heterogeneity of the partners remain visible throughout the operational phases of this process. They all work towards a common goal, i.e. replace the SCM; still in this work they maintain their uniqueness and heterogeneity. The shipping company conducts vessel operation, and ROT operations involving the replacement of the SCM is always done by suppliers like FMC. The easiest way to build a common identity in the ecology is to do it via practice and daily collaboration. The SEC must be confident that this loose organization or network knows Statoil's work processes (meaning roles, responsibilities for work tasks) and health, environment and safety standards. Without this knowledge it will be difficult for the different organizations to do a good job. At the same time the SEC is dependent on using the core competence of these niches. If he introduces a regime of strict control and regulation, it will be difficult for the niches in the ecology to do a good job, since he does not give them the opportunity to take advantage of their own identity, practice and skills. The SEC is able to balance these shared and dual loyalties. He must enroll them so that they buy the programs, meaning the values and operating procedures of Statoil. At the same time he must let them keep their loyalties to their own niche in the ecology because their identities tend to be linked to their core competence and practice.

CONCLUSION

This chapter has focused on boundary-spanning in practice in what we have called a subsea ecology in a high-tech business bracketed and described by formal procedures and a strict division of labor, or strict hierarchy and market coordination mechanisms. It presented the way in which boundary-spanning activities are dependent upon the maintenance of trust, institutionalizing boundary-spanning in human and non-human elements in order to re-establish the technical condition of subsea systems. We have shown that the trust element is the glue, or the foundation, for a flexible structure of communication and enrolment realized through boundary work. Engineers in this setting perform substantial informal boundary work to be able to do their work in an efficient manner. Their self-definition and devotion is realized through boundary-spanning, interaction with various material elements and through the extensive management of trust. The consequence of this knowledge-intensive operational practice is that the engineers have to live with paradoxes continuously. In the light of the situation of these engineers the chapter has addressed four issues related to the dynamics of collaboration and control that such professionals must cope with in today's high-tech environments.

REFERENCES

Adler, P. (2001). Market, hierarchy, and trust: The knowledge economy and the future of capitalism. *Organization Science, 12*(2), 215-234.

Aldrich, H., & Herker, D. (1977). Boundary-spanning roles and organization structure. *Academy of Management Review, 2*(2), 217-230.

Briers, M., & Chua, W. F. (2001). The role of actor-networks and boundary objects in management accounting change: A field study of an implementation of activity-based costing. *Accounting, Organizations and Society, 26*(3), 237-269.

Caldwell, D. H., & O'Reilly III, C. A. (1982). Boundary-spanning and Individual Performance: The Impact of Self-Monitoring. *Journal of Applied Psychology, 67*(1), 124-127.

Callon, M. (1999). Actor-network theory: The market test. In J. Law, & J. Hassard (Ed.), *Actor network theory and after* (pp. 181-196). Oxford: Blackwell Publishing.

Carlile, P. R. (2004). Transferring, translating, and transforming: An integrative framework for managing knowledge across boundaries. *Organization Science, 15*(5), 555-568.

Carlile, P. R. (2002). A pragmatic view of knowledge and boundaries: Boundary objects in new product development. *Organization Science, 13*(4), 442-455.

Cross, R. L., & Parker, A. (2004). *The hidden power of social networks: Understanding how work gets done in organizations.* Boston: Harvard Business School Press.

Friedman, R. A., & Podolny, J. (1982). Differentiation of boundary-spanning roles: Labor negotiations and implications for role conflict. *Administrative Science Quarterly, 37*(1), 28-47.

Hardy, C, Phillips, N., & Lawrence T. (1998). Distinguishing trust and power in interorganizational relations: forms and facades of trust. In C. Lane & R. Backman (Ed), *Trust within and between organizations: Conceptual issues and empirical applications* (pp. 64-87). Oxford: Oxford University Press.

Kramer, R. M., & Cook, K. S. (2004). *Trust and distrust in organizations. Frontiers of theory and research.* Thousand Oaks, CA: Sage Publications.

Lane, C. (1998). Introduction: Theories and issues in the study of trust. In C. Lane & R. Backman (Ed). *Trust within and between organizations: Conceptual issues and empirical applications* (pp. 1-30). Oxford: Oxford University Press.

Latour, B. (1991). Technology is society made durable. In J. Law (Ed.). *A sociology of monsters. essays on power, technology, and domination* (pp. 103-131). London: Routledge.

Levina, N., & Vaast, E. (2005). The emergence of boundary-spanning competence in practice: Implications for implementation and use of information systems. *MIS Quarterly, 29*(2), 335-363.

McEvily, B. (2003). Trust as an organizing principle. *Organization Science, 14*(1), 91-103.

Nardi, B, A., & O'Day, V. L. (1999). *Information ecologies: Using technology with heart.* Boston: MIT Press.

Orlikowski, W. J. (2002). Knowing in practice: Enacting a collective capability in distributed organizing. *Organization Science, 13*(3), 249-273.

Pawlowski, S. D., & Robey, D. (2004). Bridging user organizations: Knowledge brokering and the work of information technology professionals. *MIS Quarterly, 28*(4), 645-672.

Powell, W. W. (1990). Neither market nor hierarchy: Network forms of organization. *Research in Organizational Behavior, 12,* 295-336.

Star, S. L., & Griesemer, J. R. (1989). Institutional ecology, "translations," and boundary objects: Amateurs and professionals in Berkeley's Museum of Vertebrate Zoology 1907-1939. *Social Studies of Science 19*, 387-420.

Wenger, E. (2002). *Cultivating communities of practice.* Boston: Harvard Business School.

Williamson, O. E. (1993). Calculativeness, trust, and economic organization, *Journal of Law and Economics,* (36), 453-502.

KEY TERMS

CCR: Central control room, a room with operators that controls the production process on an oil and gas installation.

DPA: Detail pool agreement, the agreement that describes the day to day operations of the subsea pool, including its products, services and the rights-obligations of the pool partners.

FMC: A major manufacturer and service provider of subsea equipment.

HSE: Health, safety and environment.

ROV: Remotely operated vehicle, used for inspection and repair of subsea systems.

ROT: Remotely operated tools, tools used for replacement of heavy subsea components.

SAS: Safety and automation system, the system that handles the major safety functions on an oil installation.

SCM: Subsea control module, is a subsea component that contains important control functions of a subsea system.

SEC: Subsea engineering coordinator, engineer responsible for delivering availability of subsea systems.

SOW: Scope of work, is the description of the subsea intervention work to be executed.

SSA: Is a frame contract between Statoil and FMC that eases the delivery of subsea services.

Tool Pool: A pool that gives its partners eased access to subsea intervention tools and reduces the life cycle costs of tool ownership.

Chapter II
The Information Society: A Global Discourse and its Local Translation into Regional Organizational Practices

Ester Barinaga
Royal Institute of Technology, Sweden

ABSTRACT

The discourse on the information society is characterised by a democratic ideal of "general access." In this chapter, we follow the transformation of such an ideal as the discourse of the information society is translated by the Swedish parliament and implemented in a high-tech region north of Stockholm. We will see that as the discourse is being implemented, it incorporates ethnic categorical boundaries that structure the region and segregates the community where it is being implemented. The main argument of the chapter is that categorical inequalities are embedded in the economic rationality/business logic that structures the discourse on the information society, resulting in socioeconomic, geographic, and technological segregation.

During the last thirty to forty years, a new way of conceiving contemporary societies seems to have emerged. To describe the society we inhabit, scholars have started to use metaphors such as the "post-industrial society," the "post-Fordist society," (Amin, 1994), the "network society" (Castells, 1996/2000), or the "age of unworldliness" (Marquard, 1991). Conferences and seminars are organized to discuss the "image economy" (Dobers & Schroeder, 2003) and journals bearing the title "the information society" are established.[b] The last ten to fifteen years have seen an expansion of that worldview and today indexes rank states according to how well they are adapted to the "new economy" (Atkinson & Correa, 2007), politicians discuss how to make it into the "global economy," the EU launches calls for research proposals on the "knowledge society," and journalists write on the "global village."

The profusion of terms and analysis of contemporary society attest to the lively discussion on the essence of modern society and the role information technologies play in it. Neo-liberal observers praise the arrival of the information society as the coming of an open and self-regulated world market.[c] Neo-Marxist-oriented analysts warn that the information society is widening and cementing already existing socioeconomic class inequalities.[d] For some, information technologies are the means to empower and develop all groups of society; for others, information technologies are embedded into social dichotomies giving raise to unemployment, the lowering of wages, the lost of labor rights, poverty, exclusion, the deregulation of the welfare state, or migration (Bauman, 2004). Some stress the potential for a truly professional and caring society while others worry about the tightened control over the population. Often, however, and independently of political affiliations, political economist and policy makers use the term to legitimate a wide range of governmental measures—from neo-liberal policies aimed at deregulating markets and liberalizing business rules to more interventionist policies to give incentive to a particular business sector (Braithwaite & Drahos, 2000).

At another level of discussion, disagreement exists on whether today's society is substantially different to any other previous. On the one hand, some argue that what we are currently seeing is the emergence of a society that is radically different to those we have inhabited up to date. On the other hand, there are scholars who maintain that the difference between contemporary and previous societies is merely one of degree. These concede that information and theoretical knowledge play a special role in the present era but insist that current social changes are incremental, not radical in nature. Whereas proponents of the first view stress newness, discontinuity, and inevitability of the changes, the second focus on continuities, persistence, and possibility to shape the course of the changes (Webster, 2002).

The debates on the information society are thus many and varied. Yet, independently on the position one takes in what regards the goods or evils of the information society, and independently on whether society is undergoing a radical or an incremental change, one fact cannot be denied: the *discourse* of the information society has taken public discussion and scholar analysis with force.

THE INFORMATION SOCIETY: TWO EPISTEMIC APPROACHES

An interesting and, I believe, fruitful analytical distinction is relevant here. A distinction separating the information society as a descriptive term referring to a period in a larger historical process from the information society as a discourse, a way of comprehending the world, an ideology.

The information society can be regarded as a set of ideas, of language and social practices that provide knowledge on a particular subject: A discourse. As such, it defines and constructs the objects of our knowledge—in this case, "the information society"—sustaining a regime of truth (Foucault, 1980). A discourse is an *institutionalized way of thinking*, delimiting what can and cannot be said about a topic. Few question, for instance, the fact that information technologies are somehow shaping society. What is put into question is how far this shaping has gone and whether the shaping is occurring in one or both directions. None doubt the centrality of knowledge in that society for that matter, and few would dare to deny its global reach. The nuances are put, however, on how far that knowledge is reaching and on the parallel globalization of a new sort of inequalities.

"Institutionalized way of thinking"—the concepts of discourse, power, and knowledge as developed by the French philosopher Michel Foucault, are helpful at this point. Much more than a way of thinking and producing meaning, a discourse constitutes both the object of knowledge and the subject who knows. Neither is mere rhetorics. Rather, what can be talked about and how it is talked about is indistinguishable intertwined with social practices and individual identities. There is no clinic without the medical discourse, no prison without criminology, no asylums without psychiatry. Likewise, there is no patient, no delinquent, and no mad without those discourses. And, as we will see, there are no high-tech regions and no techies (and no non-techies) without the discourse of the information society. Discourse, that is, is the permitted, the authorized, and authoritative knowledge shaping and being shaped by the social context. In this sense knowledge, power, and discourse are inextricably connected.

I am not negating the reality of current historical changes. It can be argued, and many constantly do so, that the advent of the new technologies has changed how corporations make business, individuals communicate, rural communities sell their products in distant markets, or national governments relate in the international arena. But this historical sense of the information society is not the only one available. The information society may also be understood as a contemporary discourse serving certain economic and political interests. The mechanisms involved in the discourse of the information society become useful to prioritize people and goods revealing their political efficacy and economic gains.

In this perspective, the information society becomes ideological as it seeks to reshape society in accordance with a new imaginary that serves some interests better than others. It is not a coincidence that civil society organizations denounce an unholy alliance between both the corporate and the political worlds and make their discontent heard in multitudinary protests such as those in Seattle (1999), Göteborg (2001), or Genova (2003). Contradictory voices come from within academia, some denouncing the dominance of capitalism as an economic model and the pressure exerted by international institutions such as the IMF or the World Bank prompting the globalization of technologies, markets, and democracy.[e]

I choose to look at the information society as a discourse. Thus, discourse will be used throughout the chapter as an analytical concept. I will study the discourse of the information society not as mere, inconsequential talk, neutral in its effects, as if, to paraphrase Michel Foucault, the fact of speaking about the information society was of itself more important than the *forms of imperatives* that are imposed by speaking about it (Foucault, 1978/1990). Through the discourse, we will see economic policies are legitimated, corporate investments justified, and technology developments prioritized, choices most of these which tend to favor a group of people over another. It is in this sense that the information society has become the ideological production accompanying many of today's attempts to achieve economic and social development—it is used to legitimate, prioritize, and justify political, social, and economic agendas. Highlighting the discursive dimension of the information society forces us to focus on the terms under which the information society is occurring, a discussion that has been greatly neglected.

Regional development projects are particular instances where socioeconomic ideals are pursued. Projects for developing an IT region are instances to study a manifestation of the information society discourse. Kista Science City is one such project, both inspired by the discourse of the information society and with a strong Swedish sociodemocratic flavor. Its context is Kista, a region socially and economically divided along ethnic lines.

BACKGROUND TO THE KISTA CASE STUDY

When Sweden is portrayed as a highly technological society, Kista is often presented as the main example. Encouraged by *Wired Magazine*'s labelling of Kista as "the Wireless Valley" and longing for a reputation similar to that of Silicon Valley, Kista is regarded in Sweden as the region where the future of the high-tech industry is being developed. Yet, as it is the case in Silicon Valley, Kista was established long before it was dubbed the Wireless Valley.

It all started in the mid '70s. What had been a military field was transformed into a residential suburb built following the ABC-city concept; A for "arbete," work; B for "bostäder," apartments; and C for "centrum," a city center. The idea behind the concept was to create a suburb independent from the City of Stockholm, a satellite city where dwellers could live without having to commute between residence, workplace, and shopping places (Brattberg, 1977). In practice, this was carried out locating work places, center, and residences physically side by side (Johansson, 1991).

As a consequence, Kista is architecturally divided, most noticeably, by the underground line, which separates residential from working areas. Underneath the underground line, and bridging both areas, lies a big shopping center, Kista Galleria, where one can find everything from groceries and a pharmacy to jewelries, fashion shops or community services.

The Kista Borough is located some 15 kilometers to the northwest of the City of Stockholm. The underground line and a local train connect the borough to Stockholm city, while the national road E4 connects it both to Stockholm and further north to Arlanda, Sweden's international airport, and Bromma national airport. To the south west of Kista lies a natural reserve, the Järva valley, which public authorities are very keen to preserve. Bordering the south of the valley lies two other boroughs, Tensta-Spånga and Rinkeby planned following the same ABC-concept and built only a couple of years before Kista (Brattberg, 1977).

About 30,000 people live in the Kista Borough; 49% of the borough's residents have a foreign background, mainly from Asia[f] and Africa.[g] From 2002 to 2003, a period of high economic conjuncture, the percentage of open unemployment among residents in Kista increased from 3.7% to 5.1%. Such development is the most negative for the mentioned period for the whole City of Stockholm (S 2004:13, USK, 2004). Yet, the number of employees in Kista increased by 5.3 %,[h] making the negative development even more drastic; 65. 5% of residents do have employment (*cpr.* 75.3% for the City of Stockholm), 82% of which commute to work elsewhere.[i] Employment figures go down to 54.4% if only residents with foreign backgrounds are considered (SCB, 2004). Similarly, social subsidies (*ekonomiskt bistånd*) is more common among the population with foreign citizenship than among those with Swedish nationality—32.8% and 6.6% respectively (USK, 2004).

Interestingly, about 30,000 people work in Kista. Yet, these are seldom residents in the borough. People working in Kista commute daily from Stockholm or suburbs other than Kista. A mere 9% of those working in the area do live in it; 33% of Kista's working population live in other areas of Stockholm, 48% live in another municipality, while 10% live in another region (USK, 2003).

In short, Kista is a good example of a region where economic growth and urban development go side-by-side in unemployment and the ethnification of social exclusion. Architecture and the underground line are the physical counterparts of a deeper social and economic divide crossing Kista for, side by side with this typical "immigrant suburb," there is a growing business area. Public authorities and the private sector are cooperating to convert Kista into a world-leading IT-cluster or, as they prefer to put it, "the society of the future."

KISTA SCIENCE CITY: THE SWEDISH TRANSLATION OF A GLOBAL DISCOURSE

"Sweden is to become the first country to be an information society for all." The IT policy goal was made explicit by the Swedish parliament in March 2000 (prop. 1999/2000, p. 86) and was confirmed in July 2005 (prop. 2004/05:175). The underlying logic is the belief that "(a) more efficient use of IT creates a better ground for economic growth and national development."[1] Two aspects can be recognized in this goal.

First, "Sweden is to become an information society *for all*." On June 18, 2003, an IT policy strategy group was appointed. "In addition to exercising an advisory role vis-à-vis the government, the group is to play a proactive role in efforts to achieve the IT policy goal of an information society for all." That is, the group will see that the benefits of the Swedish information society reach everybody independently of gender, race, ethnicity or religion. Social background or national origin should not restrict a particular individual life chances in that new society. Traditional Swedish social-democratic ideals resonate in the policy goal.

The long-term objective of the parliamentary proposition is to achieve "economic growth, employment, regional development, democracy, justice, quality of life, equality, diversity, a sustainable society, and effective public administration." Since "IT affects all of us and changes our life conditions" the Swedish parliament reasons, the formulated strategy to achieve a democratic and equalitarian society is to focus on IT infrastructure and use (see Box 1).

Box 1. Kista Science City extended map

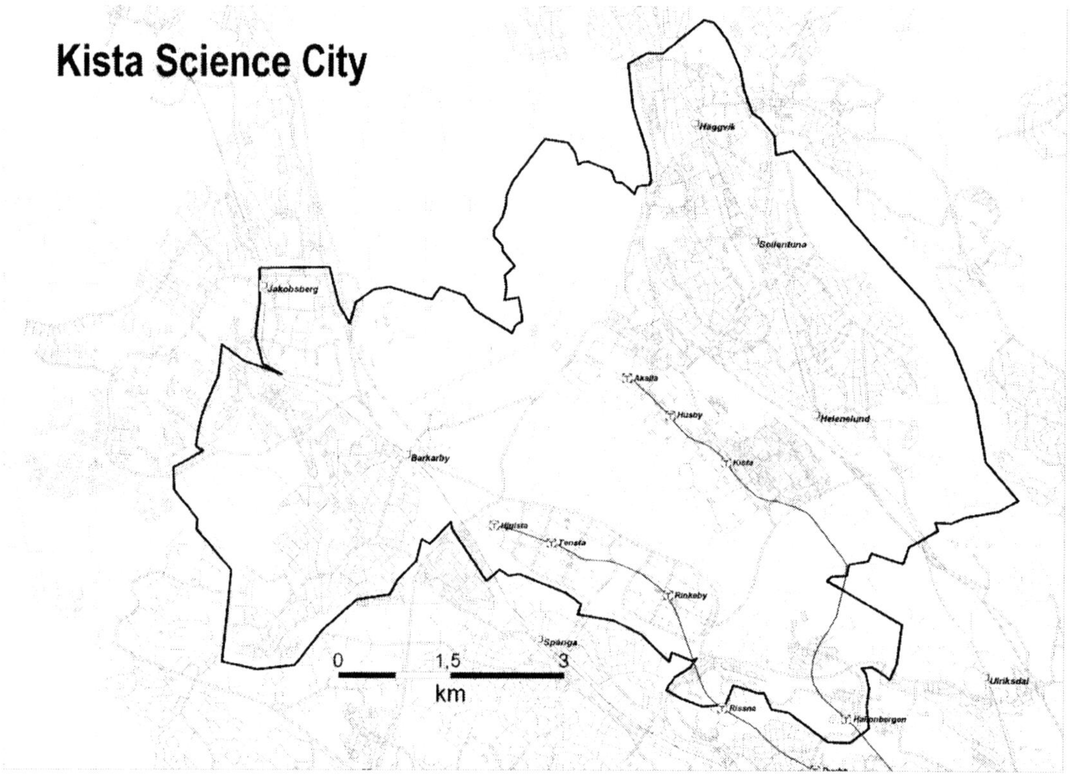

Echoing assumptions inherent to the discourse of the information society, general access to IT, and widespread IT knowledge should necessarily lead to democracy, equality, and social justice throughout society. Consequently, the Swedish IT industry and research are placed at the centre of Sweden's economic growth and social development. A strong version of technology determinism permeates the first aspect of Sweden's IT-policy goal.

"Sweden is to become *the first country* to be an information society for all." The second aspect of the policy goal puts the emphasis on being "the first country." One of the central tasks for the IT policy group is to consolidate "Sweden's leading international position at the forefront of it development."[k] An information society that benefits all is not enough. Sweden is also to take a lead in international technology development.

Whereas the first part of the policy vision has a social character, the second part has more of a technological quality. The first has to do with how the benefits brought by technology are to be distributed across society; while the second is a direct call for a technology contest. That is, national solidarity plus international competition. The conflation of both spheres, the social and the techno-economic spheres, is in line with Sweden's social-democratic tradition of juxtaposing economic growth with social security (Andersson, 2005).

The redefinition of Kista into a Science City should be regarded against this background. During the spring of the year 2000, the City of Stockholm in close cooperation with industry and university developed a vision for the future of a geographical area that includes Järfälla, Kista, Södra Järva, Sollentuna, and Sundbyberg.

The name of the vision: "Kista Science City." Its content: to "[d]evelop Kista to a living and growing Science City, with companies and a university on an international level, in an environment that attracts competent and skilled people." Furthermore, echoing the technological aspect of the policy goal, "Kista Science City will be known as a world leader in the three designated areas of growth: Mobile services, wireless systems and broadband systems." Whilst its motto captures the social aspect of Sweden's IT policy goal: "unlimited."

The responsibility to carry out the vision is Kista Science City AB, a company subsidiary to the Electrum Foundation. Striving to improve IT competence in Sweden, the City of Stockholm, Ericsson, and ABB started the Electrum Foundation in the 80s. The Electrum Foundation is, somehow, a lobby group working for the interests of the region. Kista Science City AB is its executive arm, marketing and developing Kista "to become the most attractive area for ICT companies to establish."[l]

At the time of the study, Ericsson, IBM, Telia,[m] ABB and a large Real State Proprietor were represented in Electrum's Board of Directors, as well as the Royal Institute of Technology (KTH) and the City of Stockholm—a sign of the type of interests vested on implementing Kista Science City. Similar interest groups were represented in the Kista Science City AB's Board of Directors—representatives for the private IT sector, real state proprietors, the local university, and the city administration. That is, an alliance between the corporate and the political worlds is behind the ideas shaping the future of Kista.

And indeed, much has been achieved since Kista Science City AB started in the year 2000. The IT University, a collaboration between Stockholm University and KTH, has been established in Kista attracting over 2,500 students; the Royal Institute of Music and Karolinska Institute moved their IT-related educational programs to Kista; various research centres have been established (Wireless@KTH, IMIT, Kista Photonic Research Centre, Kista Integralen); Kista high school has been transformed into IT high school; a business lab, Kista Innovation & Growth, and a risk capitalist, KTH Seed Capital, have been established; Ericsson's corporate headquarter has been relocated to Kista; Kista Galleria has undergone a major renovation; Kista Science Tower and Kista Entré have been built; road infrastructure has improved; two new bus lines have been added, one between Vällingby, Kista and Sollentuna C, and the other linking Barkarby, Kista and Danderyd; two extra wagons have been added to the blue underground line; and student housing has been built.

Undeniably, Kista Science City AB has been successful in transforming Kista industrial area into a high-tech cluster, or a "science city" as those behind Kista Science City AB prefer to formulate it. Yet, it is more questionable if the IT policy goal of creating "an information society for all" has advanced. To be sure, the Kista case shows how the very effort to place Kista in the international IT scene has contributed to enforce boundaries that further separate those working but not living in the area from those living but not working there.

Industrial and economic success as well as social failure is a reminder that discourses are not mere inconsequential words. Discourses have consequences on how we see the world and further act upon it. It seems, then, that if we are to understand the so-called information society, we need to get deeper into the workings of discourse. How does the structure of a discourse structure the region on which the discourse is being implemented? How does the economic logic inherent to the discourse on the information society (and to its locals translations) lead to power differentials among the groups present in the region where the discourse is being implemented? Is the logic of economic growth compatible with the existence of a multicultural/multiethnic suburb?[n] More particularly, what processes has the Kista Science City vision set into motion? How has a diverse population entered the plans and efforts for the growth of a regional IT cluster?

These are the questions addressed in the chapter. We will see how the economic logic embedded in the discourse of the information society and structuring the Kista Science City vision is a piece in the puzzle of understanding persisting socioeconomic differentials despite urban development and economic growth. This occurs through the discursive technique of categorical pairs. Categorical pairs, we will see, have contributed to cement socioeconomic inequalities in a high-tech region in the making.

"SCIENCE CITY": AN ORGANIZATIONAL PROBLEM

Setting boundaries to separate two groups of people in the region responds to an organizational problem.[o] It seems that not all parts involved received the label "science city" positively. Whereas some defended a more holistic vision that included social and cultural life, others seemed to have preferred a stricter definition of the area and its activity. The former, argued for a "science city;" the later for a "science park." The former wanted a new vision that set new efforts into motion; the later were content with the existing label "Kista Science Park" and its achievements. The former adopted a regional perspective; the later emphasized the importance of keeping a business cluster approach. The former stressed the sociodemocratic ideals dominant in the IT policy goal, that is, the relationship between technology development, economic growth and social justice; the later focused on technology development per se. As *bn*, managing director for the Electrum Foundation and for Kista Science Park AB until 2000, recounts:

The idea behind Kista Science City is to market a region. But, that is something much more difficult than to market a business area. It is the name Kista Science Park that put Kista in the world map. [...] My successors neglect Kista Science Park.

Competing understandings of Kista's future coexisted. The "science park" project aimed at giving the area an identity to place Kista on the international map of science parks. It focused on the business cluster. The "science city" focused, instead, on regional and social development. It aimed at coordinating the efforts of various municipalities in matters such as road infrastructure and housing. As *pah*, CEO for Kista Science City AB at the time of the interview, puts it:

What is relevant is that to think about the well-being of this area was not enough to do something good. Rather, the people here thought this area was boring. And transport service, housing, and all that presupposes that we gather everybody in our surroundings in a smart way. So we began to talk of developing this into a city instead of a business area or a science park. We broadened the concept. (*pah*)

The confusion about what Kista was to be was latent at the time of the study. In fact, when asked what the difference was between a science park and a science city, many of those involved in the implementation of the vision had difficulties to answer. mb, at the time marketing director for Kista Science City AB, and *uk*, a colleague of his,

eb: What does Kista Science Park refer to and what does Kista Science City designate?
uk: Try to answer that one!
mb: Well, that was a good question. Kista Science City is a vision for the region. If one likes putting a word to everything, one could say that Kista is 'downtown' and the rest is the suburbs. Kista Galleria is central Kista Science City. [...] Kista Science Park is only Kista. Kista is, after all, a Science Park! But Kista Science Park ab or Kista Science Park as a name does not exist anymore. The change of name was also for the other boroughs and municipalities to feel that they participated. We are after all much more than a Science Park!

The organizational problem Kista Science City AB faces can be formulated with the questions, what is a "science city"? What does a "science city" do? What makes it different than a "science park" or a high-tech cluster? How can a "science city" contribute to regional development and, by extension, advance the IT policy goals of economic growth and social justice?

Kista Science City AB has an added difficulty when defining "science city" since city inevitably brings to mind a population, a local community. However, as we saw in the previous section, a big portion of the population in Kista has an immigrant background, is unemployed, and lives on social welfare. All of these characteristics fit badly with the associations of high-status and high-income attached to a technology-intensive region. And yet, the Swedish parliament's goal is to extend the benefits brought by it to the whole population. *la*, Director for the Kista Borough from 1996 until 2001, saw in Kista's business area the possibility to achieve that goal in the whole Borough. His vision read "everyone will have a place to go at 8 o'clock in the morning." The problem of defining a science city has thus to be nuanced by Kista's local characteristics and the broader political attempt to achieve "an information society for all."

In its efforts to make sense of "Kista Science City," one of the first steps taken by Kista Science City AB has been to define who "we" are and what "we" do. Yet, a "we" presumes a "they," and hence, the beginning of a process of exclusion. It is therefore interesting to analyze how those behind Kista Science City see themselves, who is considered dissimilar, and how relations are defined between the two sets of actors. Two mechanisms moved the definition process forward: silencing, and the transformation of "the other."

Figure 1.

Facts and figures

Land area: 2,000,000 m^2

Office space: 1,100,000 m^2

650 companies
250 ICT companies
175 SMEs

28,000 employees
(Ericsson 8,000)

Main areas

Wireless Systems
Broadband Systems
Mobile Services

kista
SCIENCE CITY

SILENCING

Kista Science City AB's formal presentation of Kista Science City starts with the "facts and figures" of the region (see Figure 1).

Figure 1 talks of office space, number of employees, and amount of companies. Out of the 650 companies housing in Kista, those working with related tasks are specially high-lighted. Furthermore, technology development is preferably carried around the wireless technology, broadband systems, and mobile services. All 400 companies do not work with ICT, yet, we remain curious about their type of activity. Besides it, a second industrial characteristic being high-lighted is the number of small companies. To judge from the acknowledged facts and figures, it seems there is a desire to emphasize Kista as an entrepreneurial, dynamic, technological region.

A second aspect should be noted. All information presented in this first slide has to do with Kista's business area. This is the keynote throughout the whole presentation. The Kista they are defining as entrepreneurial and technological is a limited Kista—its size is 2,000,000 m2. Kista's boroughs and its inhabitants—Akalla, Husby, and Kista's residential side—not to forget the boroughs and municipalities encompassed by the wider Kista Science City—that is, Sundbyberg, Rinkeby, Spånga-Tensta, Järfälla, and Sollentuna—are silenced; Kista's residential area is completely absent in the presentation. In a single coup, the technological definition has shrunk Kista Science City to the IT-cluster, excluding the city's population.

In the attempt to define Kista along the technological boundary, a further step is taken. Kista Science City AB presents Kista's history. As all historical accounts, this one is written from a particular perspective, a perspective, seemingly, with an agenda.

From being a farming area and later a military training ground, Kista has transformed into a high-tech region. From the moment the City of Stockholm decided to end military training in the valley located south of Kista, the Järva Valley, up to our days, the story told follows a straight line. It is one of progressive "technologization" of the area. The 70s saw SRA, Rifa[p], and the Swedish operations of IBM move to Kista. During the 80s the City of Stockholm together with Ericsson and the Royal Institute of Technology (kth) took the initiative for planning an electronics centre in Kista. Electrum (the Electronics Centre Foundation) was thus founded in 1986. By the 90s about 20,000 people worked in more than 200 working places in Kista, cutting-edge, international IT and telecom companies had their main offices in the Borough, and numerous institutes carried out their research in close collaboration with industry. During the first years of the new millennium, the effort is extended to include the areas around the Järva Valley (Sundbyberg, Rinkeby, SpÅnga-Tensta, Järfälla, and Sollentuna). The business community, the university, and local authorities develop the vision of Kista Science City and founded Kista Science City AB for the implementation of the vision. Work with the vision starts with the renovation of Kista Galleria and the construction of Kista Science Tower and Kista Entré.

It is a story about success, about how, from there being nothing in the area, a high-tech cluster is created. Kista's historical account is built around technology, giving the impression that technology is the backbone of Kista's history. Once more, everything non-technical is left in silence. The only event mentioned that is not directly related to technology has to do with how the old economic order and its people are gone: "the last farmer leaves" Kista in 1976—but even here, the information is placed to reinforce the account of continuous technologization.

Figures 3 and 4 symbolize the modernist project of designing the forms of human togetherness, a project that emphasizes and ties together the values of economic growth and technology development with the idea of social progress, a project that has lately received the name "the information society." Furthermore, the account

Figure 2.

Some milestones in Kista's history

1905	Military training ground
1972	Construction of houses gets underway
1975	First to move in: Ericsson – SRA och Rifa
1976	The last farmer leaves
1980	IBM moves in
1985	The City of Stockholm takes an initiative to build an electronics centre
2000	The work to create a joint vision starts
2001	The building of Kista Science City starts

kista science city

Figure 3.

Figure 4.

told in Figures 3 and 4 leaves aside what is not to be considered within the particular information society project carried out in Kista, giving a hint of the exclusionary practices imposed on the borough.

The same two aspects that characterized the choice of facts and figures that introduced Kista do also structure the historical account. One, a focus on technology. Two, demarcating the geographical area to Kista's business area. These two characteristics are not independent of each other. They are related through what is left in silence: non-technological activities, non high-tech places, non-techy people.

Silence is best heard when comparing the official history with that presented by Kista's own residents. Instead of structuring their account along the technological boundary, residents' historical account is organized along ethnic and cultural lines:

At the beginning of the 80s, those moving to the area where those that lived cramped in the city centre and those who had sold their house. ... In the mid 80s, whole families, particularly from Latin America, moved from Tensta and Rinkeby to Kista. To move from Tensta and Rinkeby to Kista was to move up socially and economically. But those who were already living here, mostly Swedes, started to feel that there were far too many immigrants. It is not allowed to say so, but that's sheer racism! [...] Immigrants continued to move in, but by the end of the 80s they came from other parts of the world: Greece, Yugoslavia, Finland, and Iran. Iranians, they really wanted to adapt, they wanted and could give. The big groups of Somalis that are there today, especially in Husby, have arrived only during the last five years. (*gw*)

gw is a 76 year-old Swedish woman. She moved to Kista in 1974. Her account, like so many other accounts I have heard in the residential side of Kista, focuses on social aspects—conflict and friendship between people with an immigrant background and Swedes, unemployment, difficult economic conditions, low status jobs, segregation, and marginalization. *ss*, a young Swedish woman working in one of the local schools, says:

There are very few Swedish students [in Ärvinges-kolan]. Much has changed during the 12 years that I've been here. At the beginning there were many Swedes. It's enough to look to the schoolbooks! But gradually they moved out. Kista became a suburb, no good place to live in. Like Rinkeby or SpÅnga-Tensta. Immigrant areas. Kista was becoming a ghetto. Ericsson, they had busses for their employees, usch! They were afraid that anything would happen to them in the underground! (*ss*, ungdomsledare)

The categorical pair Swede/immigrant is manifest in the residential side of Kista. The historical account is structured along a clear ethnic categorical line—from being a middle class area mainly inhabited by Swedes, it has become a "typical" immigrant area. With "typical" it is meant all connotations that are usually attached to the category immigrant—connotations of low economic and social status, integration problems, and conflict. These connotations led many of Kista's previous Swedish neighbours to move elsewhere because, as *gw* puts it, "many felt there were far too many immigrants."

None of my Swedish friends live here any more. All moved out. (*gw*)

I wasn't the only immigrant. There were one Polish girl and another Finnish girl. But I was the only Muslim kid, and I don't eat pork, you know. They didn't know how to do, they didn't know what to give me for lunch. Today they simply don't offer pork at the school's restaurant; there are very many Muslims and they are used to those sort of things. (*ia*)

ia is a 30-year old man, born in Sweden from Turkish parents, and has grown up in Kista. His story corroborates *gw*'s account. That Swedes moved out as people of immigrant background moved in and this shift has contributed to define residential Kista as an immigrant area. Hence, not only does residents' historical account follow the categorical line Swede/immigrant. Despite Kista Science City AB's efforts to the contrary, the categorical pair gives the area a certain identity.

The organizational problem Kista Science City faces is to define who they are and what they do. This is done by using technology as the main and only defining attribute. Now, since Kista's residential side is identified with the category immigrant, the best way to move towards a technological definition of Kista is to concentrate exclusively on technology. "We" are a high-tech region. "We" do wireless, broadband, and mobile technology. "Our" history is one of continuous technologization. All that does not conform to the technological definitory requirements is left aside, silenced; it becomes a "they." Consequently, the big Kista Science City is reduced to Kista's business area. Which leads us to the second aspect of the official historical account.

Kista Science City is a vision covering the extended area around the whole Järva valley. Yet, reading the official historical account, it is only possible to read about Kista's industrial area. So much is this so that by the end of the presentation of Kista Science City a map delimits Kista's confines (Figure 5).

The map ends at the underground line, excluding the bigger Science City and, most blatantly, Kista's residential side. Defining Kista as a high-tech region implies silencing everything

Figure 5.

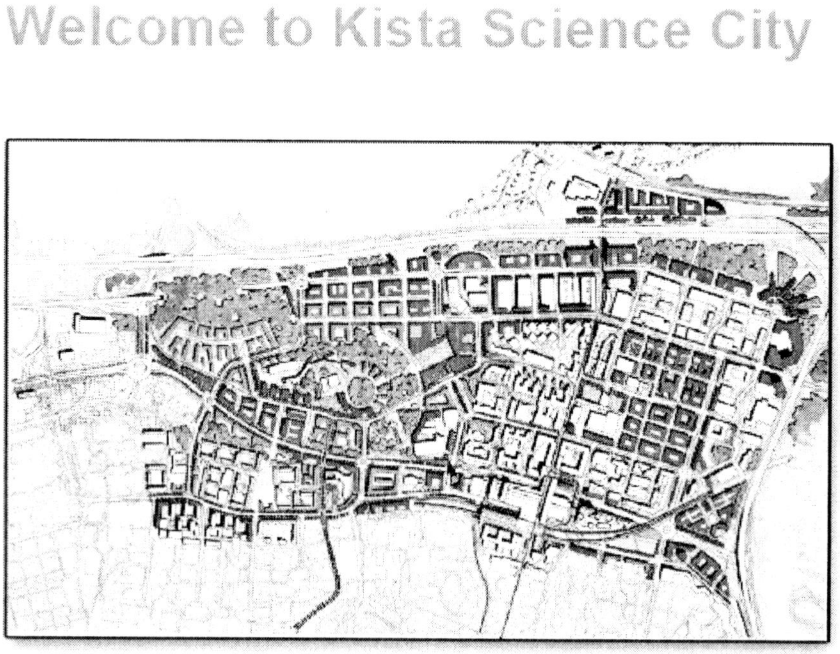

that does not conform to the technical category. And silence entails disappearing from the map. The categorical line techy/non-techy geographically separates places, splits people, and rewrites historical accounts.

Silence, thus, facilitates moving away a defining boundary—the ethnic categorical boundary. It helps to install a new defining categorical line—the technological one. It reshapes Kista's geography shrinking it to a smaller area and gives the region an identity—that of "Sweden's Silicon Valley." Silence reinforces the very logic of categorical boundaries: two sets of actors are distinguished, separated, and ranked. The set left in silence often falls into the less privileged category of the two.

A discursive trick helps Kista Science City AB to transform an immigrant area into a high-tech region. "High-tech" is often linked to terms such as "international," "high-status," or "economically strong," all of which are opposed to what the category "immigrant" is associated to. Hence, by defining Kista as high-tech, confining "immigrant" to another Kista and silencing (or failing to represent) that "other" area, that particular problem is side stepped.

One thing must be noted at this point. Discursive silence is not synonymous to un-consequential silence for the people left in silence. Discursive silence is the manifestation of power differentials between the two communities co-existing in the region. What is kept silent and what is spoken has important consequences. Economic and social policies are, among other things, designed to meet the demands of lobby groups or to answer dominant social accounts. If these social accounts are biased, as the one about Kista seems to be, economic and social policies are bound to disregard the realities of those silenced by lobby groups and by prevailing social accounts.

By keeping all but IT-related activities in silence, Kista Science City AB is actually drawing a dividing line between technology and non-technology. It distinguishes among companies and categorizes people according to how technology-intensive their activities are. Not only do words do things (Austin, 1975) but also silence does things. Silence *identifies* and *separates* between technological and non-technological activities, between "us" and "the other." It rests prominence to what is left silenced, hence *ranking*, as well as identifying and separating, different kinds of activities and people. Technology-intensive companies and people working with IT fall in the privileged "techy" category. The others, the "non-techy," are subordinated.

But separating (identifying who and what is in and who and what is out) and ranking is an inherent part of organizing (and exercising power). And so, defining Kista as a technology-intensive region is organizing the region along the categorical line techy/non-techy.

Now, it could be argued that I have been looking at official presentations, and official presentations are meant to paint a nice picture of what is being presented. However, official presentations tell about the context people inhabit, about the space for action made to individuals, and about the goals justifying unequal relations among groups of persons. In any case, in the next two sections I will analyze interviews and other sources of empirical material. We will also be able to distinguish a second mechanism contributing to embed segregation into the socioeconomic order: the transformation of "the immigrant" through the discourse of multiculturalism.

TRANSFORMING "THE OTHER"

The categorical boundary immigrant/Swede is not always kept in silence. Many are aware of the categorical divide that crosses Kista. Yet, in numerous occasions, when discussing it, the divide seems to change character. For instance, an extract from an interview with *pah*, at the time CEO for Kista Science City AB:

> *What is fun is that Kista, Husby, and Akalla have a big population coming from other parts of the world. Kista's working area and even more the IT-University have almost as big a proportion from other parts of the world. There is no-one here thinking that that is a problem at all. That's quite obvious! It is very important to bring people here that are interested and that can help these companies to grow. So that, somewhere there must be something to learn from each other, I think, at the other side of the road.* (*pah*)

pah is completely aware of the line dividing Kista. "The other side of the road," as he puts it, is inhabited by a higher proportion of people with immigrant background, unemployed, and people living on social welfare. The underground line is the physical counterpart of the symbolic, economic, and social divisory line immigrant/Swede.

Yet, redefining Kista along the technological boundary implies turning the ethnic divide upside down. And so, "what is fun" about Kista is that the population is so international. Without any other information mediating, *pah* compares the amount of foreign people living in Kista with the amount of foreign people working in the industrial area. Presenting both pieces of information side by side rubs out the fact that the relationships of these two groups to the region differ in status. The first, those residing in Kista (and positioned along the ethnic boundary), are subordinated to their Swedish counterparts. They fall into the category "immigrant." The second, those working in Kista (and positioned along the technology boundary), are not subordinated in the socioeconomic order. They fall under the label "expatriate," or rather, "technology experts," "techies." People of immigrant background and techies may fulfil similar requirements: a high-level of education; ability to speak foreign languages, most importantly English; and working experience. Yet, different boundaries mark their identities. Expatriates are valued for their educational level, professional experience, and technological expertise; people of immigrant background are equated with low-status jobs or even unemployment, and lack of technological competence (or at least, low-tech).

By juxtaposing two distinct pieces of information (one, foreign population; two, foreign workforce) *pah* draws attention to what they have in common: foreignness. This adjective has two different uses depending on whether it is used together with the technological or with the ethnic boundary. But since Kista is being redefined exclusively along the technological line, foreignness retains only the use given within that line. That is, international. Accordingly, people of immigrant background are valued for the international flavour they add to the region.

pah continues stressing the importance of "*bringing here* people that are interested and that can help these companies to grow." "Bring here." That is, expatriates. No wonder that "no-one here thinks about that as a problem." "That," expatriates. The council of the Kista Borough, public authorities, immigrant associations, government and other actors do however think of ethnic segregation as a problem. And *pah* is not unaware of this debate. My guess is therefore that he means "no-one thinks about expatriates as a problem."[q]

As a reminder of the previous map demarcating the high-tech Kista, the "people that are interested and can help [IT] companies to grow" are, to paraphrase *pah*, "on this side of the road." The road and the underground line that runs over it are the physical boundaries separating "us," the high-tech, from "them," people of immigrant background. *pah*'s last sentence, "somewhere there must be something to learn form each other, on the other side of the road" is a recognition of this divide.

One interesting thing has happened here. Two symbolic boundaries—technology and ethnicity—share a physical barrier—the underground line. As a result, the categories distinguished by the symbolic boundaries are matched, and meaning infused to each other. The high-tech region defined by the technological boundary is located at the north of the underground line. The immigrant category distinguished by the ethnic boundary inhabits the south of the underground line. The logic of opposites inherent to boundaries and categorical pairs ("we" are what "they" are not; and "they" are what "we" are not) helps the physical border to symbolically separate the immigrant from high-tech. If the north is occupied by high-tech, the south, the immigrant south, must be non-technological, or at least technologically uninformed. In other words, physical borders and the coexistence of two categorical boundaries results in matching people of immigrant background with technology unawareness.

pah must be praised for his rhetorical skill. In a few sentences he has transformed the categorical pair immigrant/Swede and allied it to the techy/non-techy dimension, granting the area a aura of internationalism. Yet, the transformation is not complete, and the meaning of low status associated to the category immigrant remains. Having matched both categorical pairs results then in an extension of the associations made to immigrant to include that of non-techy.

pah is not alone in putting different types of information side by side, making them seem of a similar kind:

Almost 30,000 people live in the Borough. More or less the same amount of people works in Kista that has a strong and expansive business life. Kista is an international city Borough with many dwellers with roots in other countries and cultures and many companies with activities all over the world. (Kista - En framtidsinriktad stadsdelsnämd i en dynamisk stadsdel)

Reading this paragraph without knowing about the socioeconomic segregation in Kista, it is easy to conclude that the 30,000 people that live in the region also work there. Yet, that is not the case. As we saw earlier, 91% of the people working in Kista live elsewhere, and a whole 82% of the local residents with an employment commute to work somewhere else.

When the people of immigrant background are mentioned, it is done in connection to an image of internationalism. "Kista is an international borough." The variety of countries of origin of Kista's population together with the existence of companies operating all over the world, serves as a proof of Kista's international atmosphere. Somehow, unemployment and segregation are left aside in favour of the exotic flavour added by a population with a diverse cultural background.

Worth noting is how the management discourse of multiculturalism has slipped into official accounts of regional development. The management argument of people from different cultures contributing to enhance a company's productivity is successfully used here to highlight the region's technological performance and suppress its social division.

The transformation of the immigrant other is, somehow, another way of silencing the voices of those living in Kista but unable to work in the higher-status jobs of the industrial side. A single characteristic is sorted out (foreignness), twisted (into internationalism), and aligned to/matched with the interests of the technical side (high-tech region of international standards). The economic preoccupations, the social troubles and fears of those at "the other side of the road" are left outside the definition of Kista. Decisions concerning Kista are thus made based on a picture of the region that relies exclusively on the voice of one single group, those at the IT side.

SEGREGATED (TECHNO-) ECONOMIC RELATIONS

Silence contributes to install a technological boundary and to redefine Kista according to that boundary. Transforming the ethnic other into an international individual is another rhetorical strategy, this time used to match the technological and ethnic boundaries. The organizational problem of defining Kista is coped by offering a biased image of the region. The image offered is that of an international area with high educational standards, an image that ignores the social reality of the region.

Yet, I believe this strategy has further consequences than ignoring a social reality. Matching ethnicity, a widely accepted categorical pair, with technology to define a region and its economic industry determines the direction of technology development. This in turn leads to entrench ethnic segregation in the very technology that is supposed to exceed it.

One of the most immediate consequences of painting a one-sided picture of a whole region is that such a picture guides technology development in a certain direction. The definition of Kista as an exclusively high-tech region inhabited by "highly-educated people" leads those with the power to influence the direction of technology development to ignore alternative paths. Let me give a couple of examples.

The first example comes from an interview with a researcher working for a research institute in Kista. They are developing a navigation service to be offered in mobile phones. The service is only possible in those phones including GPS technology. The researcher is, at the moment of our meeting, testing the service among visitors to the Kista shopping center. When describing the test group, she says:

We focus only on those who work in Kista. Weekend people do not have such an equipment. Well, we can assume that. (*ås*)

The researcher uses a euphemism to refer to Kista's immigrant population. During ordinary weekdays the shopping mall is visited by both people living in the area and those working in its industrial side. At lunchtime, restaurants are full with workers having a quick meal and the mall busts with people running their errands. During weekends though, the mall is visited mainly by those living in the area. "Weekend people" is a euphemism to refer to these, that is, people of immigrant backgrounds.

The euphemism is both a sign and a root of the social and economic segregation in Kista. It signals the already existing separation between both groups concerning their daily activities. "Weekday people," the researcher hints, work in the area, shop in the mall, and are technology interested. "Weekend people," however, spend their weekends wandering around, have less economic possibilities, and are supposed to be technology indifferent. The researcher uses day activities to point to the boundary separating two groups of people in Kista. Second, as she relies on that boundary to test the new technology, she is actively reproducing economic segregation. Since the technology is only being tested on "weekday people," the service developed will be exclusively sensitive to their preferences and everyday routines. The preferences and routines of those with other life-styles, yet also visiting the mall, are ignored. It is bound to be more difficult for them to accept the new technology.

The second example took place in the summer of 2003. An international conference was arranged at the KTH campus in town. Participants came from mostly Western countries—France, UK, Sweden, the U.S. The conference theme was digital communities. Hence, among the conference activities, organizers planned a study visit to Kista. This began with a walk through the residential parts, from the Husby to the Kista districts. I had the chance to guide that walk and, among others, told them about the social and economic segregation in Kista.

After a lunch in Kista Galleria, we continued with a visit to one of the research institutes in Kista. The director of the research institute received us. As he presented the diverse research projects carried by the institute and the various wireless products developed, the following exchange took place.

conference participant: What about wireless in Husby?
institute director: Let me be frank. Those people are not quite Swedish. They are immigrants... our products are not for them. They don't have the education or the income.

The director's answer follows an economic rationality. High-tech companies fund the research institute and these, needless to say, want return for their investment. New developed technological products are often expensive and thus, out of reach for those with poor economic possibilities. To develop inexpensive technology might not be within the interests of he institute's financial supporters. Thereof, the refusal of the institute's director to consider people of immigrant background as a niche group for technology development. As Thomas (1995) argues, developments are the result of capitalism. "Their nature is shaped by what is expected to sell, not by what is expected to help produce a utopia or egalitarian society" (p. 91).

Yet, I guess many, like me, feel tempted to characterize the previous exchange as an example of overt discrimination. The flaw of the director's reasoning is that he makes his economic calculations along the boundary Swede/immigrant. "Those people are not quite Swedish. They are immigrants." He separates two groups of customers based, not on economic aspects, but on matters of origin. He then goes on to characterize the immigrant group as lacking education and income. Hence, the director seems to imply, our products are developed for the other group—that is, Swedes—who have both the education and the income.

What the director seems to forget in his economic calculations is that people of immigrant backgrounds may turn into an economic advantageous market niche to pursue. Frissen (1995) shows how, in the early days of the telephone, major operators never thought of women as possible users. For over 20 years operators disregarded women's social use of the telephone (as opposed to the business/functional use attributed to men), thus missing an important market segment. Yet, by using the telephone as an instrument in social life, women influenced the unwilling telephone industry into accepting sociability as a major telephone use and turned it into a profitable strategy. In other words, Frissen shows how the "uses of the telephone are embedded in a set of ideas, values, beliefs and practices, expressing what it means to be a man or a woman" (p. 87). In the same way, the directors' words point to a set of ideas, beliefs, and practices expressing wider cultural notions related to ethnic social arrangements. The economic argument gives space, and even some sort of legitimacy, to an organization of society along racial boundaries. He not only disregards people of immigrant background as customers worth to be attended. He is reproducing in the very technology social segregation and cultural notions of "us" vs. "them."

Categorical boundaries rely on over-simplications of wide groups of people. However flawed such analysis, they influence decision-making and, in this case, the direction of technology development. This, in turn, further cements economic segregation. Rather than taking advantage of the so praised democratic potential of information technology, technology development becomes a boundary-maintaining practice.

DISCUSSION

The study presented here follows the discourse of the information society as it is formulated by the Swedish parliament and translated in the Kista Science City vision. We distinguished two aspects in the Swedish version of the discourse. First, a social-democratic ideal: to include everybody in the design of the new society. This will be done by investing on information technologies because, it is assumed, such technology is central in the design of a democratic and equalitarian society. Simultaneously attaining economic growth, political democracy and social welfare is possible thanks to an implicit belief in technology determinism. Second, a global technological challenge: to be at the international forefront of IT development. Both aspects were conflated in a single IT-policy formulation: "Sweden is to become *the first country* to be an information society *for all*." In Kista Science City the social element of the discourse became a motto—"unlimited"—whereas the technological element was translated into the aspiration to "be known as a world leader" in IT development.

Responsible to implement the Kista Science City vision was Kista Science City AB, a collaboration between the City of Stockholm, Ericsson and, initially, ABB. Given the conflation of economic, technological, and sociopolitical goals, it seems natural to involve in the implementation process all those actors in the position and with the power to influence the economic, technological and social spheres of society. That is, the private sector (Ericsson and ABB) and political agents (the City of Stockholm)—an unholy alliance that has been denounced by anti-globalization movements.[i]

The focus of the chapter is the process of implementing the discourse/vision of the information society. Not surprisingly given the corporate-policy alliance, Kista Science City AB's own attempts to define who they were, what they did and to what end they did it led them to focus, exclusively, on the technological aspect of the discourse. Kista was defined along technological boundaries. Silence and categorical pairs were two power techniques used in the effort, apparently erasing all traces of the discourse's social aspects. A third technique was also used: the idea of multiculturalism. This one helped to make invisible the workings of the two previous. The power effects of those techniques: the division produced in the community, the associations incorporated to the identity of "the immigrant other," and the alteration of the idea of the information society. In short, the normative ideals of the society in the making transformed from being all-inclusive to discriminate people, activities and places.

That ideas and discourses transform as they are repeated and implemented is not new. Within the field of organization studies, Czarniawska and Sévon (1996) show us how ideas travel, are picked-up by various actors, change depending on the context, and ultimately, materialize. In the same line of argument, Sévon (1996) conceives imitation as the process of translation in which ideas or practices "are transformed by chains of translators," while Sahlin-Anderson (1996) prefers the term editing to refer to a similar process. Foucault is renowned for his studies on how established cultural ideas—such as madness (in *Madness and Civilization*), illness (in *The Birth of the Clinic*), crime and delinquency (in *Discipline and Punish*), or sexuality (in *The History of Sexuality*)—transform over historical periods to serve the tactical convenience of social systems. All these researchers emphasize that the result of the transfer of ideas to new contexts or to different times in history is not necessarily even similar to the idea that was originally conceptualized.

What might be unique to this study is that it evidences the material effects of the discursive transformation on the organization of a community. In this way, the study shows the tight links existing between discourse and community structure, highlighting the conditions under which the information society is occurring.

A second interesting lesson from the case studied is the insight it provides in three discursive techniques used by the powerful for the redefinition of the region: categorical pairs, silence, and multiculturalism.

Categorical Pairs

Contrary to what is intuitively assumed, the components of a categorical pair do not describe two separate realities but one and the very same. Dichotomies characterize what is proper, but also what is improper. They demarcate a "we" through what "they" are not and a "they" through what "we" are not. The two sides of a categorical boundary acquire their meaning in relation (a relation of contrast) to each other. They help us to experience and describe the world based on contrasts, and are therefore a useful technique to any effort of defining anew a particular reality.

Given the double ambitions of the Swedish discourse on the information society (and its translation into the Kista Science City vision), Kista Science City AB started with the organizational problem of defining anew what Kista was to be and what it was to do. To their help they borrowed a categorical boundary that had already been used elsewhere to define a leading high-tech region. They travelled to Silicon Valley and imitated the technological boundary that so successfully defines that region. Being a powerful cluster of actors, Kista Science City AB quickly set out to manufacture presentations, rewrite (hi)stories, and redraw maps based on such a boundary.

But boundaries are not simply descriptive tools. Characteristic of most of them is that the relation between the elements separated by the boundary takes a hierarchical form – this being one reason why categorical pairs is so useful a technique for organizing. Often, one of the elements is subordinated to its contrary. This subordination is telling of implicit normative ideals, ideals that are imposed on that which is being defined and organized as the categorical pair is translated into presentations, stories and maps. In the technological boundary the most privileged position is clearly occupied by the high-tech part of the dichotomy, implicitly devaluing anything that may be associated to low-tech.

Kista Science City AB had an added difficulty. The region they had set out to define was historically seen as an immigrant suburb; a denotative adjective, "immigrant," that suits badly with the ambition to be in the "international forefront of technology development." In a skilful discursive movement the ethnic boundary was matched to the technological one adding to the category of "immigrant" another qualification, that of technologically unaware, low-tech. Consequently, decisions concerning the direction of technology development were made by and for the privileged group, the high-tech, the non-immigrant. (The discursive movement was made possible by the two other discursive techniques identified in the chapter, silence and the discourse on multiculturalism. Below I elaborate on how these two techniques work).

That no opposition from those excluded was met can be partly explained by the broad cultural acceptance of the ethnic boundary. Ethnicity is a sociocultural concept that organizes regions and people all over the world. It is therefore easily translated to various contexts and situations. In the case studied, the ethnic category is accepted to such an extend that it easily incorporates into the discourse of the information society, resulting in a transformation of the original ideal—from "unlimited" to divided. Discourse and the organization of community are thus linked by way of categorical pairs.

Multiculturalism

A question may still disturb some. How has it been possible to install a categorical boundary that so contradicts the democratic ambitions implied in the discourse of the information society? Surely, there are many among the privileged group that are well intentioned and engaged enough as to respond to such flagrant inequality. How come they do not seem to react?

We may find a possible answer in the slight degree of elasticity allowed to the excluding boundary. Tilly (1999) writes, "a viable category by no means entails a complete perimeter around all actors on one side of the boundary or the other; on the contrary, complete perimeters require a great deal of management and ordinarily cause more trouble than they save" (p. 66). It would be politically unsustainable, and probably economically risky, to have an all too rigid boundary dividing the region and its people.

A discourse slips in. A discourse that to unaccustomed ears may sound highly democratic; a chant to a global world, the fusion of markets, and the diversity of the workforce. The management discourse of multiculturalism. Once more, the corporate logic has sneaked in. By officially chanting to cultural diversity an appearance of tolerance is attained. It makes space for otherwise distressing biased presentations, partial stories, and shrunk maps. It transforms "the immigrant" into an exotic international specimen. The unwished badge attached to the region is changed into a longed for aura of internationalism.

The most direct power effect of this discursive technique is that exclusionary practices are effectively made invisible, thus countering many of the potential responses to them.

Silence

A second possible answer to the lack of reactions from the privileged side is silence. Silence is first and foremost a manifestation of the power differentials between the two groups co-existing in the region. It is a technique used by the powerful to avoid confronting the effects (intended or not) of their decisions. Silence comes from above, from the stronger voice, making silence louder than the potential voices that could come from those being silenced.

Silence functions in two ways. First, it *facilitates* overriding previously existing categories. Silence makes it easier to substitute those categorical boundaries for new boundaries that may be more relevant to the discourse of the information society (in the Kista case, a technological dimension replaced ethnic boundaries). Silence facilitates overriding and substitution of categorical pairs because both these practices are made less obvious, avoiding provocation of sensibilities.

Two, silence *reinforces* the very logic of categorical pairs—separate in two and rank the parts. Ranking involves giving one member of the pair a lower value, a lesser worth. Silence does this well, because what is silenced is immediately subordinated – it is not worth being talked about. The normative ideals inherent in categorical pairs are, by means of this (silenced) subordination, imposed on that which is being categorized. Foucault helps to make this point more clear.

When studying the silence that for so long has been said to envelop all that had to do with sex, Foucault (1978/1990) observes, "[s]ilence itself [...] is less the absolute limit of a discourse, the other side from which it is separated by a strict boundary, than an element that functions alongside the things said, with them and in relation to them within over-all strategies. There is no binary division to be made between what one says and what one does not say; we must try to determine [...] which type of discourse is authorized, or which form of discretion is required in either case. There is not one but many silences, and they are an integral part of the strategies that underlie and permeate discourses" (p. 27).

By means of a thick silence, categorical boundaries are transported, attached to, inserted in a variety of contexts and situations. Silence (to refer to Foucault) is not simply an innocent limit to a discourse. Rather, it is a discursive technique used by the powerful to enforce certain normative ideals. Implicit imperatives lie behind the choice of what is to be silenced by the discourse. Such normative ideals form the ground, in turn, of the classification made through the categorical boundaries.

Without negating the truth of current socioeconomic changes, the chapter shows that acknowledging the discursive dimensions of the information society is a fruitful strategy. It helps us to elicit the hidden assumptions and ideological underpinnings of many development projects inspired by the discourse of the information society. The information society is often presented as a recipe for economic and social development. Yet, considering it as a discourse highlights the conditions under which such development ideals are taking place. The following points summarize the process set in motion by the "information society":

1. Installing a categorical boundary between activities according to the amount of technological expertise required.
2. Giving lower status to the less technological activities (done by means of discursive silence).
3. Matching the internal technological boundary to an external, widely accepted, ethnic boundary (done with help of the discourse on multiculturalism).
4. Channelling immigrants into less status jobs whether or not they could perform the jobs that require more technological expertise.
5. Developing technology to suit the preferences of the prioritized social group.

Steps one and two are the actual definition of a high-tech region. Steps three and four are a way to side-step the contradiction brought by the implementation of the discourse on the information society. Segregation is up to this point a question of marking off life spaces. Step five cements segregation by incrusting it in the very technology.

REFERENCES

Amin, A. (1994). *Post-Fordism. A reader.* Oxford: Blackwell.

Andersson, J. (2005). "Mellan tillväxt och (o)trygghet: idéer om tillväxt och socialpolitik i socialdemokratisk ideology." In J. Andersson, J. Björkman, & I. Humlesjö, (Eds.), *Välfärdsstatens skräpvind: historiska spår i dagens välfärdspolitik,* (pp. 97-119). Lund: Studentlitteratur.

Atkinson, R. D., & Correa, D. K. (2007). *The 2007 State New Economy Index—Benchmarking economic transformation.* The Ewing Marion Kauffman Foundation and the Information Technology and Innovation Foundation.

Austin, J. L. (1975). *How to do things with words.* In J. O. Urmson & M. Sbisà (Eds.), *The Williams James Lectures delivered at Harvard University in 1955.* Oxford: Oxford University Press.

Barinaga, E. (forthcoming). *The power effects of discourse—inclusion and exclusion in the information society* (preliminary title). Stockholm: EFI.

Bauman, Z. (2004). *Wasted lives: Modernity and its outcasts.* Cambridge: Polity Press

Bloomfield, J., & Bianchini, F. (2004). *Planning for the intercultural city.* Bournes Green: Comedia.

Borja, J., & Castells, M. (1997). *Local and global: Management of cities in the information age.* London: Earthscan.

Braithwaite, J., & Drahos, P. (2000). *Global business regulation.* Cambridge: Cambridge University Press.

Brattberg, L. (1977). Järvastaden—en kommunal satsning. In *Sankt Erik årsbok.* Stockholm: Samfundet S:t Erik.

Brecknock, R. (2006). *More than just a bridge: Planning & designing culturally.* Bournes Green: Comedia.

Burbach, R., Nuñez, O., & Kagarlitsky, B. (1997). *Globalization and its discontents.* London: Pluto Press.

Castells, M. (1996/2000). *The rise of the network society.* Malden, MA: Blackwell.

Czarniawska, B., & Sévon, G. (1996). *Translating organizational change.* Berlin: Walter de Gruyter.

Dobers, P., & Schroeder, J. (2003). *The image economy—organizational processes and representational practices in a Swedish context.* Unpublished paper; presented at a seminar in the Stockholm School of Entrepreneurship.

Foucault, M. (1978/1990). *The history of sexuality: An introduction* (Vol. I). New York: Vintage.

Foucault, M. (1980). *Power/knowledge: Selected interviews & other writings 1972-1977.* New York: Pantheon Books.

Frank, T. (2001). *One market under god—Extreme capitalism, market populism, and the end of economic democracy.* New York: Anchor Books

Friedman, T. L. (2005). *The world is flat: A brief history of the twenty-first century.* Waterville: Thorndike Press.

Frissen, V. (1995). Gender is calling: Some reflections on past, present and future uses of the telephone. In R. Gill, & D. Gring, (Eds.), *The gender-technology relation.* London: Taylor & Francis.

Johansson, I. (1991). *Stor-Stockholms bebyggelsehistoria - Markpolitik, planering och byggande under sju sekler.* Stockholm: Gidlunds.

Khan, N. (2006). *The road to interculturalism: Tracking the arts in a changing world.* Bournes Green: Comedia.

Korten, D. C. (2001). *When corporations rule the world.* San Francisco: Berrett-Koehler.

Marquard, O. (1991). *In defense of the accidental: Philosophical studies.* New York: Oxford University Press.

Peter-Martin, H., & Schumann, H. (1997). *The global trap: Globalization and the assault on democracy and prosperity.* London: ZED Books.

Sahlin-Anderson, K. (1996). Imitating by editing success. In B. Czarniawska & G. Sèvon (Eds.), *Translating organizational change.* Berlin: Walter de Gruyter.

SCB (Statitiska centralbyrån) (2005). *Statistisk årsbok 2004*. City of Stockholm.

Sèvon, G. (1996). Organizational imitation in identity transformation. In B. Czarniawska, & G. Sèvon (Eds.), *Translating organizational change*. Berlin: Walter de Gruyter.

Thomas, R. (1995). Access and inequality. In N. Heap, R. Thomas, G. Einon, R. Mason, & H. Mackay (Eds.), *Information technology and society*. London: Sage in association with Open University.

Tilly, C. (1999). *Durable inequality*. Berkeley, CA: University of California Press.

USK. (2003). *Arbetandes och boendes uppfattning om trafiksituationen i Kista*. City of Stockholm.

USK (Utrednings- och statistikkontor) (2004). *Befolkningsstatistik 2003*. City of Stockholm.

Webster, F. (2002). *Theories of the information society* (2nd ed.). Routledge.

Wood, P. (2004). *The intercultural city reader*. Bournes Green: Comedia.

ENDNOTES

a The argument presented in this chapter is further elaborated in *Powerful dichotomies: Inclusion and exclusion in the information society* (preliminary title), Stockholm: EFI. The project on which the argument is based has been funded by Riksbanskens Jubileumsfond and by Jan Wallander and Tom Hedelius Stiftelsen.

b A short sample of journals which title reflects the trend: *The information society journal*; *Journal of International Knowledge societies*; *Knowledge society and third way*; *Network society*.

c For a good example of a neo-liberal inclined observer of the so-called "information society" see Thomas L. Friedman, 2005, *The world is flat – a brief history of the twenty-first century*. New York: Farrar, Straus and Giroux.

d For an example, see Thomas Frank, 2001, *One market under god – Extreme capitalism, market populism, and the end of economic democracy*. New York: Anchor Books.

e For an example of an academic voice that has strongly energized many anti-globalization movements, see David C. Korten, 2001, *When corporations rule the world*. San Francisco, CA: Berrett-Koehler. Other works interesting for their critique to globalization are, Hans Peter-Martin and Harald Schumann, 1997, *The Global Trap: Globalization and the Assault on Democracy and Prosperity*, London: ZED Books; or Roger Burbach, Orlando Nuñez and Boris Kagarlitsky, 1997, *Globalization and Its Discontents – The rise of postmodern socialisms*, London: Pluto Press.

f 50% of the population with foreign background comes from Asia. From USK - Utrednings- och Statistikkontoret (www.stockholm.se/usk).

g 17% of the population with foreign background comes from Africa. From USK.

h CFAR (Centrala Företags- och Arbetsställeregistret), SCB. Retrieved November 22, 2006 at www.kista.com.

i Figures refer to year 2002. USK.

j "En effektivare IT-användning skapar en bättre grogrund för tillväxt och nationell utveckling." (Verksamhetsplan – Regeringens IT-politiska strategigrupp, 2004, p. 5).

k Ministry of Industry, Employment and Communications (Background.pdf). It can be found in www.regeringen.se.

l Kista's conscious strategy for urban development is in line with the tendency, pointed out by Jordi Borja and Manuel Castells, concerning cities' competitiveness in the global economy. See Jordi Borja, and

Manuel Castells, 1997, *Local and Global: Management of Cities in the Information Age*. London: Earthscan.

[m] Telia is the Swedish National Telecom Operator.

[n] On this subject see the Intercultural City research project led by Charles Landry and Phil Woods – a comprehensive study considering the extent to which cultural diversity may be a source of innovation, creativity and entrepreneurship for the development of cities. The project has been reported in a series of books: Phil Wood (ed.), 2004. *The intercultural city reader.* Bournes Green: Comedia; Jude Bloomfield and Franco Bianchini, 2004. *Planning for the intercultural city.* Bournes Green: Comedia.; Richard Brecknock, 2006. *More than just a bridge: planning & designing culturally.* Bournes Green: Comedia; Naseem Khan, 2006. *The road to interculturalism: tracking the arts in a changing world.* Bournes Green: Comedia.

[o] Throughout the essay, I use the term "organization" in Tilly's sense (Tilly, 1999). It does not make reference to firms, governments, schools, and similar, formal, hierarchical structures. With organization I mean "all sorts of well-bounded clusters of social relations in which occupants of at least one position have the right to commit collective resources to activities reaching across the boundary" (p.9). Tilly names kin groups, households, religious sects, bands of mercenaries, and local communities as examples of organizations.

[p] SRA and Rifa are both Ericsson's predecessors.

[q] *aep*, an Spanish researcher working at the it-University, would not agree with pah. From my field-notes:

I receive an e-mail prompting researchers at the it-University to commercialize their research. Remembering his strong negative reaction a couple of months ago concerning commercialization of one's own research, I ask Alberto if he has seen the e-mail. Another strong reaction, yet not on the content but on the form of the e-mail: "I haven't read it. It's not in English. I'm fed up with the linguistic chaos of this institution. They want to appear international and send some things in English, but when it has to do with big decisions, politics and power then they send it in Swedish. (*aep*)

I have met similar reactions from other expatriates. Language is clearly a barrier hindering expatriates to fully participate in the transformation of Kista into a high-tech region.

[r] For an influential example see, David C. Korten, *When corporations rule the world.*

Chapter III
High-Tech Workers, Management Strategy, and Globalization

Jasmine Folz
Seattle Central Community College, USA

ABSTRACT

Based on qualitative interviews with Seattle area high-tech workers, this chapter explores their positioning within and reaction to globalization processes. Looking especially as cost-cutting labor strategies of contingent employment, importation of foreign workers, and the outsourcing of professional high-tech work, it is argued that these are essentially restrictive employment strategies that benefit employers at the expense of employees. While some of the interviewees more or less approved of these practices as logical from the corporate perspective, and were confident that their jobs were too complex to be at risk, most are questioning these processes and some were actively trying to organize in an effort to halt or at least slow down such trends. How and why high-tech workers accommodate or resist management policies and practices they disagree with is analyzed with attention to the impact of ideology.

INTRODUCTION

The rapidity and scale of technological innovations during the last century has contributed to a great transformation in the labor process that continues to change the nature of society in general and work in particular. Information can travel and be processed faster than ever before, and this allows for the reorganization of corporate entities with respect to both time and space. Many have interpreted these changes and shifts as the inevitable and indeed desirable result of "globalization" in which the breaking down of political and economic barriers is paired with the blurring of cultural identities. Attention to globalization has produced many valuable contributions to the anthropology of work in particular, ranging from ethnographies of how transnational production practices shape the lives of people working in the mines, factories, and "information sweatshops" of developing nations to studies of how blue collar workers in de-industrialized America have experienced these same processes (Doukas, 2003; Ferguson, 1999; Freeman, 2000; Nash, 1989, 2001). How, though, do the relatively elite workers who create the technologies that facilitate these practices experience globalization?

Drawing from qualitative interviews with eleven high-tech workers conducted in Seattle, WA in 2004, I examine high-tech work and workers as embedded in the international political economy. In their day-to-day experiences, as well as in the moments and factors that define their careers, individual high-tech workers confront and negotiate a myriad of macro level issues, including changes in technology, market fluctuations, political regulations and policies, and off-shore production. Though these issues may appear much larger and thus separate from individual high-tech workers, these macro level changes do affect the high-tech workers I interviewed in very tangible ways. To better understand this macro-micro relationship, I analyze issues associated with globalization that were brought up in the interviews, looking particularly at high-tech production and labor processes and at how the workers I interviewed accommodate or resist globalization as it is manifested in management policies and practices.

GLOBALIZATION AND HIGH-TECH PRODUCTION CYCLES

Much of what concerns contemporary ethnographers are the issues surrounding globalization, and, more specifically, the negative consequences globalization has had on many people in traditionally as well as newly marginalized spaces. Indeed, globalization is this moment's answer to what Wolf calls the "central problem" inherent to each period of anthropology (1972, p. 252). Though from an anthropological perspective, globalization, as defined by the interconnectedness and exchange of world cultures and economies, is not new, what is new is the speed and scope at which technology is shrinking the relationships between time and space. From processing information in "data factories" to working in cyberspace, location of work and workers has shifted and new sites and categories of work have emerged. The implications of these transformations for high-tech workers, their managers, and the global labor process are certainly significant, though they are also significantly contested. Within the American context, two issues exemplify the role of high-tech workers in the globalization process: flexibility and accommodation as management strategies and the high-tech production process.

Advances in computing networks, transportation, and telecommunications have allowed corporations to become more flexible and take greater advantage of economies of scale. This flexible business model, which is facilitated by neoliberal policies, favors the decentralization of production framed within correlating shifts in time and space, or shifting spaces. Harvey's analysis of the shift from the assembly line logic of industrial capitalism (Fordist) to the post-Fordist logic of flexibility and accommodation that now dominates global production is useful here for, as he explains:

> *The primary significance ... of all these changes is that the relatively privileged position of the working classes in the advanced capitalist countries has been much reduced relative to conditions of labor in the rest of the world... The secondary point is that conditions of life in advanced capitalism have felt the full brunt of the capitalist capacity for 'creative destruction' making for extreme volatility of local, regional, and national economic prospects...* (2000, p. 69)

Thus, the globalization process and, in particular, the management strategies of requiring flexibility and accommodation from employees, define the circumstances and constrictions of high-tech work/workers. While on the one hand high-tech workers can embrace unprecedented levels of personal autonomy in regards to where and when they work, they are also constrained by this same "freedom" to do work anywhere at any time. The implications being that they are seldom able to "turn off" their work selves and that their livelihood is dependent on the unstable global market. The role technology plays in enabling management to utilize such "rational" employment strategies is both obvious and debated.

Hakken suggests that technological innovations, which allow more opportunities for communication, have put the labor process on the cusp of a new transformation he terms "resocialing," or the development of a more social form of work whose sociality is both acknowledged and accepted by management (2000, p. 8). Still, the tendency for management to use technology as a means of control appears to be predominant. For most workers, new forms of technology provide management with new instruments of control through limited access to information and electronic surveillance of production and the workplace (Freeman, 2000, p. 198-9). Moreover, exploitative management strategies, such as those that favor international divisions of labor to the disadvantage of workers, are not only enabled by but premised on new technologies (ibid, p. 166-7).

Hakken maintains that with advances in communication technology, work is no longer conducted in place but, rather, in space and that online workers simultaneously construct and are constructed by a new social formation of cyberspace (2000, p. 9). Certainly, information and communication technology (ICT) has opened up new possibilities for the organization of work; however, these possibilities are predicated on access to technology and there is a "digital divide" based on regional and class barriers to computer access and literacy. Michaelson goes as far to argue that the information economy "...has not caused the gap between the rich and poor, but it has exacerbated it" (2000, p. 13). Through examination of the high-tech production cycle, we will be able to connect these debates to the high-tech industry and its workers and, in so doing, understand how these processes are perpetuated and experienced.

The high-tech production cycle is rapid, global, and cyclical. The rapid, international aspects of the production process can be directly attributed to advances in manufacturing, travel, and communications technologies whereas its cyclical nature can be traced to the booms and busts inherent to capitalist economies. The production cycle, in general, consists of product research, design, and development, followed by production. At each phase of the cycle, there is usually intensive testing to verify that the product's design, components, and function all meet in-house, industry, and government regulations. Because high-tech corporations are profit driven, they utilize the lower labor costs of contract employees and outside vendors (often in developing nations) and because they are also vying to beat their competition to the market, the cycle is continually sped up. What this means for high-tech workers is that their working lives (and often their outside lives) are dictated by what point they are in the cycle of any given product they are working on. For example, if they are preparing for the actual production of the product, there is a "crunch" time where it is common to work overtime for weeks on end, and there are also mini crunches at the end of each phase of the cycle to ramp up for the next phase. Further complicating matters, if production is taking place in another country and/or time zone it is often necessary to come in early or late to check in. In addition, it is not unheard of for individual workers to be assigned to two or more projects simultaneously.

As straightforward as the previous description of the production cycle is, because of the fast pace and reliance on multisited production teams, problems are common. Trying to beat or at least keep up with the competition entails adhering to unrealistic schedules, which can be very stressful, as Barb,[a] a software programmer, explained:

Sometimes the problems are just ongoing. You know if you're doing builds you just have to turn them out every day and just make sure they work every day. And if there's build breaks then you have to investigate those, fix them, or find a person to fix it... sometimes the developers are depending on the bits that you build for them to go on to the next step, so that's kind of a turning point kind of a thing that just goes on endlessly.

Add to this the logistics of working with team members in different time zones and, with so many variables in the production process, miscommunications and other problems are almost inevitable.

In addition to negotiating the problems inherent to the logistic and temporal constraints of production cycles, some of the people I interviewed also have to contend with government regulations. This is because the technologies their companies produce have significant health and/or safety applications. John, an electrical engineer, told me that a large part of his job is to ensure their product, which uses potentially dangerous lasers, meets federal safety standards. Similarly, Kevin, a computer scientist who works on cancer treatment technology, must ensure that his work is stringently tested to meet Food and Drug Administration requirements. Because they force corporations to conduct and document rigorous tests on their products, in some cases, strict government regulations can slow down and thus mitigate the problems inherent to production cycles.

In sum, while technology enables fast paced, multisited production cycles, even the technology must be understood in the context of the contemporary political economic environment, which favors profit driven, multinational corporations. To better understand how these changes affect those who design and work with the technologies that enable these processes we will look at how cost-cutting labor strategies such as contingency, importation of foreign labor, and outsourcing are executed and experienced in the high-tech industry.

COST CUTTING LABOR STRATEGIES: CONTINGENCY, IMPORTING LABOR, AND EXPORTING WORK

Contingent employment, which entails hiring workers on a temporary, contract basis, has been a widespread employment strategy in the high-tech industry for over twenty years (Barley & Kunda, 2004, p. 13). In fact, ten of the eleven workers I interviewed have worked on a contingent basis for part or most of their careers. While some of the interviewees prefer it because it allows for more freedom to pursue outside interests, others have never had a choice and would much prefer full time work. Whether for, against or ambivalent about contingency, everyone agreed that it is an increasingly prevalent fact of high-tech life.

Contingency is attractive to employers because hiring employees on a contract basis allows corporations to save on the costs associated with full time employees and it allows for flexibility in hiring a specialized person for one specific project or task and then letting them go when their services are no longer needed or wanted. As such, it epitomizes the flexibility and accommodation inherent to post-Fordist capitalism. Contingency also gives employers some leverage with asking contract employees to put in long, stressful hours because it's a short term, project specific request. But, as Barb explained when telling me about the demanding nature of her last job, this strategy is not always as effective as intended:

That was a very unrealistic pace. But it was a contracting job so they knew this person was just going to suffer for a certain amount of time. They're not doing the right thing with this job 'cause they really should have a permanent position—the job was so difficult, if somebody could really learn it maybe it would have been easier but they laid me off. They spent a lot of time training me on this job—so now they're gonna have to retrain somebody else—I don't know what they're gonna do, I just think they're crazy. But they're under budget constraints. That's why they do it.

So, while restrictive budgets can make contingency look good, the costs of filling a multifaceted, key position with someone who has no vested interest in the company may be very high. Be that as it may, high-tech corporations are motivated by profit and, in general, contingency is a successful and cost effective employment strategy.

From the employee's perspective, contingent work is attractive for similar reasons; it can be very lucrative and the flexibility of working on a contractual basis allows employees to sample and learn about various types of work and companies without making any commitments to stay if they don't want to. Steve, a tech writer, is especially appreciative of this aspect of contingent work, as he explained when telling me about his early days as a Microsoft temp: "I took a break and traveled to South America for seven months—and that was another thing that [was] really attractive about the temp work because when my job was done I was done—I didn't ask anyone, didn't tell anyone I'm leaving, I'm just done." That said there are some problems inherent to contingent employment.

Full time employees do not always make as much money as some highly skilled contingent workers, but they generally have healthcare benefits, access to employee assistance programs to deal with job stress and/or personal problems, stock options, access to company sponsored events, and, perhaps most importantly, job security. According to Doussard and Mastracci (2003), 96% of contingent IT workers surveyed would prefer full time employment (p. 17). To that end, many contingent employees hope contingency will be an avenue to full time employment and while it sometimes is, according to my informants, more often than not the job ends with the contract. In fact, for some contingent workers, contracting has been their only employment. Sue, a hardware designer and drafter who has worked in the industry over twenty years, told me, "In my life I think I've only had two full time jobs, the rest have all been contract." Although they are not as obligated to their employers as full time employees, contingent employees have less rights and little control over the tenure of their contracts, especially in a bear market. While embracing the freedoms of contingent employment may say something about the individuals who choose to do so, what perennial contingency among workers who would prefer full time jobs says about the state of the industry and political economy is even more telling.

Following from hiring native workers on a contractual basis is the increasingly common practice of using H-1B and L-1 work visas that enable foreign workers into the United States. These visas are premised on a lack of homegrown talent; however, some argue that this import of qualified but less expensive labor is in fact premised on the lower wages paid to and greater control over foreign employees (Doussard et al., 2003). Ben, a software developer and the only interviewee who is working on an H-1B visa, shed some light on the matter. He was brought to the United States from India several years ago by an agency that specializes in recruiting, sponsoring, and managing H-1B employees for a percentage of their wages, of course. Since that time, he has worked for several high-tech companies in California and the Seattle area, each one taking over his sponsorship for the duration of the contract. While Ben sees his experience working in the United States as an excellent opportunity to improve his skills, the context of this opportunity leaves much to be desired, as this interview excerpt illustrates:

JF: Were you lonely when you came or did you miss India or your family?
Ben: Yes, that was pretty hard but work kept me

really on my toes seven days a week. So even getting some time off on Sunday was a big issue. I didn't have much time to think about them.
JF: And so everyone was working like this and do people ever get burnt out?
Ben: Yes but we [had] no choice because we were brought in to fulfill a requirement and you cannot back out, it's nothing like I've taken a job offer and anytime I can back out. I had no other option. You are sponsored by these organizations and you are pretty much tied down to them.

Though the trafficking of highly skilled, well paid laborers who want to work in the United States doesn't inspire the same level of concern as other transnational labor migrations, still, it does resemble past exploitations and raise questions about the state and nature of the high-tech industry. This level and scope of foreign importing of labor isn't occurring in other professions, and while all contract workers are legally "tied" to their contracts, U.S. citizens are more likely to at least have knowledge of and fairly easy access to legal avenues of recourse from unfair or illegal work arrangements.

From importing foreign workers to outsourcing the work does not take a great leap of the imagination and, indeed, the specter of outsourcing is looming over the high-tech industry and there is a real sense of, if not fear, unease about its impact. All the interviewees are aware of the trend in outsourcing professional high-tech work though only four are very worried, three of whom were unemployed at the time of the interviews. The rest feel that even if professional high-tech work goes overseas that they will probably be able to get work in the Seattle area because their skills are quite good and specialized and/or they have enough contacts in the area to secure employment. That said, the interviewees did express three main objections to outsourcing: (1) that it degrades the quality of the product, (2) that it threatens their own jobs, and (3) that the practice is immoral. While most everyone agreed to the first and several to the second objection, only a few brought up the third.

Of all the interviewees, Ben was the most at ease with outsourcing as a production strategy, as he explained when I asked him if he has observed much concern about it:

Yes, you can see the feeler of people's faces the moment that a new personality comes to [the] company, especially an Indian. So the expression tells you off but it's not only India, somebody's going to do it cheaper than India... It's not a question of Indians. Typically when you look for a car in the market you're going to go to the guy who's going to give you the best deal and that's the guy who's going to survive out there so that's how the world market is going to be so it's going to happen to India as well at some point of time. Some country that's really qualified and you have people doing it much better, certainly the market is going to shift. So it's the same thing in life. You'll just have to take it philosophically, I'll be old by then.

While Ben, who plans on returning to India to continue his career, is in a good position to take the trend in outsourcing philosophically, it is only recently, with the outsourcing of some professional high-tech work, that many American high-tech professionals have begun to question the tacit approval for whatever lowers the bottom line.

Most of the interviewees have firsthand experience dealing with outsourcing of software and electronic goods and everyone who had worked with outsourcing agreed that it compromises the quality of work. This judgment is not rooted in assessments of the ability of offshore workers but in the observed difference in assumptions and inevitable miscommunications that long-distance teams encounter. In fact, Mike and Kevin, both computer scientists, told me that they aren't at all worried about outsourcing of professional software work because they have seen the compromised quality that results from miscommunications. Though not as unconcerned as Mike and Kevin, Peter, a project manager at Microsoft, gave examples of miscommunication and lack of quality he's witnessed with the outsourcing model and then went on to explain why he doesn't feel threatened by it:

In terms of it worrying me—and this is personal cause I know a lot of people that are very worried about it—but having seen the failures and the decreasing quality and things like that that happen, I'm not as worried. Eventually those kinks will get worked out and I think that a lot more of the low level jobs will probably go there but then that becomes a cascading effect that the low level jobs go but then you need some middle management over there and when before we [were] only using vendors and now there's actually a whole bunch of full time Microsoft employees, they're building campuses and everything in India and PRC, stuff like that. So, it's a bit troublesome and I feel the same way that I do about The GAP. You know, like it's ridiculous that they're willing to go over there and compromise quality and like what they're paid—they pay those people a quarter of what I get paid. So, I don't know, it's a little disconcerting 'cause I think that they're shortchanging their customers...

Peter's ambivalence is instructive. While, on the one hand, it is acknowledged that outsourcing tends to result in lower quality and has some unsavory implications about the industry in general, many high-tech workers do not perceive an imminent threat to their jobs and so they shrug it off as consumers do when they get a good deal on clothes they know were made in a third world sweatshop. And, in the same vein, just as Americans who have lost or expect to lose their manufacturing jobs to outsourcing may be more sensitive to the implications of buying clothes at The GAP than those who are secure in their job futures, for high-tech workers who have lost or fear losing their jobs the differences between offshore garment or manufacturing and offshore programming are negligible. For instance, Sue, whose job (mechanical design and drafting) has been outsourced, brought up the issue using the very tape recorder I was recording the interview with as an example:

Sue: The Sony tape recorder you are using, years ago would cost $250. Now, with rigorous competition...
JF: It was $34.
Sue: Yes, with rigorous competition from China it now cost $34, for this amazing piece of technology. But, if we turn this over I'm sure that we are going to see "Made in China."
JF: Just a sec, let's see. Made in China!
Sue: So I could tell you that this physical piece of equipment that cost you $34 in essence cost Sony company, physically, I'll be generous, $4 to manufacture. That would be for all the components and the manufacturing of this. I would add another $2 for testing and shipping and then the rest would go towards the company.

Notwithstanding debate about the differences and implications of outsourcing industrial vs. professional work, both types of outsourcing have the same motivation: the bottom line. Despite her skill and experience, because Sue's job is a relatively lower level high-tech occupation, it has been one of the first to be outsourced. Time will tell whether or not Sue represents a canary in the mineshaft but, presently, many high-tech workers are very concerned for their future.

Unsurprisingly, contingent employees are especially sensitive to outsourcing. When I asked Barb if she had witnessed much to do with outsourcing, she told me:

I've heard a lot of stories of smaller companies that have just outsourced everything and people have been forced to train their replacements. Well the companies always say they're not forced, but they're not given the extra bonus after they're kicked out. I mean I work with a lot of Indians and they're great, but... How can you compete when those people are getting paid ten dollars an hour when we're getting paid twenty to forty dollars an hour? People always say programmers get fifty to sixty, well maybe the company is paying fifty to sixty an hour, but the programmers around here, top wages are around forty an hour, if you're contracting.

Barb's frustration and concern are palatable. And for good reason: the possibility of having to train your replacement is akin to building the machine that replaces you on the assembly line and, in fact, high-tech workers can fairly and ironically be held partially responsible for the technology that allows the outsourcing of high-level work to occur.

Several permanent, full-time employees were also very apprehensive about the future. When I asked John if he was worried his job could be outsourced he said, "Yeah, actually it could be, and it might be. Some of the stuff I do, yeah, very, very possibly." When I asked if he thought the recent downturn in the industry was cyclical, as he had said earlier booms and busts were, he sighed and said no, and then told me about a friend of his who runs an internet business from Seattle but all his employees are in India, adding:

Some people think that the whole outsourcing thing creates work here, which I find very hard to understand. A lot of other people just say oh no all of our jobs are going overseas and that's it, we're all going to be out of work. I think this friend of mine might be an example; he didn't have enough money to do what he's doing without the Indian programmers. So, I think it kind of goes both ways, but I don't know. I think it's probably a net [of] jobs going away than creating his business. ...Maybe we'll create a few jobs here because he's able to get the start and get the work he needs done from India.

As John's friend illustrates, to stay competitive, it's necessary to utilize "emerging markets." And, if he is successful with this company, he may very well hire some people in Seattle but, there are a lot of contingencies in this scenario and, for the present, he cannot afford and is overlooking American workers.

Even for seasoned high-tech workers the increased level competition brought on by the dot.com bust in late 2000 and outsourcing is daunting, as Sue told me:

There are more people competing for the same job. I just recently went to a job fair for Boeing, and they just said that they received a contract that they will be hiring four thousand people. Two months prior they laid off maybe six thousand people and they recovered some of those people and then they had open interviews to fill positions. I have no idea how many people there were, but I know that when I got in for the interview process I was amongst maybe two thousand people. And I saw a line trailing out of the building around the corner, and all of those people were [told], after waiting for two or three hours in the sun, 'We have too many people, please go home.' They did not accept resumes, they just said go home.

The image of thousands of workers waiting in line to apply for a few jobs is not unknown of in the United States, but the image is associated with relatively uneducated members of blue-collar occupations. When that line is comprised of educated workers waiting to apply for jobs that their education and experience led them to believe they could expect as the reward for completing the socially prescribed avenue to economic security, the very foundation of the American achievement ideology becomes suspect.

For some high-tech workers, outsourcing has changed their perceptions of the socioeconomic framework they work in. Their criticism of outsourcing is based not only on the tendency for poorer quality products and its threat to their own livelihoods, but also on a moral judgment of the role their industry plays in the globalization process. For some, outsourcing represents a shift in society; when I asked Mary, a tech writer, how she felt about outsourcing she told me, "There was something in the paper just a couple days ago about community and the automatic checkers at the grocery store and how, with the outsourcing, [we are] losing our sense of community. And that's part of my objection to outsourcing." Mary's connection between outsourcing, automated checkouts at the grocery store and sense of community speaks to a fundamental criticism of how technology is utilized. The implication being that when jobs that were formally performed by people are done by machines or by people far away there is a loss of community because the daily interactions between the people that produce and consume goods and services are circumvented or eliminated altogether.

For Sue as well, her objections to outsourcing are not merely based on the personal cost she has paid but also on her indignation at how it affects the Americans who are left without jobs and the foreigners who now perform them for less money:

Well, my personal opinion is that there are people in my country and other countries who have money and I think the object is to maintain that money and to continue making money—never losing. They have found a resource—it was probably like the industrial revolution, how the industrial revolution changed the world—things could be mass-produced. And so now we have found the next phase, we have found a way to produce twenty-four hours a day and in the last ten years it's extremely obvious to me how offshore manufacturing is affecting America. To the point that yes, some people are making money, they're making good money, but now there is another issue—there are more and more people who are not making any money. So the rich are getting richer but now the middle classes are plummeting to the poor. I've noticed less and less opportunity here, but more and more in third world countries. And I guess I could say I'm happy for them, but I'm not because there is sacrifice. There are reasons things cost money. In America we have rules; there are stringent environmental and safety issues that companies have to comply to. And in, for example, China, Malaysia, Viet Nam... people suffer because there are no regulations or constraints.

From her comparison of contemporary outsourcing to the industrial revolution to her assertion that there are reasons things cost money, Sue is articulating a critique of the relationship between technology and the economy. That jobs are going away as a direct result of outsourcing and that high-tech workers are aware of the trend is clear, what is not clear is whether this will prompt an organized response from high-tech workers in an effort to mitigate or halt the process.

ACCOMMODATION AND RESISTANCE

Historically, engineers have had very little reason to resist management, with whom they assumed shared interests (Noble, 1979, p. 41). In fact, though Veblen (1964) had hopes engineers would organize the overthrow of business, he conceded this was improbable because, "Engineers and industrial experts are a harmless and domicile lot, well fed on whole, and somewhat placidly content with the 'full dinner plate' allowed by the vested interests" (p. 135). Nevertheless, as high-tech corporations ask for higher levels of commitment from their employees while simultaneously offering less job security, causes for resistance, if not revolution, abound.

Freeman (2000) notes that due to the hegemonic nature of corporate power, resistance is not a clear-cut issue and many workers have conflicting and/or ambiguous feelings about the work they do. Kunda (1993) finds that in the contemporary high-tech corporation, management has consciously blurred the lines between work and play, organization and self, in an effort to control and extract commitment from employees. The result, from the high-tech workers' perspective, is that more and more of the self is monopolized, and as Hochschild (1983) posits, in order to salvage personal autonomy and identity outside of work, personal feelings must be "managed" and the management of feelings in response to work constitutes a new form of labor: emotional work. The consequences of this monopolization of the self by the organization can be devastating and often leads to serious personal problems such as divorce or alcoholism and psychological breakdown, generally referred to as "burnout," which all of the interviewees have witnessed and several have experienced.

Whenever the workers I interviewed described a management policy or practice as inherently unsatisfying, I asked them how they and their coworkers dealt with it. While all the interviewees found at least some management tactics disagreeable, there was very little resistance beyond complaint: accommodation was the most common reaction. For instance, when I asked Tim, a game developer who frequently remarked that management only got in the way of technological work, if he or anyone he worked with who was unhappy with management ever did anything about it he replied, "Nah, you don't wanna piss off management. It's never a good move." So, even though Tim was the most vocally anti-management worker I interviewed, he was no more likely to actively resist policies or practices he disagreed with than workers who were happy with or indifferent to management.

When workers did provide stories of resistance to management, they discussed three basic tactics: humor, slowing down, and labor organizing. When I asked Peter about how workers at Microsoft were reacting to compensation cut backs, he gave an example of how humor is used as protest:

There's been jokes lately, a lot of joke memoranda showing up by elevator bays about bathroom usage and how Microsoft employees, blue badges, will get to use the bathrooms and they'll have to use their cards whenever they go.[b] *They can go in twice a day, and every time they go into a stall they're allotted three sheets of paper per swipe and no more than a maximum of four swipes per day. And the non-native English speakers will sometimes grab me or come to my office and say 'Have you seen this?!' and I'm like, 'I'm gonna tell you this is 100% positively a joke.'*

Humor can be a very powerful form of critique, and because it is often used in hidden and public transcripts, there is room for different interpretations of its meaning(s). However, whether it, alone, can be considered a form of resistance as conceived in labor studies is not clear.

While discussing how hard the recent trend in outsourcing has been for her and her coworkers, Sue told me about how some employees do resist such practices:

If I think of some people, since their jobs are being taken away from them, since portions of their jobs are being done in other countries, they are making their functions last. So, let's say there was somebody testing a piece of equipment. If it took two hours to test that piece of equipment, all of a sudden it's going to take four hours. They're trying to stretch their work out, trying to make themselves valuable. But actually it's harmful to themselves and to the company. It does not help the company really—it doesn't matter if it helps the company or not, they are going to change—but for you to slow yourself down, you have to think, is this going to be helpful, this skill set, is this going to be helpful to me in another company? Or, will I just be perceived as a slow worker?

Sue's assessment that slow downs aren't good for the company is certainly valid. However, the fact that she even mentioned what is good for the company, as a worker who has been laid off from her own job due to outsourcing, is perhaps a telling clue as to why so few high-tech workers resist management in significant ways.

According to my informants, high-tech workers are not likely to actively resist management policies and practices they find unsatisfying and if and when they do, they are likely to do so in non-confrontational ways. Though many workers disagree with many management decisions, they are usually able to find other sources of job satisfaction that balance the effects of management. Or, if they feel they are easily employable, they will simply quit. To understand why high-tech workers are unlikely to participate in organized resistance we must examine their occupational history.

The professional identity attached to formal education generally and high-tech work specifically, in combination with the distinctive labor history of engineers, combines to produce a seemingly contradictory worker identity. Throughout the twentieth century, the occupation of engineering grew with industrialization and as it grew in scope and its ties with the military-industrial complex became ever stronger, it evolved from a craft formally associated with blue-collar work into a vaunted profession on par with medicine and law. This ascension to the rank of profession was accomplished as both industry and the need and niche for management grew. In fact, as Shenhav (1999) shows, it is the influence of engineering rationality in the management field that laid the foundations for subsequent and contemporary management theory as a "naturally" rational endeavour. Nevertheless, among high-tech workers there is an inherent distrust of management. Significantly, this distrust is not based on criticism of management per se but, rather, on an assumption that managers, by definition, are too far removed from the design and production process to provide accurate or efficient guidance to those who actually work with the technology. Management isn't bad because it's management so much as because it's *not* engineering. Thus, while high-tech workers have not historically respected management, nor have they questioned its priorities.

Though everyone interviewed was at least aware of if not worried about the trend in outsourcing high-tech work, only about half were aware of efforts by high-tech workers in Seattle to unionize in an effort to mitigate, if not stop, the trend. Five of the eleven interviewees were aware of the attempt to organize high-tech workers; three of those five are pro unionization. For the interviewees who are pro union, they see unionization of high-tech work as a viable and necessary means to secure their financial ends. Opposition to high-tech unionization fell into two basic categories: (1) that unionization is undesirable for ideological reasons and (2) that the grievances of professional high-tech workers do not warrant unionization in the ways those of blue collar workers do.

For the interviewees who were aware of but not supportive of high-tech unionization, Mike and Peter, their opposition to unionization is based on their experiences in and perceptions of the high-tech industry. For Mike especially, unionization represents a part of the high-tech industry that he doesn't want to feel connected to:

I've never been a part of a union, and I've always kind of viewed them as an evil thing. I probably got it from my dad, from my family when I was a kid. At Boeing, half of the company is unionized and there was always an animosity between upper management and the unions; it just seemed really odd to me to be working in this environment where your boss is your enemy. And when the unions would go on strike, you know that was very much the feeling was it's us against them to the point of vandalism and actual physical violence and who would want to work in an environment like that? So I guess my experience is that if you're good at what you do you can make good money. And maybe that's just because I've been lucky in my career or whatever, I've been at the right place at the right time and maybe that's not true across the board and there's people with valid grievances that a union could solve but it hasn't been my experience.

Mike's discomfort with the confrontation inherent to union politics at Boeing, as well as his assertion that if you are good at what you do that you will be successful, are both rooted in his experience which, in turn, is rooted in his acceptance of rational management and the American achievement ideology. Both are also open to question, if not refutation, by his acknowledgment that his experience may not be representative. In fact, his experiences are representative of issues that prompt some high-tech workers to resist management. Earlier in the interview, he spoke of feeling like a second-class citizen when he worked on a contingent basis and of his prolonged, unplanned unemployment when the dot.com bubble burst. His resistance to organized resistance, then, can be understood as ideological in nature.

When I asked Peter if he had heard of unionization efforts by local high-tech workers, he said he had and articulated a more nuanced but ultimately anti-union argument:

*The people that wanna unionize, I don't think they understand the situation of the world economy anymore. I mean [if] you look at South Korea and how it's economy completely fell apart it was because of the over unionization of industry in South Korea, they couldn't compete with North Korea or China or any of these other places. So, when you look at Microsoft or you look at the technology industry and you look at unionization, I can understand from the perspective of temporary workers, why they would want to do that... I think [it] will do more harm than not, it'll expedite outsourcing, and it's actually frankly kind of ridiculous for technology people to do it. When you're talking about meatpackers that sustain permanent injuries because they're working in freezing conditions and using things that cause major vibrations in their bodies and they're only making $8 an hour and have been working there for years, there is significant more push behind that. When you're talking about someone who is making $85,000.00 a year... Ok, maybe if you wanna talk about in relation to how much money the company makes, you can talk about that. But you know what? It's not a relational system and we're not a socialist society. It's hard though, because I do see a lot of people get f***ed, but it's a relative f***ing compared to those people that work at McDonald's.*

That Peter is also one of the few interviewees with significant experience in project and personnel management must be considered a significant factor when interpreting his perceptions of the situation. Peter's opposition to unionizing the high-tech sector is based on his assessment that it renders industry uncompetitive and that while some high-tech workers are taken advantage of, relatively speaking, they are far better off than most workers. His analysis, then, is based on what he perceives as the irrationality of unionization. While no one else couched their opposition or ambivalence to unions in the competition argument, several other interviewees made this second observation.

Among the interviewees who were not aware of the efforts to unionize high-tech workers the reaction to such efforts was one of general ambivalence. Like Peter, many see their situation as imperfect but do not think their grievances are significant enough to warrant union intervention. Ben's reply to my question of what he thought of high-tech unionization implies that, by definition, high-tech workers do not need unions: "A union is very often [for] when people are neglected and people are not cared for. So I think IT is a highly paid job and typically seems to be going fine so no one has any complaints." Kevin, too, while aware that there are some valid causes for complaint, does not see parallels between high-tech work and traditionally unionized labor:

I guess it could make sense, like I've always had great contracts at my companies. The Microsoft type, I wouldn't call it exploitation, but what ended up in the lawsuit was probably as close as you'd get to a need to have a union.[c] *There's ACM, there's software programmers and software developers [who] band together on an intellectual level but also on like what their benefits are and usually if you're getting screwed with a company and you're a computer software engineer and you're good at what you're doing, then there's so many other companies that will take you. You know, it's like opposite of being a laborer being screwed over. But the outsourcing might actually cause some more union like activities, but we'll see. Not right now, I never really thought about it, or I never felt any need for it, yet, but it's interesting activity.*

Kevin's acknowledgement of the instigating factors for high-tech unionization, combined with his assertion that, as Peter said, any exploitation experienced by high-tech workers is relatively benign, suggests, at best, an ambiguous class-consciousness. Also of note is Kevin's reference to a professional organization (the Association for Computing Machinery) that provides intellectual and career support. Most professions, including doctors, lawyers and university professors, belong to professional associations, many of which perform several of the same functions as labor unions such as grievance representation and contract negotiations. That professional high-tech workers would belong to such organizations is not so interesting as is their simultaneous eschewing of unions.

Of the six interviewees who were not aware of attempts to unionized high-tech workers, only one, Tim, was decidedly supportive:

Well I guess it depends on how they unionize but unions in general I think are a good thing. I know a lot of temporary employees get forced to work ridiculous hours, like that was one of the problems with High-Tech-Corp-A was you had to work just amazing hours. I guess the Washington laws have been written such that they can require you to work up to twelve hours a day and they can fire you if you don't want to or something like that. So there's a lot of contract employees getting no benefits and that kind of stuff.

Tim's support for unionization and derision of management can, at least in part, be attributed to his experience growing up working class. Conversely, Peter, who also grew up in a working class environment, is against efforts to unionize high-tech workers and sees management as rational, if not ideal. Despite their similar class backgrounds, age, and years in the industry, Tim's support for and Peter's derision of high-tech unions provides an excellent example of how individual experience and agency can shape class-consciousness.

When I asked Sue, who has lost her job and drawn connections between her industry and global disparity, if she thought other people were dealing with the same issues as her, she told me:

I thought I was the only one against this, but I'm meeting more and more people who are recognizing that the WTO, without the right constraints or regulations, that it is a Pandora's Box. That it provides opportunity, but does it provide equal amount of opportunity as much as it provides disparity? I think the disparity is greater, both here and abroad.

Though the previous statement may indicate that Sue is ripe for joining an organized resistance to the processes she finds so ideologically reprehensible, when I asked her if she was aware of the efforts of high-tech workers to unionize she answered, "I was not aware of it. To unionize, I don't think it would matter." Sue's ambivalence raises the question of how or even if it is possible for unions to effectively recruit and represent a group of workers who are not aware or supportive of unionization. Even Steve, who has a history of political activism, is overwhelmed by the present situation, telling me, "It looks like things might change radically you know? But if they do there's nothing I can really do to stop it. If there's really smart people in India that can do these jobs really well and if they can they're gonna and it's gonna go."

Yet for some high-tech workers, like Nancy, a tech writer, making a connection between their work in the high-tech industry and the global disparity in wealth and power has led to active resistance. Six months before she retired from Microsoft Nancy joined a union working to organize high-tech workers in the Seattle area, WashTech, which was started by contingent workers at Microsoft and is affiliated with the Communication Workers of America (CWA). [d] Before joining WashTech it hadn't occurred to her that Microsoft was a part of the globalization process. And, in fact, initially WashTech wasn't focused on globalization, but, rather, was concerned with the plight of contingent workers and occupational safety issues. But, as the interviewees have made clear, the contracting, importing, and outsourcing of labor is becoming an issue of great concern and its connection to the globalization of the production process has galvanized some high-tech workers to become at least interested, if not active, in unionization. Despite all of this, Nancy acknowledges that it is very hard to organize high-tech workers, noting, "Tech workers are passive in a way."

At the time of the interview, WashTech represented five bargaining units in the Seattle area; all in the public sector. On average, about thirty people attend the quarterly meetings but the membership has shot up from 250 to 450 since the union started placing off shoring ads in newspapers and there are 15,000 people on the mailing list. While Nancy is the most actively involved in labor organizing of the interviewees, she is also the only research participant who has been able to retire from the industry and thus has the time and resources to devote her energy to the cause.

After Nancy, Barb is the most adamantly and actively supportive of high-tech unions. But, perhaps because she is working on a contingent basis, her assessment of the situation is bleak:

I think it's gonna take a lot of people being out of work or not being able to retire, having to work forever, for people to realize maybe a union might be a good thing. But, we're not there yet. There's people from all over the world doing this work and the jobs are becoming shorter and shorter, more contractual, you agree to only work for a certain amount of time and then you're laid off, and that was the result of the lawsuit with Microsoft... And there's no big retirement plans and you don't get health care when you get laid off; it's bad. But I don't think people are desperate enough yet. Or, people get laid off and they don't have any job—they lose their house they lose everything, but they're not involved in the union because they're just trying to survive...

When I asked Barb if she ever tried to recruit and organize coworkers, she told me that she thinks she would be fired for being too vocal about unionization at work, adding: "...I've *occasionally* mentioned it to other workers but nobody's ever that interested. So I don't really make an issue of it, you know, yet." Barb's support for WashTech may be strong, but so is her concern that her politics could cost her jobs.

CONCLUSION

Letting the workers speak for themselves, it is obvious that although their connections to the global economy are strong, they are also fraught with insecurity. While the daily stress of working to keep up with production cycles that are, by definition, too rapid and dependant on too many variable to avoid problems is offset by generous compensation, this too is offset by the insecurities associated with the increasingly global nature of the high-tech industry. Despite rhetoric about free agents and markets, contingency, importing labor, and outsourcing work must be seen as essentially constrictive employment strategies that, in general, favor employers at the expense of employees. This raises the question: If high-tech corporations continue to utilize contingency and reliance on foreign labor pools in an effort to stay competitive in the global economy, to what extent will American high-tech workers still identify with management's interests? The interviewees' assessments of their contribution to and positioning within globalization processes demonstrate that they are uniquely situated in terms of critiquing these processes. While most of the interviewees actively questioned management policies and practices that impact their job security, whether this will lead to active, organized resistance is not clear. While there are emerging labor organizations focused on the globalization processes that concern many high-tech workers, it remains to be seen if these efforts will affect how or to what extent high-tech workers who are willing to construct an identify against management are also willing to identify *with* organized labor. Finally, though the present findings are inconclusive in regards to if or how high-tech workers are likely to actively resist globalization processes, the power of technology and the workers who create it in enabling these and other changes cannot be underestimated. As such, as an occupational group, high-tech workers have a great potential to exert a meaningful and effective critique of the production and labor processes inherent to late capitalism.

REFERENCES

Barley, S. R. & Kunda, G. (2004). *Gurus, hired guns, and warm bodies: Itinerant experts in a knowledge economy.* Princeton, NJ: Princeton University Press.

Doukas, D. (2003). *Worked over: The corporate sabotage of an American Community.* Ithaca, NY: Cornell University Press.

Doussard, M. & Mastracci, S. (2003). *Uncertain futures: The real impact of the high-tech boom and bust on Seattle's IT workers.* Center for Urban Economic Development: University of Illinois at Chicago.

Ferguson, J. (1999). *Expectations of modernity: Myths and meanings of urban life on the Zambian Copperbelt.* Berkeley, CA: University of California Press.

Freeman, C. (2000). *High tech and high heels in the global economy: Women, work, and pink collar identities in the Caribbean.* Durham, NC: Duke University Press.

Hakken, D. (2000). Resocialing work: The future of the labor process. *Anthropology of Work Review, XXI*(1), 8-10.

Harvey, D. (2000). *Spaces of hope.* Berkeley, CA: University of California Press.

Hochschild, A. R. (1983). *The managed heart: Commercialization of human feeling.* Berkeley, CA: University of California Press.

Kunda, G. (1993). *Engineering culture: Control and commitment in a high-tech corporation.* Philadelphia, PA: Temple University Press.

Michaelson, K. L. (2000). Laboring in cyberspace: Internet realities and the future of work. *Anthropology of Work Review, XXI*(1), 11-14.

Nash, J. C. (1989). *From tank town to high tech: The clash of community and industrial cycles.* Albany, NY: SUNY Press.

Nash, J. C. (2001). *Mayan visions: The quest for autonomy in an age of globalization.* New York: Routledge.

Noble, D. F. (1979). *America by design: Science, technology, and the rise of corporate capitalism.* Oxford, UK: Oxford University Press.

Shenhav, Y. (1999). *Manufacturing rationality: The engineering foundations of the managerial revolution*. Oxford, UK: Oxford University Press.

Veblen, T. ([1921] 1964). *The engineers and the price system*. New York: A.M. Kelley.

Wolf, E. R. (1972). American anthropologists and American society. In H. Dell, & H. Hymes (Ed.), *Reinventing anthropology* (pp. 251-263). New York: Pantheon Books.

ENDNOTES

[a] Pseudonyms have been employed to protect the anonymity of the research participants. As well, with the exception of references to Microsoft and Boeing, I have employed pseudonyms for all other high-tech companies discussed in an additional effort to protect the interviewees' anonymity. Because Microsoft and Boeing each employ tens of thousands of workers it is highly unlikely that interviewees who discuss working at those companies will have their anonymity compromised.

[b] At Microsoft, there is a colour coded ID card system in which FTEs and contingent employees are differentiated by the color of their badges. FTE badges are blue, whereas contingent badges are orange. Thus the fake memoranda are lampooning the company's recent cutbacks *and* its stratification system.

[c] Kevin is referring to the precedent setting class action lawsuit, *Vizcaino v. Microsoft*, which was filed by perennially contingent (or "permatemp") employees who felt they were being treated as "common law" employees and thus should be eligible for inclusion in company pension and stock options plans. The case was decided in favour of the plaintiffs and, since then, Microsoft and other high-tech companies have gone to great lengths to ensure contingent employs cannot legally be perceived as employed by the company. At Microsoft this has resulted in three month "breaks" between contracts for contingent employees so they cannot depend on or claim Microsoft as their sole employer, and the increased use of employees who have "vendor" status and who, in theory at least, can or do work for other companies in addition to Microsoft.

[d] Please see: http://www.washtech.org for more information on this organization.

Chapter IV
Language Norms and Debate in Triple Helix Organizations

Kathryn J. Hayes
University of Western Sydney, Australia

J. Anneke Fitzgerald
University of Western Sydney, Australia

ABSTRACT

Commercialization activities combining the discoveries of one occupational group, such as scientists, with the commercial skills of managers involve interactions across occupational cultures. This chapter considers how dissent can be interpreted as a sign of dysfunction or cause for concern. The context of the study is temporary Australian hybrid industry-research organizations composed of academic, government, and industry personnel. Semi-structured interviews of twenty scientists, engineers, and managers focused on their experiences and perceptions of occupational culture, including styles of debate. Distinctive patterns of argumentation were identified as typical of commercial and research occupations. Extended argumentation contributed to knowledge creation, and played a role in maintaining a hierarchy among research institutions. Members of research and commercial occupational subcultures working in Australian CRCs reported frustration and reduced effectiveness of argumentation due to different norms for dissent. Initial expectations of similarity, built upon identification of occupational hierarchies, heighten the impact of these differences.

INTRODUCTION

From the 1980s, governments of industrialized economies have looked to innovations involving the generation and reconfiguration of knowledge as a path to achieving a competitive advantage (Gibbons et al., 1994; Lehrer & Asakawa, 2004; Nowotny, Scott, & Gibbons, 2001; Premus, 2002). From a commercial perspective, the attractions of entering collaborative research partnerships include access to complementary physical and intellectual assets, reductions in product development time, cost, and risk, and increased organizational and financial flexibility (Senker & Sharpe, 1997). The creation of hybrid research organizations that use resources and/or governance structures from more than one existing organization (Borys & Jemison, 1989), has been a feature of the response to government and business pressures for increased innovation. The evolution of "triple helix" organizations, in which private firms and publicly funded research groups collaborate with the support of government funding, has been traced across Europe, the USA, Latin America, and Asia (Etkowitz & Leydesdorff, 2000). In addition to a rapid geographical spread, the number of hybrid industry-research centers in operation has also quickly increased (Gray, Lindblad, & Rudolph, 2001; Premus, 2002) as has the percentage of industry and government funding they attract (Gray et al., 2001). Clearly, the operation of hybrid organizations composed of government, business, and research groups within national systems of innovation is a subject worthy of attention.

Hybrid industry-research organizations share some of the management challenges encountered by joint-venture partners from the same industry. Boundary issues between the partner organizations, threats to the hybrid organizations stability from the sovereignty of member organizations, and the challenge of reconciling disparate goals, technologies, and organizational cultures can occur even when commercial organizations in the same industry combine (Borys et al., 1989). The merger of HP and Compaq (Burgelman & McKinney, 2006) provides a recent and instructive example of the difficulties that can be encountered even between commercial organizations from the same industry segment. Temporary collaborations between industry and research partners present even more complicated structural, managerial, and organizational cultural challenges than those that have made mergers and acquisitions between comparable commercial organizations into acknowledged high-risk endeavors (Barringer & Harrison, 2000).

In temporary hybrid industry-research organizations, the initial stages of innovation, including the identification, capture, and evaluation of new knowledge, are increasingly conducted across organizational *and* occupational boundaries, and authority structures can be unclear. The experience of interoccupational contact in hybrid industry-research organizations forms the focus of this work. This chapter considers in depth one management challenge inherent in the operation of high technology, hybrid research organizations; that presented by the debating norms of research and commercial occupational subcultures. Specifically, members of commercial occupations can interpret ritual dissent, considered a normal and constructive mode of interaction by members of research occupations, as a sign of conflict and an issue that needs to be controlled. Commercial occupations include commercially oriented engineers and business managers. The context for the study is four hybrid industry-research organizations in the Australian Cooperative Research Centre (CRC) program.

The Australian CRC program provides an example of government intervention to stimulate innovation through encouraging collaborative arrangements. Australian government funding policies changed in 1990 to encourage research institutions, commercial organizations, and government bodies to work closely together to commercialize promising ideas and technologies. CRCs are composed of research, government, and industry members working together to bring an invention to market and operate as trans-disciplinary, temporary organizations intended to link discovery, application, and use. Additional policy objectives include the maintenance of Australian ownership of intellectual property, provision of training positions for postgraduate students and the inclusion of regional organizations, beyond the major cities (Department of Education Science and Training & Howard Partners, 2003). In 2006, 71 CRCs were receiving funding, with $925.9 million being provided for administered grants between 2006-7 and 2010-11 (Department of Education Science and Training, 2006).

The program has predominantly been supported by public money in a public to private ratio of approximately three to one. While the CRC program has produced more than 70 spin-off companies with a total turnover expected to reach up to $300 million by the end of 2007 (Young, 2003), it is viewed as a qualified success by politicians and has received much "sympathetic criticism." In the early years of the CRC program, over 80% of CRC professionals had a research background (Liyanage & Mitchell, 1993), with a 13% minority drawn from industrial and commercial sectors. It seems reasonable to expect at least one subculture, associated with the researchers who worked together in the "pre-history" of the CRC and guided the discovery through early commercial feasibility studies, to exist in the early stages of CRC development.

Innovation generally involves the practical application of new knowledge. This requires at least two types of occupational cultures; the exploratory roles performed by "pure" researchers, and pragmatic, systematic roles filled by a variety of commercial occupations. Commercialization activities combine the discoveries of one occupational group, such as scientists, with the commercial skills of engineers and managers (Steiner, 2000). Therefore, bringing innovative ideas to markets involves interactions across occupational cultures and interoccupational tension may be implicit in innovation. Previous work (Hayes & Fitzgerald, 2005) has explored some of the boundaries between occupational cultures working together in commercialization activities, but the influence of language norms has not been considered in the literature.

We argue, based on interviews with scientists, engineers, and managers that the styles of argument used in occupational groups within hybrid industry-research organizations may act as a barrier to commercialization. Analyzing different argumentation styles extends existing research into occupational norms for debate to include innovation, commercialization processes, and the context of triple helix organizations.

Argumentation

In triple helix organizations, argumentation, which we define as the action or process of reasoning systematically in support of an idea, commonly occurs in groups composed of more than one occupation. For an extensive review of argumentation theory see Sillince (2002). Rational or scientific argumentation uses logical reasoning about competing ideas, for an extended period, until the topic has been exhaustively analyzed. It is important to distinguish between the form of argumentation identified as rational by Sillince (2002) and the rational school of epistemology. Rationalism refers to the Cartesian approach to the philosophy of knowledge, based upon the use of logic and mental constructs to deduce truth. Rational argumentation, as referred to in this chapter, is not limited to Cartesian notions of truth and knowledge. It is equally applicable to empirical approaches based upon induction and sensory experiences (Nonaka & Takeuchi, 1995).

Far from being an outmoded form of rhetoric at risk of extinction, scientific argumentation plays a vital role in the creation of knowledge. Programs to measure and increase the teaching of argumentation in school science are being implemented in the UK (Erduran, Simon, & Osbourne, 2004). Although scientific and academic groups are skilled in and accustomed to using constructive dispute to produce creative ideas and outcomes, occupational norms that proscribe dissent may restrict their effectiveness in cross functional groups. In addition, the value of minority opinions and constructive dissent in increasing group creativity while guarding against "groupthink" is well documented (Nemeth, 1997). However, members of business cultures, which in the main value cohesion and conformity (Pech, 2001), may view signs of disagreement between researchers as alarming and predictive of failure.

Scientific forms of argumentation have been identified as poorly adapted to public debate of trans-scientific social issues such as global warming (Ziman, 2000). The challenges of engaging in argumentation across the boundaries of national cultures (Liu, 1999), and rhetorical traditions (Dolinina & Cecchetto, 1998) have also been acknowledged. The use of language, dialogue and argumentation is an occupational culture construct, policing acceptable communication within the group. As McKerrow noted (1980, p. 31):

... communities are typified by the specific rules which govern argumentative behavior, by social practices which determine who may speak and with what authority, and by their own "display" of these rules and social practices in response to challenges from within and outside the community.

Furthermore, McKerrow asserted that when individuals from two communities with different standards for judging the strength and appropriateness of arguments argue, the choice is to select one set of argumentation standards or "face a contradiction that cannot be easily resolved." However, little attention has been directed to the potential impact of conflicting norms for argumentation upon the commercialization phase of innovation in either traditionally structured, or triple helix, industry-research organizations.

Culture and Innovation

When organizational culture is considered in studies of innovation, integrationist assumptions of consistent values and beliefs are often made (Martin, 2002). For example, entire organizations are identified as exemplars of innovation culture, despite known subcultural differences amongst its members. Moreover, systems of innovation are compared and contrasted across nations and continents (Lehrer et al., 2004), assuming organizational cultural homogeneity exists within the boundaries of nations or even economic communities.

In contrast to monocultural representations of organizations, the three perspective view of organizational culture (Martin, 2002; Meyerson & Martin, 1987) identifies three cultural patterns co-existing in an organization's culture at any one time. These are the integration, differentiation and fragmentation perspectives. At any time, one pattern may be dominant and more easily detected than the other two. However, over time or due to environmental changes, the organization's culture will be dynamic; the currently dominant perspective will recede and other frames will become more visible.

An *integration* perspective focuses on cultural manifestations that have consistent interpretations, and regard organizational culture as a clear consensus without ambiguity. The *fragmentation* perspective claims relations between cultural manifestations are neither clearly consistent nor inconsistent and views ambiguity, not clarity, as the core of culture. Viewed through the frame of fragmentation, consensus is transient and issue specific.

The *differentiation* perspective focuses on cultural manifestations that have inconsistent interpretations, and result in organizational groupings. The differentiation perspective considers consensus to exist only in subcultures. Subcultures may exist in harmony, independently, or in conflict; however, within a subculture there is clarity. Recent research has pointed to the existence of subcultures within hybrid industry-research organizations (Hayes et al., 2005; Siegel, Waldman, Atwater, & Link, 2004).

Subcultures and Innovation

Subcultures are defined as groups that have unique patterns of values and behaviors, providing a distinctive identity. Subcultures can be consistent or inconsistent with the dominant culture of the organization (Trice, 1993). Every large organization consists of potential subcultures. For example, subcultures may exist based on characteristics of gender, ethnicity, profession, age, functional division or geographic location, wage levels, and employment status (Kunda, 1992). Hybrid industry-research organizations are likely to import occupational subcultures along with the individuals recruited from research, commercial and government institutions. In addition to other group behavior, professional and occupational affiliations influence argumentation styles (Sillince, 2002).

Hence, while existing literature acknowledges the functions of argumentation in knowledge creation, it does not necessarily address the difficulties of occupational cultural differentiation within hybrid organizations. This research explores business and research modes of argumentation and gives insight into their potential to act as a barrier to commercialization.

METHODOLOGY

A qualitative, semistructured approach was employed. Australian CRCs typically consist of individuals from a variety of professions aligned with a particular industry, located within a broad context of Australian society and public research policy decisions. The ability to effectively identify, isolate, and accurately measure both dependent and independent variables within a natural setting is limited. In addition, the scant literature on occupational cultures within industry-academic research partnerships would not support attempts to define dependent and independent variables. As social activities occurring within organizational and national cultures, commercialization processes in a complex social system suit holistic investigation using qualitative methods.

A total of twenty scientists, engineers and business managers were recruited from four CRCs. Two of these CRCs were from the Information and Communications Technology sector and two were from the Biomedical sector. The organizations' maturity ranged from newly formed with only a few years of operation to twenty years of operation for a publicly listed company that developed out of the CRC program. The participants' experiences and perceptions of differing subcultural norms, including styles of debate, were recorded and transcribed verbatim. QSR N-Vivo® software was used to aid detailed coding and analysis of the collected research material, facilitating the interpretation process. Member checks, in which the data and interpretations were provided to participants for correction, verification and challenge, were used to increase the credibility of the research.

Through the analytic phase of the project, the research material was found to cluster around a number of core themes. Through a reflective, iterative process, we interrogated theme content to explore relationships between and within the themes.

Despite the suitability of the selected methodology, the research findings depend upon the memory, insightfulness, and honesty of the interviewees. The findings are also constrained by time, place, and the changeable nature of individual perspectives. Consequently, the findings cannot be readily extrapolated into other contexts.

However, as the following section illustrates, the insights provided by the interviewees extends existing theory to consider the potential impact of interoccupational argumentation on the functioning of hybrid industry-research centers. Therefore, further research into occupational subculture, occupational identity, and commercialization processes in the context of CRCs in Australia is appropriate.

RESULTS AND ANALYSIS

The interviewees demonstrated clear familiarity with the research topic. They described visible signs of occupational groups working in commercialization, the values and beliefs of particular cohorts and perceived boundaries between commercial and research occupational subcultures. Such ability to rapidly grasp and articulate the notion of occupational subculture might be attributed to acumen and education. However, the emotionally charged responses, often marked by frustration, suggest that the impact of different styles of occupational argumentation in commercialization extends beyond mere intellectual awareness.

An alternative to occupational subcultural differences as an explanation for the friction reported between occupational groups is that participants' experiences were colored by stereotypes, beliefs that all members of a particular group are similar and think and behave in the same way (Trice, 1993). However, frequent contact between individuals of similar status, cooperating in working towards common goals is demonstrated to reduce stereotyping based on age (Hale, 1998), race (Dixon & Rosenbaum, 2004) and nationality (Stangor, Jonas, Stroebe, & Hewstone, 1996). These conditions apply in CRCs. After extensive periods of work contact, in some cases more than twenty years, the CRC members' reports of occupational difference are based upon observation and engagement, not simple stereotypes.

Argumentation as an Occupational Norm

... we wonder if all of those lovely, extrinsic rewards will cause us to become arrogant, overconfident, or lazy, or – worst of all – to stop arguing. We believe the quality of our work will suffer if we stop bickering and are left with only our optimism. (Sutton & Rafaeli, p. 127, in Frost and Stablein (eds), 1992, Doing Exemplary Research.)

Scientists, engineers, and managers from government and private sectors described distinctive patterns of argumentation as typical of research and commercial occupations. Descriptions of scientific and research group debates were consistent with earlier reports (Nemeth, 1997; Ziman, 2000). Specifically, restrictions on the amount of time allocated for discussion, the likelihood of reaching a clear conclusion and expectations of action resulting from debate were very different for commercial and research communities. Both commercial and academic groups agreed that researchers enjoyed debate and viewed it as an important part of their identity. As stated by a scientist, who moved to a management position:

The important aspect is inquiry into "What's the best focus?" It's not inquiry for the sake of "Let's muck around." And that's often the criticism of science going into commercialization. Inquiry is a necessary process before you arrive at an agreed way forward. That's very stimulating and something I really, really miss in shifting from that culture ... difference of opinion is part of the creative process.

The importance of dissent as a tool to hone solutions through a verbal contest of ideas and as a symbol of scientific identity was evident. When asked if the academic and research members of the CRC enjoyed debating amongst themselves, a public sector manager responded that it was a core part of their professional identity, stating:

They'll [researchers] consider [debate] very much part of their reason for existence and if they didn't have that debate I think they would see themselves as not representing their profession. .... The debate may well air a range of options and different opinions and approaches and thoughts and so forth but it doesn't crystallize an outcome. And for the private sector all that debate is good, but if it doesn't bring you closer to an absolute outcome, an absolute decision, well then all it is, is debate. It may be very worthwhile from the academic point of view but time is money when it comes to the private sector.

Sharing the view that business communication should be action-oriented and produce a clear conclusion a product development engineer remarked:

There were meetings that only scientists used. They had a university common room meeting culture. They would sit and talk about the latest discoveries in the field; it was just a big chinwag. And that was work. Because they were scientists...they'd talk about the latest things in the field and if it didn't relate to the thing I was trying to make I couldn't care less. I used to get dragged along to them occasionally but, operations point of view, meetings to me were a waste of time. I do the communication I need to do as I need to do it.

Similar comments by other participants indicated a clear understanding and acknowledgement of differences, both in the acceptability of dissent and the desirability of reaching a clear conclusion in a specified time. Researchers appeared comfortable with, and even proud of their ability to use "talking as thinking" in a mono-cultural group, in contrast to commercial interviewees who preferred talking to be preparation for action. Comments made regarding time limits for debate link argumentation norms to different temporal orientations towards pace and punctuality held by the two broad occupational groups (Hayes, 2007).

Research Hierarchies and Argumentation

While researchers generally focused upon the function extended argumentation performed in knowledge creation, interviewees acknowledged that argumentation also played a role in expressing and maintaining a hierarchy among research institutions. Managers reported that universities in the "top tier" preferred to deal with others of similar status and did not give as much time to universities perceived to hold low rankings. One manager with an academic background explained:

This is all about the prestige or precedence of the universities who have a pecking order. To collaborate with someone who is below you in the pecking order is going to pull you down.

Researchers generally described their debates as impersonal, driven by intellectual challenge and focused on reaching the best possible solution, but several research and commercial participants identified extended debate as a vehicle to demonstrate and consolidate academic hierarchies and express conflicts within the research community. The role that "contests of ideas" play in establishing and maintaining a research hierarchy was clear. One researcher commented,

The idea of science being a tra-la-la freedom to act, it isn't! It's hierarchical beyond anything you would expect.

This researcher continued to express the view that the hierarchy in research communities was more vigorously contested that that of industry:

Have you ever been to an international [research] conference where people are fighting over the stakes and seen concession and kind hearted behavior? It is absolutely back biting beyond belief and beyond anything that Wall Street can turn up.

The existence of a research hierarchy was confirmed by a commercial engineer from the same organization:

There was just a pyramid of scientists. Scientists are like that you know ... you get to be a scientist, a senior scientist, a super-super scientist, an extra-super-duper scientist and a king scientist.

Additionally, one manager in a biomedical CRC explained that linkages between universities and hospitals represented "power blocs" in his field, and certain institutions were unable to join a CRC if existing members regarded them as inferior.

The above demonstrates that members of both research and commercial subcultures identify hierarchies within research communities. However, expectations of how a hierarchy will modify debate are very different for research and commercial groups, and present a challenge to communication between managers and researchers in high technology, triple helix organizations. This challenge will now be examined in detail.

Research Hierarchies, Argumentation and Violated Expectations

To reiterate, confusion and frustration can result from contact between research and commercial occupational subcultures. Managerial occupational culture features a competition based work hierarchy, in which senior managers act in decisive and controlling ways to express their status (Schein, 1995). Hence, both research and commercial occupational cultures have strong and visible hierarchies, but differ markedly in their attitude to dissent (Nemeth, 1997; Pech, 2001). Members of commercial and research occupational cultures in the CRCs studied recognize a hierarchy of authority in their own, and in the other occupational community. This leads to expectations of similar attitudes towards deference and dissent. However, the norms of the two occupational cultures vary markedly in their use of dissent. Language norms associated with dissent in research and commercial occupational cultures violate initial expectations of similarity built upon a common recognition of occupational hierarchy, as is illustrated in Figure 1.

False impressions of similarity, based upon use of a common language, can reduce the ability of participants to attend to the substance of an argument and can lead to suspicion and hostility between members of different national cultures (Dolinina et al., 1998). This research has found that members of research and commercial occupational subcultures working in Australian CRCs report frustration and reduced effectiveness of argumentation due to different norms related to the acceptability of dissent. Initial expectations of similarity, built upon identification of occupational hierarchies, heighten the impact of these differences. In research communities a contest of ideas builds and sustains the hierarchy; open intellectual dissent is considered normal and constructive. In contrast, dissent is viewed as divisive and destructive in commercial organizational cultures (Nemeth, 1997; Pech, 2001). While briefly acknowledging the opportunities for creativity that can be presented by dissent, Levine clearly views conflict as an expense to be avoided through seeking resolution as quickly as possible, cautioning "It is very important to understand the magnitude of the transaction cost of remaining in conflict" (Levine, 2006, p. 379).

It is beyond the scope and design of this research to comment upon whether business or research communities have more pronounced hierarchy. However, both groups operate within hierarchies, leading to expectations that communication within the hybrid, tripe helix organization will follow familiar patterns. These expectations are routinely violated. A commercial manager identified the following instance in which research dissent norms conflict with those of business:

The academic world is about critical debate and a lot of people go to [research] conferences to generate critical debate so they stand up and they expect to get criticism about their work and have to defend it. But if you go to a conference and it's an industry conference it's more about promoting your product and if people were to be critical of it you'd be offended.

Figure 1. Hierarchy and dissent norms in research and commercial occupational cultures

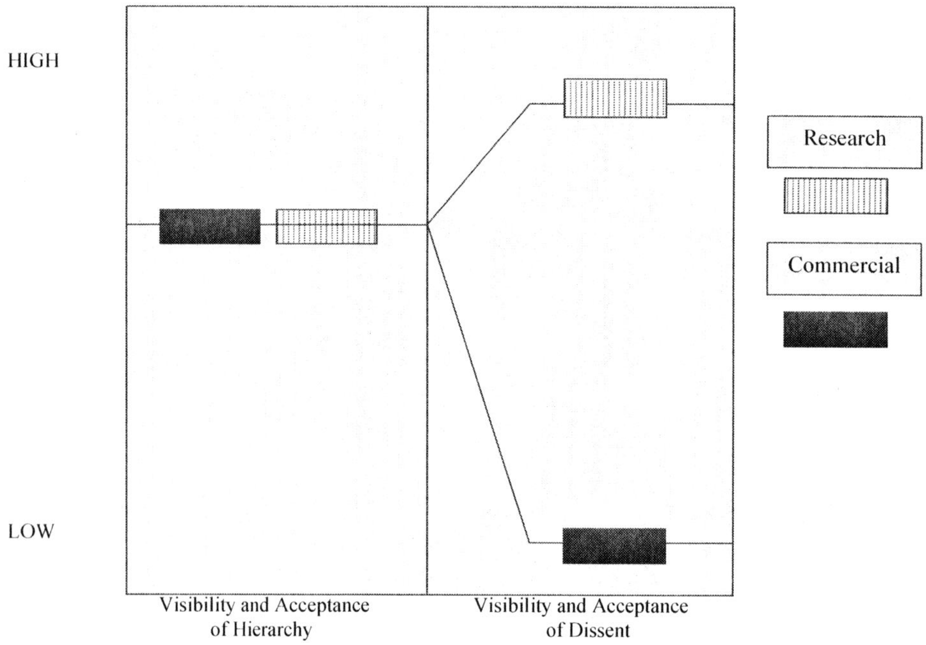

In counterpoint, a researcher reported that while criticism of his work was to be expected as a normal part of being a researcher, assessments of his work by managers, and the use of managerial authority to remove him from a project was intolerable and resulted in his resignation:

For me that was quite a significant impact on my life because just emotionally, that was something I had worked on for a number of years and suddenly I was taken out of this and it was not really fair I thought, and not really based on my value as a founder and as a scientist ... I could have avoided the conflict with this person but the methods that he used were just inappropriate and I couldn't sustain insults and things like that and I just had to respond. Although maybe if I had a different culture and ... was not that focused on the company succeeding, if I was focused more on keeping my position then maybe I would behave differently.

In a different CRC, a manager reported a similar response from a researcher who had funding removed from his project after it had been publicly criticized by managers from government departments, but not actively debated by mangers from industrial organizations:

... he was a very senior academic and thumbed his nose a little bit by resigning a position on the CRC. The research he was doing wasn't going to result in something the private sector could take to market. So that's probably why the private sector weren't active participants.

In summary, while both research and commercial groups in the CRCs explored are comfortable operating in their own occupational hierarchies, the two groups hold opposing norms related to the acceptability of dissent, and especially the public expression of dissent. In addition, judgments made by people outside the boundaries of each specialized research community are unwelcome, consistent with the importance of peer assessment *within* scientific and research communities (Merton, 1957; Ziman, 2000). Managers of organizations that are temporary alliances between research and commercial groups can be perceived as lacking the professional and organizational legitimacy required to judge and control the work of researchers.

Cross-Functional Team Leadership, Participation, and Argumentation

In addition to signaling membership of research or commercial subcultures in hybrid industry-research organizations and constructing a research hierarchy, commercial informants identified the use of rational argumentation as a handicap to researchers assuming cross-functional group leadership. Clearly stating all assumptions and relying upon logic to win an argument were viewed as incompatible with the political demands of balancing the needs of multiple public and private stakeholders. As explained by a manager:

... the current CEO is extremely political in [his] behavior in order to satisfy the conflicting interests of board requirements or stakeholder environments vs. scientific requirements. Getting a scientific person to step into that role and try and manage is near impossible. [laughs] That's what they'll try and do. They'll get in there and say "I'm just going to be straight." And of course you need to be able to balance those two conflicting worlds.

The manager was quick to praise the skill of the scientist and the quality of their logic and reasoning, but continued:

...there is no question he uses logic and reasoning to rule the day. That won't necessarily work in a political environment where one of the conflicts would come up is his employer is sitting around as one of the stakeholders... But he'll keep going back and use logic and reasoning as opposed to political processes.

Some of the interviewees managing research groups clearly understood that engaging in scientific argumentation could alarm industrial members of the CRC. After saying that he enjoyed engaging in rational debate with other experts in his field, a manager of scientists volunteered:

I'm just conscious that if I'm having a debate with scientists, even scientists from [industry member of hybrid research organization], I'm happy to be quite open. Whereas if I'm having a discussion with the business people from the company then I'm going to be much more careful in how I say things so as not to overly worry them about things which indeed are a very small worry, but could be taken out of context.

Consequently, the style of argumentation employed by members of research or commercial occupations does more than identify group membership. It can impede communication across occupational subcultural boundaries. Furthermore, the interviewees revealed that occupational membership, not organizational affiliation, was the source of differing norms of argumentation.

Participation in Argumentation and Occupational Subcultural Boundaries

Participation in extended bouts of rational argumentation was reported to be restricted to members of groups involved in knowledge creation, whether located in academic, public or private sector organizations. Interviewees spoke of extended and heated, but congenial debates occurring between researchers while commercial representatives present did not participate. Individuals from non-research groups only became involved to stop the debate if they thought it had gone on for too long. A public sector manager described the motivations of business and government representatives present during the debates as follows:

Private sector, their role tends to be very vested interests, hence narrow focus and short and sharp, not bringing in a whole range of otherworldly thoughts into the process but short, sharp, and focused. From the government perspective, the government input is more generalized than the private sector, whether it's going to be useful from a government perspective. Not as focused and as to the point as the private sector but still somewhat focused relative to what the academic debate might be.

There was general agreement that research debates were closed to non-members, indicating that styles of argumentation may help to create and reinforce occupational boundaries. Tolerance of discussions without time limits and with ambiguous outcomes was identified as an important difference. As a manager commented when asked if rational arguments reached final and lasting conclusions:

Well sometimes they do. Sometimes they just peter out in the context of "Yes, there's differing opinion on the subject" and if there's no categorical resolve well then everyone can go back to their corners agreeing that there's differing opinions which could be revisited at a later date.

The argument repertoires reported are consistent with the norms of scientific argument, which include a preference for a written format, a well-informed audience, no imposed time limits, and little expectation that a "yes" or "no" conclusion will be reached to support a particular course of action (Ziman, 2000). On the other hand, business decisions have been reported to be based upon patterns of preparation that do not conform to the typical pattern of argumentation used in research communities (v. Werder, 1999).

Occupation Specific Argumentation as a Barrier to Commercialization

In addition to functioning as an occupational identity construct, rational argumentation may constitute a barrier to commercialization. As a scientist explained:

The clash occurred where ... the mode of working within the culture of science is one that's driven by inquiry, which is driven by brainstorming. The commercial product managers wanted to focus on one approach only and pin it down, no divergence whatsoever. Even the mode of discussion, conversation, engagement, was so different and the commercial guys found it quite threatening and kept saying, "Stop arguing!" ... But it's viewed as the dysfunctionality of the [scientific] group, and sort of a threatening aspect. And a couple of times where I've sort of gone into "science mode" of really wanting to understand "Why and how", has been seen as threatening. Here as well [referring to new employer].

A manager from the same CRC offered the following rejoinder when asked about the need for scientists to engage in rational argumentation to identify the best possible solution:

I generally think that there is a, sensible way forward with a number of options that you can discuss but you should be aware of the options beforehand. And the thing is scientists always say, "You don't know," and therefore just stop planning for it, which does drive me nuts. ... And that's not really thinking far enough ahead. And that I've seen many, many times, that scientists focus on the next few experiments, and being completely unable to handle uncertainty and complexity which is odd given that's what they should be doing, and I think that is what business does all the time. If you haven't told somebody something might go wrong and it does, it might be blindingly obvious to a scientist but if you haven't told your supporters and investors of the possibility then they are certainly going to see vigorous debate as a failure.

Researchers recognized the value and complementary potential of contributions from both research and business cultures, but saw managerial language norms intruding on their professional judgments of the "correct" way to conduct research, and degrading the quality of work produced. For example:

And that difference in culture, the way a scientist will approach a problem is to look at all the possibilities and then choose, and pursue a particular path. The way these guys were doing it was choosing the path and not deviating from it and just going down further and further into that. Now both of them absolutely constructive because you have that broader engagement to come up with a better solution that you then feed into the very necessarily so, straight-jacket type focus. ... What that did then, by quelling that inquiry, it actually reduced the standards. The sort of consensus approach reduced the standards of work created. There was a mediocrity that came into the work. It was very interesting, an absolute culture change that occurred.

From the perspective of the researchers, attempts by business people to limit debate were dismissed as "micro-management" and indicative of a desire to "dictate" technical solutions to problems they were not qualified to address. Researchers reported dismay and distrust upon hearing business people publicly announce project timeframes based upon what they viewed as a naively optimistic view when dealing with non-routine discovery work; that everything would run to schedule. Both commercial and research informants reported business people viewing rational argumentation as a sign of occupational dysfunction, evidence of inadequate forward planning, or suggestive that the proposed solution will fail.

At least three points emerge from the previous discussion. First, argumentation styles can act as occupational and subcultural norms. Second, the use of distinctive occupational argumentation repertoires may not facilitate communication in cross-functional meetings and thirdly, argumentation style may influence perceptions about project statuses and likely outcomes. Undoubtedly, the diverse forms of argumentation favored by research and commercial occupations provide a sense of group identity and are suited to their differing tasks and organizational environments. However, different argumentation styles may reduce the ability of interoccupational contributions to commercialization to be accepted and recognized as constructive.

CONCLUSION

How Can Research and Commercial Groups Talk to Each Other?

Argumentation patterns typical of commercial and research groups reveal their occupational values and assumptions. In commercial settings action and results are valued, particularly as focus upon speed as a competitive strategy has increased over the past fifteen years (Vinton, 1992). This leads to argumentation patterns favoring quick decisions oriented toward taking action to achieve short-term organizational goals. This commercial argumentation is not easily reconciled with scientific values of accuracy and thorough understanding. Furthermore, popular management texts praising "strong" business cultures that display consensus and conformity (Pech, 2001) may have exacerbated perceived differences between the two groups' use of argumentation, contributing to commercial perceptions of scientific argumentation as an occupational and organizational problem.

Research or academic styles of argumentation are characterized by vigorous dissent to create and continuously improve ideas and are not timed while business debates are reported be action oriented, timed and managed to a definite conclusion. In commercial groups, dissent may be perceived as a threat to group cohesion, group hierarchies and the efficient execution of business plans. Members of research groups report enjoying intellectual dissent, consider it to be constructive and value it as a key part of their professional identity. When the groups are separate, they can follow their own norms in parallel. When combined in cross-functional teams, encountering seemingly foreign standards for debate can create frustration and may jeopardize commercialization outcomes. Members of each community generally regard the debating norms of the other group as unusual and an obstruction to equal participation (even to the extent of applying labels such as "dysfunctional"), creating barriers to communication and commercialization. At the same time, members of both groups maintain their occupational subcultural boundaries and identity by continuing their distinct argumentation style in a combined forum, even when doing so impedes communication.

The argumentation rituals used by research and commercial groups may reflect their underlying reward systems and motivation towards producing knowledge or making money. Mutual conflict exists between business pressures for speed and punctuality and research emphasis on thorough understanding to protect and enhance scholastic reputations. Moreover, the different forms and norms of debate and rhetoric fit two key roles in innovation: imagining the as yet non-existent product or service and making it concrete. Scientific or rational argumentation plays a complementary role to business modes of debate through the creation and testing of ideas vital in the early stages of the commercialization process. Focusing on a specific form of a new product or service, and organizing the team to fund, test, and manufacture in volume to meet set deadlines provides the corresponding applied role.

In contrast to Levine's (2006) assertion that conflict is detrimental to an organization's culture, Roberto (2005) champions the benefits of conflict in improving the quality of decisions, provided that it is paired with a consensus to execute decisions once they are made. In one CRC evidence existed for cross-pollination of researcher culture, with it's willingness to question and debate, with a commercial focus on planning and execution essential for the survival of the participating small and medium enterprises (SMEs). This did not occur serendipitously, but only by the conscious design and action of the CEO, working through the CRC's Communications Manager. The CEO commented on the issue of using "creative dissent" in meetings of commercial and research members, and the fact that some of the industry partners had adopted rational argumentation as a technique for use inside their own organizations:

I found it productive then [in a public sector organization] and still do, when used sparingly. So it has wider application than just in the academic domain. Yes, I think some SME's do use it, especially in their innovation activities, because this where the employees are given a little more latitude to question and explore.

Nevertheless, it is unlikely that there will be wide-spread adoption of research forms of argumentation by industrial participants in triple helix organizations, or that research participants will willingly accept arguments based on consensus or formal organizational authority. The debating norms of each occupational culture govern that community, are acquired through training or socialization, and their display indicates that an individual chooses to remain a member of that group (McKerrow, 1980). Hence, the differing argumentation styles of research and commercial groups will be resistant to change. Nevertheless, if members of triple helix organizations understand and are aware of the norms and debating rituals of business and research occupational subcultures it may be possible to reduce misunderstandings and support successful commercialization. To obtain the benefits promised from the formation of hybrid industry-research groups, it is important that each occupational subculture be free to employ its own subculturally appropriate rhetorical forms in homogenous, single occupation meetings. However, a consciousness of differences in styles of debate and modified use of research and business argumentation when occupational subcultures meet, are likely to be advantageous.

When viewed from the traditional, integrationist perspective of organizational cultural research, hybrid industry-research organizations face the challenge of operating within an organizational cultural paradox; the beliefs, values and assumptions held by groups within the organization are seemingly contradictory and yet valid. The divergent bodies of knowledge and skills that make the parties attractive to each other, and provide reason to collaborate, simultaneously create difficulties in communication and cooperation.

However, recognition of the dominance of the differentiation view of organizational culture in hybrid industry-research organizations may allow managers of innovation to resolve the apparent paradox through acknowledging, respecting and facilitating productive interactions between dissimilar, but ultimately complementary occupational cultures. Clear agreement and enthusiasm for high-level goals can co-exist with fundamental assumptions and values that are in conflict.

Commercial and research occupational subcultures have a history of working separately. They occupy separate thought worlds, use different norms and have disparate assumptions about goals and methods of working. The distance between the assumptions, beliefs, values, and norms of the subcultures associated with research and commercial communities necessitates acknowledgement, and conscious effort to accommodate their differences. Both are efficient and effective in their own contexts, but thought, effort and action are needed to improve interoccupational working patterns, communication, and outcomes. Whether or not a hybrid industry-research organisation thrives and becomes productive will depend to some extent on the management of occupational cultural differences between its research and commercial members.

ACKNOWLEDGMENT

The researchers would like to thank the participants for their generosity in allowing us to benefit from their time and experience.

NOTE

Based on "Business and Scientific Forms of Argumentation in Commercialization: Dictators and Chinwaggers by Kate Hayes and Anneke Fitzgerald, which appeared in the Proceedings of the 2006 International Conference on Management of Innovation and Technology.

REFERENCES

Barringer, B. R., & Harrison, J., S. (2000). Walking a tightrope: Creating value through interorganizational relationships. *Journal of Management, 26*(3), 367-403.

Borys, B., & Jemison, D. B. (1989). Hybrid arrangements as strategic alliances: theoretical issues in organizational combinations. *Academy of Management Review, 14*(2), 234-248.

Burgelman, R. A., & McKinney, W. (2006). Managing the strategic dynamics of acquisition integration: Lessons from HP and Compaq. *California Management Review, 48*(3), 5-27.

Department of Education Science and Training. (2006). *CRC Directory 2006*. Retrieved October 1, 2006, from https://www.crc.gov.au/HTMLDocuments/Documents/PDF/CRC_Directory_2006.pdf

Department of Education Science and Training, & Howard Partners. (2003). *Evaluation of the cooperative research centres programme*. Canberra, Australia: Howard Partners Pty Ltd.

Dixon, J. C., & Rosenbaum, M. S. (2004). Nice to know you? Testing contact, cultural, and group threat theories of anti-black and anti-Hispanic stereotypes. *Social Science Quarterly, 85*(2), 257-280.

Dolinina, I. B., & Cecchetto, V. (1998). Facework and rhetorical strategies in intercultural argumentative discourse. *Argumentation, 12*, 167-181.

Erduran, S., Simon, S., & Osbourne, J. (2004). TAPing into argumentation: Developments in the application of Toulmin's argument pattern for studying science discourse. *Science Education, 88*(6), 915-933.

Etkowitz, H., & Leydesdorff, L. (2000). The dynamics of innovation: From national systems and "mode 2" to a triple helix of university-industry-government relations. *Research Policy, 29*, 109-123.

Gibbons, M., Limoges, C., Nowotny, H., Schwartzman, S., Scott, P., & Trow, M. (1994). *The new production of knowledge: The dynamics of science in contemporary societies*. London: Sage Publications.

Gray, D. O., Lindblad, M., & Rudolph, J. (2001). Industry-university research centers: A multivariate analysis of member retention. *Journal of Technology Transfer, 26*, 247-254.

Hale, N. M. (1998). Effects of age and interpersonal contact on stereotyping of the elderly. *Current Psychology, 17*(1), 28-48.

Hayes, K. (2007, Sept 6-7, 2007). *Triple helix organisations, knowledge-stewarding communities of practice and perceptions of time: The hunters and gatherers of commercialisation.* Paper presented at the 8th European Conference on Knowledge Management, Consorci Escola Industrial de Barcelona, Spain.

Hayes, K., & Fitzgerald, A. (2005, Sept 4-7). *Preliminary findings of an investigation into interactions between commercial and scientific occupational cultures in hybrid research organisations.* Paper presented at the Proceedings of the 6th International CINet Conference, Brighton, UK.

Kunda, G. (1992). *Engineering culture: Control and commitment in a high-tech corporation.* Philadelphia: Temple University Press.

Lehrer, M., & Asakawa, K. (2004). Pushing scientists into the marketplace: Promoting science entrepreneurship. *California Management Review, 46*(3), 55-76.

Levine, S. (2006). High performance organizations: Creating a culture of agreement. *Handbook of Business Strategy, 7*(1), 375-380.

Liu, Y. (1999). Justifying *my* position in *your* terms: Cross-cultural argumentation in a globalized world. *Argumentation, 13*, 297-315.

Liyanage, S., & Mitchell, H. (1993). Organizational management in Australian cooperative research centres. *Technology Analysis and Strategic Management, 5*(1), 3-14.

Martin, J. (2002). *Organizational culture: Mapping the terrain.* Thousand Oaks, CA: Sage Publications Inc.

McKerrow, R. E. (1980). *Argument communities.* Paper presented at the 1st Summer Conference on Argumentation, Annandale, VA.

Merton, R. K. (1957). Priorities in scientific discovery: A chapter in the sociology of science. *American Sociological Review, 22*(6), 635-660.

Meyerson, D., & Martin, J. (1987). Cultural change: An integration of three different views. *Journal of Management Studies, 24*(6), 623-647.

Nemeth, C. J. (1997). Managing innovation: When less is more. *California Management Review, 40*(1), 59-74.

Nonaka, I., & Takeuchi, H. (1995). *The knowledge creating company: How Japanese companies create the dynamics of innovation.* New York: Oxford University Press.

Nowotny, H., Scott, P., & Gibbons, M. (2001). *Re-thinking science: Knowledge and the public in an age of uncertainty.* Malden, MA: Blackwell Publishers Inc.

Pech, R. J. (2001). Termites, group behaviour, and the loss of innovation: Conformity rules! *Journal of Managerial Psychology, 16*(7), 559-574.

Premus, R. (2002). Moving technology from labs to market: A policy perspective. *International Journal of Technology Transfer and Commercialisation, 1*(1/2), pp. 22-39.

Roberto, M. A. (2005). *Why great leaders don't take yes for an answer: Managing for conflict and consensus.* Upper Saddle River, NJ: Wharton School Publishing.

Schein, E. H. (1995). *Organizational and managerial culture as a facilitator or inhibitor of organizational transformation* (MIT Sloan School of Management Working Paper). Cambridge, MA, USA.

Senker, J., & Sharpe, M. (1997). Organizational learning in cooperative alliances: some case studies in biotechnology. *Technology Analysis and Strategic Management, 9*(1), 35-51.

Siegel, D. S., Waldman, D. A., Atwater, L. E., & Link, A. N. (2004). Toward a model of the effective transfer of scientific knowledge from academicians to practitioners: Qualitative evidence from the commercialization of university technologies. *Journal of Engineering and Technology Management, 21*, 115-142.

Sillince, J. A. A. (2002). A model of the strength and appropriateness of argumentation in organizational contexts. *Journal of Management Studies, 39*(5), 585-618.

Stangor, C., Jonas, K., Stroebe, W., & Hewstone, M. (1996). Influence of student exchange on national stereotypes, attitudes, and perceived group variability. *European Journal of Social Psychology, 26*, 663-675.

Steiner, C. (2000). Teaching scientists to be incompetent: Educating for industry Work. *Bulletin of Science, Technology, and Society, 20*(2), 123-132.

Sutton, R. I., & Rafaeli, A. (1992). How we untangled the relationship between displayed emotion and organizational sales: A tale of bickering and optimism. In P. J. Frost & R. E. Stablein (Eds.), *Doing exemplary research* (pp. 115-128). Newbury Park, CA: Sage Publications Inc.

Trice, H. M. (1993). *Occupational subcultures in the workplace.* Ithaca, NY: ILR Press, Cornell University.

v. Werder, A. (1999). Argumentation rationality of management decisions. *Organization Science, 10*(5), 672-690.

Vinton, D. E. (1992). A new look at time, speed, and the manager. *Academy of Management Executive, 6*(4), 7-16.

Young, E. (2003). Joined-up thinking. *New Scientist, 179*(2408), 46-52.

Ziman, J. (2000). Are debatable scientific questions debatable? *Social Epistemology, 14*(2/3), 187-199.

Section II
The Knowledge Worker

Chapter V
High-Tech Meets End-User

Marc Steen
TNO Information & Communication Technology, The Netherlands

ABSTRACT

One challenge within the high-tech sector is to develop products that end users will actually need and will be able to use. One way of trying to match the design of high-tech products to the needs of end users, is to let researchers and designers interact with them via a human-centred design (HCD) approach. One HCD project, in which the author works, is studied. It is shown that the relation between interacting with end users and making design decision is not straightforward or "logical." Gathering knowledge about end users is like making a grasping gesture and reduces their otherness. Making design decisions is not based on rationally applying rules. It is argued that doing HCD is a social process with ethical qualities. A role for management is suggested to organize HCD alternatively to stimulate researchers and designers to explicitly discuss such ethical qualities and to work more reflectively.

HUMAN-CENTRED DESIGN

Many organizations, both private and public, need to or want to innovate. Not for the sake of innovation itself, but in order to create new products, services, or processes that will create added value for their customers, for the end users of their products or services or for citizens. Developing innovations that match end users' needs or wishes is especially (but not exclusively) problematic in the high-tech industry where many innovations are driven by technology push. A risk of technology push is that researchers and designers invent some product or service that nobody needs or nobody can use. One way in which researchers and designers try to match their innovation efforts to end users' needs and wishes is to *interact* with them during an innovation project. They try to learn from them, to be informed or inspired by them. This can be seen as an attempt to narrow the gap between researchers and designers in their high-tech ivory tower vs. end users "out-there."

My current interest is in researchers and designers activities. However, it is acknowledged that their work is only one half of the innovation process: end users also play crucial roles in adoption, domestication, and appropriation processes (Oudshoorn & Pinch, 2003, p. 11-16). Users and technology are "co-constructed" (*ibidem*): people influence technology and technology influences people, both designers *and* end users shape an innovation. However, I am currently more interested in how researchers and designers think and speak about end-users "out-there" (Latour & Woolgar, 1986), rather than being interested in any "real" existence of end-users or their "real" characteristics.

There is a broad variety of methods available for researchers and designers[a] to involve (future, potential, or putative) end users[b] in their projects, for example: participatory design (e.g., Schuler & Namioka, 1993) where people who will be using the system that is being developed are invited to cooperate during development, evaluation, and implementation of that system (such efforts are often related to workers' emancipation); the lead-user approach (e.g., Von Hippel, 2005) where innovative users are seen as a source of innovation and are invited to help develop or improve a product (similar to participatory design, but with less emphasis on emancipation); fieldwork, inspired by ethnography or ethnomethodology, to study the social and cultural aspects of what people do, in order to design applications (often combined with participatory design, for example in the field of computer supported cooperative work) (Crabtree, 2003); contextual design (Beyer & Holzblatt, 1998), a method to observe people doing tasks in their natural context, with attention for their physical surroundings, the artefacts they use as well as their activities, communication, power and culture, and to articulate system requirements based on this; empathic design (e.g., Koskinen, Battarbee, & Mattelmäki, 2003), where researchers or developers try to get closer to end users' lives and experiences, for example by observing their daily life or work, or role-playing some of their activities, and apply what they learn from that in the design process; codesigning (Sanders, 2000), a kind of participatory design where end users make things together with researchers and designers (the focus is on *making* things, and doing that jointly, rather than on *saying* things in interviews, or on being observed *doing* things); and usability engineering, a range of methods to evaluate and improve a product's usability together with end users.

Such approaches are known in more general terms as human-centred design (HCD), which is characterized by four principles: (1) the active involvement of users for a clear understanding of user and task requirements; (2) an appropriate allocation of functions between users and technology; (3) iterations of design and evaluation processes; and (4) a multidisciplinary approach (ISO/IEC, 1999). One goal of HCD is to involve future or potential end users as early as possible: preferably from the start of a project, when problems are articulated, when the design brief is formulated, when directions for searching solutions are chosen so that end users' input can optimally help to steer the project—rather than, for example involving end users at the end of a project for usability testing. Kujala (2003), in a review of various approaches for early user involvement, concludes that early user involvement has "positive effects on both system success and user satisfaction." She is positive about ethnography-inspired field studies as a way to learn about end users' needs and wishes and formulate system requirements (rather than asking them in interviews), and she is critical about whether designers have the necessary skills to conduct and interpret such fieldwork.

STUDYING ONE PROJECT

In many accounts of HCD projects, the relation between interacting with end users and making design decisions is *not* seen as problematic. It seems like taken for granted that such interactions lead to better decisions: "after the observations it was decided to prioritize problem x and make it central in the design process" or "during the workshop solution y was chosen as a basis for further product development." Such phrases seem to cover up what happens in-between interactions with end users and making design decisions. The process of HCD is often presented as a "black box" (Latour, 1987): a container that only shows what goes in (interactions with users) and what goes out (design decisions) and which covers up what happens in-between. However, it is easy to imagine that *something* happens in-between. Some filtering process happens, because researchers and designers do not straightforwardly follow everything that they see and hear from end users, nor do they do nothing with what they see and hear.

My current interest is to open-up the "black box" of HCD. I am interested in how researchers and designers who work in the high-tech ICT industry, for example in research labs, seek interactions with end users, and how this informs or inspires their design process and decision making: how these interactions relate to the process of exploring and articulating of problems and of solutions. Furthermore, I focus on research and design as a *social* process (Bucciarelli, 1994). I focus on the interactions between people, rather than on the technological or economic aspects of research and design.

Such a study requires *access* to researchers and designers doing their work, and therefore I chose to study one project in which I work and of which I coordinate one part, and to conduct the study as "participant observer" (Easterby-Smith, Thorpe, & Lowe, 2002, p. 110-4). This project is sponsored by the Dutch Ministry of Economic Affairs and by the participating organizations, and its goal is to create concepts that will help businesses and organizations to develop and apply innovative information and communication technology (ICT). The project is positioned in the "fuzzy front end" of innovation (Koen et al., 2002). It is about articulating opportunities and creating concepts. This means that the project team members have a relatively large amount of freedom and responsibility concerning what problem to address or what direction to look for solutions—more than in regular product development, implementation, or market introduction projects.

People from different organizations and with various backgrounds work in this project. My study is confined to one part of the project in which we design and evaluate a new kind of telecommunication applications: *we-centric* applications. At the start of the project, there was only this vision of what we wanted to achieve with *we-centric* applications: to stimulate and to facilitate people to communicate and cooperate better with each other, especially among people who currently don't communicate and cooperate enough with each other. During the project we further developed this concept, partly based on interactions with end-users, so that by the end of the project we were able to characterize we-centric telecom applications as follows. A *we-centric* telecom application is typically meant to be used on a mobile phone or *smart phone,* a handheld device that combines phone and basic computer functions, similar to a personal digital assistant. Furthermore, a we-centric telecom application would provide suggestions to communicate or cooperate with others, and has two key functions. These functions were not clear at the start of the project, but they emerged during its course: (1) it composes and presents a dynamic list of potentially relevant people, which is meant as a suggestion to communicate or cooperate with these people—this list is created automatically based upon searching for similarities in data on participants' contexts and interests or preferences; and (2) it presents information about these other people's contexts, which is meant to actually help communicate or cooperate with them—this "context information" may contain their "presence" (one can be "available," "online," "busy," "away," etc., similar to instant messaging) and a short explanation why this person is supposed to be relevant for you currently. Furthermore, we envision that a *we-centric* telecom application is reciprocal: if A receives a suggestion to talk with B, then B receives a similar suggestion about A, and if A can see something of the current activities of B, then B would be able to see something of A's likewise.

The project's goal is to develop and evaluate *we-centric* application together with, and for, two target groups: (1) with/for police officers; and (2) with/for informal carers. These efforts are described in the next two sections.

DESIGNING WITH/FOR POLICE OFFICERS[C]

At the start of this part the project, Albert—who works at an ICT expertise center, which is associated to the police organization and who is a former community police officer—proposed to support and improve the work of community police officers. The work of those kind of police officers consists largely of talking with various people, for example with shop owners, school administrators, people from the municipality, and building corporations as a way to serve citizens and to prevent crime. Their work is rather different from the work of *emergency* police officers, who react to emergencies and who would, for example, drive around with lights flashing. Albert wished to empower community police officers and to create tools for them to better communicate and cooperate with other people for their various tasks. This idea matched the goal to design a *we-centric* application.

Then Mandy, Dirk, and me—we work at two ICT research labs and the three of us are responsible for the design and evaluation of a *we-centric* application—and several fellow project team members accompanied one or more police officers during one working day. We did this to learn hands-on about police work, rather than from documents or from interviews. We made individual notes of their observations and then summarized these in the form of "personas" and "storylines" (e.g., Cooper, 1999). Based on our observations we created three personas, three typical police officers Ad, Bert, and Theo, and described, in the form of storylines, three typical working days of these police officers—see below for an excerpt:

About community police officer Ad

Ad is 45 years old and has been working for 20 years as a community police officer. Ad is married and has two children: a daughter of 18 and a son of 16. Five years ago, Ad and his family moved from Amsterdam to Haretown [in the countryside]. Amsterdam was becoming too hectic for him and he preferred quieter surroundings. [...]

Monday 11:00o'clock.

Ad is walking around in "his" area. He decides to go to the swimming pool to have a quick look how things are. The weather is beautiful and it will probably be very busy in the swimming pool. The last few days a group of teenagers has been causing a lot of trouble. Ad arrives at the swimming pool and looks for his contact person, John. John however appears to have a day off.

We portrayed Ad as relatively old and preferring quiet work, which is typical for some community police officers—so we were told. This is in contrast to a typical emergency police officer, who is younger and enjoys the thrills of emergency police work. Furthermore, we are suggesting that his work may be improved by a *we-centric* telecom application, which would send a notification to Ad that his contact person at the swimming pool is currently absent, so that Ad may decide to visit the swimming pool another time.

We then organized a workshop with the police officers whom we spent a day with, to discuss our observations and interpretations. During that workshop the project's goal shifted from improving communication between police officers and other people *outside* the police organization, to improving communication and cooperation *within* the police organization, mainly because opening-up communication and cooperation towards people outside the police is not desirable from a security perspective. Furthermore, the idea was developed to help community police officers to share their "implicit" knowledge with emergency police officers. Community police officers typically have a lot of knowledge about people and locations in their heads—not everything is made explicit in reports or databases—and such implicit knowledge may be helpful for emergency police officers who often lack such knowledge. A typical "use case" would be this: Emergency police officer Bert is sent to an address because of domestic violence. Without the application, he would enter this situation unprepared, but *with* it he would receive a suggestion to contact community police officer Ad, who was at that address for a similar incident just two days ago and who made some arrangements with these people. If Bert calls Ad, Ad can update Bert, quickly, of course. This idea also builds upon observations of the researchers at the Police Academy that community police officers and emergency police officers rarely communicate with each other on the street, and that their work could improve if they would.

The police officers' manager wanted to participate in the workshop, which had several effects. At the start of the workshop, he became irritated about the storylines that we created. He was irritated because we portrayed his daily practice and problems as "children's stories." During the workshop he probably influenced what other police officers were willing or dared to say. And at the end of the workshop he pointed out that we "learned nothing at all about police work" in one-day observations. He suggested that we should do observations over a longer period.

Mandy, Dirk, and I then made sketches for a telecom application, which we dubbed *PolicePointer*. This application is envisioned to run on a *smart phone*. Its main function is to provide suggestions to contact other police officers who may have information that is useful for you, given your current task or context.

We organized two workshops for which we invited both community police officers and emergency police officers. We discussed the topic of knowledge sharing and our ideas and sketches for receiving suggestions to communicate with other police officers. During these workshops, the *PolicePointer* was further developed. It was discussed that emergency police officers may be reluctant to "waste time" contacting other people; their job is to react quickly. Therefore, a notification function was added: when Bert receives a suggestion to contact Ad, Ad simultaneously receives a notification, so that he may contact Bert proactively. Furthermore, the police officers mentioned that emergency police officers have knowledge what happens outside office hours and what happens in different areas and that this kind of knowledge may be useful for community police officers. We adopted that idea so that the project's focus then became to support *both* kinds of police officers to communicate and share knowledge with each other.

With the help of fellow project team members these sketches were developed into a prototype (see Figure 1). This prototype was evaluated by five police officers during a field trial of two days, during which they used a *smart phone* with the *PolicePointer* on it. During this trial, very few incidents occurred, and, partly because of that, only few situations occurred in which the *PolicePointer* was perceived as having added value. Nevertheless, the participating police officers were positive about the *PolicePointer*, especially when they would receive suggestions for "unstructured" situations, situations for which there are no procedures and in which they must improvise, and when they would receive suggestions when they were in a relatively unknown area in which they don't know many colleagues (yet). Interestingly, both project team members and police officers

Figure 1. A prototype of PolicePointer, running on a smart phone

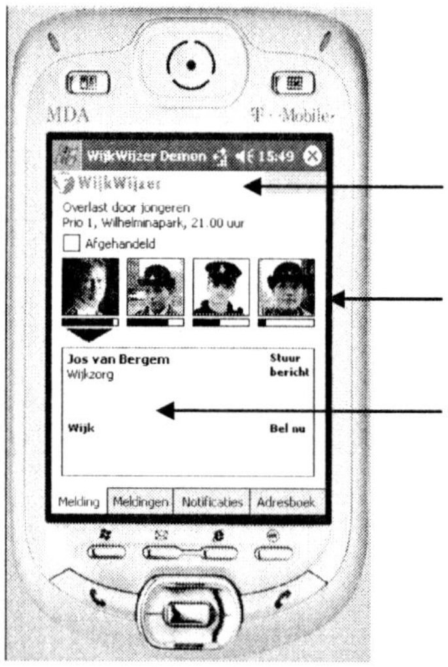

downplayed the prototype's shortcomings during this trial: the hardware was too bulky, the login procedure was too long, and the system responded too slow. We explained that these issues would be improved if the prototype were developed into a working product.

Looking at this design process, we see a mixture of participatory design (inviting end users during problem definition and solution finding), of the lead user approach (the police officers who participated in our project were selected based on their innovative behaviour or attitude), and of empathic design (project team members joined a police officer for one day). Furthermore, we can see that each interaction with police officers resulted in some shift of the project's focus. This is in line with the goal of HCD: to let end users have influence upon the research and design and design process. The *PolicePointer* started as a tool for communication between community police officers and people outside the police, then became a tool for community police officers to share their knowledge with emergency police officers, and then became a communication tool between these two sorts of police officers.

However, we may also say that our focus on designing a *we-centric* telecom application gave us eye-flaps. In the process of going from an idea via sketches into a prototype, the *PolicePointer* gradually solidified. But in each step of the research and design process, the end users—the police officers—became less "real" and more "ideal." In each interaction with them, we further modelled real police officers into ideal users of our product. We started with the goal to design a *we-centric* telecom application, a solution "in our heads" looking for a problem "out-there," and we were happy that Albert found community police officers as a target group. We observed some police officers and wrote storylines about them, storylines that irritated one police sergeant, storylines in which we implicitly suggested that they should work differently, and that they should use our *we-centric* application. And we created a trial in which we managed to neglect the prototype practical shortcomings and instead focused on the *PolicePointer*'s theoretical, potential added value.

DESIGNING WITH/FOR INFORMAL CARERS[D]

Project team members Catherine and Edith, who work at a university's medical department and who have years of experience in working with and for people with dementia and their informal carers, articulated this goal: to help informal carers who provide care to people who suffer from dementia and who live at home (*not* in an institution). In many cases, these people provide care to their husband, their wife, or one of their parents and such care is often needed fulltime. The ultimate goal is to improve the quality of life for both the person with dementia and the informal carer, by supporting the informal carer, so that he or she can better provide care and will feel less burdened or will be less likely to suffer from burn-out.

In order to better understand the needs of people with dementia and of their informal carers, Pauline, together with her colleagues Catherine and Edith conducted a large scale survey. Over 300 "dyads"—a person with dementia and his or her "primary" informal carer—were interviewed using several standardized questionnaires. This fieldwork took several months and Pauline presented their preliminary results several times during meetings. Typically, she would show a table with the "most frequently reported (unmet) needs of people with dementia and their informal carer": "memory (40.5%); daily activities (17.9%); information about health and treatment for informal carer (17.3%); company (12.7%); psychic need of informal carer (12.3%) …."

The other project team members, Annelies and Martin, who work at a new media lab, and Rachel, who works at an ICT research lab, are responsible for the design and evaluations of a *we-centric* application. Dementia and informal care are relatively new topics for them, and therefore they chose to do some additional observations and interviews. This caused friction within the team. Pauline suggested that they first study the literature and the results from their survey, and to do additional research only if they *miss* specific data. Furthermore, there were discussions about method. Pauline and her colleagues do a survey to obtain a statistically representative overview of people's needs, whereas Annelies and Rachel want to get acquainted with a small number of people with dementia and their informal carers and to be inspired by them. This difference is also described by sociologist Haddon and designer Kommonen (2003), who characterize social scientists as being concerned with existing knowledge and studying and documenting reality, and designers as being concerned with originality, imagining alternatives and changing reality. There were several conflicts within the project team about doing a large, statistical representative survey vs. doing "interviews with only one or two people," and about which method can be a basis for conclusions or decisions. Only after project team members had actually cooperated in interpreting each other's data, were they able to cooperate constructively.

Based on the survey and on the additional interviews, a design focus was agreed upon to alleviate the burden of the informal carers, to support them in their "work." In many interviews informal carers told that they experience taking care of a person with dementia as very demanding, especially if he or she is your husband, wife, or parent. Furthermore, the situation of a person who suffers from dementia cannot improve but only become worse. Such informal carers have to do everything alone: there may be nobody who offers help, or they may not dare to ask others for help.

Annelies, Martin, and Rachel, helped by Pauline, then started a user study. Four informal carers were invited to participate in a series of three interviews in their homes. In the first interview the project team members got acquainted with them and learned about their situation. Based on these interviews they created personas and storylines, in the form of "a day in the life of…" – see below for an excerpt:

About Ans and Simon

Ans (73 years) has dementia. Her husband Simon (76 years) is her "primary informal carer." They have been together for 51 years. They have two sons, Johan and Pieter, and four grandchildren. They were born and raised in a rural part of the country. Simon had his own shop where he manufactured and repaired sails. He retired 12 years ago. Ans used to do the financial administration and took care of their two sons. She used to be very active; she did volunteer work and used to go for a bicycle trip once a week with the woman from next door.

Time: 5:30.

Ans is stumbling around in the house, which awakens Simon. He sighs and tries to sleep again. He went to bed late last night, because he wanted to finish reading his book. Ans is calling Simon. Simon gets out of bed and goes to help her.

Time: 06:15.

Simon brushes his teeth and dresses. Then he chooses clothes for Ans. He helps Ans to shower, then he helps her dry herself and dress. Helping her with the sleeves of her dress takes a while. Then Ans is dressed.

Ans is portrayed as a woman who used to be active and social. This shows us the woman she used to be before she got dementia and became passive and isolated. Furthermore, the situations of waking up early and helping Ans shower and dress suggest that Simon "suffers" from Ans suffering from dementia. He has to help her fulltime. This also suggests that we could help Simon, for example with a *we-centric* application, which may help to alleviate informal carer's social and emotional needs.

In a second round of interviews, Rachel and Annelies read these storylines aloud and discussed these. Together with the informal carers, they identified several situations that the informal carers found relevant. After the interviews, Rachel, Annelies, Martin, and Pauline selected three situations to be used as input material for a concept development workshop with six project team members and with six additional "external" creative designers (not project team members). In this workshop, three groups worked in parallel, each on one of the three situations. One group developed an idea for an ICT application that matches people who look for help and people who offer help, which is meant to let informal carers support each other, practically and emotionally. The other two groups came up with a "domotica" application which monitors what the person with dementia does in the home and assists him or her; and a jewelery product with localization technology, which helps the person with dementia to return home if he or she gets lost outside.

The first idea was selected to be further developed because: (1) it is closest to the goal of designing a *we-centric* telecom application, it is about communication and cooperation; (2) it is a "mobile, context-aware and adaptive" application, which fits the project's scope and goal; (3) it is similar to ideas developed previously by project team members and to the *PolicePointer*, which makes the design more comfortable; (4) and it seems most feasible to build a prototype of it within planning and budget. Making this decision was difficult because some project team members thought that the other two ideas were more interesting and creative. Furthermore, it was noted that the people behind the first idea were involved in the research process, they were "experts on informal carers' needs," whereas the people who came up with the other two, more creative ideas were people with little or no experience with dementia or informal carers.

In a third round of interviews, Rachel and Annelies discussed this first concept with the four informal carers who participated in the first two rounds of the interview, and with seven other informal carers. Questions that were discussed were How do you wish to invite informal carers to participate? How would you like to form or control such a group of informal carers? How do you think participants can be motivated to offer and accept help? What kind of help requests would people ask of each other, for example structural or incidental? And should there be mechanisms to monitor how much help a person offers and receives? Based on these interviews, the following functional requirements for a we-centric telecom application—dubbed *WeCare*—were articulated: (1) an informal carer can ask for help by putting a "help request" in her online calendar; (2) an informal carer can offer help by filling in her online calendar (when she is available to help) and her profile (the kind of help she wishes to offer); (3) the system automatically matches the "help requests" in with other participant's calendars and profiles; and (4) messages are sent to the person asking for help and the person offering help—these messages can be sent by e-mail or SMS or as entries in participants' online calendars. Additionally, participants can post on a bulletin board: "help requested" and "help offered."

The current plan is to build a prototype of *WeCare* and to evaluate this together with several informal carers, similar to how the *PolicePointer* prototype was evaluated. The prototype will include a database, which can be accessed via a webpage on their computer and also via their *smart phone*. The idea is that people will be able to do complex tasks, such as scheduling, while sitting at their computer, see Figure 2, and use a subset of the functionalities via their mobile phone or *smart phone*, for example when they are travelling.

In this design process we saw a mixture of participatory design (discussing possible problems and solutions with potential end users) of the lead user approach (the informal carers who participated were selected based on their use of computer, internet and mobile phone), and of usability engineering (discussing concepts and sketches with end users). If we look at how the interactions with informal carers influenced the design process, we see that the initial idea—to support informal carers to do their "work"—remained relatively stable. The interviews with informal carers helped to make design decisions and to articulate functional requirements. However, during the interactions with the informal carers the concept was not critically questioned,

Figure 2. A mock-up of WeCare: A webpage on a computer

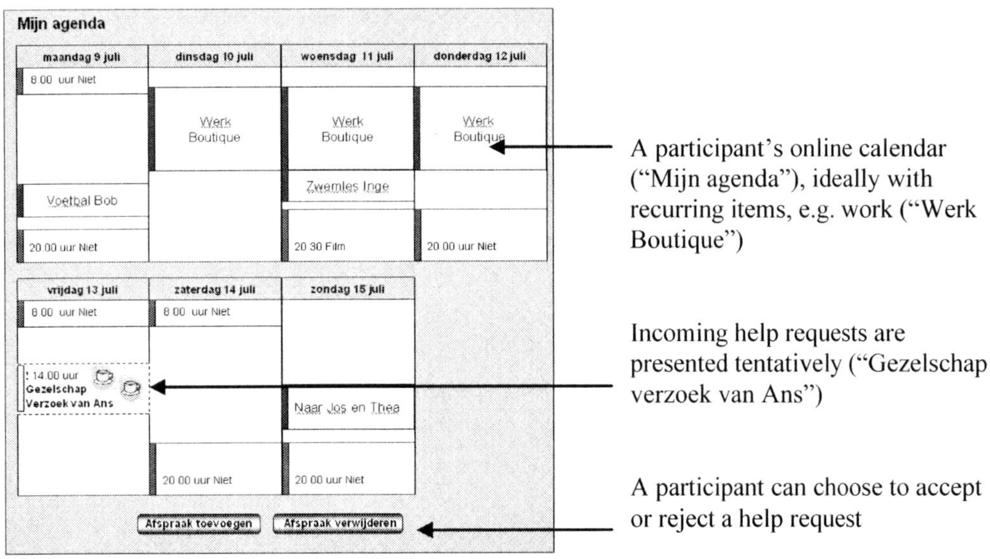

and an attempt, in the concept development workshop, to come up with alternative ideas was not welcomed.

Furthermore, we can see that there was friction *within* the (relatively large and heterogeneous) project team: about which research methods to use (a systematic survey, which produces facts vs. informal interviews for inspiration) and about choosing a concept (a conservative concept vs. a more innovative concept). This contrasts with police officers' case, where there was friction between the (relatively small and homogeneous) project team and the police officers, for example when the officers' manager was irritated about "our" storylines about "their" practice. This draws attention to the importance of cooperation *within* the team in a human-centred design project: not only the interactions with end users, but also the interactions between people in a multidisciplinary team are important (cf. principle 4 of HCD).

GATHERING KNOWLEDGE

In the next two sections, I will critically examine two assumptions that seem to be key in human-centred design (HCD): that researchers and designers can gather knowledge about end users' needs, wishes and preferences; and that they can apply this knowledge to make appropriate design decisions.

These assumptions can be criticised from various angles: one can argue that end users are not aware of their needs, cannot articulate these, or do not want to or cannot speak about these (van Kleef, van Trijp, & Luning., 2005); that zooming in too much on a small group of end users will result in an over-customized product that will interest only a few (Stewart & Williams, 2005); or that paying too much attention to end users erodes the role of the designer, whose vision and creativity are essential (Hekkert & Van Dijk, 2001). Such effects can indeed be seen in the cases, for example when some of the police officers seemed to talk less freely with their manager present in the workshop; when the validity of making design decisions based "on interviews with only one or two" informal carers was disputed; and when the experts on the informal carers' needs came up with a relatively conservative idea.

What is probably most striking is that the end users—the police officers and the informal carers—were most often absent in the process. The police officers were invited for several workshops and informal carers were interviewed, but they did *not* participate in design decision making. In project meetings, where design decisions were made, they were not present, but they were *represented*. Project team members portrayed them via statistics, which they constructed out of their survey, and via storylines, which they constructed out of their observations and interviews. The storylines are meant to portray the current situation ("is"), but in it were suggestions that these situations are problematic from a very specific angle, and that these situations can be improved by using our *we-centric* telecom application ("ought").

Furthermore, the project team members acted as end users' spokespersons, for example, when they discussed their needs, wishes, and preferences, and when they made design decisions, decisions which are meant to influence the product, which is meant to influence end users' life and work. Representing end users has resembles the "configuring" (Woolgar, 1991) of end users, and the creation of "scripts" (Akrich, 1992) about end users. Parallel to the product which they create, researchers and designers create an image of what an end-user should look like and how he or she should use their product. Moreover, the gathering knowledge about and from end users and representing them is *not* a neutral activity (Rohracher, 2005, p. 16):

Representing users in design is by no means a simple and straightforward process, but continuously reshaped and negotiated by actors involved in the design process. [...] User representations are constructed and shaped by the interests, specific discourses and traditions of actors involved and often are also entrenched in material infrastructures or methods to investigate demand.

The process of representing end users is a process of mobilising resources to influence others through communication (Latour, 1987). When Pauline talks about her interviews with informal carers or when Mandy quotes what police officers said in a workshop, they mobilize these end users to make their point.

Gathering knowledge about the world around me, including other people, was also examined by Levinas. He wrote that when I gather knowledge I almost automatically reduce everything to concepts that are already familiar for me: I transmute the *other* so that it matches to my *self* (1996a, p. 11-12):

The knowing I is the melting pot of such a transmutation. It is the same par excellence. When the other enters into the horizon of knowledge, it already renounces alterity.

When I gather knowledge about another person, I make a gesture of grasping: I grasp what I see and hear of that other person, and pull him into my own world. Levinas uses words like "the con*cept* or the Beg*riff*" (1996b, p. 152, emphasis in original) to exemplify the "concreteness of the grasp." The project team members could not escape this tendency of drawing others into their "melting pot." Their own interests, ambitions, intentions, methods, and their creativity—their "selves"—make them filter what they heard, saw, and understood of "the other" during their observations, workshop and interviews. It is possible to imagine a HCD project in which researchers and designers do their observations, workshops and interviews according to the book, and then follow their own interests, ambitions, intentions and do what they had in mind already. The gesture of the grasp becomes poignant when constructing storylines or personas turns into a way of "involving users by simply excluding them. The users are instead represented by an archetype of a user" (Blomquist & Arvola, 2002).

Researchers and designers who study end users are not engaged in some neutral fact-finding activity, but in negotiating between different people's interests and values. Explicitly or implicitly they are taking sides, they are making decisions all the time.

MAKING DECISIONS

A research and design project is about making decisions. There are many options open, many uncertainties, but at the same time, there are only few criteria or requirements to ground decision making. Nevertheless, one has to make decisions in order to proceed: create ideas, choose between ideas, turn an idea into concepts, choose between concepts, turn a concept into a prototype, and into a product. One proceeds by reducing "design space" gradually until one has one product. Explicitly or implicitly one must make decisions about what problem to address, about criteria to choose between problems, about what direction to search for solutions, about criteria to choose between solutions, etcetera. This makes research and design different from for example a study in social science or an engineering project.

Roozenburg & Eekels (1995) argue that "design thinking" is different from other "logical" ways of thinking: *de*duction starts with premises and then one draws a conclusion (in mathematics); *in*duction starts with several observations and then one speculates about a pattern (in natural science); and *ab*duction starts with observing an effect and a process, and then one reasons backward to a possible cause (in history). Contrastingly, design thinking or *inno*-duction starts with imagining a problem that seems worthwhile to try to solve, one simultaneously imagines possible solutions, and also criteria to choose between solutions, one may reformulate the problem or go looking for solutions in other directions, etcetera. Articulating the design problem and formulating the design brief are part of the design process. Imagining the problem and imaging the solution are intertwined. "The problem and solution co-evolve" (Cross, 2006, p. 80)[e]. A designer will try-out and play with solutions as a way of exploring the problem. This play with problem and solutions can be found especially in the "fuzzy front end" where problems and solutions are discussed iteratively.

This is different from how one (supposedly) conducts a study in social science—one starts with a question, develops a method, does the study, and interprets the data—or how an engineering project is (supposedly) done—one starts with a brief and then proceeds via steps of generating and choosing alternatives until one has an optimal solution. A research and design project starts with a bet: "No innovation, no invention develops without this initial bet" (Akrich et al., 2002, p. 219). Albert betted that designing a telecom application to support community police officers to cooperate better with others would be worthwhile. The project team members betted that designing a telecom application to support informal carers in their "work" by requesting and offering help would be sensible.

Derrida has a special way of looking at the making of decisions. He argued that only in a situation *without* rules, where one cannot use knowledge, logical rules, or moral rules can one make a decision (Derrida, 1995, p. 147-8):

The only decision possible is the impossible decision. It is when it is not possible to know what must be done, when knowledge is not and cannot be determining that a decision is possible as such. Otherwise, the decision is an application: one knows what has to be done, it's clear, there is no more decision possible; what one has here is an effect, an application, a programming.

Making design decisions about what problem to focus upon, what end users to interview, how to interpret what they say, about which direction to look for solutions, and how to choose between alternatives can be called "impossible" decisions. Furthermore, Derrida said that only when one makes such an "impossible" decision, freedom and responsibility become possible (Derrida, 2001, p. 28):

A decision, as its name indicates, must interrupt, cut, rend a continuity, the fabric or the ordinary course of history. To be free and responsible, it must do other and more than deploy or reveal a truth already potentially present, indeed a power or a possibility, an existent force.

This freedom to choose may not be there in every project, but this freedom, which comes together with responsibility was there in the two cases studied. If we organize a research and design project in such a way that logical or moral rules steer the decision making, new possibilities are not possible—no innovation is possible. Innovation can *only* happen if we make "impossible" decisions, without rules. And the making of such "impossible" decisions make it possible to act freely and responsibly.

SELF AND OTHER

A key idea behind HCD, behind interacting with end users and behind multidisciplinary teamwork is that "other people" should be able to influence the project. What researchers and designers do in a HCD project can be seen as making moves towards the *self* or towards the *other*. They move towards the *other* when they listen to end users and try to understand their needs, wishes, and preferences or when they listen to another project team member and try to understand what he or she says. Conversely, they move towards the *self*, when they view end users as raw material for their own creativity, when they study end users with their own methods and represent them within their own argument, or when they remind others about the project goal and focus and mobilize this project goal and focus to make their own point.

Researchers and designers are making movements in both directions, at different moments, in different situations. During an interview or workshop, they try to move toward the other and when they discuss their findings within the project team, they move toward the self. They even make these two movements simultaneously, for example when they decide that other people are in need and must be helped: they move towards the *other* who is (supposedly) in need and they offer help which fits within their own expertise and ambition, within the *self*. This movement is salient when these products are meant to *empower* end users, to help them change their behaviour in a (supposedly) beneficial direction: the *Police-Pointer* was developed to empower police officers to become more self-steering, rather than following hierarchical lines of command; and *WeCare* was developed to stimulate informal carers to share their tasks with others, rather than doing everything themselves.

Project team members made these movements in order to accommodate different demands upon them. I see at least three forces at work here.

Firstly, there is the *self* that demands them to use their expertise, their methods, their creativity. This is what they bring into the project. Dirk is clear and positive about team members having their own pet subjects: "Everybody has of course his own ideas which he wants to introduce" (Transcript 31 May 2006, p. 6). He provides examples of Albert who wishes to experiment with location technology and maps and of Barry who likes *smart phones* and who thinks of applying such *smart phones* almost regardless of the target group or the problem at hand.

Due to their ambition to develop a *we-centric* application the people working on the *Police-Pointer* learned slowly about what police work is about. It was only after several workshops with police officers that they were able to create something that is both interesting for police officers and for their own project. This focus on telecom brings the risk of missing the larger context of police work. During lunch, after one workshop, the police officers talked heatedly about their uniform's trousers. They currently wear cotton trousers, and their managers want them to wear woollen trousers. But they do not want woollen trousers. They told us how easily you get blood or other stains on your trouser and that they are responsible for keeping their uniform clean. They can easily wash their cotton trousers, but they must bring such woollen trousers to the dry cleaner. Listening to such stories we could have learned more about power and culture and about how innovation works in a police organization. Moreover, there may be interesting parallels to draw between the introduction of woollen trousers and the introduction of something like the *PolicePointer*.

It was only during the field trial of the prototype that we were discussing in detail how the *Police-Pointer* may fit in the current police organization and culture. Before that, we had paid attention to communication and information as functional processes, which can be supported by ICT. If the police is currently organized top-down and reacts to incidents, how will police officers think and feel about the *PolicePointer*, which is meant to stimulate working more proactively and more self-steeringly?

The people working on *WeCare* used different methods to study and represent end users. Their focus on their own respective methods and their discussions about methods may have blurred their view on the end users. During one such discussion, there was confusion about *whose* problem we are trying to solve. The table that summarizes the survey's results has four columns: "reported needs" and "reported unmet needs" of people with dementia, and "reported needs" and "reported unmet needs" of their informal carers. Confusingly, many of the informal carers' needs are "caused by" the needs of the people with dementia, for example needs related to "memory." And trying to solve the informal carers' needs is, of course, meant to help solve the needs of the people with dementia. During one such discussion, somebody remarked "Our need is to do something about that problem." This draws attention to the researchers who are concerned with their own methods and interpretations, to designers who are concerned with their own creativity and ideas—and to the risk of altogether forgetting the end users. Another risk is that a focus on technology—often the *self* of researchers and designers—leads to a technological view: as if organizing the exchange of help between informal carers is a technical problem which needs to be solved through a technical solution.

Secondly, there are the end users, the *others*. The project team members tried to move towards the *others*, and the interactions with them influenced the project. Looking critically at the moves made towards the *others*, we saw that the end-uses were often not present, but represented. They are made part of one's argument, and drawn towards the *self*. The project team members—probably unintentionally—reduced the otherness of end users.

In workshops the police officers talked about problems in their work, for example about how they balance conflicting roles or identities, such as being a "spider in the web," being a "go-getter" on an emergency call, and being servant to larger bureaucratic processes simultaneously. Although these utterances appeared in meeting minutes, they were rarely explicitly discussed during decision making within the project team. The police officers' otherness was also reduced when project team members turned their field notes into storylines. The notes described vividly how the different project team members experienced situations like arresting a thief, driving with lights flashing, wearing a bulletproof vest, or interviewing suspects. However, they omitted such descriptions when they made their storylines, which consequently became relatively sterile. They constructed the storylines with the project's focus in mind and focused on situations where a *we-centric* application may be of value.

The informal carers' otherness was reduced when the project team members applied questionnaires with pre-defined concepts to interview people and summarized their findings into one table. Interestingly, Pauline mentioned relatively late in the process of doing the survey fieldwork that almost all the interviewees were crying during the interviews. This fact was not told before nor was it asked about. Crying respondents was maybe not an apparent aspect of conducting and reporting a survey. The informal carers' otherness was also reduced when we constructed storylines based on the interviews, privileging what we are interested in and overlooking other aspects. Rachel talked several times about how it moved her that an elderly lady, who was informal carer, found it hard to call upon others for help, but these emotions of Rachel and of this lady were lost in the translation into storylines.

Human-centred design is an attempt to move toward end users. But, since such projects are often multidisciplinary, and with different organizations participating, it is also an attempt to move towards other team members: people who have different backgrounds, positions and interests. Many attempts were made to move towards the *other*, but often one easily moved (back) to the *self*.

And thirdly, there is the context of the project. There is a project plan, on which the participating organizations had agreed upon, and in it is the goal to develop and evaluate a *we-centric* telecom application. I tried, on various occasions, in my role as coordinator of a part of the project, to influence my fellow project team members to focus on needs or problems that can be related to telecom (and to neglect other problems or needs) and to develop a telecom application (and to disregard other kinds of solutions). This can be illustrated by the observation that, during the development of the *PolicePointer*, we only became interested relatively late in the project in how police officers input and edit reports in their database. We became interested when and because we realised that we needed to understand "their" information process in order to make "our" telecom application work: the *PolicePointer* must access and match these reports in this database in order to make automatically generate its suggestions for the police officers to communicate. The development of *WeCare* was similarly steered by mobilizing the project's scope and goal, for example, when during and after the creative session these criteria were put forward: it must be a *we-centric* application; it must be "mobile, context-aware and adaptive"; it must be similar to current ideas within the project team; and it must be feasible technically.

Although the project plan is text in a paper document, it can be brought alive. It can be mobilized when one team member refers to it, interprets it and makes it into a part of his or her own argument. This mobilizing is similar to how end users are represented and mobilized. The project plan is not part of any one of the project team members' *self*, but it can be drawn towards the *self*.

In summary, project team members are moving in a space which spans between *self* and *other*. And in this space they move towards end users when they interact with them, or they pull them towards their selves when they represent them to make their own point. Similarly, they move towards fellow project team members or pull them into their own argument. And they move toward the project's goal and scope, or pull it towards themselves to make their own point.

SPEAKING ABOUT ETHICS

Talking about *self* and *other* and making references to Levinas and Derrida is my way of arguing that human-centred design (HCD) is a social process which has ethical qualities [f]. Furthermore, I would like to suggest a role for management in bringing these ethical qualities more to the fore.

In the two cases I described how researchers and designers interacted with end users and how they made design decisions, and I described their activities in terms of making "grasping" gestures, of making "impossible" decisions and as moving between *self* and *other*. I would like to argue that such actions have ethical qualities. My argument is *not* concerned with evaluating whether what researchers and designers do is *morally* good or evil, with suggesting that they should behave more "morally" or with suggestions to improve HCD.

The point which I wish to make is that ethics are happening in HCD—all the time, *already*. But these ethical qualities are rarely discussed explicitly. We act as if they do not happen, we marginalize such ethical qualities. My point is also *not* that moves towards the other are better or worse than moves towards the self. The two movements stand for the two "faces" of HCD: it is about meeting another person and trying to create something for this other person and about deploying one's own creativity. Both moves are made, and both moves must be made.

I am interested in the ethics, which happen in interactions between people when researchers and designers interact with end users and with fellow project team members—whom they listen to, how they listen—and when they make design decisions – when they choose to focus on a certain problem or choose to develop a certain solution. I would like to ask questions like How much participation do I allow in "participatory" design? How much empathy do I have in "empathic" design? How central are people in *human-centred* design?

My suggestion can be seen as an attempt to "deconstruct" (Critchley, 1999; Derrida, 1991) HCD, where I understand deconstruction as a "radical form of questioning text(s)" (Letiche, 1998, p. 125): providing a critical reading of what happens in a project and providing an alternative reading of that project. What would happen if these ethical qualities were brought more to the fore? What if managers, researchers and designers become more aware and explicitly discuss their *selves*, their own respective positions, their own expertise, methods and creativity? What if they become more aware and explicitly discuss how they relate to *others*, to what they hear and see from end users and from fellow project team members? What if they become more aware and explicit about how they move between *self* and *other*?

This suggestion can be positioned in a debate about the relation between ethics and science and technology studies (STS). Many social constructivist studies of technology in the field of STS can be criticized for *not* speaking (enough) about ethics, for their "lack of and, indeed, apparent disdain for anything resembling an evaluative stance or any particular moral or political principles that might help people judge the possibilities that technologies present" (Winner, 1993, p. 371). Recently Poel and Verbeek (2006) stimulated a debate between STS and ethics. They argue that, traditionally, engineering ethics has been concerned with studying the ethical *consequences* of developing or applying technology, and advocate that STS might help to conduct more situated studies of what people involved actually *do*, to "open the black box of technology" (rather than building theoretical arguments), and that, vice versa, engineering ethics "might help STS to overcome its normative sterility."

My suggestion is in line with theirs: I suggest that researchers, designers, their managers, and others involved in a HCD project can be stimulated to reflect upon their practice and upon the ethical qualities of their practice. This could be a way of letting the other put into question my *self*, my spontaneity, my creativity (cf. Critchley, 1999, p. 5; Levinas, 1996a, p. 17), and it could be a way to help researchers and designers to act more freely and more responsibly. Responsibly in the sense that they can question their design decisions more explicitly and ask each other to *respond* to such questioning. And freely in that they can more consciously and explicitly choose between options. I think that many researchers and designers are currently attempting to do HCD, but that they are doing their projects as if they are doing social science or an engineering project. My suggestion is to conduct HCD differently, to conduct it more accordingly to what it *already* is—a process between people, with ethical qualities—to continue doing HCD, but more reflectively.

ACKNOWLEDGMENT

This text was written as part of the Freeband Communication research program (http://www.freeband.nl). I would especially like to thank my fellow project team members for their cooperation while conducting my study and for their permission to write about them.

REFERENCES

Akrich, M. (1992). The description of technical objects. In W. E. Bijker & J. Law (Eds.), *Shaping technology/Building society: Studies in sociotechnical change* (pp. 205-224). Cambridge, MA and London, England: MIT Press.

Akrich, M., Callon, M., & Latour, B. (2002). The key to success in innovation—Part 2: The art of choosing good spokespersons. *International Journal of Innovation Management, 6*, 207-225.

Beyer, H., & Holzblatt, K. (1998). *Contextual design: Defining customer-centred systems*. San Francisco: Morgan Kaufmann.

Blomquist, Å., & Arvola, M. (2002). Personas in action: Ethnography in an interaction design team. In *Proceedings of the 2nd Nordic Conference on Human-Computer Interaction* (pp. 197-200). Aarhus, Denmark: ACM Press.

Bucciarelli, L. (1994). *Designing engineers*. Cambridge, MA and London, England: MIT Press.

Cooper, A. (1999). The inmates are running the asylum: Why high-tech products drive us crazy and how to restore the sanity. Indianapolis, IN: SAMS Publishing.

Crabtree, A. (2003). *Designing collaborative systems: A practical guide to ethnography*. London: Springer-Verlag.

Critchley, S. (1999). *The ethics of deconstruction: Derrida and Levinas* (2nd ed.). Edinburgh University Press.

Cross, N. (2006). *Designerly ways of knowing*. London: Springer-Verlag.

Derrida, J. (1995). *Points: Interviews, 1974-1994*. Stanford, CA: Stanford University Press.

Derrida, J. (2001). Deconstructions: The impossible. In S. Lotringer & S. Cohen (Eds.), *French theory in America* (pp. 12-32). New York and London: Routledge.

Derrida, J. (1991). Letter to a Japanese friend [original 1987]. In P. Kamuf (Ed.), *A Derrida reader: Between the blinds* (pp. 270-276). New York: Columbia University Press.

Easterby-Smith, M., Thorpe, R., & Lowe, A. (2002). *Management research: An introduction* (2nd ed.). London: Sage.

Haddon, L., & Kommonen, K. H. (2003). *Interdisciplinary explorations: A dialogue between a sociologist and a design group*. Helsinki: University of Art and Design.

Hekkert, P., & Van Dijk, M. (2001). Designing from context. In P. Lloyd & H. Christiaans (Eds.), *Designing in context* (pp. 383-394). Delft: Delft University Press Science.

ISO. (1999). ISO 13407: *Human-centred design processes for interactive systems*. Geneva, Switzerland: International Organization for Standardization.

Koen, P., Ajamian, G., Boyce, S., Clamen, A., Fisher, E., Fountoulakis, S. et al. (2002). Fuzzy front end: Effective methods, tools, and techniques. In *PDMA toolbook for new product development* (pp. 2-35). John Wiley & Sons.

Koskinen, I., Battarbee, K., & Mattelmäki, T. (2003). *Empathic design: User experience in product design*. Helsinki: Edita Publishing.

Kujala, S. (2003). User involvement: A review of the benefits and challenges. *Behaviour and Information Technology, 22*, 1-17.

Latour, B. (1987). *Science in action: How to follow scientists and engineers through society*. Milton Keynes: Open University Press.

Latour, B., & Woolgar, S. (1986). *Laboratory life: The construction of scientific facts* (2nd ed.). Princeton, NJ: Princeton University Press.

Lawson, B. (1997). *How designers think: The design process demystified*. Architectural Press (an imprint of Butterworth-Heineman, a division of Reed Educational and Professional Publishing).

Letiche, H. (1998). Business ethics: (In-)justice and (anti-)law. In M. Parker (Ed.), *Ethics and organizations* (pp. 122-149). London: Sage.

Levinas, E. (1996a). Transcendence and height [original 1962]. In A. Peperzak, S. Critchley, & R. Bernasconi (Eds.), *Emmanuel Levinas: Basic philosophical writings* (pp. 11-32). Bloomington and Indianapolis, IN: Indiana University Press.

Levinas, E. (1996b). Transcendence and intelligibility [original 1984]. In A. Peperzak, S. Critchley, & R. Bernasconi (Eds.), *Emmanuel Levinas: Basic philosophical writings* (pp. 149-159). Bloomington and Indianapolis, IN: Indiana University Press.

Oudshoorn, N., & Pinch, T. (2003). Introduction: How users and non-users matter. In N. Oudshoorn, & T. Pinch (Eds.). *How users matter: The co-construction of users and technology* (pp.

1-25). Cambridge, MA and London, England: MIT Press.

Poel, I. v. d. & Verbeek, P. P. (2006). Ethics and engineering design. *Science Technology & Human Values, 31,* 223-236.

Rohracher, H. (2005). The diverse roles of users in innovation processes. In H. Rohracher (Ed.), *User involvement in innovation processes: Strategies and limitations from a socio-technical perspective* (pp. 9-35). Munchen/Wien: Profil Verlag.

Roozenburg, N. F. M., & Eekels, J. (1995). *Product design, fundamentals and methods.* Chichester, UK: John Wiley and Sons.

Sanders, E. B. N. (2000). Generative tools for codesigning. In Scrivener, Ball, & Woodcock (Eds.), *Collaborative design.* London: Springer-Verlag.

Schuler, D., & Namioka, A. (1993). *Participatory design: Principles and practices.* NJ: Lawrence Erlbaum Associates.

Steen, M. (2006a). *Our need is to do something about that problem—Studying how researchers and developers interact with informal carers during an innovation project.* Presented at EASST 2006 Conference (European Association for Studies of Science and Technology), August 23-26, Lausanne, Switzerland.

Steen, M. (2006b). We don't want woollen trousers—Studying how researchers and developers interact with police officers during an innovation project. In R. ten Bos & R. Kaulingfreks (Eds.), *Proceedings of SCOS 2006 (Standing Conference on Organisational Symbolism)* (pp. 644-666), July 13-15, Nijmegen, The Netherlands.

Stewart, J., & Williams, R. (2005). The wrong trousers? Beyond the design fallacy: Social learning and the user. In H. Rohracher (Ed.), *User involvement in innovation processes: Strategies and limitations from a socio-technical perspective* (pp. 39-71). Munchen/Wien: Profil Verlag.

van Kleef, E., van Trijp, H. C. M., & Luning, P. (2005). Consumer research in the early stages of new product development: A critical review of methods and techniques. *Food Quality and Preference, 16,* 181-201.

Von Hippel, E. (2005). *Democratizing innovation.* Cambridge, MA: MIT Press.

Winner, L. (1993). Upon opening the black box and finding it empty: Social constructivism and the philosophy of technology. *Science Technology & Human Values, 18,* 362-378.

Woolgar, S. (1991). Configuring the user: The case of usability trials. In J. Law (Ed.), *A sociology of monsters: Essays on power, technology, and dominations* (pp. 57-102). London and New York: Routledge.

ENDNOTES

[a] I use "researcher" and "designer" for roles or for activities—not for people or for occupations. One person can at one time do research, "study something carefully and try to discover new facts about it", and at another time do design, "decide how something will look, work, etc. especially by drawing plans or making models" (Oxford Advanced Learner's Dictionary, 7th ed.). I am keen not to say "R&D" because that word is associated with an organizational job, function or department.

[b] I use "end-users" as a more readable alternative for "future, potential or putative end-users." Nevertheless, it is strange to talk about end-users when there is not yet a product or service that can be used – it is currently being developed. Furthermore, I use "end-users" rather than "users," to refer to people for whom a product or service is meant primarily or ultimately, and not to e.g. a repair-person, who may also use the

product, or e.g. a decision-maker who may decide to install or implement the service.

[c] More details about this project are in a conference paper (Steen, 2006b).

[d] More details about this project are in a conference paper (Steen, 2006a).

[e] Similarly, Lawson (1997, p. 121-7) argues that design problems cannot be comprehensively stated and require subjective interpretation, that the number of possible solutions is inexhaustible and no one is optimal, and that the design process is involves finding as well as solving problems and is a prescriptive activity.

[f] I intend to talk about the ethical qualities of research and design. Not about the ethical qualities of applying or using the products that come out of that process.

Chapter VI
Professional Dimension of IT Specialists' Social Role

Agnieszka Postuła
University of Warsaw, Poland

ABSTRACT

This chapter gives the opportunity to describe and analyze how IT specialists construct their professional reality and how they perform their roles. Nowadays companies (or top management) influence its employees in multiple aspects. Thus, professional relations should be investigated in the terms of mutual interactions within the organization, which is the main topic of the chapter. Often, top management tries to conclusively define (consciously or unconsciously) what the organization is. It means that workers of lower levels in such companies are manipulated by their higher-level colleagues. Sometimes this influence, albeit both hierarchical and controlling, leaves some space for the development of more equal bonds. In other organizations, manipulation is not a part of the standard work climate. The relations between the employees are, in such organizations, based on rituals and direct communication. The most important detail of this construction is to take care of every interaction between all organization's actors. In this chapter, I intend to describe work relations and other aspects of everyday work life of IT professionals. Empirical material is derived from the field research conducted during 2002-2004 in two stages with different intensity of data gathering. The first stage of the research concerned a software company creating customized business software "to meet the requirements of clients." The second stage was based on interviews with IT professionals from different organizations including software companies, international corporations, and also IT departments within companies operating in different business branches. Complete analysis of gathered material is presented in a doctoral thesis (Postuła, 2007). Therefore, in case of some discontinuity of considerations in this chapter I would refer to the full text.

SOCIAL ROLE OF IT PROFESSIONALS

Modern humans are involved in organizations qua roles, rather than qua persons (Kallinikos, 2003, p. 595).

In classic sociological concept, IT professionals play different social roles (Goffman, 1959/1990). Social actors play roles connected to specific everyday activity and in reference to the audience. IT professionals take their roles in project teams or other groups (like information technology departments). By the term "social role," the author means:

the pre-established pattern of action, which is unfolded during a performance (playing the role) [...] Defining social role as the enactment of rights and duties attached to a given status, we can say that a social role will involve one or more parts and that each of these parts may be presented by the performer on a series of occasions to the same kinds of audience or to an audience of the same persons. (Goffman, 1959/1990, p. 27)

There are two social groups involved in playing social roles. In this case, it would be IT specialists who take roles and play them and observers who watch IT specialists' performance and evaluate it (Goffman, 1959/1990). My aim is to show how Polish IT professionals play their roles in different organizations: if they are happy with these roles, how they feel about their professional creations, and what are the implications of constructing[a] roles for management processes.

IT professionals act according to scripts, which means procedures and regulations existing in his or her organization as well as cultural norms and values.

Taking up a job imposes duties, liability (responsibility), and obligations, which (who) is related with professional role, which (who) catches on (Bowie & Duska, 1990, cited in Kostera, 1996, p. 124).

Expectations of the actor's performance could be expressed in two ways: in the written form and as the unwritten rule accepted by a culture (Kostera, 1996). Each society and organization has its own principles and procedures concerning detail activities and workers' behavior. They are different on low level of generality. On the higher level, these rules are much more similar between societies (i.e., attitude of young people toward older).

Formal organizations emphasize written rules. It facilitates executing regulations, as well as introduces order. Thus, each worker knows all the scripts of a role, which he or she is taking at the moment (each person can play different roles simultaneously), and can behave according to observers' expectations. A set of rules could also be an instrument for motivating employees as well as an effective way of creating a good atmosphere at work (i.e., culture of efficiency). However, under one condition, the roles of one actor and his or her position in the organization cannot be mutually exclusive. In other case, it can evoke social conflicts.

Many authors (Berger & Luckmann, 1966/1983; Czarniawska-Joerges, 1992; Kostera, 1996 and others) stand that:

social roles are cultural phenomena, network of interpersonal communication, and structuralization of meanings. They are also a link in a process of constructing sense, element of social process of culture. (Kostera, 1996, p. 126)

IT professionals' roles are being defined repeatedly in the process of constructing organizational and social reality. Everyday actions allow new interpretations of principles, behavior, and workplace processes. Constructing relations, the actors build indissoluble network of meanings, which defines them as a professional group attributed with specific features. Each role consists of three elements. Kostera (1996) enumerates dimensions of manager's social role, which could be transmitting on the ground of IT specialists' profession. They are not explicitly separated, but mutually infiltrate each other. These are:

1. Social expectations (dominated scenario)
2. Constructing role by IT specialists (professional dimension)
3. Institutional context (organization, where IT specialists play their roles)

Three dimensions of IT professionals' role are seized in Table 1.

Societal Dimension

This dimension is the most important from the point of view of close environment because it describes how outside actors perceive IT professionals. It is possible to single out three key groups involved in interactions with IT specialists: customers, strangers (outsiders), and IT professionals' families. The first group mainly consists of purchasers of IT systems and services. The second group are people not connected with the IT profession (i.e., people of different professions particularly those not of exact sciences, connected anyhow with IT professional's work). And the last one consists of all members of families taking part in informal organizational meetings.

Social norms and values shared and obeyed by IT specialists as well as a prestige of this profession are the most important factors for people from the environment. High respect for the IT specialist profession boosts barrier of access to the occupation[b] on the one hand, but builds its attractiveness on the other. It causes wages increase in this sector and also effects demand on labor market. Presently, many students in Poland who graduate from information technology studies (or some related faculty like mathematics) decide to work in different IT specializations. A big part of professional employees consider financial motivation as the most effective. But there are also other reasons to work in this profession. It seems like Polish IT specialists consciously create job image as an inaccessible and of high level of expertise. Of course, partly it is a matter of a body of profession. However, the second factor of barrier building is active influence on creating and maintaining this image.

Professional Dimension

Professional dimension is the most important from the perspective of profession analysis. General elements of this dimension are an ethical code, an education, and career paths, which are components of a professional community. IT specialists are in particular the occupational community (Van Maanen & Barley, 1984), which creates expectations and standards inside professional group. They compose hermetic occupational com-

Table 1. Dimensions of IT professionals' social role

Dimension	Social group (community)	Meaning
Societal role: Social expectations	Social stakeholders	How should / shouldn't IT professionals act? What is important, valuable, undesired?
Professional role: Professional dimension	Professional community	"Game rules," how should professional / IT specialist act?
Organizational role: Institutional context	Organization (participants)	Norms and values of the organization referring to the way IT professionals should / shouldn't act in their organizations.

Source: Based on Kostera (1996, p. 128)

Table 2. Social groups related to IT specialist profession (professional role)

Dimension	Social group (community)	Description
Professional role: Professional dimension	IT specialists' co-workers	All people staying in friendly relationships cooperating with interlocutors, staying also in professional relationships and exchanging professional experience
	IT professionals' community in wide sense	Total number of IT professionals working in Poland, actively creating professional role

Source: Based on Kostera (1996)

munity that protects its interests from external influences. It allows claiming that IT specialist is a pure profession (Freidson, 1994; Hall, 1986). IT professionals could be investigated from two perspectives:

- As co-workers in project team or in organization as a whole
- In a wider depiction: As a social occupational community

Professional dimension describe how IT specialists construct their reality as the part of the process of professionalization (Larson, 1977). Therefore, analysis could begin with daily tasks and actions of IT professionals as well as procedures for education and occupational ethics. Particularly, this dimension concerns efficiency, norms, and values related with education, experience, and ethical standards (Kostera, 1996). Conscious creation of professional image on labor market and in wider environment, which also regards to societal dimension, has further implications in professional dimension (inside this professional group). Low accessibility to the profession creates its standards, which determine outsiders' perception of IT specialists. Thus, we deal with closed process, which interrupted can cause change of certain values and norms. There is such possibility, because not all rules of IT specialists' professional life belong to the sphere of consciousness. Part of them is profoundly unaware. Aspects of these norms can harm the profession's prestige, lower its meaning, and thus, negatively affect already accepted standards. Following the translation process of IT specialist actions gives a picture of relations of all actors and shows a role of a main actor (Callon, 1986; Latour, 1987).

Two factors have a major influence on definition of IT professionals' role. First, concerns a form of organization where IT specialist works. It is a matter of organization profile, which determines IT professional's position and all the tasks related to occupational duties. Education, self-development, and possible career paths depend on a form of organization likewise. The 2nd factor influencing professional role is management. Members of management trace certain standards of behavior (ethics), range of tasks, quality of work, products, or services. IT specialists in Poland have a specific attitude to define their profession and they perceive it in many ways. Particularly, they differently identify themselves with the profession. Mainly, there are two groups: those that don't assign their profession to IT specialist (i.e., saying that they are programmers, software designers, etc.) and the other groups, which have strong identification with the profession. The main problem concerns naming profession and sub-professions, which is

Table 3. Elements of professional dimension of IT specialist social role

Element of professional dimension of IT specialist social role		**Expectations and values**	**Definitions of scenario elements**
Task		Efficiency: What is considered as an achievement?	What should be done?
Standards	Education	"General rules" of IT professional's work: What their knowledge and skills should be?	How should it be done? By whom?
	Ethical code	Ethics: How should IT specialists act?	

Source: based on Kostera (1996, p.150)

very wide and heterogeneous. IT professionals divide into number of different specializations, which causes divergence previously mentioned.

Organizational Dimension

The last element of a social role is organizational dimension, which completes the picture of IT specialists' occupation. It is the greatest exploration area, however, it doesn't concern profession itself as much as the former dimension. It took a lot of time for interlocutors to describe their daily activities, main problems, and favor tasks in projects. They usually like their work and they are proud of it. Thus, they like to talk about it which is very useful for a researcher. Although, it is not so easy to get a full description of a workday from an IT specialist. They tend to concentrate on details, which could be not so explicit for a researcher and does not establish a clear picture of IT specialist work. Moreover, their organizational role is very complex and it consists of many various elements difficult to unequivocal defining. Under consideration are: everyday work, specific tasks, IT professionals of different specializations, their place in organizational structure, norms, and values of culture characteristic for certain company etc.

NOTES FROM THE FIELD

Ethical Code and Profession's Definitions

In the first place, it is necessary to explain how IT professionals perceive their occupation. Many of them take their profession very seriously, but there are workers who treat the profession as a temporary source of earnings. They have a different attitude toward computers, which are the main IT specialists' tools. Computers are also one of the sharp dividing lines of professional models. For some IT specialists, computers are almost objectives of worship and devotion while for some of them (usually older persons) it is rather regular working tool. These groups of professionals share different work activities. Younger IT specialists spend all day working in spite of a daytime (and treating informatics like life priority) while the oldest stay at work only as long as it is required (and treat profession as one of their interests and regular source of earnings).

So for sure I'm different to young boys coming here because I can see the attitude of people my age, let's say, not adult but not of this first fresh passion to computers. And I'm not fascinated with it that much. I treat it as a craft. But I can see that for these young boys, the job is a real passion. So when they write a program, they keep talking about it and discussing it with friends. I treat it only as the craft and it's not a kind of a job I would do with passion anymore. It is just a job that lets me earn a living. Just that. (BTK)

An interesting part of the field material concerns definitions of profession. Polish IT specialists define their job in many different ways. The problem lies in too capacious term for the name of profession. That's why IT professionals stand for separating their occupation for specializations. Interlocutors spoke about wide and narrow definitions.

Unfortunately, Polish language is littered with the word informatyk *(IT specialist) and generally speaking, representative of IT professionals, we might say, is every person who writes on the keyboard faster than an average user. This is an IT professional.* (LMI)

Polish problems with generating homogeneous definitions for IT specialists could rise from huge dynamics of development of this profession. Every day brings new specializations and it becomes extremely difficult to group them and name properly. It results in weak identification of IT specialists with their profession in a wider sense. Therefore, they identify mainly with particular specializations.

I avoid the word informatyk *for defining my job. My kids used to name my job like this but I was disgusted with it. For me,* informatyk *is a worker of an IT department in the company from outside the IT branch, for example bank or so.* (KMI)

Some interlocutors underlined elitism of their profession. Thanks to this statement, it becomes clear why IT specialists use technical language and name specializations in the way incomprehensible for the outsiders. They try to establish borders for new potential members of the community and to hamper the access to the profession.

In some way I have this sense of superiority in reference to IT professionals. Because I have this feeling that IT specialists that I know think that they are better than average person. [...] Is it right or wrong? In some way it's not because it is rather racist. (NZA)

On the one hand, elitism and emphasizing separation of particular specialization leads to closing profession from strangers' influence. It determines very strict ethical code, which is comprehensible only for members of close society. Ethical code is rather not written but interlocutors claim that it is precisely executed. On the other hand, it causes organizational conflicts because outsiders have plenty of difficulties with understanding IT professionalists' behavior and language[c]. Only representatives of the profession are able to explain details of each specialization and its deeper chracteristics. However, even IT specialists' point of view often differs.

Someone who fixes printers or monitors can be named an IT professional. Notice that there is a position helpdesk or someone who administrates network and they are administrators. Someone who codes is a programmer. Someone who plans projects is a designer. You can define almost every activity related to IT department. But IT professional? Who is he? A person anyhow related to computers? (RPK)

Some interlocutors agreed that IT professional derive from engineer occupation. Interlocutors when asked about their job without computers specified engineers. Interlocutors claimed that former engineers took up activities, which are a domain of today's IT professionals. Work tools are the elements that become changed during the time and are much more advanced now. Such point of view makes understanding IT specialists' work easier and marks out boundaries of profession.

One more important issue considering the professional attitude of IT specialist to their job is attachment to positions in organizations, or rather lack of it. IT professionals pay attention only to specializations, not to positions and precise places in organizational hierarchy. An average software development company has flat structure and it's not very complicated.

Firm structure is quite strange. I think it is a very bad organization for making a career. There is a boss on the top of the structure, cappo di tutti cappi, *lower there are few project managers and programers and that's all.* (RBL)

Great corporations hiring IT professionals have a precisely determined structure and it usually cannot be modified. Every person in such an organization is strictly assigned to his or her position. Organizational structure affects level of positions' fluency. In smaller IT companies, the worker is responsible for many different kind of tasks and that gives him or her great experience on one position for a long time. Inversely, in the international corporations, workers usually tend to get promotion and change position quite fast. IT professionals consciously choose organizations that they prefer.

Education

Education is one of the factors strongly determining IT specialists' profession. The process of education shaping people interests and graduation has great influence on further career paths. Moreover, education is one of the traditional determinants of the profession (Abbott, 1988; Hall, 1986; Larson, 1977; and others). Research results gave five factors describing process of education:

1. School
2. University
3. Knowledge and practical skills
4. Fast learning
5. Professional experience

Interlocutors state that IT specialists have to be educated in an appropriate way, which means graduating from an information science faculty or mathematics (exact sciences).

Mathematics gives an opportunity for perceiving reality logical way. (RRR)

Many interlocutors chose their profession at a very young age. Already in school they delimit interests quite precise. Many of them spent most of the time by the computer; however, these tools were much different from present ones. Most of the interlocutors studied specializations in accordance to their current profession.

I graduated physics at the Warsaw University and it was a great bonus for me that there was a lot of mathematics. You can get to know different methods of thinking. They represent strongly structuralized approach to different issues, rules, etc. And it was great benefit for me. (NLL)

Education is only the first step to becoming a professional. Knowledge and skills count as much as education, in some cases even more. Interlocutors take common positions on deriving knowledge from practice. They try to improve their qualifications by performing everyday tasks. It is an essential condition that enables them to maintain a high place in the IT specialists' organizational hierarchy. That is why education knowledge and work experience count the most and are almost an obligatory element for every professional's career.

IT professionals should know how to write code. Definitely. [...] Programming is an ability to use analytical approach. It is the ability to see something as a whole and in details as well. It is not about applying solutions straight as they are. It is about the ability to perceive relations. (NZA)

Interlocutors emphasized that the most important thing about their work is not quantity of knowledge but particularity of investigating the subject. Knowledge in one narrow subject decides about IT specialist's professionalism. That is why cooperation between representatives of different specializations is one of the most important factors for effective organization. Each of the IT project teams prepare their part to connect it in the specific moment. Interlocutors claim that strong specialization is a matter of years of work and getting knowledge.

It is not so difficult to start. But entering more specific issues and getting to know details from a particular field means a lot of work and devoted time spent on learning. [...] On the level of writing some procedures or packages, you don't have to have specific knowledge. You should have only well defined catalogue. (ILR)

Apart from specific knowledge, IT specialists on the higher organizational positions should have general knowledge about project procedures. In each project team there is at least one person who can embrace all the problems encountered in the IT project. They are project managers and they are usually former programers. As interlocutors claim, the best bosses were programers once. Such a boss can understand all the problems occurring in the project and is able to fix it immediately. It is a matter of relations between workers and management, too.

There is one more aspect of education and it is the ability to learn fast. IT branch is one of the most dynamic and IT specialists should keep up with the quickly changing environment. It seems that the one and only stable element is employment and it depends on individual IT specialist's needs, not on the employer's decision. In Poland there are employees markets (not employer) in the IT branch, although IT professionals have a choice to work abroad (and they are valuable manpower in many countries).

In spite of the quite comfortable market situation, IT specialists should be up to date with new technologies, any news appearing in their field. They should care about improving qualifications, practice skills, and getting knowledge of specialization. It is a rather individual task, however, older, more experienced colleagues are the best teachers who can give younger IT specialists support.

Here, you can get great knowledge, you can meet people who are extremely knowledgeable about different specializations. In the same way, I can see a huge gap between myself and people who are new here or those who are from the beginning [in the organization]. Some of them don't evolve at all and others change in weeks. (RRR)

Some companies have their own way of hiring new workers. Young IT specialist (students or very young graduates) get to specific departments in organizations where they can learn fast on the particular field. Managers with IT experience lead such a students to careers showing them required parts of skills, experience, and knowledge.

At first I got to bank department. Bank department in our firm is such a big bag which gathers all new employees who are actually not skilled, know not too much. And they can learn there a lot. (RRR)

Thus, IT specialists who can learn very fast get a lot experience in a short term. Interlocutors said that this is one of the most valuable elements of the profession. IT specialist-theorist is an unwanted person in software company. The real value is his or her knowledge and experience. That's why IT professionals start working at a very young age. Moreover, interlocutors argued that studies do not give them proper knowledge. They claim that a university is a place that doesn't give practice—the most desirable knowledge useful for solving real problems occurring in IT projects.

At the university, we made small projects and that's all. It seemed to be great projects, then. But here [in the firm] I can see the immensity of problems. I'm just getting used to it. And I can see chances to learn a lot. (JAA)

Specific knowledge is a key to IT specialists' profession and at the same time an element differing workers for specializations. Having great knowledge is a determinant of IT professional's status in the organization and his value on the market. The most precious achievement is practical knowledge, which IT specialists can get in the company, not at the university. Although higher education is very important in this profession, experience is still on the first place. That is why IT professionals start to work so early and pay great attention to fast learning.

Professional Career Criteria

Determinants of professional role of IT specialists' previously described influence career paths within this occupation. Education affects the direction and quality of worker's professional development. IT professionals don't attach much importance to the occupation as a way to make a living. Interlocutors state that the most important thing about their job is the essence of the profession, which means programing, designing, system analysis, etc., depending on a particular specialization. Therefore, many IT specialists don't plan their career path. They follow their temporary interests and try to get to know as much as possible. The only choice they made is about the university. As it was mentioned, graduating from a specific faculty is sufficient element of planning future. There are three identified in the field factors determining career paths of IT professionals:

1. Motivation to work
2. Success and failure
3. Professional burn-out

Usually, young IT specialists' motivation to work in this profession grows gradually in the course of time. Parents play important role in stimulating interests and making children aware of professional determinants. They are helpful with buying tools (computers) and organizing place and space for developing interests of their children.

I made the decision about being an IT specialist quite early, actually already at 4-5 class of primary school. [...] My parents are both IT professionals and we always had computers at home, father was engaged in programing, soldering circuits, because those were the old times. And I always liked it. (BLA)

First decisions about choosing a profession were made at young age and were affected by different factors. Later, IT specialists made their decisions more consciously, and one of the main motives for changing jobs was financial. Changing a job usually meant leaving the work place not a profession. Majority of interlocutors state that being an IT specialist is a profitable profession and there are only few causes, which can make them change occupation.

And it finished with my admitting to Warsaw University on information systems faculty. I think it was motivated by potentially high incomes because it seemed to be a good job in Poland that time. And it appeared to be a profitable branch in the closest future. (RBL)

I had a gap in my career right after graduating [...] I started to work at school [as a teacher], then my children were born and I had to raise them, and after that it appeared that I must earn some money to make a living so I started to look for possibilities to get back to informatics. And my friend worked here... (LHS)

Besides financial motives, interlocutors mentioned concept of creating in this profession. As it was previously described, IT specialist is a very capacious occupation, containing different specializations. Many graduates of IT faculties may find something interesting to work with as an IT professional (programers, designers, administrators, webmasters etc.) because almost each specialization requires creating.

As an advantage, interlocutors named possibility to work in selected part of the day. Especially software companies allow for working at free hours. Therefore, IT professionals may choose the most efficient time of the day for work.

Success and failure are directly connected with creating. Such specialization as a programer requires solving problems that demand knowledge, skills, and experience. Usually it means great challenge for IT specialist. At the same time, solved problems or well-designed systems are a success for IT professionals. Even promoting on higher positions in organization is not as valuable as a satisfaction from well-done job.

And vice versa, useless system, "put on the shelf" as interlocutors said, means deep failure and one of the most unpleasant moments at work for an IT professional.

Designing something for the future is a great fun. Sometimes even after one or two years systems appear to be useful and efficient. Sometimes clients have to grow up to understand that. There was a case when a client some time after buying our IT system decided to put some modifications in it. And our application was ready for that, we designed it earlier. I think you have to be "professionally twisted" to understand how much it means in this job. (UCZ)

An IT professionals' career is often affected by factors already enumerated. Interlocutors choose a working place on the basis of standard motives. First there is usually the possibility of creation. Free working hours are on the second place.[d] Changing positions in organizational hierarchy does not play an important role in IT professionals' career path. IT specialists know that their profession usually brings sufficient earnings so this motive was rarely mentioned.

Usually everyone starts with programing. Then they become system designers and pass from writing a code to designing work, which still means working with computers but it is something different and requires new skills, which you get with experience. (NNI)

Sometimes IT specialists think of getting supplementary education. It could be related to issues like management, team working, projects planning, and so on. IT professionals perceive the labor market in positive way. They believe in high flexibility of employers and thus possibilities to changing workplace easily. Even though they are optimists, IT professionals tend to plan career paths according to their own interests. In order to do so they learn new programing languages and get knowledge from new fields. This is the way to build competitive advantage on the market. But financial factor is only partial motivation of this behavior. To understand IT professionals, it is important to realize that they learn constantly not to earn money but to develop knowledge and get experience. Moreover, IT professionals' own (intellectual nature) interests are of the first importance.

For sure, I would like to start some postgraduate studies, soon. Maybe something about managing projects, something like that. Frankly speaking, I'm aware that in about five years there will be lots of young IT professionals on the market and they are going to replace me. (SPT)

The last issue mentioned by interlocutors was professional burnout. As a researcher, I wasn't prepared for asking this question directly. One of the interlocutors, young programmer was interested in older IT specialists' opinions to this topic. It turned out that IT specialists' professional burnout is nothing unusual comparing to other occupations. Interesting were answers of older interlocutors who claim that it is almost impossible to get burnout in this profession. They state IT specialist is so interesting and capacious profession that everyone may find interesting and challenging work for himself.

I would interpret burnout as a phenomenon of every occupation. When a worker is doing something for a long time, especially in a big company [...] it becomes less interesting, enthusiasm disappears. [...] To a low degree, burnout occurs with small long-lasting projects. When you are working on the same project for months, you wish it to end quickly. Thus, routine and boredom may cause small burnouts. Bigger burnouts may occur when you do the same things for a long time. Then you have to change department, a job, or so. (KIA)

Concluding, the best way to stay a fresh and satisfied worker, IT professionals should care about constant delivering challenges. It should be based on new technologies or any activities which IT professionals are interested in. Thus, they are particularly effective and bring income to their organizations. However, this is a problem of IT managers to hire workers on the right place and to find them proper job.

IMPLICATIONS FROM THE FIELD MATERIAL

An IT specialist is a particular technical profession. It should be considered as an occupational community (Van Maanen et al., 1984). IT professionals are characterized by focusing on their work, not caring about organizational matters, and sharing common values. The most important value is getting a status of full member of the community, which means developing knowledge in definite specialization. To reach this goal, they share information in everyday practice and assist in each other projects if needed. In many cases, IT professionals do not aspire to any promotion (i.e., project manager). It often means leaving the community and becoming a member of a top management. Especially employees of international corporations working with programers in their community rather than leaving for other positions. This tendency is strictly connected to organizational role of IT professionals. They prefer working in favorable environments, which is staying with closest co-workers of similar specialization rather than becoming a boss. It means new challenges, but also new obligations, like contacts with clients. One of the interlocutors told me a story of degradation from position of project manager to programer to stay away from any contact with clients. For the interlocutor it was better to choose a lower position and keep desired atmosphere at work instead of getting a promotion and making deals with client. This behavior is close to described in a technician reality.

The only real career option, promotion to management, means leaving the community, and most technicians would rather remain a technician-hero than become an organization manager. Then too, the technicians are in some ways more involved with their customers than with their own corporation, so even though they are always working in a space that is not theirs, it makes sense to remain within the triangular relationship of technicians, machines, and customers. (Orr, 1996, p. 76)

As it was mentioned, a characteristic feature of any IT profession is concentrating on work, not on the organization. The most desirable value is to become a qualified engineer by means of getting the status of full member of community. The way of pursuing this goal is by sharing information, mutual support at work activities, and competition in extent of occupational skills. Thus, a promotion of profession outside of the community is not rewarding. Occupational community shares same values, which does not concern organization. These are values characteristic for a professional role of IT specialists.

Clients and organizational units in need of IT products (i.e., marketing or financial department) are the source of IT professionals work. This separates IT specialists into groups:

1. IT specialists working in software companies where programers are representatives of one profession.
2. IT specialists working in organizations of other branches (not IT); IT specialists are the minority of employees.

This classification determines IT professionals' relations with the environment. The first group implies having a professional and a societal role, in the second one, an organizational role. The one and only career path of young IT professionals in organizations other than software companies is becoming a manager, which means leaving colleagues and a favorite job (programming). Consequence of such an approach is concentration on self-development within the same position and doing work as good as possible without being promoted. This makes customers the closest environment of IT professionals. That is why much stronger relations connect IT specialists with clients (although this relation has not only positive effects) than with any other social actors. IT specialists depend on their both organizations and organizations of their clients. First one makes up a technical background relevant to knowledge of each IT specialization and it is a source of incomes. Second organization creates a field for practicing skills and building professional experience. Organization of a client is so important, because it is satisfying one of the strongest needs of IT specialists—pursuing for professional development.

If the client is the most important social actor in the work environment of IT specialist, the organization hiring him or her is not the first place of importance in his or her professional career. Interlocutors confirmed this statement in the interviews. IT professionals spoke about their organizational life very rarely. Instead, they told stories about current work, project currently in progress, and all related subjects. Therefore, the professional role is the one the IT professionals identify with. It seems to be a common feature for technical workers (Barley & Kunda, 2001; Orr, 1996; Shenhav, 1999). That is why relations within the occupational community (which gives technical background) are so strong and any organizational community is not so important.

In spite of wide spread relations within their occupational community, IT specialists are strong individuals. It does not mean that they are not capable of teamwork. It is a peculiar paradox that IT professionals value individual achievements and base their work on community (common) knowledge and experience at the same time. Both individual and group matter counts. In the creative work process, IT specialists try to maintain essential balance between these reverse tendencies. It will bear fruits of individual actions supported by common knowledge and experience. New client's order means new task to handle individually or with his or her team. Work becomes a priority as it makes such a great challenge for an IT professional and gives satisfaction. In fact, a client with his or her satisfaction is much lower in this hierarchy. It emphasizes importance of work with reference to other occupational activities.

Combination of an individual challenge at work and a support from occupational community could be the key for understanding the attractiveness of this profession for IT specialists. They have the opportunity to take part in common projects as an individual worker (entrepreneur) and it gives them the desirable independence. All these elements put together allow IT professionals to manage their own time and incomes freely to a great extent. Moreover, most IT specialists work as contractors (Barley & Kunda, 2006). In Poland they are hired on short-time contracts (often just for one order) and they are not obliged to stay at work fixed hours. These are first of all the work contracts with organizations outside the IT branch. Other organizations, especially software companies, give their workers more freedom. This allows IT professionals are able to choose comfortable times for work and also gives an opportunity to work as much as it is possible (of course if there is such a need in organization) for the purpose of getting enough income. Thus, effective work time brings out an anticipated salary. In addition to the pay off, free work time gives a chance for IT specialist to work at his or her favorite hours of concentration. A lot of IT professionals prefer to work at night time. IN average software company IT specialists have choice to work in their most comfortable hours. Sometimes it means spending nights at the work place. Companies also allow IT professionals to work at home. Interlocutors agreed that this is one of the most important arrangements with their employer.

Flexible hours are not only a consequence of personal needs of employees but also the result of specific profession's characteristics. The work is sporadic and unpredictable and therefore cannot be scheduled (see Janssens & Steyaert, 1999). It depends on the quantity of client's orders every day or week. It is impossible to construct a schedule for a long time ahead. Moreover, IT professionals, especially working in software companies, do not like to submit to deadlines. In their opinion, there should be the only one deadline and that is finishing project.

One more element of IT specialists' approach to playing the social role (which is conscious or unconscious action), distinguishing it from other technical professions, is a distance to their work. IT specialists tend to do their job professionally but they do not limit their work to meet the client's needs. Quite a big part of their work satisfies needs of self-development what differs in actions depending on IT specialization and organization. A common feature of this activity is some kind of entertainment (Hjorth, 2004; Strannegård & Friberg, 2001). This action can be discussed from the perspective of second organizational dimension—an informal organization. IT professionals have a strong tendency to create social groups independent from the company that hires them. Occupational communities previously mentioned are the first unit of IT specialists' identification, not the company as a whole. Other informal groups also have the advantage over companies. Friendly relations in organizations have special value and are one of the strongest elements of professional role.

Conception of collective identity of IT professionals previously discussed is linked with Van Maanen et al.'s definition (1984) which states that:

occupational communities represent bounded work cultures populated by people who share similar identities and values that transcendent specific organizational settings. (Van Maanen et al., 1984, p. 314)

Such a work order enforces self-control of workers in their community, although it is not easy condition to make. Van Maanen et al. (1984) claim that matter of control in occupational communities is a high challenge for management. In particular, in the organizations outside the IT branch, division of labor (what is an effect of deep specialization process) results in increasing control at the expense of occupational community values. New approach to management may rely upon creating an effective career (development) path, which links workers closer to the organization that to the occupational community. This strategy mostly depends on real promotion possibilities in organization. Nevertheless, IT professionals are a specific group where other values count more than changing organizational positions even if it would be connected to rising wages. That is why communication with project managers (only of computer sciences education) could be a much more effective instrument of managing control in organizations[e]. Identifying IT project managers' needs may facilitate integrating interests or finding new convergent goals of both organization and occupational community. Engaging directly members of IT department in these changes without mediation of project manager may offer workers' resistance. It concerns particular changes in the field of work process in regards to increasing dependence of worker on organization or its environment as well as limitation of his or her practice possibilities (it especially concerns programers). This restriction may offer the greatest resistance of workers.

Knowledge, which is the basic element of IT professionals work, is developed by occupational community too, not by organization (so it is an element of professional role, not organizational one). Abbott (1988) claims that professionals have to deal with increasing knowledge systems and its complexity. IT specialist's profession is a good example of this tendency. The number of new specializations is connected with creating new systems of knowledge (constructed mainly verbally). It is one of the elements of constructing competitive advantage between occupational communities. Abbott attributes particular place to abstract knowledge in process of creating new fields of professional work. However, he maintains that knowledge should not be too abstract and concrete because it may interfere with building this advantage.

Approach establishing competition of professions and discovering strategies used by different groups to achieve a certain social status as a profession rises from theory of professionalization (Larson, 1977). From this perspective, main research stream constitutes consensus, which is the base for professional knowledge in society. "Professions are knowledge-based occupations." (MacDonald, 1995, p. 160) Thus, each profession should tend to decide the question of body of knowledge, socio-cultural development of knowledge, and strategy of maintaining this knowledge base (MacDonald, 1995). Professions operate in the specified system (Abbott, 1988). Effective increasing of power of profession in this system causes permanent change in mutual relations. It is an obvious advantage of IT specialists because of high barriers of access to this profession (Benoit,

Table 4. Elements of professional dimension of IT specialist social role

Element of professional dimension of IT specialist social role		**Expectations and values**	**Definitions of scenario elements**
Task		*Efficiency: What is considered as an achievement?* Satisfaction of a client or any other receiver of IT product or service. Operational system or other IT service ready to use.	*What should be done?* Main goal of IT professional work is well-done and implemented system or other service, which become useful for receiver (client). Entertainment is a goal for IT professional almost simultaneous to work but irrelevant for client.
Standards	Education	*"General rules" of IT professional's work; what their knowledge and skills should be?* Graduating ITS studies or other science studies (i.e., mathematics). Experience in particular specialization. Pursuit of professional self-development counts.	*How should it be done?* *By whom?* Individual gaining qualifications is required. IT professionals should care about work quality standards and a way of providing service.
	Ethical code	Ethics: How should IT specialists act?	

Source: Based on Kostera (1996)

1989). Therefore, abstract knowledge of profession becomes the most effective tool of competition. The strongest professions struggle of power not only in categories of work but also well developed system of profession (the dominant element of advantage between professions). This tendency is a base of emerging new professions (i.e., predomination of medicine over nursing). These days various IT specializations behave similarly collecting new fields of knowledge.

The most important dimension of IT professionals' social role may be summarized in Table 4.

REFERENCES

Abbott, A. (1988). *The system of professions. An essay on the division of expert labor.* Berkeley, CA: The University of Chicago Press.

Barley, S. R., & Kunda, G. (2006). Contracting: A new form of professional practice. *Academy of Management Perspectives, 20*(1), 45-66.

Barley, S. R., & Kunda, G. (2001). Bringing work back in. *Organization Science, 12*(1), 76-95.

Benoit, C. (1989). The professional socialisation of midwives: Balancing art and science. *Sociology of Health and Illness, 11*(2), 160-180.

Berger, P. L., & Luckmann, T. (1966/1983). *Społeczne tworzenie rzeczywistości.* Warszawa: PIW.

Callon, M. (1986). Some elements of sociology of translation. In J. Law (Ed.) *Power, action, and belief. A new sociology of knowledge.* (pp. 196-233). London: Routledge & Kegan Paul.

Czarniawska-Joerges, B. (1992). *Exploring complex organizations. A cultural perspective.* Newbury Park, London, New Delhi: SAGE Publications.

Freidson, E. (1994). *Professionalism reborn. Theory, prophecy, and policy.* Berkeley, CA: The University of Chicago Press.

Goffman, E. (1959/1990). *The presentation of self in everyday life.* London: Penguin Books.

Hall, R. H. (1986). *Dimensions of work.* California: Sage Publications Inc.

Hjorth, D. (2004). Creating space for play/invention—concepts of space and organizational entrepreneurship. *Entrepreneurship & Regional Development, 16*, 413-432.

Janssens, M., & Steyaert, C. (1999). The inhuman space of HRM: Sssssssssssssssensing the subject. *Organization, 6*(2), 371-383.

Jemielniak, D. (2004). Kultura: Zawody i profesje. *Prace i Materiały Instytutu Studiów Międzynarodowych SGH*, 32.

Kallinikos, J. (2003). Work, human agency, and organizational forms: An anatomy of fragmentation. *Organization Studies, 24*(4), 595-618.

Kostera, M. (1996). *Postmodernizm w zarządzaniu.* Warszawa: PWE.

Larson, M. S. (1977). *The rise of professionalism: A sociological analysis.* Berkeley, CA: University of California Press.

Latour, B. (1987). *Science in action. How to follow scientists and engineers through society* Cambridge, MA: Harvard University Press.

MacDonald, K. M. (1995). *The sociology of the professions.* London, Thousand Oaks, CA, New Delhi: Sage Publications.

Orr, J. E. (1996). *Talking about machines.* Ithaca, NY, London: Cornell University Press.

Postuła, A. (2007). *Rola informatyków w organizacjach.* Doctoral thesis. Warszawa: Uniwersytet Warszawski.

Shenhav, Y. (1999). *Manufacturing rationality. The engineering foundations of the managerial revolution.* Oxford University Press.

Strannegård, L., & Friberg, M. (2001). *Already elsewhere. Play, identity, and speed in the business world.* Raster Forlag, Sweden.

Van Maanen, J., & Barley, S. R. (1984). Occupational communities: Culture and control in organizations. In L. L. Cummings, & B. M. Staw (Eds.), *Research in organizational behaviour* (Vol. 6, pp. 287-365). Greenwich, CT: JAI Press.

KEY TERMS

Culture: Generally refers to patterns of human activity and the symbolic structures that give such activity significance.

IT Specialist (IT Professional): Very generalizing term meaning someone whose work is connected with computers. It may be a person teaching information technology at school as well as very specialized IT worker (programmer, webmaster, IT architect etc.). That is why the term IT specialist (or IT professional) is rarely used by members of the occupational community.

Occupational Community: Occupational communities represent bounded work cultures populated by people who share similar identities and values that transcendent specific organizational settings.

Organizational Role: Dimension of social role which considers norms and values of the organization refering to the way IT professionals should or shouldn't act in their organizations.

Profession: An occupation, vocation or career where specialized knowledge of a subject, field, or science is applied. It is usually applied to occupations that involve prolonged academic training and a formal qualification (IT professional). Professions are usually regulated by professional bodies that may set examinations of competence, act as an licensing authority for practitioners, and enforce adherence to an ethical code of practice.

Professional Role: Dimension of social role which considers "game rules" of workers' professional life. This part of social role explains how should professional (IT specialist) act referring to members of professional community.

Programmer: Or software developer is someone who programs computers, that is, one who writes computer software. The term computer programmer can refer to a specialist in one area of computer programming or to a generalist who writes code for many kinds of software. Programmist knows required in his work computer language (Java, C++, etc.).

Social Role: A set of connected behaviors, rights, and obligations as conceptualized by actors in a social situation. It is mostly defined as an expected behavior in a given individual social status and social position. In organizational anthropology "role" is seen as the set of expectations that society places on an individual.

ENDNOTES

[a] Within the Berger's and Luckmann's (1966/1983) meaning of social constructing of reality.

[b] Jemielniak (2004) enumerates differences between occupation and profession. In this chapter I would use both terms interchangeably.

[c] It is rather a part of organizational role of IT specialists. Wider it is analyzed in whole elaboration (Postuła, 2007).

[d] This data is not based on quantitative research. These are conclusions from interviews.

[e] This situation particularly concerns organization outside of IT branch, which has IT department.

Chapter VII
Employee Turnover in the Business Process Outsourcing Industry in India

Aruna Ranganathan
Cornell University, USA

Sarosh Kuruvilla
Cornell University, USA

ABSTRACT

In this chapter, we explore the problem of high turnover in the high-tech BPO sector in India, where relatively well-educated employees are performing a variety of primarily low skill, low cost jobs. We highlight the various approaches employers are taking to solve the turnover problem. As we will argue, some of these strategies are fairly traditional, focusing on various instrumental incentives to promote employee retention, while some others are new and rather radical, particularly the articulation of an organizational and work culture tailor-made for the particular demographic profile of BPO employees: young, upper middle class, well-educated graduates. Based on anecdotal evidence and interviews with industry personnel, we sense some ambiguity regarding the effectiveness of these strategies. We argue that this ambiguity is a function of (a) the recent and rapid growth of the industry and the fact that firms are experimenting with a whole variety of retention strategies, and (b) the inability of firms to develop an integrated organizational culture that permits a focus on both longer term organizational performance, as well as retention.

INTRODUCTION

The IT (information technology) industry in India (commonly referred to as the outsourcing industry) is touted as being crucial to the country's economy. The industry has grown its revenues tenfold in the past decade, from U.S.$ 4.8 billion in 1997-98 to U.S.$ 47.8 billion in 2006-07.While it currently (2006) contributes only 5.4% of India's GDP, this figure is expected to rise to 12.3% by 2012 (NASSCOM, 2007). Thus, the IT industry is expected to be a key driver of India's economic development via its contribution to GDP growth, employment growth, and poverty alleviation (Srinivasan, 2006).

There are three main segments in the IT industry: IT services and software, business process outsourcing (BPO), and hardware (NASSCOM, 2005). Though IT services and software accounts for the bulk of the industry at present, this segment has matured to some extent. On the contrary, the BPO sector is newer and is experiencing explosive growth. Projections indicate that the BPO sector will substantially overtake the software sector in value by 2012, other things being equal (Knowledge@Wharton, 2004b). And it is the BPO industry that has earned India the reputation of being the outsourcing capital of the world. Further, consolidation in the industry has resulted in several large software firms (e.g., Infosys, Wipro, and IBM) starting or buying out BPO units, thus blurring the distinction between software and BPO. We focus exclusively on the BPO sector in this chapter.

Although low costs and a seemingly endless supply of skilled human resources have been the key to India's leadership role in software and business process outsourcing, we argue in this chapter that continued growth of the BPO sector is contingent on it overcoming its biggest human resources problem (i.e., high turnover, which is posing a serious threat to growth and profitability in this sector[a]). Average turnover rates in the industry range from 25-40% (NASSCOM, 2005), imposing a significant cost on firms as they attempt to replace 40% of their employees per year in a very competitive labor market. Thus, despite possessing one of the world's largest supplies of skilled manpower, there is a desperate shortage of people in the industry.

Our goal in this chapter is modest, for example, to understand the causes and consequences of high turnover in the industry, and to document the variety of ways in which human resource departments in the BPO industry are attempting to address the critical issue of employee retention. In an industry where costs are a key source of comparative advantage, paying higher salaries to retain employees is self-defeating, as it threatens the longer-term viability of the industry. As it is, salary growth rates of 15-20% a year over the last few years has resulted in the movement of BPO jobs from India to China and other locations. Thus, BPO firms have to find other means of employee retention.

This situation therefore presents an interesting case study of a unique management problem, for example, the management of relatively well-educated employees in the high-tech sector who are doing a variety of primarily low skill, low cost jobs (although there is a growing high skills based research process outsourcing sector in India as well). Our study, based on interviews and participant observation of the BPO industry in India during 2005-2006, suggests that employers are experimenting with a variety of strategies to combat the turnover problem. As we will argue, some of these strategies are fairly traditional, focusing on various instrumental incentives to promote employee retention, while some others are new and rather radical, particularly the articulation of an organizational and work culture tailor-made for the particular demographic profile of BPO employees: young, upper middle class, well-educated graduates.

In the next section, we provide a (very) brief overview of the BPO sector in India. Thereafter, we discuss the nature and causes of turnover in the industry. We then describe various human resource (HR) strategies that companies are implementing in attempts to retain their employees. In the final section (although we do not yet have clear empirical evidence on which of the HR strategies are successful), we discuss these initiatives against the broader research background of turnover and organizational culture.

THE BPO SECTOR IN INDIA

There were various catalysts that led to the boom in business process outsourcing in India in the mid-1990s (see Dossani and Kenney, 2003 for a comprehensive discussion). The growth of the BPO industry capitalized on India's reputation as a software services provider, a reputation that first developed during the Y2K crisis. This reputation, alongside a large English speaking workforce and low costs (Indian costs are about 20% of U.S. costs in business processes), coupled with the time differences (India works while the U.S. sleeps), resulted in the establishment of call centers and back office services. The pioneering efforts of some MNCs in attempting to re-engineer business processes for cost savings paved the way for many others to outsource to India. Specifically, General Electric, American Express, and British Airways were quick to capitalize on the huge salary cost differentials between similarly trained employees in India and the United States. These MNCs also demonstrated that relocation could be undertaken with minimum disruption. And improvements in the enabling environment—notably the availability of telecommunications bandwidth at increasingly lower rates, the ability to digitize documents, and the usage of standard software platforms in corporate information systems—played a key role too.

Table 1 provides an overview of the BPO sector in India. While the $7 billion sector currently accounts for only 23.8% of the total IT industry revenue, it has seen an average growth rate greater than 55% during the 2000-2005 period. It is this dramatic growth rate, its relatively high contribution to exports and the accompanying escalating employment potential that has raised the possibility that this sector will be the most important "development engine" for India.

India has a dominant position in the global market for offshore outsourcing, expected to account for 51% of the global aggregate by 2008 (NASSCOM, 2005). This market share growth is driven largely by India's numerous advantages (costs and availability of skilled manpower), including the newly acquired reputation of Indian firms' ability to "ramp-up" operations quickly based on demand. Indian also benefits from an English language advantage given that the U.S. and the U.S. are the predominant adopters of offshoring. Consulting firms such as AT Kearney rank India as the most favorable location for BPO activities in the world (see the data in Table 2) and highlight India's advantages in both compensation costs and skills.

Currently, there are over 400 large BPO firms in India (DTA Consultants, 2004), and this number is growing. Within the BPO segment, there are two kinds of firms - captives and third party firms.

Table 1. Overview of BPO sector in India

Year	Total IT Industry Value (U.S.$ billion)	BPO Segment Value (U.S.$ billions)	BPO Exports (U.S.$ billion)	Growth Rate of BPO Segment	Contribution to Employment
00-01	12.1	1.0	0.9	66%	70,000
01-02	13.4	1.6	1.5	60%	106,000
02-03	16.1	2.7	2.5	68%	180,000
03-04	21.5	3.9	3.6	44%	253,500
04-05	28.2	5.7	5.1	46%	348,000

Source: NASSCOM, 2005

Captives are fully owned by major MNCs (e.g., Dell, GE Capital, EDS, HP) and provide business process services only to their owners. Third party firms (both Indian-owned and foreign-owned) provide BPO services to any client, and normally service a large number of clients simultaneously. Table 3 lists notable firms in both categories. Captives thus have a relatively stable market because their fortunes increase or decrease strictly based on their parent company's fortunes. Third party firms, on the other hand, are dependent on securing contracts with a variety of clients, and thus face a higher degree of uncertainty, which we argue will impact their ability to introduce policies to retain employees. Some captives such as GECIS (previously owned by GE) have now become third party firms. Further, the industry is undergoing a bout of consolidation, with large software firms (e.g., Infosys, IBM, Wipro, and TCS) entering the business process outsourcing industry.

What is the nature of work in the BPO industry? Customer care and support services (call centers) are the largest "service" lines, accounting for close to 38% of the BPO sector's employee base and a third of its revenues (NASSCOM, 2005). For example, Delta Airlines outsources parts of

Table 2. Comparing alternate destinations for offshore sourcing of BPO

Factors	India	China	Malaysia	Czech Republic	Singapore	Philippines	Brazil	Canada	Chile	Poland
Financial Structure (Scale: 1-4)										
Compensation	3.19	3.00	2.50	2.24	0.91	3.14	2.84	0.42	2.56	2.45
Infrastructure Cost	0.23	0.23	0.33	0.27	0.22	0.22	0.17	0.32	0.22	0.26
Tax and Regulatory	0.30	0.09	0.26	0.13	0.34	0.23	0.16	0.26	0.21	0.17
Sub Total	3.72	3.32	3.09	2.64	1.47	3.59	3.17	1.00	2.99	2.88
People Skills and Availability (Scale: 1-3)										
BPO Experience	1.03	0.48	0.19	0.23	0.61	0.42	0.41	0.82	0.25	0.19
Size & Availability of Labor	0.47	0.6	0.02	0.01	-	0.08	-	0.04	0.01	0.05
Education	0.25	0.21	0.27	0.33	0.33	0.19	0.11	0.44	0.08	0.28
Language	0.21	0.07	0.14	0.17	0.25	0.14	0.13	0.4	0.15	0.14
Employee Retention	0.13	-	0.11	0.18	0.17	0.11	0.21	0.19	0.21	0.22
Sub Total	2.09	1.36	0.73	0.92	1.36	0.94	0.86	1.94	0.70	0.88
Business Environment (Scale: 1-3)										
Country risk	0.83	0.68	1.03	1.06	1.41	0.57	0.91	1.26	1.03	0.91
Country Infrastructure	0.20	0.15	0.24	0.28	0.40	0.13	0.25	0.41	0.29	0.19
Culture Adaptability	0.10	-	0.28	0.38	0.43	0.05	0.01	0.40	0.11	0.24
Security of Intellectual Property	0.18	0.10	0.22	0.30	0.39	0.17	0.24	0.41	0.25	0.23
Sub Total	1.31	0.93	1.77	2.02	2.63	0.92	1.41	2.48	1.68	1.57
Total	**7.12**	**5.61**	**5.59**	**5.58**	**5.46**	**5.45**	**5.44**	**5.42**	**5.37**	**5.33**

Source: NASSCOM, 2005

its worldwide reservation services to an Indian third-party vendor called Wipro Spectramind. Wipro now manages Delta's reservations from its Mumbai call center and saves Delta $26 million per year (Kalakota & Robinson 2004). Similarly, Amazon.com outsources some of its customer service operations to Daksh (now owned by IBM) in order to respond quicker to e-mail from customers. Today, several hundred Daksh employees work exclusively for Amazon.com and they reply to 95% of e-mail received within 24 hours and 100% within 48 hours. This saves money for Amazon.com too since a full-time customer service representative would cost Amazon.com $2000 per month in the U.S., while an equivalent representative costs $150-$200 per month in India (Kalakota et al., 2004).

Note that these customer service center jobs are similar to the customer service jobs elsewhere that have been studied extensively (see Batt, Doellgast, & Kwon, 2006 for detailed comparative evidence). As such, work organization in this industry follows closely the "tayloristic" model (see Remesh (2004) for a description of the average customer service center operation). Although the work is done in a "high-tech" environment with the latest in telecommunications and videoconferencing technology, it is relatively low skilled

Table 3. Notable captive and third party outsourcers in India

Captive	Third party
Banking and Financial Sector	WNS Global Services
Fidelity , HSBC, Deutsche Bank	WIPRO BPO
American Express, Goldman Sachs	HCL Technologies BPO Services Ltd.
Bank of America, Standard Chartered	IBM – Daksh
JP Morgan, ABN Amro, Goldman Sachs	ICICI OneSource Ltd.
Prudential, Morgan Stanley,	EXL-Services
Lehman Brothers	Mphasis BPO Services
Professional Services	Intelenet Global
Mckinsey & Co, Accenture, Bain	GTL Ltd.
Deloitte, Ernst & Young, Reuters	Progeon
Technology and Telecom	24-7 Customer
HP, IBM, EDS, DELL	Datamatics Technologies
Samsung, HONEYWELL	Hinduja TMT
Automotive & heavy Machinery	TransWorks Information Services
General Motors, Ford	Tracmail Group
Daimler-Chrysler, Hyundai	Convergys India Services Private Ltd.
Pharmaceuticals	Zenta India Private Ltd.
Eli Lilly, Vision Health Source	vCustomer
Pfizer, Astra-Zeneca	Sutherland Technologies
Others	
AOL	
TESCO	

Source: NASSCOM, 2005

work. However, since the market is primarily in the U.S. and the UK, a knowledge of good English is key. And therefore, Indian BPO companies are forced to hire post-graduates and graduates from a variety of disciplines (a graduate from a good university guarantees, generally speaking, some English language capability). Thus, these are highly educated workers, perhaps over-qualified for the jobs that they do, that are in demand in the BPO industry.

Other high growth segments in the industry include human resource outsourcing and legal and medical transcription services. Big human resource consulting firms such as Exult and Hewitt have realized that much of their lower skilled payroll and benefits functions could be relocated to India. Legal transcription services constitute the conversion of interviews between clients and lawyers into documents to be presented in court. There has also been a long established tradition of the outsourcing of medical transcription. These are the typical, mundane, fairly low skill, limited complexity jobs that form a majority of the work in the BPO industry.

However, the composition of work being undertaken in the BPO segment is in transition and there is a shift toward "high-end" outsourcing involving more advanced skills. There has been growth in higher-level technical support, account management, customer data analysis, and hosting of technology that allows for interaction with the customer. Further, new high-end BPO services are emerging and are growing rapidly, though from a small base. Over the last two years, a number of highly skilled jobs in the West have been outsourced to India. These jobs are mostly concentrated in the financial, research and pharmaceutical industries, leading to the terms financial process outsourcing (FPO), research process outsourcing (RPO), and knowledge process outsourcing (KPO).

In the FPO area, equity research, actuarial analytics, data modeling, risk assessment, management of foreign commodity accounts, and the processing of home loans and mortgage services are being outsourced by the big U.S. and European companies. For example, Smart Analyst Inc. and OfficeTiger employ financial analysts in India who can scrutinize personal credit histories, access corporate public financial disclosures, and troll oceans of economic statistics by mining databases over the Web ("New Global Shift" 2003). In the RPO arena, the kind of work being outsourced includes developing data search tools for medical research, medical research and services (clinical trials), drug development research, and engineering design. The U.S. Dept of Labor estimates that over 36,000 jobs in life sciences will move to India. A number of research and development centers have already been established (e.g., Siemens, Snecma Aerospace, Sun Microsystems, Samsung, Phillips, Delphi, Daimler, and Bosch). Finally, in the catchall KPO arena, we find the outsourcing of a variety of activities including publishing services, remote education (such as providing math tuition to U.S. school children in the Southwest), animation and simulation, and general research and development. Many of these jobs require advanced degrees. Table 4 shows the spectrum of BPO work being undertaken in India.

THE TURNOVER PROBLEM

It is clear that the BPO industry will not be able to maintain its current growth rate without solving its biggest human resources problem (i.e., employee turnover, a problem that has become apparent since 2004). The dimensions of the turnover problem are large by any standards. A wide variety of turnover estimates are currently available. Most reports talk about an average of 30-40% (per year). Commercial and popular reporting tends to focus on the extremes. There have been reports of turnover being as low as 12% and turnover being as high as 90%. Mccue (1995) cites one large BPO company which claimed that high staff turnover rates had forced it to replace

Table 4. The spectrum of work done by Indian BPO Firms

Customer Interaction Services	Data Processing	Skilled Support Services	Knowledge Process Outsourcing	Financial Process Outsourcing	Research Process Outsourcing
-Product Care	-HR administration	-Publishing services	-Knowledge management	-Equity research	-Life sciences research
-Billing services	-Insurance claims	-Graphic design	-Business Intelligence	-Data analysis	-Clinical trials
-Tech support	-Medical transcription	-Accounting	-Internet based tutoring	-Market research	-Engineering design
-Marketing	-Legal transcription	-Tax	-Medical databases	-Customer data mining	-Semiconductor design
Low-End ---→**High-End**					

90% of the 14,340 employees in it's largely call center-focused BPO business within a year.

A 2004 study of turnover in seven large "third party" call centers showed a wide variation in turnover rates, ranging from as low as 12% to as high as 62%, with a mean turnover of 30% (Remesh, 2004). However, turnover in "captive" BPOs is typically much lower than in third party BPOs. In our field research, too, we found turnover rates in five captive firms as being between 15-20%, although some captives reported turnover rates as high as 30%. There is job-based variation in turnover as well. Average turnover in "voice-based processes" is between 45-50%, while in "non voice" operations it is 15-20%. Further, "infant mortality" (turnover in the first 45 days of employment after training) in voice based processes is about 20% (although anecdotal evidence suggests that this figure may vary to a significant extent).

There are two types of turnover. The first kind is when workers leave the BPO sector altogether. The second is when they leave one BPO operator to go work for another, often lured by promises of higher compensation and superior incentives. Again, we do not have clear data as to which type of turnover presents a bigger problem. HR managers of well-established leading firms in the BPO industry tend to argue that much of their turnover is due to people leaving the industry altogether. However, at the same time, many other HR managers complain that there is a significant amount of job-hopping in this industry. Some argue that the cost of switching employers is really small (all one has to do is to walk down the street to the next BPO and if trained, one would be hired immediately). Several reports suggest that people would be induced to move with relatively small improvements in salary even as little as 500 rupees per month (approximately $10). Clearly BPO firms see this type of turnover as being significant (despite what HR managers at top firms tell us). Evidence that this kind of turnover is a problem can be seen in the efforts (largely unsuccessful) of large firms in the industry to establish a "no-poaching" pact. The key elements in these pacts are that (a) firms will not poach employees and (b) firms will refuse to hire employees from other firms within 3 months of their leaving their last employment. Large BPOs including IBM-Daksh, Dell, and EXL have openly declared the collapse of such pacts suggesting that they are unlikely to work in today's volatile BPO environment (The Economic Times, 2006a).

The problem of turnover and increased costs are interlinked obviously. The shortage of people has by itself resulted in an increase in salaries above the normal range in India (average BPO salaries are increasing at between 10 and 15% per year). Although salary costs are "taken out of competition" since it affects the whole industry, recruitment and training costs (some of which are firm specific) are also a matter of significant concern. It takes the average high-end call center about 14-16 weeks of training before an employee is ready to work. Moreover, the current average recruitment conversion rate (the number of people who are actually hired as a percentage of the total number interviewed) is really low (4%) in the big metros where these BPOs are located. Therefore, turnover is a critical problem for BPO firms.

CAUSES OF TURNOVER

Interviews with human resource managers highlight a variety of reasons why employees leave their jobs. Since the BPO industry is less than 10 years old, and has grown dramatically in the last five, high turnover has only recently become apparent, and there has been relatively little analysis as to which causes are the most important. We also do not know whether turnover as a result of people leaving the industry altogether is higher or lower than people moving from one company to another within the industry (anecdotal evidence suggests that the latter is the more critical problem, but we need more analysis to make sure). Evidence from the HR departments of various BPO firms suggest the following reasons for high turnover.

Job Related Reasons

There are several job related reasons, but this category is by far the most important cause of turnover. One important issue here is dissatisfaction with the immediate supervisor (the "lousy boss syndrome"). The BPO segment is less than 10 years old years old and most of the employees are between the ages of 19 to 27. Given the high turnover, people with only 6 months of BPO experience are often promoted to the position of supervisor to manage teams of 10 to 15 people. However, they receive insufficient training due to rapid growth and turnover, and lack significant managerial expertise, thus rendering them ineffective at supervision. Human resource professionals label this the "managerial bandwidth" problem and liken it to attempts at creating a "10 year old scotch in 2 years."

A second job-related problem is inherent in the nature of the work itself. Unlike other high-tech work, BPO jobs are not intrinsically interesting. In fact, employees in this sector are disparagingly called "techno-coolies.[b]" The nature of call center jobs has been the subject of widespread research since it is the primary reason for turnover in this sector all over the world (Batt et al., 2006, Deery & Kinnie 2004). It is widely acknowledged that call center jobs are highly repetitive and tayloristic in nature, with tightly regulated lunch and restroom breaks, stringent call targets, and extensive monitoring. Indian call centers are apparently no different, as Remesh (2004) suggests from his extensive research on call center jobs in India. These aspects, in combination with working the night shift in order to service the U.S. market by day, have resulted in an outgrowth of what the industry calls BOSS (burnout stress syndrome). There are numerous reports of call center workers using counseling and psychological services (Pandey & Singh, 2005). Besides, dealing with customers can be tedious in itself. And in this way, employees soon begin to see call center work stripped of its glamour and leave within a couple of years.

Third, avenues for career development in many BPO firms are limited. This is supported by our research and other existing evidence. Batt et al.'s (2005) survey of 60 call centers revealed that while 15% of call center workers were promoted to higher positions within the call center setup, only 1% was promoted to upper management levels beyond the call center section of the organization.

Our research provides more nuanced insights. We find that captive call centers are able to provide promotions to higher levels than "third party" call centers. The market stability that captive call centers enjoy permits them to do so through an envisionment of long-term career paths for their BPO employees.

Demographic Profile of Workers

A second category of causes relate to the segment of the labor market that BPO work attracts. The workforce is young (primarily aged between 18 and 25), and tend to view BPO jobs as transitory (i.e., their goal is to earn good money for a couple of years in the BPO sector before moving on in life) (the average BPO and call center salary of Rs.10,000 per month represents about 2.5 times what a graduate could earn in any other occupation) (Knowledge@Wharton, 2004a). Remesh (2004) presents a more nuanced picture by studying a cross-section of workers in a large third party BPO. His survey of 277 workers suggests that 75% of BPO workers are graduates of convent schools (private schools that are more expensive but noted for their superior English-medium education, typically run by catholic or protestant churches). Ninety-four percent of their fathers and 63% of their mothers are graduates, many of whom work for the government. And they are predominantly female. This reflects a slice of the Indian middle class that has no incentive to settle for a call center job. Many aspire to high-level professional or government jobs, or further education at the graduate level, while others leave to get married and start a family.

Psycho-Social Factors

A third category of causes are psychological factors that are in part related to the demographic profile previously noted. Working the night shift leads to chronic fatigue and insomnia, and is also linked with behavioral changes such as smoking and a poor diet among young graduates. Some workers gain weight because of the sedentary lifestyle the job promotes, involving working on a computer at night and sleeping during the day with little exercise in between. Contact with family and friends is also restricted. These young people work at night, go back to their families during the day, but find no one at home as they have all gone to work. Although there is a long tradition of night shift work in blue collar occupations all over the world, young upper middle class Indian graduates are less able to deal with the absence of a "social life" in the evenings. Further, Pradhan and Abraham (2005) argue that using a different name and adopting a different persona during work hours can cause a questioning of one's identity and can lead to what they term the "multiple personality disorder." (Several client firms from the U.S. require Indian call center workers to adopt American names and American accents). For example, Anjali becomes Angie during the night and talks with an American accent to American customers, leading to cultural self-alientation and a sense of dissonance. All these are sacrifices that a young graduate is either not willing to make or has difficulty dealing with.

To compound matters, some middle-class parents disapprove of BPO jobs, leading to an increase in stress for these young graduates. Parents for example, fear that the BPO lifestyle is a corrupting influence with too much emphasis on fun and western forms of life. Pradhan et al. (2005) suggest that "the new working habits and patterns of living (i.e., working late into the night, the pub culture, the consumerism) are in complete contrast to ways of living together, value systems, traditions, and beliefs still nurtured by the middle class." Other parents disapprove because they feel that their children are lured by easy money and will forgo the chance to pursue higher education. Yet others worry about the reputation of their daughters.[c] Consequently, young people face a variety of different stressors that lead to their decisions to leave the company.

The relative importance of the various factors noted above is still unknown. Turnover as a function of having a lousy boss tends to involve people leaving for other jobs within the industry. Those experiencing the stress syndrome and multiple personality disorder tend to leave the industry completely, and in many cases, those wishing to pursue graduate education and marriage also tend to leave the industry completely. Firms are in the process of putting in place a range of policies designed to combat turnover, which we discuss in the next section.

HUMAN RESOURCE POLICIES TO REDUCE TURNOVER

Indian BPOs are experimenting with a range of strategies to reduce employee turnover. Our goal in this section is to describe the various approaches. In the discussion section, we will take a more analytical perspective regarding these strategies, using the prior literature to gain a perspective regarding the relative efficacy of these approaches. At this stage, we classify the responses of firms into two broad categories (i.e., "traditional" approaches and "new" approaches).

Our category of "traditional" includes a variety of strategies by firms that correspond broadly to the existing academic literature on employee turnover. In the field of human resource management and organizational psychology (as represented by journals such as the Academy of Management Journal, the Journal of Applied Psychology, the Journal of Management, etc.), the problem of turnover has been studied almost incessantly for almost 50 years, beginning with the work of March and Simon (1958), and there have been several meta-analyses that have reviewed past studies (see Griffeth, Hom and Gaertner, 1998). Broadly this voluminous and cacophonic literature suggests (as we gleaned from a reading of several "meta-analytic" studies) that the most important predictions of turnover are variables such as job satisfaction (including satisfaction with various dimensions or facets of the job), organizational commitment (which is predicted by both job satisfaction and a number of other variables), and the perceived ease of finding alternative employment. Therefore, any action by firms to promote job satisfaction (or satisfaction with various aspects of one's employment), we categorize as "traditional" approaches. As the discussion below shows, this includes a variety of different initiatives.

Our category of new approaches below is much smaller, yet more interesting, as they do not fit easily into the prior turnover literature. Rather, these approaches appear to conform more to the notion that it is not the job that matters as much, but the culture of the organization and a variety of "non-job" aspects that are important in employee retention. There is, for example, an emerging literature in management that focuses on "job embeddedness" which highlights an employee's non-job related links to the workplace. Clearly these "new" approaches are consistent with this view, but they go beyond it in many ways, and appear to be tailored to the peculiar characteristics of the BPO workforce in India (i.e., young, upwardly-mobile, highly educated English-speaking middle class people). Next, we describe the strategies.

TRADITIONAL STRATEGIES

Increasing Career Opportunities

In response to perceptions that BPO jobs do not offer career growth, (a widely held perception amongst BPO employees) employers are going to great lengths to convince their employees that there are sufficient career growth opportunities. According to a NASSCOM-Hewitt survey, there was a 14% increase in the number of BPO companies offering a formalized career path to their employees in 2003 alone (NASSCOM 2004). BPOs are actively encouraging employees to move up the ladder and are consciously filling their front-end leadership roles from the talent pool available within the organization. To facilitate this, employees are being cross-trained in multiple domains and are being provided opportunities to work in various lateral departments. Both vertical and lateral movements are important in career development. Thus, BPOs are highlighting that they can offer many career options and functional specializations including project management, IT, systems migration, business process consultancy, training, HR, finance and accounts, quality, business analysis, sales, marketing, proposal development, and the like.

As an example, Intelenet Global Services (one of the larger BPOs) offers career growth options to its agents after two years. The company offers three career paths: vertical (agent/team leader/ team manager/operations manager), horizontal (across functions) and progress to parent company (TCS or HDFC). Further, Intelenet has made it clear to its employees that 70% of its higher level vacancies will be filled from within, ensuring sufficient promotional opportunities for its entry level people. Moreover, they have emphasized that if a vertical movement is not desirable (such as not wanting to be a team leader), a person could move laterally to another function such as HR/ Training and then move vertically upwards from there. Prudential Process Management Services (PPMS), the captive BPO arm of Prudential UK, also offers vertical as well as lateral movements to its employees. The career development streams here are domain specialists, support function specialists and global exchange program (Dev 2004).

These formal processes are buttressed by individual career counseling and the creation of individual road maps to career advancement. Most BPOs have established in-house counseling teams that offer employees a variety of career advice. Through face-to-face counseling as well as psychometric profiling, these counselors are devising customized action plans for employees based on their aspirations and strengths. In addition, they can assess skill sets and recommend specific career options for workers. Finally, they also offer to work with employees to build skills sets and bridge gaps that would enable them to develop their careers effectively.

For example, at Infowavz International, a relatively newer BPO that operates about three call centers, a career planning sheet is given to all employees at the recruitment stage itself. This outlines the organizational structure and the skill sets required for the various functions at various levels of the enterprise. Apart from this, each individual's competencies, strengths, and areas of improvement are evaluated under a comprehensive 'training needs analysis' program for both current as well as future roles (Dev, 2004). Similarly, Progeon (now called Infosys BPO) has recently started a career planning committee to guide young employees on their future prospects (Mukherjee, 2004). These efforts try to contradict the widely held perception that there are limited career growth opportunities in the business process outsourcing industry, especially by highlighting growth opportunities when people join the industry.

It is too soon to gauge the success of these efforts. Earlier, the "captive" BPOS were better able to provide career development, largely because they had a stable market for their business. But today, the big third party BPOs such as Infosys BPO and IBM have seen their business grow to an extent to which they can essentially guarantee the same career opportunities as captive BPOs can.

Providing Educational Opportunities

Recognizing that most young upper middle class Indian graduates have higher educational aspirations (after working at a BPO firm for a couple of years after finishing college), BPO operators are beginning to sponsor executive education for employees. This is quite an extreme strategy since they are providing access to educational opportunities that will almost certainly result in the employees leaving after they complete those opportunities. At best, these efforts will result in prolonging an employees' decision to leave by about 2/3 years. On the other hand, the "war for talent" is so acute that even 2/3 years of stability on the employee front is important in this extremely "tight" labor market. Thus, many firms have entered into agreements with premier Indian business schools (e.g. Indian Institutes of Management, Xavier Institute of management) to offer part-time MBA programs for their employees. Employees also have the option of pursuing technical certifications like MCSE and CCNA to gain domain-specific skills in the technology sector (InfotechOnline, 2006). The course content and class timings are tailored as much as possible to the specific students' needs. Mostly, the classes are virtual, offline or held at the educational institutes. But in some cases, they are held over weekends on the BPO firms' campus.

Wipro Spectramind, India's largest BPO company with 9,300 employees, has recently started such an "earn and learn" program. "We have tied-up with educational institutes like Symbiosis for MBA and BITS Pilani for computer courses so that the employees are able to work and study at the same time. We are even ready to reimburse the course fees once the employees enter into an agreement with us for an assured number of years of service," states the VP Talent Engagement and Development at Wipro Spectramind (Shrinate, 2004). This initiative not only "up-skills" workers, but helps Wipro retain their A-level employees, and undoubtedly offers an opportunity to create future industry leaders. Since the idea is to the retain employees, Wipro is not selective about which employees they sponsor (Shrinate, 2004).

Other BPOs, such as TransWorks, Sitel, ICICIOneSource, and BNKeInternational also offer their employees the opportunity to continue studying whilst in full time employment. Some firms conduct internal "entrance tests" to shortlist the promising employees while some others consider past academic performance or performance on the job. At Sitel, any employee can apply for the scholarship program, although preference is given to high performers. At Transworks, the applicant must have been employed for a minimum of six months and should have performed well on the service level agreements. ICICIOneSource also insists on the six-month rule and encourages only the top performers to apply. There are a variety of post graduate diplomas in subjects such as marketing, finance, and management. There is also diversity among firms as to whether or not they pick up the tab for these educational opportunities. Some firms pay the whole fees, while other provide a subsidy or arrange educational loans. Sitel only picks up half the fees of the student, while TransWorks often reimburses the entire course fee. ICICIOneSource has an upper limit of Rs.25,000. Further, some BPOs require employees to commit to a minimum period of employment. At Wipro Spectramind, it is a year, while at ICICIOneSource it is six months (Shrinate, 2004).

This is a germane strategy. It is well known that the typical BPO employee is interested in further professional education. Our interviews with various employees suggest that they value this "perquisite." And although we do not yet have firm data, anecdotal evidence suggests that BPOs who are establishing these programs tend to report lower turnover than those without similar programs.

Leadership Development

Allied to career development, BPOs are also investing heavily in leadership training. While the motivation here is clearly one of self-interest (i.e., BPOs would like to increase "managerial bandwidth" and thus simultaneously reduce turnover), there is an element of development and longer term career planning here that is of mutual interest to both employer and employee. From the employer's perspective, leadership development is key to establishing a strong middle management cadre, but also increases the loyalty of employees who go through the training. From the employee's perspective, management training complements the career development process and binds the employee more closer to the organization in the longer term.

These training programs are quite intensive and require considerable organizational resources. BPOs are training their employees to motivate their colleagues, instill team spirit, conduct performance appraisals and take on a battery of other responsibilities. At ITC ClientLogic (CLI3L), for example, there is considerable emphasis on training beyond that required for the current job, with a wide span of skills that are covered, from workforce management, coaching and feedback, communication skills and rewarding and recognizing staff to understanding of business metrics and Six Sigma (The Economic Times 2005).

Employee Stock Option Plans

Providing employees with stock options has had a long history of building loyalty and forcing an alignment of employees' longer term goals with the organization's longer term goals. Thus, using stock options as a motivating tool is not new, but in most organizations, this tends to be focused on senior management. Introducing stock options for lower levels of employees, as the BPOs are now doing, is more clearly tied to a retention objective (although 52% of BPOs provide stock options to their employees, most organizations have a vesting period of between 3-6 years) (NASSCOM 2005). Among higher end BPOs operating in the financial industry and in bio tech, stock options are common as well.

Job Enrichment

One of the defining features of a call center job is the repetitive nature of the work. As such, observers of the customer service industry note that these are "lousy" jobs, wherever they are. The intrinsic nature of the work does not seem to vary across countries. However, HR managers in Indian BPOs, facing high levels of turnover in a tight labor market, argue that call center workers can "add more value to a company than just being an outlet for customer frustrations." (InfotechOnline, 2006) Hence, BPOs are outsourcing work like data entry and seasonal work to smaller companies so that their employees get to do high value-added work. For example, E-Edit Solutions, a BPO in e-publishing, outsources basic composition work and XML coding work to multiple vendors (IndiatimesInfotech, 2006). This way, BPO operators get to work on new technologies. This is not possible in all cases, but the principles are transferable.

These various new initiatives are in addition to traditional strategies of good pay and benefits (which do not vary much across the industry) and good human resource management practices (which varies quite a lot) in the industry. Indian BPO firms are also engaged in a number of other strategies, such as targeting older workers who might be more willing to see this as a long-term career.

NEWER APPROACHES

Developing a Fun Culture

A key development is the recent focus of several BPOs in positioning themselves as "fun" places to work in to retain their predominantly young workforce. In general, the approach here is to sponsor fun-filled activities, parties, invest in recreational facilities, construct trendy interiors, encourage employees to watch movies, play games, and relax (when workload is light). The goal of these activities is to create an ambience in which the young educated and middle class workforce will enjoy the work atmosphere and hopefully will not leave. As the Director of Business Development at

Motif a BPO suggests "We understand that salary is only one part of what motivates employees... an agreeable working environment is what really allows us to retain talented individuals." Thus, BPOs are increasingly setting aside a budget for "fun" so as to keep their employees "happy."

At Wipro Spectramind, "everything revolves around fun" according to the VP Talent Engagement and Development (Doke 2003). Fun is a core, written-down, defined value and hence, permeates all aspects of life at the organization. Wipro Spectramind's mission statement reads "Together we create a fun-filled work environment, encouraging all to enjoy pushing their individual and original roles." The BPO holds that fun is critical to enhancing mutual trust and building dynamic relationships amongst employees. Further, they believe that fun helps to mitigate stress arising from BPO work and allows employees to maintain high energy levels in their interactions with customers. Thus at Wipro Spectramind, fun is not a reward for performance but a core aspect of the organizational environment. It is even linked to the performance management system whereby employees are rewarded for living up to the value of fun in the workplace. As a senior executive warns, "make no mistake, fun is serious business here [as it] helps reinforce the values of the organization" (The Economic Times, 2004).

24/7 Customer has developed a similar company motto: FITO (fun into the organization). By budgeting $10 million per year for these activities, they ensure an engaging time for all those on their roster. The BPO believes that employees bond through such activities and then go back to work recharged. And strong bonds of friendship certainly help to retain employees in an organization. Fun activities can help foster bonds with customers too. For example, at 24/7, the procurement of an important contract was celebrated with a 30 feet long cake with the customer's logo on it (The Economic Times, 2004).

At HCL BPO and Contech, a balance between fun and work is obtained by pushing the "work hard, party hard" concept. The BPOs recognize that their employees are fascinated with western ways of living and they devise programs that capitalize on this. And as the director of business development at Motif observes, "an agreeable work culture is what allows [BPOs] to retain talented individuals" (O'Driscoll, 2005).

Creating a balance between work and fun in the workplace requires a significant organizational commitment and a variety of activities as the Wipro Spectramind previous example indicates. Thus, there is focus on the *physical environment* with BPOs making sure that their physical layouts appeal to their trendy workforce. EXL, for example has three state-of-the-art facilities spread over 150,000 square feet at Noida, near Delhi. These facilities have a bright, colorful, and lively design on the shop floors, while exuding elegance and classiness in "common areas" and on the corporate floor. The color schemes are bright and visually appealing, given that BPO work is deskbound, repetitive and sometimes stressful. Bright reds, pinks, and yellows have been used on the shop floor, with a textured finish to counter the glare of the bright walls (Business India Intelligence, 2003). Similarly, there is a focus on *recreational facilities,* which appeal to the workforce. For example, Infosys' campus houses a state-of-the-art gym, golf course, sauna, 50,000 square feet swimming pool, lake with paddling boats, grocery store, pool tables, ping pong tables, and a dance floor (Infosys, 2001). 24/7 Customer has invested in a complete range of music equipment where their seven in-house rock bands practice during work (The Economic Times, 2004).

Food is also an important aspect of the environment. IBM Daksh offers free breakfast buffets to all their employees (The Times of India, 2006). The Infosys campus contains food courts that serve Chinese, North Indian, South Indian, and Western cuisine (Infosys, 2001). Then there is the focus on *wellness.* The negative aspects of the call center environment, including night shifts, sedentary work, talking on the phone for hours together, and their associated ailments are now well known. Therefore, specific initiatives are being taken by some BPOs to address nutritional and stress-related concerns that employees might have.

For example, KGISL recently hired a physiotherapist to teach employees exercises for flexing their fingers and toes (Revathy, 2004). While the initiative is small, it is nevertheless an attempt at protecting employees' health and spirit. Other BPOs have initiated various programs like yoga classes, neck and shoulder massages, dietary and nutrition advisory sessions and workshops on managing sleep habits and achieving work-life balance.

To complement the physical attributes of the workplace catering to the young is a focus on an *informal atmosphere at work.* Many BPOs try to keep the work atmosphere quite informal so that employees loosen up, feel comfortable, and have fun at work. With this in mind, Wipro Spectramind has no dress code. Therefore, torn jeans worn to work are just as acceptable as traditional Indian clothing (Doke, 2003). And there is a focus on *"fun-filled activities" at the workplace.* Events like treasure hunts, voice impersonations and rangoli competitions are common. At HCL BPO, frequent contests and team dinners are a way of life. At Contech, it is common to walk into the cafeteria and witness a round of musical chairs. Even a one-minute-game-show was organized recently during which employees had 60 seconds to stick as many bindis (Indian adornment) as possible on their colleagues' faces. At Effective Teleservices, each time an agent secures a sale, he or she spins the office's wheel of fortune with the chance of winning a prize. On Halloween Monday, the employee with the best costume is awarded a gift. Infowavz International celebrates Rose Day, Friendship Day, Valentines' Day, etc. with great fervor in a way that involves all their employees (O'Driscoll, 2005). Sometimes employees put up cultural shows too. Thus, at these BPOs, at any given point in time, there is something or the other going on, with never a dull moment. And after work, there are *parties.* Many BPOs sponsor parties for their employees at a nightclub or a discotheque, with unlimited food and drink to ensure a night of partying and fun with friends and colleagues. BPO workers enjoy this immensely as it legitimizes partying by making it 'official' and therefore allows them to elude late night curfews imposed by strict parents. And there are *trips.* BPOs are organizing regular outings too, as an opportunity for their employees to bond with one another and with their company outside of work. There are day trips to malls, weekend trips to resorts, and so forth. And finally, various Indian *festivals* are celebrated on company time and at company expense.

To coordinate all these activities, BPOs have hired a new cadre of employees called "fun officers." For example, Progeon (now called Infosys BPO) has hired a "chief fun officer" (CFO) who articulates new "fun" strategies (Mukherjee, 2004). In other BPOs, fun teams are in charge of spur of the moment activities including pizza parties, outings, and picnics. At Wipro Spectramind, each supervisor is rated for his or her efforts to encourage team members to organize and participate in events and group activities (Shahane, 2005). Clearly there is an organized effort to promote fun as a central element of one's employment at a BPO in India.

Developing Family Friendly Practices

A second key element that is new is the effort to integrate the employee's family into the enterprise in some way. The underlying notion here is that if the family is engaged, then there is greater likelihood of employee retention, as the employee's family is also in some way "embedded" in the organization. Infosys BPO and Transworks have instituted periodic *family days* for their employees to show off their workplaces to their parents, siblings, partners, and kids. It is also an educational experience for family members as they learn about the various facets of life in the BPO industry and are introduced to team leaders. In the words of a VP of Human Capital at Transworks "The objective is to familiarize families with the nature of the industry and the work done, so that they feel proud of their children" (Ganesh, 2006).

The family is also integrated into the *reward systems* of the enterprise. "When an employee has performed well, we call up the parents or the family members concerned and tell them about the employee's achievement," claims a lead executive of Accenture's India Delivery Centre (Ganesh, 2006). On similar lines, when an employee recognition ceremony took place earlier this year at Wipro's headquarters in Bangalore, all awardees along with their spouses from locations across India, were flown down exclusively for the ceremony. And then there is this notion of "*sharing the gains*." When Cognizant reached its $ 1 billion revenue milestone, it celebrated by doling out the latest 5th generation iPods to 15,000 employees who had served the organization for over a year (Chatterjee & Ramnarayan 2006).

Integrating the family at the time of hiring is also common. In their own way, BPOs are addressing parental queries about their ward's job profile, career opportunities, and the workplace environment. Accenture's BPO arm has even re-engineered its hiring process to involve employees' families right from the interview stage to the post-recruitment phase. During the advanced stages of an interview, the company asks the prospective candidate to bring his or her family or spouse along. "We want them to have a personal feel about our workplace and deal with their apprehensions, if any, then and there," says the VP HR, Convergys India, who invites parents of new recruits to the firm (CiteHR, 2004).

And specifically geared to retention, many BPOs offer fully paid family vacations for longer service. Wipro Spectramind gives out two-three day holiday packages to an employee's entire family for a destination within India in appreciation of the employee's long stint in the organization (Chatterjee et al., 2006).

Finally, there is the notion of work-family balance, a response to the problems expressed by young people. For example, at Wipro Spectramind, there are times when an employee needs temporary leave from work when his or her partner has been posted abroad. Such requests are accommodated as far as possible. When a husband and wife pair works for Wipro, the company tries to ensure that both of them are always placed at the same location. GlobalLogic BPO bears the travel expenses of their employee's spouse and dependant children if the employee is required to travel abroad for more than three months on B1/L1/H1 visas (Shahane & Savitha, 2007).

Perhaps the most radical effort at retention is the complete *embedding of the family* at the workplace: BPOs are encouraging entire families to work together in the company. One pioneering New Delhi based call center, Vertex, with over 900 employees popularized this concept. "They are several couples employed here, a mother and son duo, and an instance of 3 brothers but this is just the beginning," according to the CEO of Vertex India. The idea was explored when some employees who were leaving said they were doing so because they hardly spent time with their families as a result of working night shifts while their spouses were asleep. But now, families can travel together to work, spend time at work and also while commuting back home. Vertex is hoping that the emotional bond to the company will be stronger now that family ties are involved (InfotechOnline 2004b). Ness Technologies, too, always makes a job offer first to an employable spouse before moving to the general pool of applicants. "We also retrain the person if necessary," claims the VP HR at Ness Technologies (Shahane et al., 2007).

In sum, the "new" strategies category appear to focus on the one hand on non-work related aspects of the organization that serve to make their BPO an attractive environment to work, while simultaneously attempting to increase the commitment of employees family's to the organization in order to increase employee retention. These strategies have not received much attention in the voluminous research on turnover.

DISCUSSION

As noted, it is too early to comprehensively evaluate the effectiveness of these ongoing retention strategies. From a more analytical perspective, it is clear that the "new" strategies are qualitatively different from the traditional variables highlighted in prior turnover research. Removing the causes of job dissatisfaction is important, as demonstrated by prior turnover research. But the interesting aspect of the "fun" and "family integration" strategies is that they tend NOT to focus on the most fundamental issue (i.e., the job[d]). Thus, the new strategies give great emphasis to the creation of an organizational culture that provides some avenues for "fun" and "family integration" without addressing fundamental issues with the job itself. This approach appears to turn decades of turnover research on its head.

On the other hand, it is possible that decades of turnover research has not focused appropriately on the importance of such approaches. However, there is a new strand of research on turnover that seems consistent with the new approaches described in this chapter (Mitchell, Holtom, Lee, Sablynski, & Erez, 2001). They focus on the concept of "embeddedness," which refers to three inter-related notions: the extent to which people have links to other activities at work, the extent to which their jobs and communities are similar to or fit with other activities in their life spaces, and the ease with which these links can be broken. As Mitchell et al. (2001, p. 1104) note, "we can describe embeddedness as like a net or web in which an employee has become stuck." The higher the number of links between the person and the web, the more the person is "bound" to the organization. Thus, it is possible that an employee can become enmeshed or embedded in many different ways. For example, an employee will consider how well he or she fits in the work community and the organization, and therefore issues like amenities, outdoor activities, social and entertainment activities all play a role in this notion of fit (Mitchell et al., 2001) that leads to organizational attachment.

This notion of embeddedness, is quite consistent with the concept of organizational culture (in fact, most HR managers used this term), although culture has NOT been an important variable in turnover research (thus far). Clearly Indian BPOs are 'engineering' a particular type of culture that increases the embeddedness of the employee in the work environment. Whether the focus on increasing employees' embeddedness through the creation of a particular type of culture apparently will have a positive effect on retention remains an open question; more systematic research is necessary to firmly establish its effectiveness.

We do however, have some preliminary anecdotal evidence regarding the effect of both classes of strategies. First, on average, we find that turnover levels at captive firms tend to be lower than those at third party BPOs (in our sample of firms, turnover ranged from 15-20%), consistent with our expectations that captive firms have a more stable environment. Second, we know that at Sitel, the attrition rate is down to 30% from 37% in just 8 months as a result of the part-time educational opportunities (Shrinate, 2004). Also, "new" strategies at 24/7 Customer has reduced "infant mortality turnover" to less than 2% (Menon, Sinha, & Urs, 2004). Finally, in a recent Dataquest-IDC survey measuring the satisfaction level of BPO workers, IBM (previously Daksh) emerged on top as a result of high satisfaction scores on a variety of human resource dimensions, including work culture and image. This is despite their low rank on "employee satisfaction with wages." (InfotechOnline, 2004a). As a result of Motif's fun at work strategies 97%

of Motif employees said in the same Dataquest-IDC survey that they "look forward to coming to work in the morning."

By creating a young employee-friendly work environment, KGISL reports that it has managed to contain their attrition rate at between 5-8% (Revathy, 2004). Similarly creative human resource management initiatives geared towards fun at Effective Teleservices have helped the company maintain an impressive 5.2% attrition rate (O'Driscoll, 2005). Thus, the preliminary evidence suggests that both types of strategies appear to be delivering results.

However, there is also some recent evidence of a dissatisfaction and disillusionment with these strategies as well. The chief personnel manager of 24/7 notes that "some BPOs have gone overboard with the fun theme" (The Economic Times 2006b). With the "fun" approach, the seriousness related to doing business had gone, particularly as BPOs moved up the value chain and started doing more value added work. At 24/7 the company has newly introduced a formal corporate dress code, and is attempting to draw a line between work and fun (The Economic Times, 2006b). And as the chief delivery officer of the company notes, they are trying to now convey to employees that working for a BPO is not fun, but a serious career (Borthakur, 2006).The Chief Executive of Progeon, for example, notes that the concept of "having fun at work" was necessary in order to attract young people to work in the industry, but now employees have realized that working in the outsourcing industry involves serious work. Similarly, a senior vice president at another BPO argues that "during the initial phase of the BPO industry, branding BPO work as "fun" was a way to attract and retain these young kids to work" (Borthakur, 2006). But now, with maturation, it is necessary to re-brand the business. Bakshi (2005) suggests that "although the fun factor has not been thrown out of the window, it has taken the backseat" and that career planning and development are replacing the "fun" factor as the key to retaining employees. Recruitment advertisements for BPOs do not emphasize the fun factor as much anymore, while highlighting the career growth possibilities in the industry. The VP HR of Tracmail argues "it is high time the industry realizes that showcasing the BPO space as a fun extension of college life is grossly incorrect" (Offshoring Times, 2005). The president of NASSCOM (the industry association) notes that companies are now shifting to growth prospects as a prime motivator rather than concentrating on fun initiatives to retain talent (Bakshi, 2005).

Although the evidence above is anecdotal, it is clear that some BPOs highlight the importance of the culture of fun, while others question its effectiveness. Clearly, there is some ambiguity here. What explains these differing perceptions on the importance of the "culture" strategies? We offer two tentative and inter-related explanations here. First, we argue that the given the recent evolution (the last 8 years) and the rapid growth and development of the industry (average annual growth rates exceed 50%—see Table 1) under tight labor market conditions, firms were following a "scattergun" approach, experimenting with a variety of strategies to solve the turnover problem. Many HR managers told us that all they seem to do is "recruit, recruit, and recruit." Thus organizations have just not had the time during this period of breakneck growth to establish coherent human resource policies. As some of the comments above seem to suggest, the last year has seen some maturation in the industry, where some of these culture strategies are being re-evaluated.

Second, we argue that it is probable that strategy of creating a "culture of embeddedness" is too narrow in its focus. Kunda (1992) suggests that an organizational culture is "engineered" (i.e., researched, designed, developed, and maintained) to attain certain strategic goals. The culture defines the social characteristics of the company and acts as a guideline for company action. Over time, the grasp of this culture runs even deeper to control the "mindsets" and "gut reactions" of the employees by creating in them a strong commitment to the company and identification with company goals (Schein, 1985). There is considerable evidence that companies with strong corporate cultures are successful (Deal & Kennedy 1982). Organizations with a strong culture create clear and coherent values (Chatman & Cha, 2002; Saffold, 1988), increase organizational performance via a better understanding of organizational objectives and increase inter-personal ties (Deal et al., 1982; Pottruck & Pearce, 2001; Tushman & O'Reilly, 1997). And there is evidence from studies of several organizations (e.g., Southwest airlines) that companies with strong cultures attain a variety of strategic advantages (Friedberg & Friedberg, 1996; O'Reilly & Pfeffer, 2000a). Further, strong cultures create organizational stability (Denison & Mishra, 1995), which appears to be the need of the hour for Indian BPOs.

The debate about the usefulness of these "fun culture" strategies in Indian BPOs is, in our view, a reflection of the narrow focus of these strategies. Indian BPO strategies on culture appear to concentrate on the creation of an environment that promotes retention, and less on the creation of a culture to accomplish broad organizational goals. Their HR initiatives, for example, focus on social engineering in the BPO workplace to create a fun, nurturing environment. Through the provision of various social activities, these initiatives establish communal ties among employees to the extent that the workplace becomes the focal point of social interaction. This is further buttressed by various policies that link employees' families to their workplaces, further binding employees to the organization. This is a human resource initiative, not an effort to create a particular shared culture by officially propagating "values" and having employees interpret them (Hochschild, 1997).

We think that what is really going on is that BPOs have failed to integrate the two categories of strategies effectively and have been as yet unsuccessful in creating a strong organizational culture that promotes not just retention but overall organizational performance. For these efforts to be successful, a more integrated corporate culture approach such as that suggested by Kunda and others may well be necessary. What appears to be required is an integrated strategy that develops a strong organizational culture, one that in the long term aligns individual and organizational goals in a more serious way. We leave it to future research to investigate our tentative hypothese further.

CONCLUSION

In this chapter, we briefly described the problem of high turnover in the high-tech BPO sector in India. We highlighted the various approaches to solving the turnover problem (which are currently ongoing). Although we do not have evidence yet of the effectiveness of these strategies, based on anecdotal evidence and interviews with industry personnel, we sense some ambiguity regarding the effectiveness of these strategies. We argue that this ambiguity is a function of (a) the recent and rapid growth of the industry and the fact that firms are experimenting with a variety of strategies to limit turnover, and (b) the inability of firms to develop an integrated organizational culture that permits a focus on both longer term organizational performance, as well as retention.

The chapter is focused on highly educated workers, many of whom are working in low skilled jobs in the high tech environment of the outsourcing industry in India. There is a contradiction here. These employees are way too overqualified for the low skill jobs that they do. It is not surprising that turnover is high, given the nature of the work itself. Yet the employees are critical to the growth of India's BPO industry. Perhaps the answer in the longer term is to recruit employees with lower qualifications who will be more likely to see these jobs as providing long-term careers. However, this would require the Indian education system to graduate more English speakers.

Alternatively, maturity and consolidation in the industry will provide a more stable basis for more effective human resource policies to contain attrition. Until then, Indian BPO employees are "having fun."

REFERENCES

Bakshi, M. (2005, March 28). *Heard you, loud and clear.* The Hindu Business Line. Retrieved March 9, 2007, from http://www.thehindubusinessline.com/ew/2005/ 03/28/stories/2005032800200300.htm

Batt, R., Doellgast, V., & Kwon, H. (2006). Service management and employment systems in U.S. and Indian call centers. In L. Brainard & S. M. Collins ((Ed.), *Offshoring white-collar work: The issues and implications.* The Brookings Trade Forum, 2005. (pp. 335-361). Washington, DC: The Brookings Institution.

Batt, R., Doellgast, V., Kwon, H., Nopany, M., Nopany, P., & da Costa, A. (2005). *The Indian call centre industry: National benchmarking report strategy, HR practices, & performance* (CAHRS Working Paper #05-07). Ithaca, NY: Cornell University, School of Industrial and Labor Relations, Center for Advanced Human Resource Studies. http://digitalcommons.ilr.cornell.edu/cahrswp/7

Borthakur, M. (2006, Nov. 9). *BPO: Only business.* The Times of India. Retrieved March 9, 2007, from http://timesofindia.indiatimes.com/articleshow/374855.cms

BusinessWeek Online. (2003). *The new global shift.* Retrieved December 20, 2006, from http://www.businessweek.com/magazine/content/03_05/b3818001.htm

Chatman, J. A., & Cha, S. A. (2002). Leading though organizational culture. In S. Chowdhury (Ed.), *Next generation business series: Leadership.* Financial Times. Prentice Hall Publishers.

Chatterjee, M. B., & Ramnarayan, A. (2006, April 10). *For you and your family.* The Hindu Business Line. Retrieved March 10, 2007, from http://www.hinduonnet.com/ thehindu/thscrip/print.pl?file=2006041000010100.htm&date=2006/04/10/&prd=ew&

CiteHR. (2004). *BPOs encourage parents day out.* Retrieved March 10, 2007, from http://www.citehr.com/bpo-india=bpo-indusrty-vt7396.html

D'Cruz, P., & Noronha, E. (2006, May 27). *Organising call centre agents: Emerging issues.* Economic and Political Weekly. Retrieved February 27, 2007, from http://www.epw.org.in

D'Cruz, P., & Noronha, E. (2006). Being professional: Organizational control in Indian call centers. *Social Sciences Computer Review, 24*(3), 342-361.

Deal, T., & Kennedy, A. (1982). *Corporate cultures.* Reading, MA: Addison-Wesley.

Deery, S., & Kinnie, N. (2004). *Call centres and human resource management.* Hampshire: Palgrave Macmillan.

Denison, D. R., & Mishra, A. K. (1995). Toward a theory of organizational culture and effectiveness. *Organization Science, 6*, 204-223.

Dev, S. (2004, Jan. 26). ITES Players offer long-term careers. *IT People, 8*(18), 1-2.

Doke, D. D. (2003, April 29). What makes Indian call centers tick. *Personnel Today,* 6-8.

Dossani, R., & Kenney, M. (2003). Went for cost, stayed for quality? Moving the back office to India. *Berkeley Roundtable on the International Economy.* Berkeley, CA: University of California.

DTA Consultants. (2004). *The Indian IT and IT-enabled services Industry.* Retrieved January 20, 2007, from http://www.dtainternational.com/ITES_in_India11.pdf

Economist Intelligence Unit. (2003). Finding staff and keeping them. *Business India Intelligence, 10*(1), 5-8.

Friedberg, K., & Friedberg, J. (1996). *Nuts! Southwest Airlines' crazy recipe for business and personal success.* Austin, TX: Bard Press.

Ganesh, V. (2006, December 1). *Want to retain staff? Keep it in the family, say BPOs.* Hindustan Times. Retrieved from http://www.hindustan-times.in/news/ 181_1857532,00020021.htm

Griffeth, R. W., Hom, P. W., & Gaertner, S. (1998). A meta-analysis of antecedents and correlates of employee turnover: Update, moderate testes, and research implications for the millennium. *Journal of Management, 26,* 463-488.

Hochschild, A. R. (1997). *The time bind: When work becomes home and home becomes work.* New York: Metropolitan Books.

IndiatimesInfotech. (2006). *BPOs outsourcing work to smaller cos.* Retrieved January 5, 2007, from http://infotech.indiatimes.com/BPO__ITES/Outsourcing/ BPOs_outsourcing_work_to_smaller_cos_/articleshow/2011023.cms

Infosys. (2001). *Management of human assets at Infosys.* Retrieved March 10, 2007, from http://www.infosys.com/about/cases/INFOSYS6%20case%20withchanges.pdf

InfotechOnline. (2004a). *BPO workers more satisfied now.* Retrieved November 10, 2006, from http://infotech.indiatimes.com/articleshow/msid-919912.cms

InfotechOnline. (2004b). *Invited! Your family for the BPO job.* Retrieved January 5, 2007, from http://infotech.indiatimes.com/articleshow/msod-940751.cms

InfotechOnline. (2006). *BPOs take care of your career too.* Retrieved September 5, 2006, from http://infotech.indiatimes.com/articleshow/msid-1456697.cms

Kalakota, R., & Robinson, M. (2004). *Offshore outsourcing: Business models, ROI, and best practices.* Alpharetta, GA: Mivar Press.

Knowledge@Wharton. (2004a). *A view from the developing world.* Retrieved January 25, 2007, from http://knowledge.wharton.upenn.edu/article.cfm?articleid=860

Knowledge@Wharton. (2004b) *Why corporations pursue BPO.* Retrieved February 25, 2007, from http://knowledge.wharton.upenn.edu/article.cfm?articleid=860

Kunda, G. (1992). *Engineering culture: Control and commitment in a high-tech corporation.* Philadelphia: Temple U.P.

March, J. G., & Simon, H. A. (1958). *Organizations.* New York: Wiley.

McCue, A. (2005, January 21). *High staff turnover hits Indian offshore firms.* Offshoring Special Report. Retrieved December 15, 2006, from http://www.silicon.com/ research/specialreports/offshoring/0,3800003026,39127243,00.htm

Menon, S., Sinha, S., & Urs, A. (2004, August 1). Tug of talent. *The Economic Times,* p. 7. Kolkata, India: The Times Group.

Mitchell, T. R., Holtom, B. C., Lee, T. W., Sablynski, C. J., & Erez, M. (2001). Why people stay: Using job embeddedness to predict voluntary turnover. *Academy of Management Journal, 44,* 1102-1121.

Mukherjee, W. (2004, June 17). *Most BPO workers change careers.* The Economic Times. Retrieved October 6, 2006, from http://wwwl.economic-times. indiatimes.com/articleshow/msid-742807.cms

NASSCOM. (2007). *NASSCOM Strategic Review 2007.* New Delhi: NASSCOM.

NASSCOM. (2005). *NASSCOM Strategic Review 2005.* New Delhi: NASSCOM.

NASSCOM. (2004). *NASSCOM Strategic Review 2004*. New Delhi: NASSCOM.

O'Driscoll S. (2005, November 8). *BPOs bring fun back to the workplace*. The Times of India. Retrieved March 10, 2007, from http://www.timesofindia.indiatimes.com/ articleshow/msid-1288762,prtpage-1.cms

OffshoringTimes. (2005). BPOs-Those who look for a serious career. Retrieved January 9, 2007, from http://www.offshoringtimes.com/Pages/2005/BPO_news330.html

O'Reilly, C. A. & Pfeffer, J. (2000) Cisco systems: Acquiring and retaining talent in hypercompetitive markets. *Human Resource Planning, 23*, 38-52.

Pandey, A., & Singh, P. (2005). *Women in call centres*. Economic and Political Weekly. Retrieved February 10, 2006, from http://www.epw.org.in

Pottruck, D. S., & Pearce, T. (2001). *Clicks and mortar: Passion-driven growth in an Internet-driven world*. San Francisco: Jossey-Bass.

Pradhan, J. P., & Abraham, V. (2005). Social and cultural impact of outsourcing: Emerging issues from Indian call centres. *Harvard Asia Quarterly, Summer*. Cambridge, MA: Harvard Asia Center.

Rediff News. (2005). Salary hike, a threat to Indian BPO. Retrieved January 25, 2007, from http://www.rediff.com/money/2005/apr/20bpo2.htm

Remesh, B. P. (2004). Cyber Coolies' in BPO: Insecurities and vulnerabilities of non-standard work. *Economic and Political Weekly*. Retrieved June 20, 2006, from http://www.epw.org.in

Revathy, L. N. (2004, July 19). *Live it up in office*. The Hindu Business Life. Retrieved March 6, 2007, from http://www.hinduonnet.com/thehindu/thscrip/ptint.pl?file-2004071900040100.htm&date=2004/07/19/&prd=we&

Saffold, G. S. (1988). Culture traits, strength, and organizational performance: Moving beyond "strong" culture. *Academy of Management Review, 13*, 546-558.

Schein, E. H. (1985). *Organizational culture and leadership*. San Francisco: Jossey-Bass.

Shahane, P. S. (2005, August 11). *Wanna have fun sans sex? Come, join our BPO*. Infotech Online. Retrieved March 10, 2007, from http://infotech.indiatimes.com/ articleshow/msid-1197310,prtpage-1.cms

Shahane, P. S., & Savitha, V. (2007, January 6). *It's family first and tech cos and BPOs*. Infotech Online. Retrieved March 11, 2007, from http://infotech.indiatimes.com/ articlesehow/msid-1067797,prtpage-1.cms

Sheikh, Z. (2004, Nov. 1). *Want a free international trip? Work for a BPO*. RediffNews. Retrieved from http://in.rediff.com/getahead/2004/nov/01ga-zia.htm

Shrinate, S. (2004, June 6). Truly, a learning experience. *Business Today*, 156-157.

Srinivasan, T. N. (2006). Information technology and India's growth prospects. In L. Brainard & S. M. Collins ((Ed.), *Offshoring white-collar work-the issues and implications, The Brookings Trade Forum, 2005*. (pp. 203-240). Washington, DC: The Brookings Institution.

The Economic Times. (2007). How to bell the BPO cat. Retrieved March 10, 2007, from http://economictimes.indiatimes.com/articleshow/msid-1736650,prtpage-1.cms

The Economic Times. (2006a). BPOs' no-poaching pacts fall apart. Retrieved November 26, 2006, from http://economictimes.indiatimes.com/BPOs_nopoaching_pacts_fall_apart/articleshow/498749.cms

The Economic Times. (2006b). BPOs get serious, axe "fun job" stage. Retrieved March 9,

2007, from http://economictimes.indiatimes.com/Hot_Links/BPO/ BPOs_get_serious_axe_fun_job_tag/articleshow/1486707.cms

The Economic Times. (2005). Why BPO execs are given training. Retrieved October 5, 2006, from http://www.bpo.nasscom.org/artdisplay.aspx?art_id=4491 &cat_id=608

The Economic Times. (2004). BPOs make staffers boogie woogie. Retrieved March 5, 2007, from http://www.247customer.com/newsandevents/Article -56.php

The Times of India. (2006). Two Knights @ The Call Centre. Retrieved March 9, 2006, from http://www-935.ibm.com/services/us/gbs/bus/html/dakshnews_21apr06.html

The Tribune. (2006). Future on the line. Retrieved March10, 2007, from http://www.tribuneindia.com/2006.20060111/jobs/main1.htm

Tushman, M., & O'Reilly, C. (1997). *Winning through innovation: A practical guide to leading organizational change and renewal.* Boston: Harvard Business School Press.

ENDNOTES

[a] Note that the software sector is also experiencing high turnover, although not to the same extent as the BPO sector.

[b] Coolie: A colonial Indian word meaning porter.

[c] There is a stigma associated with women working at night. The idea of a girl staying out at night, all dressed up, and returning in the early hours of the morning looking tired raises eyebrows.

[d] Even some of the "traditional" strategies noted in this chapter, e.g., providing non-job related education opportunities, do nothing to improve the nature of employees' current jobs.

Chapter VIII
Old and New Timings in a High-Tech Firm

Pauline Gleadle
The Open University, UK

ABSTRACT

We use a lens of organizational time and timings in a recent historical study of high-tech engineers in one leading U.S. firm, Techco. Our aim is to arrive at a nuanced account of whether such engineers emerge as being either highly privileged knowledge workers or as being "caught in the middle" between management and labor with no shelter from the harshest dynamics of capitalism. At the time being studied, senior management had introduced a range of measures, the effect of which was to disrupt existing organizational timings and so to threaten both the work and the self-identity of engineers. However, we argue that such disruption does not originate solely from management action. Instead the speed of technological change itself threatens to swamp the engineers, the very architects of such progress at the same time as new notions of organizational space render them more visible to senior management gaze.

INTRODUCTION

There are at least two diametrically opposed views of hi-tech engineers. On the one hand, as knowledge workers (Alvesson, 2000; Knell, 2000) they have been viewed as highly privileged, with substantial autonomy and as able to influence greatly their terms and conditions of work. However, there is also an older literature on the position of engineers within U.S. and UK firms (e.g., Creighton & Hodson, 1997; Whalley, 1986), which views these workers as caught in the middle between the shop floor and management, and so as subject to the harshest dynamics of capitalism. This chapter will attempt to arrive at a more nuanced view by conducting a recent historical study of the position of a group of high-tech engineers ("techies") in the UK R&D centre of a leading U.S. firm. In order to achieve this, we present a longitudinal case study exploring engineers' identity work using a lens of time and organizational timings. This is justified on the basis that not only has time and timings been little researched in the management literature (Lee & Liebenau, 1999), but also because time is argued to be central to understanding identity and power.

We find that there are two major aspects of techies' identity work. The first is their desire to differentiate themselves strongly from managers whom in the words of one of the lab directors, they regard as "primeval slime." Secondly, a common desire of techies is to want to change the world in some way through what they have achieved at work. Given this, techies talk emotionally of needing to "survive" in the workplace and of how difficult this is when projects are "killed." We argue that under the changed timings, this has significant implications for intra-organizational power relations and so for techie identity.

The chapter is structured as follows. We first explore the key role of high-tech engineers as knowledge workers and the extent to which they vest issues of identity in their work. This raises issues of existential insecurity and its relationship to time and organizational timings. After discussing methodology and method, we introduce the case study, setting out how Techco is perceived to be breaking with its past in important ways. We chart how the changes have been introduced in the name of rationality in contrast to the perceived "messiness" of the past. Under the changes, engineers employ a range of tactics to limit what they see as the damage wrought by the introduction of a new planning time. In the discussion and conclusion section we argue that organizational timings have adopted an altogether new character in contrast with the remembered cosiness of the past. While this has significant implications for both intra-organizational power relations and techie self-identity, other threats also arise. These include the sheer speed of technological change, which techies fear as leading to the obsolescence of their skills as well as the advent of new organizational spacings, rendering engineers more vulnerable to senior management gaze. Such threats cause considerable anguish to techies who are united in wanting to make a contribution to the world in some way.

LITERATURE REVIEW

The Key Role of High-Tech Engineers as Knowledge Workers

Given the key role of technological change in generating profits and the perceived increase in the rate of innovation in recent years (see Cooper, 2001; Tidd, Besant, & Pavitt, 1997), knowledge workers such as the Techco hi-tech engineers are the subject of much debate within the management literature and more generally. Indeed, Creighton et al. (1997) argue that with the increasing importance of technical work, there is potential for a return to levels of pre-Industrial Revolution craft autonomy. Such a view is supported by Knell (2000) and Alvesson (2000) who argue that knowledge workers such as engineers working in R&D enjoy greater discretion, voice, and responsibility than other employees. Indeed Alvesson stresses the need to manage the unwanted exit from the firm of such employees in order to avert real damage being done to the organization. According to such views, hi-tech engineers therefore emerge as being a highly privileged group largely immune for example from the negative effects of job insecurity characteristic of other workers in the U.S. and the UK (Heery & Salmon, 2000).

However, there is another literature on the role of engineers within capitalist economies. In view of the tricky relationship between technological advance and commercial constraints, Whalley (1986) for example, argues that this needs to be actively managed. Promising projects from a technological viewpoint do not guarantee profits but engineers are frequently unwilling to subordinate their technical and scientific orientation to constraints arising from the search for profits.

According to this perspective, engineers therefore find themselves caught in the middle, with no shelter from the dynamics of capitalism, neither identifying with capital nor with labor.

In summary, the position of high-tech engineers is subject to dispute. They may be viewed as highly privileged knowledge workers or else as engineers caught in the middle and so subject to unique pressures of capitalism. What is common to the two views is that it is agreed that such hi-tech engineers are in many respects different from the majority of the workforce. Also subject to agreement according to the two perspectives is that such employees invest much of their personal identity in their work. This therefore raises the issue articulated by Knights and Willmott (1989) that insecurities can be reinforced further by individuals' attachment to particular notions of self. Viewed in this manner, hi-tech engineers rather than being perhaps exempt from the negative effects of job insecurity according to Knell (2000) and Osterman, Kochan, Locke, and Piore (2001) instead would be particularly vulnerable to another type of insecurity. This would relate to their identity-very real existential insecurity. In this connection, timing is argued to play a key role in the coming about of existential insecurity.

Existential Insecurity and Time

For Giddens (1991), discontinuity in temporal experience is often a basic feature of existential insecurity. Under such conditions, time may be perceived as a series of discrete moments, each of which severs prior experience from subsequent ones in such a way that no continuous "narrative" can be sustained. By way of contrast, self-actualization implies the control of time. Essentially, this involves the establishing of zones of personal time, which have only remote connections with external temporal orders or the routinized world of time-space governed by the clock and by universalized standards of measurement. So far as possible, the future is to be organized by exactly those active processes of temporal control and active interaction on which the integration of the self's narrative depends. Part of this temporal control includes life planning as a means of preparing a course of future actions mobilized in terms of the self's biography. According to this view, the reflexive construction of self-identity depends as much on preparing for the future as on interpreting the past. Sennett (1998) however comments on individuals' feelings of weakness because of the sheer passage of time under which past experience no longer counts in the world of the flexible organization. In the flexible organization, rapid change of skills is assumed to be the norm so that complex skills in particular, such as those possessed by Techco engineers are no longer additive, permitting individuals to build ever higher on the same foundation. Given that hi-tech engineers vest much of their personal identity in their work, the increasingly non-additive nature of their skills would appear to present very real threats to their existential security. Knowledge and identity therefore emerge as intricately intertwined under the power relations of the flexible organization.

Power Relations, Knowledge, and Time

Indeed, this intertwining of knowledge, identity, and power is present in Dubinskas' (1988) classic study of biotechnologists and other R&D specialists. Each group has a different sense of temporality or of culturally constructed understandings of what time is in the different contexts of their working lives. These understandings inform and are shaped by different ways of patterning their work activities across a wide range of contexts. Differences in the social construction of times are seen as crucial factors in doing scientific and technical work and in shaping the communities of professionals who do this work. In this schema managers follow planning time whilst biotechnologists for example recognize as valid only developmental time tied to scientific life cycles.

Biotechnologists' developmental time therefore provides an arena for conflict, the scientists seeing their being as intimately connected to their image of science as a uni-linear, infinite progressive movement of knowledge. According to this perspective, these scientists view managers as cases of arrested or frozen development in contrast to how they see themselves as being subject to continuous maturation and development. Biotechnologists' model of nature as an open-ended infinite system matches their cultural pattern of time in opposition to short-termist planning time.

In this connection, Zarubavel (1979) sees such temporal asymmetry as occurring when actors exist co-terminously but subscribe to different temporal arrangements. In this manner, Zarubavel views temporal orders as supporting patterns of social integration and differentiation and thereby influencing dynamics of cohesion and conflict. Similarly, Barley (1988) in following Zarubavel, views every work world as carrying the brand of a socio-temporal order, a structure mapped by sequences, durations, temporal locations, and rates of occurrence. In Dubinskas' terms, developmental and planning times must somehow accommodate each other. Following Dubinskas and Zarubavel, time therefore emerges as a key articulation point around which issues of power relations and individual and collective identity cluster.

More recent work has further explored the world of knowledge workers from a timing perspective. Barley and Kunda (2004) have investigated the world of hi-tech contractors in the United States, finding that such workers needed to learn to treat their time as a form of capital. In so doing, they had to regularly make calculated decisions about its allocation and use. Shih (2004) uncovered similar phenomena in Silicon Valley where a new project time was emerging leading to erratic, intensified, and speeded up rhythms. Against this background of changing timings, employees were held responsible for their own performance at work by persistently gaining new skills and so remaining marketable. How employees chose to invest their time at work therefore emerged as key. Timing has also been investigated by Perlow (1999) in her study of software engineers and the temporal rhythms of their work. She draws upon Weick's (1979) concept of enactment whereby individuals in organizations act and in so doing, create the conditions that become the very constraints and opportunities they face. Using the concept of enactment, Perlow suggests that the behaviors of software engineers, which are glorified and actually act as the source of the need for engineers to devote so much time to their work in the first place. In this way, the behaviour of software engineers becomes self-perpetuating. In summary, knowledge workers such as hi-tech engineers have been the subject of debate, with much of the literature viewing them as highly privileged whilst another strand views them as caught in the middle of firms with no shelter from the dynamics of capitalism. More nuanced work which addresses the position of knowledge workers from a timing perspective suggests a more complicated picture. The possibility of reaching such subtler understandings provides support for the contention that adopting a timing perspective is useful in exploring the position of such knowledge workers. The use of a time lens has certain implications for methodology and research method, which are discussed next.

RESEARCHING TECHCO

The research approach is qualitative and inductive with a focus on the generation of subjective experiences located within an organizational context in order to analyze processes of identity construction. Indeed, the analysis aims to be holistic and to explore phenomena through the lived experiences of those who lived it. The ultimate aims of the analysis were to explore the underlying themes and contexts that accounted for this lived experience. The stress is therefore upon producing a holistic analysis of phenomena, with resultant findings being context-bound. While this would be viewed as a limitation within more positivist approaches, the production of such context-specific findings resonates with Collinson's call (1994) for detailed examination of the conditions, processes, and consequences of workplace resistance. Collinson argues that few studies have attended to the way that resistance is a condition and consequence of particular "knowledges" and subjectivities.

This historically focused chapter draws on two longitudinal research projects exploring individuals' perceptions of continuous change. It is multi-method building on earlier research in Techco carried out by a colleague; press searches, interviews, focus groups and surveys conducted over the period from the late 1990s to early 2005. Background interviews were conducted with the overall company HR manager, the lab strategic planning executive, and members of the lab IT support function. There was therefore substantial triangulation of these various different sources of evidence against the main interview data. While this material provided background supplying evidence of perceived fundamental changes within Techco over the period, the main data analyzed for this paper consists of 17 transcribed semi-structured interviews. These were conducted in early 2001 and 2002 with technical staff within the lab and explored perceptions of continuous change. Follow-up was by means of a further seven interviews in 2005 with key interviewees and with the lab strategic planning manager and the previous lab HR manager. Moreover, initial case studies based on these two sets of interviews and background material were fed back to the HR and strategic planning managers within the lab for comment, again providing triangulation of the findings.

Like Thomas and Linstead (2002), this chapter presents a theoretical framework that attempts to accommodate voice and the individual's position as both subject and object. Identities are viewed as not only embedded in the present, but also as constructed in terms of the conjunction of past and future, crucially given the timing focus of this chapter. Following Gabriel (2005), employees are seen as creating niches that are unmanaged and unmanageable. So, forms of resistance are seen to have changed to more subtle and nuanced acts of disobedience to create spaces sheltered from continuous exposure.

The results of this analysis are therefore context-specific and represent our construction of individual actors' constructions of their lived experiences. Whilst we do not argue that our analysis is generalizable, its aim is to generate a rich portrayal of a key dimension of life within the lab for R&D staff. This relates to the importance of time and timings which emerged as a much more emotive underlying theme than the many individual changes introduced by management over the period of the study within the Techco R&D centre in Swindon which we introduce next.

CASE STUDY

Techco: A Break with the Past

Techco, a leading multi billion U.S. dollar high-tech engineering firm, was founded in 1962 by Fred Smith, who stayed as head of the company until into the 1990s. Traditionally, Techco had been celebrated for its unique culture characterized by a high measure of respect for the individual employee. However, from the 1990s onwards, Techco had been subject to many pressures. These included the sheer speed of technological change and the dot.com boom and bust which led to both significant investor unrest and also to major industry consolidation in order to exploit economies of scale. With these pressures, Techco culture came to be regarded as too bureaucratic and slow-moving so that in time a new CEO was appointed to introduce the radical changes seen as necessary.

Under the leadership of the new chief executive, Techco management style had become tougher with shock waves reverberating throughout the whole organization when two board members were sacked on failing to "make the numbers." These sackings were regarded as highly symbolic of Techco changing from its previously almost cosy management style to a much tougher regime under which a conscious decision was made by very senior management to expose the corporation more to the discipline of external market pressures. Further changes followed swiftly:

There has been lots of gung-ho talk from the new CEO about troops and how important we all are but then the next thing there is a voluntary redundancy program, which was compulsory in the U.S. where 6,000 people were made redundant. (Engineer)

These redundancies were perceived as all the more shocking because now in contrast to the more recent past, it had become more difficult to find a job outside Techco. Previously, Techco engineers would have other employers lining up to recruit them.

The new CEO signaled that further changes were imminent:

So when the new CEO came on board that was the fundamental change he made. He realized that the company had a lot of things that were simply not working and we lacked a strategy, we lacked a clear identity as a company. Within days he had changed this and he started to put some flesh on those bones. (Manager)

The new CEO viewed Techco, with its previous almost cosy management style, as lacking a clear identity. On going through Techco's financial results, he adopted an externally focused perspective by looking also at the figures of the company's competitors:

The new CEO on going through Techco's results which were good-he said "Let me go through the numbers of some of our competitors." So it is an example of moving us away from having internally oriented measures of whether we are doing well enough. The reason I mention this is that by starting to benchmark ourselves externally, a lot of things that are at the root of some of the changes we have seen in the company are based on that. (Manager)

The focus of this case study is on the Swindon R&D site which employs around 180 technical staff including management engaged in highly technical corporate R&D as opposed to more applied new product development. There are three labs at Swindon, which report now directly to the U.S. with no overall UK head of R&D. This is a change from previously when the three lab directors reported directly to a UK Head of R&D who acted as something of a buffer between the UK and the U.S. Indeed, the whole of Techco corporate R&D which is spread across five different sites worldwide had been insulated by company senior management from the vagaries of the market so that engineers could concentrate on the real business of innovation. The particular focus of this case study is upon the effect on engineers' identity of management changes, emanating primarily from the centre in the U.S. but also locally in the case of portfolio management.

Measures Adopted By Techco: Centralization of Very Senior Management in the U.S.

The centre-led reforms tended in the direction of increased centralization, with very senior management coming to be concentrated in the centre in the U.S. Interviewees commented on how reorganizations of any type now very much had the flavour of top management coming over from the U.S. and telling Swindon staff what had been decided. With the loss of the UK head of R&D, the Swindon site had lost any sense of cohesion amongst its different groups. To quote one manager:

There was a big reorganization of the structure and the management of how the research labs in Swindon worked...we used to be a relatively independent centre. There was someone in charge overall of the three Techco labs in Swindon and they had separate lab managers reporting to them and a hierarchy beneath that. Then it was decided they were going to divide it up much more along business lines and the people who were going to be in charge were going to be in the U.S. ....the entire lab that I was part of disappeared overnight. Many of the projects ceased to exist completely...that was typical of the way things ended up, there were lots of different bits in Swindon reporting to different bits in the U.S. ....so this was all a bit concerning because it was all split into little bits with managers in the U.S. having a little bit in Swindon and most of their work in the U.S.. It struck me that it was an ideal situation for these U.S. managers to say that when push came to shove and money is tight, then this little bit over here in Swindon will go...so what has been lost in Swindon is any sense of cohesion between the groups. (Manager)

This centralization led to a lack of cohesion in Swindon, one engineer commenting:

Once your reporting lines kind of go across the water...then maybe it just makes it harder to feel as though you're a part of a group where you can just go and say "Hey, what do I do? I'm new to this area and so on." (Engineer)

As part of these changes emphasizing a more centralized U.S. managerialist approach, patenting was given a heightened priority within the organization.

Moves to an Increased Stress on Patenting

This new stress on patenting reflected stock market pressures as financial analysts had come to regard a high-tech firm's number of patents as an important indicator of financial health. Indeed, league tables of firms by number of patents were common, with comments being made that Techco was falling behind such firms as GlaxoSmithKline. Additionally, it had also become common practice for competitor firms of Techco to sell patents to each other with the aim both of generating income and of allowing others to take on the risks of detailed new product development. So under the new regime within Techco where patent writing was encouraged, this in effect represented a commodification of techie time whereas in the very recent past, such activity had been regarded as a waste of time and in fact an obstacle to the real business of innovation. However, in those days Techco policy had been to insulate the labs from the outside environment including those pressures emanating from the stock market. This encouragement of patenting was attributed to the new CEO and other new very senior managers:

When Bob became CEO there has been a lot more willingness to say, 'Well you know, if we're about inventing things that means patenting things.... let's actually say we've got this large pot of money, rather than just sitting on it, let's, say, spend it on putting patents through the process'.... We've got some other people coming in from outside Techco into management positions over the last 18 months. A lot of those messages have been, we need to patent more, we need to have this explicit and obvious measure of inventing things, which also has other means of exploiting it. And people have been looking outside and saying, what are other people doing? One of the things we do is look at IBM who have been top of the U.S. patent list for at least the last 6 years, they're number one...I think they are twice the number of the next highest...they have an aggressive licensing campaign. (Middle manager)

Perhaps predictably, a sizeable group of engineers resisted these changes. To quote one middle manager:

People don't like patenting at all...I think the problem is that there's a very vociferous group who are very hostile to the idea of social property.... So there is a problem with people who are having to work with that community, if your name pops up as inventor on something they think will be patented. Even if you're saying something sensible, someone will point out that you've actually got a patent with your name on it so a lot of people will say therefore you are an un-person. (Middle manager)

So, there is a vociferous group active in voicing opposition to the encouragement of patenting and Techco's new stress on the quantity of patents issued. This opposition stems partly from resistance to heightened transparency of work undertaken by techies, one middle manager commenting that previously, with less stress on patenting.

...so before (i.e., before the current stress on patenting) there was very little visibility of what was going on. (Middle manager)

Another major change included the introduction of forced ranking.

Forced Ranking

Techco had used ranking in performance appraisal for a long time. However, two things were new. Firstly, ranking was changed to forced ranking based on the normal distribution within statistics so that a given number of individuals had to be classified unsatisfactory. If this happened for two appraisals in a row, the individual concerned might be fired. Forced ranking was therefore introduced as a means of introducing market pressures inside Techco, providing a mechanism for reducing the workforce and so reflected management's new tough-mindedness. Secondly, the desired behaviors of promoting the Techco brand, teamwork, speed of execution, and risk-taking were defined for the first time whereas in the past, these had not been documented. Importantly, in the new environment of rapid technological change, speed of execution was prioritized and risk-taking defined whereas planning horizons for technical work had been longer in the recent past. Following McGovern (1998), this represents a formalization of practice that had previously been in the informal domain, with the apparent aim of this formalization being to encourage entrepreneurial behaviour by techies.

Most of the interviewees were strongly against forced ranking. To quote one engineer:

We are also in competition because ranking means that once you get to a certain level there is very little co-operation between engineers.... Why would you discuss an idea with someone who might steal it? (Engineer)

Forced ranking was therefore seen as discouraging collegial ways of working and in any case, was totally inappropriate given the distinctiveness of the individuals working in R&D at Techco. The same engineer comments:

Forced ranking is an utter nonsense, a nightmare in this environment. If it were a call centre it might work because they are all doing the same job but here everyone is as different as chalk and cheese... You can do tremendously well by sheer luck and be ranked highly as a result. Alternatively you can bust a gut doing your best work and be ranked low because of external circumstances. Most people eventually completely ignore ranking and do what they think is right, or they leave. It's a management tool for encouraging competition between employees.... It discourages certain personality types and gets everyone's back up. (Engineer)

So, in line with the nostalgically remembered old culture under which individualism was valued, the engineer quoted above is adamant as to the extreme individuality of the Techco R&D engineers making comparison virtually impossible unlike in more homogenous workplaces. Moreover, given the indeterminacy of the work and so its unpredictability of outcomes, forced ranking is viewed as extremely unfair and as provoking instead a destructive type of competition among employees. Most techies are motivated by a higher calling, of doing what they judge to be right and so they eventually decide to ignore ranking altogether.

Under the changes, market pressures from the external environment were allowed to permeate down to the level of individual engineers.

Rationality of the Present vs. the "Messiness" of the Past: Portfolio Management

These moves by the centre were echoed at the local operational level with one of the three individual labs at the Swindon site introducing a system of portfolio management. Under this policy, physical assets such as the Techco laboratory, were repositioned as financial resources and so as amenable to the workings of options theory within financial economics. Portfolio management involves a 'modern' view of risk, as something to be managed at given times at so-called stage gates when progress is reviewed. This represented a substantial change from the old regime under which essentially highly creative engineers had enjoyed working conditions of responsible autonomy (Friedman, 1977). The introduction of portfolio management by local executives therefore represented an attempt to micro manage the time of engineers. Managers were no longer to be supportive technically oriented department heads but instead rational businessmen using latest management theory to manage a portfolio of projects.

The introduction of portfolio management was seen as necessary by senior management because they viewed the previous organizational structure as "irrational." Under the old departmental structure, collaboration between departments was difficult. To quote the lab director interviewed:

How do you break down the silos (referring to the strong divisions between the different departments before the introduction of portfolio management)? How do you get people to respect each other's work, how do you get them to collaborate and co-operate and share and synchronize with each other rather than just saying "Well, my stuff is all right but theirs is crap?" (The lab director)

So, the move to portfolio management was undertaken in order to discourage the essentially tribal behaviour current under the old departmental system. Instead of departmental heads ruling the roost and jealously hanging on to their staff, portfolio management was to introduce a regime of rationale utilization of resources, primarily staff. Importantly, technical staff were to be tamed and so rendered into politer beings who would be happy to accept common timings as their work is synchronized with that of others. The lab director continues:

So what we wanted to do was to adopt a portfolio management approach where instead of there being winners and losers, everybody was involved in making decisions about what the best use of our resources were. So you know, it was not a case of he gains two heads, he loses two heads, it was overall, "What is the best thing to do with our resources, and what roles do we need to play?" (The lab director)

The lab director uses the new managerial language in stressing the supposed coherence of practice under portfolio management in contrast to the fragmentation and politicking of the past.

However, there were conflicting views of the past. To quote the lab director:

So the old structure of the lab which was a very hierarchical structure so you have a lab director, departments with departmental managers, projects with project managers and the rest and from my point of view this really encouraged the silo organization...also there was very little quality control over what went on and there were no good processes for you know, managing resources, making trade-offs and priorities. It was mainly a political fight. (The lab director)

The old system was therefore seen as subject to the ills of bureaucracy, being hierarchical with many layers of staff, thereby encouraging the old silo mentality. The lab director again uses the language of the new managerialism in talking about processes, managing resources and making trade-offs and setting priorities. During the bad old days of hierarchical management, quality control was poor.

These views are in line with those of the portfolio manager interviewed, who has never been a departmental manager under the old system. A recent recruit, a project manager, who has also had little experience of the old ways of working comments:

As I understand it, the labs tended to suffer from compartmentalization at the department, even at the project level. So you would get teams of people that would be almost in complete isolation, going off and doing work in a particular area and it didn't seem to have any relevance to what other people were doing. It was a very insular activity. (Recently recruited project manager)

The new management in this lab highlights the compartmentalism and insularity of the old system in contrast to the new rational ways of working. However, what he omits to mention is the advantage in terms of autonomy enjoyed from a techie perspective.

One middle manager praises the new coherence and rationality of strategy under portfolio management:

Where previously we had departments within Techco which were quite political so you were fighting over budgets, headcount etc. and so you wouldn't always be that productive... in the department structure departments were allowed to be a group of individual projects with not necessarily any connection between them whereas portfolios with, headcount is more structured so there appears to be relationships because the portfolios are layered and there appear to be relationships between the different layers suggesting that you have to organize your progress within a portfolio....we have a strategy for the lab and then the portfolio fits around that strategy. (Middle manager)

Another project manager who has known the previous regime, voices a more nuanced view of the past:

One of the changes is in the management style. Before, Techco was always made up of a number of fiefdoms...The culture was very much grass roots, the power was in what you knew. The style of leadership was not top down but supportive of the engineering staff. (Project manager)

So while this project manager acknowledges that the past was not perfect as in his use of the word "fiefdoms" signalling a pre-modern way of working, the culture was perceived as being very different from the present. While there may have been hierarchy, there was also power residing in the knowledge of technical staff who would be supported by their managers even while protecting their own interests, and so acting irrationally from a managerialist viewpoint

This support of technical staff was seen as emanating from the very highest levels, going back to the philosophy of the founder of Techco. One engineer reflects on Techco's supposed golden age:

People do hark back to Fred Smith, the golden age of Techco.

Interviewer: What do they(people) say about that?

The founder is very fondly remembered. There have been editions of the Techco newsletter dedicated to his wit and wisdom.

Interviewer: Doesn't it all feel a bit corny?

No, he did seem like a good guy. He had a very different attitude to the workforce than the current management. He really valued employees.

Interviewer: How do you feel he did that?

The classic one is that he always preferred to get people to take pay cuts rather than make redundancies-whereas now we are doing both. (There were various other home-spun stories)....Fred Smith used to walk around the U.S. lab and fix the central heating! (Engineer)

According to this engineer, the founder embodied all that was good about the old culture in his support of the workforce and his directly caring nature in busying himself with the central heating, thereby showing himself to be an engineer at heart. So while the old Techco was characterized as being almost medieval in operating as a series of fiefdoms with insular managers following only their own interests, from the viewpoint of some engineers they enjoyed protection and high status within a supportive culture. Now however there has been disruption of the old culture, thereby breaking the previously much valued link with the firm's glorious past. Given techies' feelings about their work, this was important however irrational the past might appear in retrospect.

Life/Death/Survival at Work

When engineers talk about their work, they do so often in terms of life, death, pain, and survival. This is understandable in terms of the fragility of their work projects:

The nature of research is that you are doing it because you really get into it and so once you have bought in emotionally to it, then it is a very painful period when it is killed. You feel that you have wasted that part of your life, all that energy. As professionals, as researchers, that matters a lot. (Project manager)

Time is important because when a project is killed, you have wasted that part of your life and the energy expended. Importantly, it takes time to become an expert, as long as four to six years but technology is changing at a faster rate. To quote another engineer:

I had become an expert I think in an area to the point where I was able to go and look at new areas of contribution and immediately contribute to them. Also, I think what I realized was that it takes 4, 5, 6 years working in an area before you become so expert that you really have the confidence that you can contribute and then when you get a change in a field, coupled with a rapidly changing environment that you're working in, both internally and externally, that doesn't happen overnight. (Engineer)

Speed is stressed in that the techie is able to look at new areas and immediately contribute to them. But such knowledge is again viewed as fragile and subject to onslaught by both changes within the field as well as by changes by management, disrupting the internal environment.

Another engineer agrees that it takes a significant period of time to develop expertise and contribute. However, planning time asserts itself in that managers keep changing things. In the face of planning time, engineers have to remain true to their own technical goals:

You go crazy. In this environment, it is going to take you three years to do anything of value and the managers keep changing things on a six-monthly or yearly basis. If you allow yourself to be deflected by those changes, you will never finish anything or achieve anything. (Engineer)

Temporal asymmetry comes about as techies need years to create work of value, to achieve closure, whilst managers persist in disrupting this through their changes under planning time.

Another engineer speaks of the destruction wrought instantaneously on occasion by planning time:

The entire lab that I was part of, disappeared overnight. **Many** *of the projects ceased to exist completely.* (Engineer)

Faced with the reality of projects being subject to axing with little warning, one project manager describes typical reactions to this by engineers:

Interviewer: What do you and other people around you tend to do, typically, if a project gets axed?
Some people go into a state of semi-depression, some rapidly move onto the next new thing. Most people find that it takes quite a while to get to a position of doing something that will make a meaningful contribution again... But if you are going to do that(Pushing back the frontiers of knowledge), you cannot switch from A to Z and work in a completely different area and suddenly be pushing back those frontiers again... You never get to actually pushing back the frontiers....so it is all about change and in this environment it can be difficult for people to really feel that they have made that contribution and sense of fulfillment from having delivered something which has made a difference to a product or to the world. It's hard to do. (Project manager)

The project manager agrees with the engineers quoted earlier that once a project is axed, it usually takes engineers a considerable time to make a meaningful contribution on another project. To quote one engineer, on a project being axed:

You're throwing everything away-it's the equivalent of being made redundant. (Engineer)

In the face of such threats at work, one engineer worries that he cannot keep up to date with the internet so that he is at risk from ending up on the scrap heap 10 years after graduation. So engineers face not only what they perceive to be the threat of planning time but also obsolescence of their highly valued skills seen as having a potential sell by date. In the words of one engineer:

Once I've read and digested this then I know the stuff, well it's like by the time you've read that book, it's probably out of date. (Engineer)

One tactic adopted faced by such threats is for engineers to seek to differentiate themselves strongly from management. In the words of one engineer:

Everyone talks quite openly at lunchtime about the silly things that managers do. A techie would not propose something unless it had a real impact and a real benefit. I look at these things that come through from management sources with changes and I wonder what the hell it is all about. It doesn't really achieve anything. You get a manager come along and stand up in a big room an make a presentation, you ask questions and he doesn't have the answers. Either the strategies are half baked or he is not prepared to tell you. (Engineer)

Techco engineers are therefore very strident when faced by managers who lack real answers or else who are not straight-talking like the techies they manage. Indeed, even visitors are not spared if techies feel that they are putting forward a weak argument:

All Techco people are argumentative, they are not afraid of criticizing a view. We get used to that but we often find that visitors find it rather off-putting. They are used to coming in and giving a presentation and everybody being meek. Whereas here they are challenged, even senior people will be challenged by oiks! ... Techco people are extremely argumentative, if there is a flaw in your argument or a corner case, they will find it for you and some would take great delight in pointing it out to you-whether or not it is relevant to your argument. (Engineer)

Given techies' argumentative nature, which they celebrate, and their lack of respect for managers, most engineers would not countenance the very idea of switching to a managerial career within the firm. In the words of one techie when posed the question as to why anybody would choose to become a manager:

Interviewer: So why would someone choose to stop being a techie and become a manager? Because they have run out of ideas. They feel they can not do it(make technical contributions) anymore. They are drying up-or because they are fed up with it(Trying to make a technical contribution) for some reason. Generally, it(why people become managers) seems to be because people want to play the power games and politics.... Really as far as change is concerned the biggest problem is that managers have to be seen to be doing something, generally they cannot find anything useful to do, so they just do something.....so you learn that most of the changes don't really matter at all. (Engineer)

Managers therefore lack meaningful knowledge and may be "dried up" in contrast to the knowledge, ephemeral though it often is, possessed by techies. Instead, managers have to be seen to be doing something, being concerned with appearances. In the view of this engineer most management changes ultimately do not matter at all. Another engineer emphasizes his distance from management when he once went to a management meeting:

I went to a couple of these (i.e., management meetings) but felt that it was very difficult for me to contribute to setting out what is basically a strategy for a research laboratory. Because we are very down in the decision-making process so all we can say is what our technical interests are ... But in the end setting up the strategy and agenda for the lab is a job for management. (Engineer)

Whilst many engineers are quite emphatic as to the nature of their distance from managers whom in the words of the lab director, they regard as "primeval slime," techies engage in certain tactics in order to protect their interests.

Tactics Employed by Engineers in the Face of Management Action

One engineer comments that on occasion he will describe his project differently in order to fit current circumstances and play the game:

You might have to be politically correct and change the name or emphasis of something slightly ... people soon pick up how to play the game. Most of these management fads are just that as far as the technical community is concerned. (Engineer)

He justifies such action on the basis that most management action constitutes no more than fads in contrast to his own work which addresses critical weaknesses and gaps in technology. Other reactions in the face of management intrusion include just ignoring such irrelevant change:

People were very cynical. The way to survive is to just ignore it and get on with what you are doing. (Engineer)

This engineer again uses the language of life and death in talking of how to survive. The necessity for such a strategy is echoed by a fellow engineer who justifies it on the basis that in any case, management strategy will have changed again in another six months time.

My strategy has generally been to keep my head low down on these things because you know that in six months time it will have changed again. (Engineer)

Despite the new management language of managerialism and rationality, this engineer sees planning time as essentially capricious in nature.

A third engineer in his "survival strategy" comments that he may go as far paying lip service to management proposals but otherwise, does not allow himself to be deflected from his chosen path:

You might pay lip service to what they propose but other than that you ignore as much as you can. That is my survival strategy. (Engineer)

All these engineers are "buying time" in broadly similar ways so as to stave off the situation of their projects being axed, and so ensure their personal survival.

Finally, in order to survive at work, engineers talk about the importance of the project manager role in acting as the interface between themselves and senior management:

Interviewer: You have talked about the project manager as spin doctor...
It's a thankless role. The level above is trying to get you to produce slides for dissemination to their public and meanwhile you have your employees trying to give you the straight message and you have to somehow take those conflicting demands and produce something that looks plausible to a higher level. You can spend your whole life doing PowerPoint. Sometimes there are things presented that depend whether a project lives or dies.... Sometimes a whole lab has just been axed and everyone re-assigned. Sometimes whole departments have been completely re-jigged... Other people moan about it all the time. (Engineer)

Techies are again presented as giving "the straight message." However, they are prepared to engage in cooperating with the project manager in order to change the presentation of their project. This is necessary because, using the language of life and death again, presentation can dictate whether a project lives or dies.

One manager summarizes the current situation as he sees it:

People join places like Techco because they are bright and want to make a difference. .....So there is no real authority in that the management of change is not command and control... I think there are always the people that probably would get characterized by the techies as being bean counters that are constantly trying to systematize research and pin it down. I don't think that any of those things have ever succeeded. So, the support structures that try and institute those changes are clearly being thwarted by technical staff refusing to play ball. (Manager)

In his view, a standoff persists between managers trying to systematize research and techies who resist this.

Finally, in summarizing the effect of the changes, one engineer comments as regards the new organizational culture:

(Talking about a session with culture consultants) ... They (the culture consultants) were looking for a kind of sense of what we thought the Techco culture was and the thing that came out was that actually we don't know anymore. We know what it used to be.... it used to be this high degree of trust based in individuals, kind of everybody was on the same side, always willing to help and a stronger sense of community.... and people kept saying, "Oh yes, this is what Techco's like," but often it was "Well, it used to be (like this) but I'm not sure anymore.. " (Engineer)

SUMMARY

In summary, that techies recall with nostalgia Techco in "the old days" when management's planning timings and those associated with the project, developmental time meshed. This has changed as Techco's new senior management has quite consciously moved from a policy of insulating the lab from outside market pressures to a position where these are transmitted inside the organization through a variety of measures. These include trends towards the centralization and concentration of senior management power in the U.S. compared with the recent past when there had been a greater degree of local autonomy, the forced ranking of staff appraisal, the introduction of portfolio management within one UK lab and the incentivization of writing patents whereas this had previously been regarded as time-consuming and even as a barrier to the real business of innovation.

However, not only do management's planning and techies' developmental timings collide, but other issues connected with time arise. Techies are keen to make a difference over the lifetime of their careers but they are confronted constantly by internet time, which threatens to swamp them although they themselves and their colleagues are the very architects of the technological change creating internet time. These threats mean that they live in constant fear either of being unoccupied when not employed on a project and as a result, ending up on the scrapheap. In order to avert these outcomes, techies engage in a variety of strategies as they resist but also at the same time, accommodate management moves with the aim of prolonging the life of their project and so continuing to feel truly alive in their aim to change the world in some way through their work.

DISCUSSION AND CONCLUSION

From the case study, organizational timings are seen to have adopted an altogether new character. As viewed nostalgically by techies, previous planning horizons had been longer term and were moreover sustained by the sense of history which was important in the organizational culture. Following Zarubavel (1996), the past is part of a social reality shared by the Techco engineers which although far from being absolutely objective, transcends individual subjectivity and is commonly shared. However currently engineers feel that all that matters is the current moment which is unpredictable, severing the past from the present, and subject to managerial fad. Narratives of the past have therefore been disrupted as the old regime rather than being celebrated is instead decried as irrational by the CEO and very senior management. A new type of temporal asymmetry therefore arises in the starkly contrasting views of the past, viewed as glorious by techies. However, the new CEO sees Techco as previously lacking a clear sense of identity and subject to a situation where a lot of things did not "work." Moreover, in the supposedly idyllic past, timings had been more individual, subject to departmental manager discretion in contrast to the present when there are attempts under portfolio management to synchronize the work of engineers. Not only however are there attempts to synchronize the work of individual engineers but through forced ranking, speed of execution is prioritized and moreover, defined for the first time as is risk-taking whereas engineers experience day to day their own version of risk. This is the risk that their projects might be axed and so they will have to start from scratch again in probably an altogether new field where their existing knowledge counts for nothing. The different times, management's planning time and techies' development time therefore represent two very different views of reality as the two are seen to no longer mesh. Temporal asymmetry has come about which shapes the dynamics of conflict between techies and management as engineers contest what they see as the new organizational regulation of time.

Planning time has now become aligned to the external market and moreover, attempts have been made to ensure corporate-wide uniformity in organizational timings. From being protected in the past within individual departments, organizational timing is perceived as being stop-start and subject to reversal as axed projects are on occasion resurrected. However, despite the new language of managerialism with its talk of rationality and individuals as "resources," techies claim to see through the changes in describing them as management fads. Under the changes, following Ancona, Okhuysen and Perlow (2001), time is viewed increasingly as a resource that can be measured, standardized, used, bought and sold under the new regime of economicity of time. This economicity of time is particularly evident under forced ranking with its prioritization of speed of execution and with the new stress on patenting. Under the latter, time invested in a particular project has to be recouped through its commodification in the form of legally enforceable patents.

In the face of such threats including the decrying of the old culture as bureaucratic and inefficient, engineers experience discontinuity in their experience of time. Following Giddens (1991), they experience problems in their control of time in terms of life planning and their organization of the self's biography. As a result, they experience threats to their identity as they struggle to sustain a narrative of themselves as technical experts making a real difference to the world in some way. They engage in differentiating themselves strongly from management whom they characterize as politicking and as taking action simply in order to be seen to be doing something, rather than undertaking any actual meaningful change. However, whilst techies are keen to distance themselves from management, their apparently straightforward resistance is heavily tinged with compliance as they are not seeking radical change within Techco. Instead, their actions are largely directed at ensuring the survival of their projects so that as techies they can make a difference. As commented upon by Fleming and Spicer (2003), despite techies' apparent hostility to managers, as engineers they are doing their very best to make a contribution and indeed they see themselves as the true organizational heroes. This heroism arises from the risks they undertake in how they choose to invest their time in a particular project. Power, knowledge and identity emerge as inextricably intertwined as techies engage in heroics in seeking to protect and extend the life of their fragile projects, subject at any moment to axeing by an apparently capricious management. Like Barley et al.'s (2004) contractors, Techco engineers treat time as a form of capital and regularly make calculated decisions about its allocation and use. They sink time into activities that maintain, enhance and build their skills. Indeed, without investing time in this manner, they might wake up one day to find their time had no value at all, a persistent worry facing Barley et al.'s contractors and Techco engineers. Both groups of worker have to make choices about what to learn, making risky choices and so exhibiting their hypersensitivity to the very real opportunity costs of time. However, such knowledge is ephemeral as techies have to be able to invest enough time in honing their expertise under constant threat from managers' planning time. Such temporal threats do not arise from planning time alone. Instead, technology has its own momentum, internet time, which engineers struggle to master. Following Shih under such conditions of rapid technological change, project time becomes of paramount importance as erratic, intensified, and speeded up.

Following Perlow (1999) and Weick (1979), enactment arises as the process by which individuals in organizations act, and in so doing, create the conditions that in turn become the constraints and opportunities they face. Viewed in this light, temporal rhythms arise from the way people interact and the way people interact "enacts" or further generates temporal rhythms that dictate these work patterns. So, in Techco increased control by managers results in resistance by engineers which perhaps then intensifies attempts at control by management. However, a certain accommodation is present as in many ways managers accept techies' role as the perpetual outsider. They appear to do so in accepting the "rudeness" of engineers which techies themselves celebrate. Nonetheless, there is evidence that very senior management is encouraging changes whereby Swindon managers become less supportive of techies than they had been in the past as they try to control and manage the process of innovation. This arises because the environment is perceived as having speeded up so that they in turn are being managed more closely in contrast to the more recent past. However, despite these changes, techies still need perhaps four to six years to develop an area of expertise so that they resist attempts at closer management of their work. In turn, management reacts by attempting still more detailed management so that the process becomes self-perpetuating.

However, it is not timing alone that poses a problem to engineers as organizational spacings also constitute a threat. Under the new regime in Techco, there is less tolerance for the fragmentation of the past as instead, projects are to be managed across the whole portfolio in one lab and moreover, are more immediately subject to senior management gaze in the U.S. Previously, there had been a UK wide head of R&D and departmental managers acting as buffers between the Swindon engineers and the corporate centre. Because of the existence of these roles, despite the apparent fragmentation of the Swindon site characterized by separate fiefdoms, there had been a sense of cohesion locally. Moreover, while management had been regarded as irrational under the system of departmental as opposed to portfolio managers, projects were sheltered behind the safety of silos and so technical work nurtured. It is therefore not timings alone that have altered but also organizational spacings. Indeed, the reconfiguration of organizational spacing has exacerbated the perceived problem of the new temporal asymmetry under the changes.

In summary, under the new timings, it becomes more difficult for engineers to develop their expertise in the longer term and so to develop their careers according to their own definitions. Tolerance for individual project timings under the old regime allied to what was perceived as a supportive grass roots culture had been altogether more nurturing of engineers' careers and more tolerant of their individuality. So, despite their own continuing lack of respect for authority, they prize the recognition they felt was accorded to them as highly individual researchers under the old regime.

While the new organizational arrangements have been put in place in the name of rationality in contrast to the muddle and perceived bureaucracy of the past, the engineers are vociferous in viewing the changes as detrimental to themselves in their pursuit of making a real difference to the world in some way. Particularly given the new harshness of the external job market in contrast to the recent past, techies emerge largely as engineers caught in the middle with no shelter from the dynamics of capitalism rather than as the privileged knowledge workers portrayed by writers such as Knell and Alvesson.

REFERENCES

Ancona, D. G., Okhuysen, G. A., & Perlow, L. A. (2001). Taking time to integrate temporal research. *The Academy of Management Review, 26*(4), 512-529.

Alvesson, M. (2000). Social identity in knowledge-intensive companies. *Journal of Management Studies, 37*(8), 1101-1123.

Barley, S. (1988). On technology, time, and social order: technically induced change in the temporal organization of radiological work. In F. A. Dubinskas, (Ed.). *Making time: Ethnographies of high-technology organizations.* Philadelphia: Temple University Press.

Barley, S., & Kunda, G. (2004). *Gurus, hired guns, and warm bodies: Itinerant experts in a knowledge economy.* Princeton, NJ and Oxford, UK: Princeton University Press.

Collinson, D. L. (1994). Strategies of resistance, power, knowledge, and subjectivity in the workplace. In J. M. Jermier, D. Knights, & W. N. Nord (Eds.). *Resistance and power in organizations.* London, UK: Routledge.

Cooper, R. G. (2001). *Winning at new products: Accelerating the process from idea to launch* (3rd ed.). NY: Perseus Publishing.

Creighton, S., & Hodson, R. (1997).Whose side are they on? Technical workers and management ideology. In S. R. Barley & J. E. Orr (Eds.), *Between craft and science: Technical work in U.S. settings.* Ithaca, NY: ILR Press, Cornell University.

Dubinskas, F. A. (1988). *Making time: Ethnographies of high-technology organizations.* Philadelphia: Temple University Press.

Fleming, P., & Spicer, A. (2003). Working at a cynical distance: Implications for power, subjectivity, and resistance. *Organization, 10*(1), 157-179.

Friedman, A. L. (1977). *Industry and labour: Class struggle at work and monopoly capitalism.* London, UK: Macmillan.

Gabriel, Y. (2005). Glass cages and glass palaces: Images of organization in image-conscious times. *Organization, 12*(1), 9-27.

Giddens, A. (1991). *Modernity and self-identity: Self and society in the late modern age.* Stanford, CA: Polity Press.

Heery, E., & Salmon, J. (2000). *The insecure workforce.* London, UK: Routledge.

Knell, J. (2000). *Most wanted: The quiet birth of the free worker.* London, UK: The Industrial Society.

Knights, D., & Willmott, H. (1989). Power and subjectivity at work. *Sociology, 23*(4), 535-558.

Lee, H., & Liebenau, J. (1999).Time in organizational studies: Towards a new research direction. *Organization Studies, 20*(6), 1035-1058.

McGovern, P. (1998). *HRM, technical workers and the multinational corporation.* London, UK: Routledge.

Osterman, P., Kochan, T. A., Locke, R., & Piore, M. J. (2001).*Working in America.* Cambridge, MA: MIT Press.

Perlow, L. (1999).The time famine: Toward a sociology of work time. *Administrative Science Quarterly, 44*(1), 57-81.

Sennett, R. (1998). *The corrosion of character: The personal consequences of work in the new capitalism*. New York, & London, UK: W.W. Norton & Company.

Shih, J. (2004). Project time in Silicon Valley. *Qualitative Sociology, 27*(2), 223-245,

Thomas, R., & Linstead, A. (2002). Losing the plot? Middle managers and identity. *Organization, 9*(1), 71-93.

Tidd, J., Bessant, J., & Pavitt, K. (1997). *Managing innovation: Integrating technological, market and organizational change*. Chichester, NY: Wiley.

Weick, K. (1979). *The social psychology of organizing* (2nd ed.). Reading, MA: Addison-Wesley.

Whalley, P. (1986). Markets, managers, and technical autonomy. *Theory and Society, 15*(1-2), 223-247.

Zarubavel, E. (1996). Social memories: Steps to a sociology of the past. *Qualitative Sociology, 19*(3), 283-299.

Zarubavel, E. (1979). *Patterns of time in hospital life*. Chicago: Chicago University Press.

Chapter IX
Trustworthiness as an Impression

Dominika Latusek
Kozminski Business School, Poland

ABSTRACT

This chapter discusses current developments in the theory of trust and distrust drawing upon a field study between solution providers and their clients in the IT industry in Poland. It addresses the question of how the suppliers (trust-takers) establish the image of being trustworthy, and how their trustworthiness is in turn examined by their potential customers, when cooperation takes place in the context of the openly expressed lack of trust (or distrust). Trust in the chapter is seen as a tranquilizer that suspends the feeling of vulnerability and enables action regardless of uncertainty involved in the situation. Cooperation is understood in dramaturgic terms--the supplier is to produce an impression that will make them appear trustworthy as perceived by the customer.

INTRODUCTION

The idea of the research project that forms the basis for this chapter originated from two reflections. On the one hand, the researchers usually agree on the crucial role of trust in business contracting practice: it is seen as "vital for the maintenance of cooperation in society and necessary as grounds for even the most routine, everyday interactions" (Zucker, 1986). It seems to me that it is particularly relevant for the high-tech business, where—in terms of knowledge—a substantial asymmetry between the supplier and the client may exist; this asymmetry is accompanied by a relatively high disclosure of key information about the organization on the part of the client. This, along with the oftentimes-particular character of the service (e.g., software design) that eludes precise evaluation, forms an environment conducive to the emergence of trust. The significance of trust for enabling mutually advantageous relations of exchange and cooperation between organizations has been largely acknowledged (e.g., Sako, 1992, 1998). Especially in the case of buyer-seller relations, trust plays an important role because even if both parties are interested in continuing the relationship, there is still vulnerability and the potential hazard of opportunism (Möllering, 2006).

Regardless of various virtues attributed to trust, we still have only limited understanding of its functioning in market processes (cf. Lewicki & Bunker, 1996; Tillmar, 2002). The reason for that may be twofold. First, according to Möllering (2006), it is because empirical studies of trust have been dominated by quantitative endeavors. Taking this into account, the field project, the excerpts of which I am partially presenting here, was originally intended to address the need of local, interpretative studies of trust pointedly expressed in the recent literature (e.g., ibid, p. 151-154). Second, as Beckert (2005) claims, our problem in understanding how trust works, specifically in market relationships, originates in a bias that is characteristic of most of the existing research. It is assumed that in dyadic market relationships, e.g. between buyers and sellers, the buyer is the trust-giver (i.e., initiator of the relationship). The general problem is that the majority of research so far has been focused on the role of trust-giver, and somewhat ignored the likely contribution of the trustee to forming the successful relationship, especially by appearing trustworthy even before the actual business contact is even initiated. It is not to mean that the role of the trust-taker has been completely neglected, but rather that the parties were not seen in a balanced mode, with a clear bias towards the role of the trust-giver. Through the research project that I draw upon in this text, I wanted to address this issue and look at the specific cases of market cooperation from equal perspectives of both partners.

An interesting framework for conceptualizing the role of the trustee in the market relationships is offered by Beckert (2005), who claims that the key to understanding the initiation of market exchange based on trust is the "performative acts" of the trust-taker. "Performative here means the production of the impression of trustworthiness, especially through acts of self-presentation with which the trust-giver tries to persuade the trust-taker of the sincerity of his intention to cooperate" (Beckert, 2005, p. 5). I believe that drawing upon the dramaturgic metaphor (Goffman, 1959) and concept of "performative acts" proposed by Beckert may cast a new light on the cooperation between business partners. Furthermore, in my case, the parties declare openly that they are rather distrustful, or at best wary towards each other, so establishing grounds safe enough to start cooperation constitutes here an additional challenge.

The question I am going to address throughout this chapter is how the suppliers (trust-takers) establish the image of being trustworthy, and how their trustworthiness is in turn examined by their potential customers, when cooperation takes place in the context of the openly expressed lack of trust (or distrust). In the first part of this chapter, I will clarify the notion of trust and discuss its role in market relationships. Trust will be seen here as a *tranquilizer* that suspends the feeling of vulnerability and enables action regardless of uncertainty involved in the situation. Cooperation, in turn, is understood in dramaturgic terms, according to the metaphor of drama (Goffman, 1959). The supplier is to produce an impression that will make them appear trustworthy as perceived by the customer. This section is most heavily based on Möllering's (2006) conceptualization of trust, and the view of market exchange set out in the article by Beckert (2005).

What follows is the presentation of excerpts of empirical studies that are to address two issues: (1) the perception of the phenomenon of trust and (2) the practical methods for verifying trustworthiness of the supplier before making the decision about entering the project together. The process of cooperation will be presented from the viewpoints of two parties engaged in cooperation: the suppliers and the customers. This is to avoid bias towards the trust-giver and fulfill the advice concerning qualitative research on trust. "In dyadic relationships, both sides of the dyad should be interviewed (Huemer, 1998)—not so much as a means of confirmation or triangulation (Altheide & Johnson, 1994) but rather as a means of taking in multiple perspectives that allow reflection on the idiosyncrasy of trust experiences" (Möllering, 2006, p. 153).

In the concluding section, I will offer my interpretation of the empirical research against the theoretical background presented previously. I should include there both the positioning of research within the theoretical framework as well as practical implication that this study may offer for managing IT projects.

METHOD

The project I would like to draw upon in the chapter is the research conducted in Poland, at three software and IT services corporations and their four client organizations (both private and public sector were represented). The main aim of the study was to get a picture of relationships between high-tech corporations and their customers, who work on relatively complicated, long-term projects (tailor-made solutions that are designed, produced, and implemented by the IT company); from the perspective of both parties engaged in the corporation. The first part of the study involved in-depth, non-structured interviews conducted at three IT companies in Poland, and subsequently at four organizations representing their clients (together 31 people were interviewed, the interview usually took 1.5-2 hours). The interviews were tape-recorded (with two exceptions) and transcribed according to the guidelines provided by Silverman (2001). This in-company phase of the research project was then complemented by direct observation of everyday interaction between the client and the supplier as well as the analysis of selected professional literature and professional press. The choice of specific sources was guided by the interviewees themselves, as they mentioned their usage in their professional life.

Hence, the presented research is ethnographical (Rosen, 1991), and the aim of the article is to present the understanding of the problem in the eyes of organizational actors, rather than apply the authors' own view of it.

TRUST IN MARKET RELATIONSHIPS

Trust as a Leap of Faith

The notion of trust may be understood intuitively and is a kind of umbrella term that covers a variety of experiences (Gambetta, 1998). Both in everyday life and in a major part of academic literature, it is assumed that trust is understood by itself and therefore there is no need to define it (Barber, 1983; Gambetta, 1988; Hardin, 2004). Trust is also a part of everyday human experience—this determines that it is on the one hand a relevant and appealing topic for academic inquiry, but on the other hand a true challenge for empirical investigation. Trust is also one of those terms that is broad and abstract enough to attract interest from many distinct academic disciplines; it is also widely present in our everyday language. In what follows, I adopt Möllering's (2006) view of trust, where it is seen as a way of dealing with vulnerability and uncertainty in interaction. There is a fine, but important difference between the notions of risk and uncertainty: trust is risky in the sense that it is always a kind of hazard, but it is far from a calculative view of risk as a probability of failure or success (Möllering 2006, p. 8). When we talk about trust, we have to take into account the trustor's positive expectation (upon which actual actions are undertaken) that the partner will not exploit the trust-giver—so will not fail the trust vested in him (Baier, 1986). Möllering (2006, p. 105-121) also indicates the element that constitutes the crux of the experience of trust. Referring to the works of Georg Simmel he states that trust is "a state of mind which has nothing to do with knowledge, which is both less and more than knowledge" (Möllering, 2006, p. 109). The essence of the concept of trust is, hence, the element of the "leap of faith" (also called "suspension") that ultimately makes the concept of trust unique and powerful. The "leap of faith" helps us to cross the gap of uncertainty and does not reduce it, but enables us to act regardless of it. Hence, when we undertake measures that are intended to reduce uncertainty, we actually also reduce the need for trust, and finally, we apply mechanisms that enable action without resorting to trust.

Market Relationships Based on Trust

The buyer-seller relations are one of the common types of interorganizational cooperation. They are, indeed, thick relationships, where "trust plays an important part because the threat of opportunism and vulnerability is acute even when there is a mutual interest in continuing the relationship" (Möllering 2006, p. 156; see also Lane & Bachmann, 1996; Sako, 1992; Zaheer, McEvily, & Perrone, 1998). However, trust is not indispensable for such cooperation to happen—as there exist a wide array of alternatives that may secure the (subjective) confidence in transactions. Thanks to them, uncertainty that otherwise would refrain a party from cooperation may be reduced to the level that enables action—and the need for trust (leap of faith).

According to Beckert (2005, p. 11), four mechanisms that automatically reduce uncertainty, make us able to act and reduce the need for trust. These are tradition, identity, power and norms, institutions, and calculation.

In situations when one exchange partner can (subjectively) reduce uncertainty and enter the relationship by one of the previously identified mechanisms, there is no need for trust. Market relationships that depend on trust (or, more accurately, on the perception of trustworthiness of the prospective partner) are characterized by the following features:

- The quality and value of products or services cannot be completely assessed by the trust-giver. This uneven distribution of information between the trust-giver and the trust-taker contributes to the general feeling of uncertainty in the relationship.
- The market is competitive (i.e., the trust giver has a choice between various trust-takers)
- The relation happens "for the first time," or is "short term" (so, that the mechanism of tradition and identity cannot develop)

Considering these features, the Polish IT market seemed to me the perfect research environment, taking into account the intangible character of software services and the highly specialized knowledge on the part of the supplier coupled with the high disclosure of information about the core business on the part of the client. The potential role of trust becomes salient when we consider the high cost and long duration of IT projects. The market is also quite competitive; although the customer's choice is limited by the technological conditions (earlier choices regarding core technologies).

Self-Presentation in Market Relations

In trust research, the focus in most cases is on the trust-giver, who vests trust in the prospective exchange partner, and thus initiates the relationship. The act of placing trust is to precede the action of the trust-taker (Beckert, 2005, p. 8). However, this stance seems to miss an important element of the concept of trust in the context of market relationships. It is of special interest in the buyer-seller setting. The buyer can choose between various sellers—the latter must persuade the customer that they are worth cooperating with. The goal of the trust-taker is then to produce the *impres-*

sion of trustworthiness (Beckert, 2005, p. 9) that can influence the actions of the trust-giver (the customer). From this perspective, the trust-taker is the initiator of a relation. According to Beckert (ibid.), the trustee tries to evoke the impression of trustworthiness through so-called *performative acts*. If the seller succeeds, that may begin a relation that is based precisely on trust (in the sense of the leap of faith), not other alternative mechanisms (tradition, control, etc.). The concept of performative acts heavily draws upon the notion of *dramaturgic action* introduced by Goffman (1959). Borrowing from his terminology, I could say in relation to the research problem here that, when there are no other available measures of reducing uncertainty, the supplier aims at producing the impression (it is about perception) of trustworthiness; based on the trust customer places and enters (it is about action) the cooperation. Beckert (2005, p. 19) writes that we can call it "a socially constructed fiction of trust" (see also Berger & Luckmann, 1966).

In the following section, I present excerpts from the field study that refers to the motives and reasons given by the suppliers and the clients for starting the cooperation. Based on their opinions, I distinguish the strategies for producing the "impression of trustworthiness" that are applied by the suppliers (in the initial contact) and accepted by the customers. The key concept in the opinions of my interviewees appears to be "rationality" or "logic"—however, it comes in various guises. Both parties in the exchange stress repeatedly that any decision and action they are to undertake must be not only understandable and justifiable to themselves, but must also appear as "rational" and logical" for other people in their immediate environment.

THE ROLE OF TRUST (FIELD STUDY)

When the word trust is brought up by my interviewees, it initially struck me that from the perspective of both the providers and the customers, it was directed only to the supplier. From the perspective of the buyer, the issue of trust usually boils down to the catalogue of features and behaviors of a trustworthy supplier. In not one interview did the topic of the interviewee's own company or them, as a trustworthy partner, come up. The providers are quite clear about it.

DL: And your customers, do you need to trust them?
PC – provider: No, not at all. There's no trust at all. No. What for?

Moreover, it definitely seems to be a business of the supplier to win trust of the prospective partner. It is needed, first, to secure the quality of contacts between both parties. The well-functioning IT solution is a product of long-term cooperation and cannot be worked out by the provider in isolation. Trust may help to build good rapport between both partners that is crucial in this context. However, it appears to require trust only on the part of the customer - building up of that is seen as a major task of the supplier.

LS – provider: In the relation between the client and the provider, there is trust, but only in one direction. Me... to the customer... there is simply no room for trust, you know? It's out of the question. Being frank with you... I just want them to give me cash and bugger off. But the customer has to trust me. It's something else.

It looks like from the point of view of the interviewees, trust neither has to, nor maybe even can, at all imply reciprocity. Even when they talk in abstract terms about a customer, the providers seem to take for granted that it is their job, not the customer's, to work on trust. From their point of view, customers are not interested in trust building, as they, as providers claim, will not benefit from it. The IT firm sees a clear profit in building-up trust. This viewpoint was strengthened further by the representatives of customers themselves, who claimed that trusting your provider might be the first step towards inescapable disaster. As they see it, when trust is developed, they will lose their alertness, diligence, and ability to mind their business with the IT partner.

According to several of my interviewees representing both parties, there also are situations, where they cannot talk about trust at all, on either part, and yet they still work together. Having mutual interest in the collaboration, coupled with some possibilities to exert pressure on the partner seems to be a more certain guarantee against failure in cooperation than simple trust.

AP – provider: Having the goods on each other is more important than mere trust. We have something to catch them, they have something on us and somehow we manage to progress .... One has to cooperate in spite of the lack of trust.

Even if the customers talk about trust towards the providers, they immediately apply restrictions to it. Trust in the market seems to be always limited to the field where customers and suppliers share common interests. Any other expectations would be seen as naïve, irrational, against the taken-for-granted rules of the market. It seems that one of the rules of business is about manipulation and, as one of the interviewees told me, in business there is "no room for taking offence" (KK - customer). Still, when the lack of trust may be an impediment in the relationship, it does not stop parties from cooperation. However, it may be reflected in the cost and duration of the projects, the quality of the final product, and satisfaction of the collaborating parties.

PZ – provider: Well, my personal opinion is that it's naturally the key to successful projects, but it's neither sufficient nor necessary. But I think that without building trust over time, in projects like those that we run that are long... then building trust at least at the level of team members who actually work together... that is crucial. I'm not saying that it can't be done, but any project can go smoothly and with expected profits for both partners without it.

VERIFICATION OF TRUSTWORTHINESS (FIELD STUDY)

Knowledge

Talking about starting cooperation with a particular supplier, the customers claim they make the decision based, most notably, on their own knowledge and competence in the field. They maintain that even during the initial talks with the prospective supplier, they are able to verify his[a] words against their own understanding of IT. However, it is irritating to the customer that the suppliers "waste his time" underestimating his competence, and they dare to present roughly drafted offers counting on his gullibility. According to the buyers, the sellers underrate the fact that their partners are professionals, quite often field-educated and experienced in IT, who have knowledge and experience enabling them to verify the credibility of an offer.

JN – customer: I have already met suppliers who promised me the moon. Such company was submitting a solution proposal but it wasn't completely meeting our needs; yet they were telling me it was perfect for us. But I can identify such dishonest sellers by asking them a few questions .... For example, when [a company] comes to us and I ask these questions and I learn that "we don't have that, but we'll be ready for tomorrow," then I know that since they don't have it, then it would take them dozens, hundreds of man-hours to implement it, to make it really work... having asked several questions of this kind, I know very well where the baseline is. So, an honest seller, after few days, would withdraw this offer. But still, there are those who are either less intelligent or more determined and they try to trick us into their game, to sell us things they simply don't have, they've no clue about, they simply don't know how to conduct business. But they have no chances here, in our place, because it is not only me who is competent, there is my team as well. So, I'd say that if there is knowledge here, then it's out of the question that anybody could get here feeding us rubbish.

Even if the customer engages into talks with the prospective supplier, they claim that they are perfectly able to distinguish between sales smart talk and the clue of the offer. The offers are usually big and comprehensive, yet only a few pages contain information that is considered crucial by the customer. This information usually refers to the specification and cost details of the proposed solution. The customers see themselves as very competent in the field of IT, and they claim they know exactly what they want. However, from the perspective of the supplier, it looks like the skills and even the general notion of IT is, in particular in the public sector, still rather limited. As several representatives of the IT companies said, the major factor for building the strong image in the perception of the client is to appear proficient in the field. A person who enjoys a reputation of being an expert in IT is more likely to be received as credible. They are usually able to explain the intricacies of an offer and provide satisfactory rationale behind a specific solution. However, it has been emphasized in the interviews that the competence of the provider should not only be evident in direct face-to-face communication, but also in one's personal professionalism (e.g., through credentials, diplomas, etc.).

To me, it seems worth noting that, although I have never mentioned the theme of trust during the interview, it usually came up, brought up by the interviewees. Somehow, they were using the word trust, but never reflecting upon it. However, the idea that trust of the client must be won, and this process is strongly tied to reason, was prevalent in all interviews. Trust is never to be granted unconditionally, but must be strongly footed in logically justifiable motives.

JK - provider: Trust is based on ... the content... it means that I know what I'm doing and that when working with me you always succeed. That I know... that I have the best knowledge, that I know my job and that such task may be granted to me.

Two points raised in the interviews become particularly salient. Both customers and providers perceive themselves as experts in IT, which is at odds with their views of each other as ignoramuses or laymen. The expertise in IT that both parties claim to possess plays a twofold part. For the client it provides a feeling of self-confidence of being able to control the partner's actions. For the IT firm it is a chance to win interest of potential customers through the image of true professional.

Corporate Procurement Policies

In the decision-making process in the buyers' organizations, quite an important role is played by the foreign stockholders, or so-called corporate procurement policy. It concerns mainly big companies that also dedicate major funding to IT infrastructure. Such organizations form a particularly important customer group for the suppliers, as their orders constitute a major source of revenue. In such cases, the decisions are made at the corporate headquarters' level, not at the company in Poland where the solution is eventually implemented and intended to work. The criterion for selection, from the perspective of an employee of the Polish subsidiary is unambiguous: only reputable firms can become suppliers.

JN - provider: The rule in our corporate group is that only reliable suppliers may be contracted, and only those about whom the biggest consulting companies recommend they will have not breathed their last by the end of next decade. It is corporate policy that only the most well-known make it on the procurement list.

Being reduced to the level of merely "users" of a certain software program is sometimes regarded as limitation by employees of the Polish subsidiaries, but most of my informants were ultimately finding this policy reasonable. When IT is considered a component of basic infrastructure that shapes daily operations of the whole corporation and provides a common platform of communication and information exchange, the standardization is clearly a desired feature. As the customers were saying, too, the process of pushing the decision-making process to the upper level of their international headquarters is even welcomed by them, as it relieves them from the burden of responsibility they would have to assume for their own choices.

The Opinion in the Community

The customers also say that in choosing the supplier they rely on the so-called "hearsay in the community"—the community meaning professional colleagues and friends working in the industry. The world of Polish IT firms, in the words of my interviewees, appears as rather small and hermetic.

EI - provider: Well... one changes jobs so frequently that everyone knows what is going on at each of the organizations.

Good connections within the professional circle give access to informal knowledge, for instance about failures and successes of specific IT projects. Having this knowledge means being a true part of the IT market community. Actually, during our meetings, my interviewees seemed to enjoy telling stories "from the inside": about their competitors, former employers or subcontractors, usually stressing that as private as they may seem, there are really "open secrets" within their everyday professional environment. And, as the customers also participate in this community, such information by them is considered crucial in making the decision about their suppliers. Every piece of information, even a gossip heard somewhere in the community may be considered quite precious.

DC - customer: You should remember that much of the gossip that goes around in the market is in fact true. Someone may hear some astonishing news about some supplier and then it turns out... I don't know... That they run some project in Dubai. But they've never ever gone abroad! But it turns out that through some channels... they managed to win a contract... only not in Dubai, but in Egypt... but still, there's information and it's even quite precise. It's that they proved they are good, and even in a tough and new environment. Information of this kind cannot be rightly appreciated.

Opinions acquired from people in the industry about the supplier constitute a worthy indication of the value of a prospective partner, but is never taken unquestioningly. It seems recommendable to have in mind who conveys information, or a gossip, and what the current interests of that person are. Information seemingly passed accidentally or by the right of "old friendship" may be manipulated by the competition. One customer coined even his "golden rule" of dealing with information circulated informally, which he called "the rule of limited trust." According to it, to be on the safe side, one should always try to find out exactly who the source and the transmitter of the information is, and what their motives may be. A colleague from your former company may happen to work for your current competitor and that is enough to be suspicious about information he gives. In business, as the field material seems to convey, even when you have known a person for a long time, their current professional status (e.g., employer) may totally change the way you would perceive information given by them.

Another reason for being careful as far as opinion within the industry is concerned is a high individualization characteristic of IT projects. Their eventual perception and rating is also very much led by knowledge and individual competencies of their users. Solutions that work in one organization and enjoy great reputation may prove completely useless in another place. There were several examples of specific IT software systems that located them on the opposite sides of the rating continuum: either as a wonderwork or as a complete disaster. In the same vein, a team of programmers and managers responsible for interactions with a customer may be rated very high by one organization and get negative opinions from the other. That is why, as customers claim, one should always be able to form one's own opinion and own competencies.

Furthermore, one of the interviewees also talks about how the rumor in the community may be used during negotiations, how it may be turned into a tangible advantage. Not always is it wise to turn down somebody who is disrespected by the "general opinion" in the market. Trusting our own opinion that we can watch such a company, is just good business.

JN - customer: The truth is, in fact, that really... relations with colleagues may give you some notion about some concrete supplier, but I also know how to use it for our benefit. It is quite simple, like that. When a firm enjoys rather bad reputation, but still constitutes a promising business ... then sooner or later, they will need solid recommendations. And it's happened several times in my life already hat I bought systems that were nominally worth 100000 USD, and I paid 5000 USD for it.

Hence, among the customers, there is an informal circulation of opinions about projects and suppliers. However, it was quite interesting to me that the suppliers rarely mentioned this community. There are opinions about customers, but of a totally informal nature—while the opinions about suppliers are partially institutionalized in the form of recommendations. Although the suppliers maintain that they are not allowed to talk about their clients in an informal setting, opinions of that kind still find their way to the community. There are organizations with a long-lasting black mark of a "whining" or "tough" customer. Those buyers are then treated with highest wariness and distance. JH told me the story of a customer who - through constant filing of complaints—made the company work for several (sic!) years beyond the schedule from the contract. He is sure, however that at the end of the day it does not pay off for the customer either.

JH - provider: He reports the next problem, and our condition is getting worse, as we do not progress, but we rather take a step backward. It turns out that he reports another hundred problems and we thought that we were finished and it suddenly turns out that we still have something to do. Naturally, when I look at it, I think they are completely aware of what they're doing... people who work on the other side are perfectly aware of what they are doing. And they know that in this way, they extend the warranty that initially was contracted for a year, for three or four years. So, it is their conscious action, and they get as a gift a piece of a system that cannot be copied, but somebody has to work, to write it, doesn't he? So, this is deliberate. ... And naturally, we don't have good memories of such customer. ... But people who work there are then recognizable on the market. Because it is also those people who work here in the company, when they work weekends because they have a job to get done for a certain customer, because some manager had a wish and blackmails us again. Even in private, they will never talk in good terms about such customers.

The IT market appears in the interviews as a rather small community where information is circulated easily and quickly; however, it may take numerous forms: gossip, rumor, or provocation. While the interviewed customers certainly declared limited trust in such community opinions, they also acknowledge that it is usually an important criterion in screening a potential partner.

Formal Recommendations

In the IT market, an important role is played by a system of formal recommendations—opinions about work done for a customer by an IT company. For other buyers, it is one of the identifications of the reliable provider. Other sellers, knowing about the important role played by the recommendation system in making the choice of the customer, pay attention to it.

Under "recommendation," my interviewees understood a "positive opinion." Formal recommendations with negative opinions are simply not given, but the customers who are dissatisfied with a certain supplier usually refuse to give a recommendation. One of the customers, DC, told me a story of such situation:

DL: Have you ever refused a customer to give a recommendation?
DC - customer: I'd have done it once, but that company never asked for the recommendation. No... perhaps I refused then.... In such cases, I try not to make the decision by myself, but I go to my supervisor with the recommendation. And I went to him with the completely negative recommendation and my supervisor confirmed it, accepted it fully. He didn't hesitate; he also said that we shouldn't give them the recommendation.
DL: But why did you refuse?
DC - customer: As I told you, at that instant I felt cheated.

Recommendations are important to the point that they may be even subject to extra verification. Due to malpractice by the suppliers, they are not reliable by themselves, but they must be additionally "vetted," usually at the source of information.

JN - customer: When it comes to talks, to negotiations, I have a wide range of arguments. Very often, it is so that the word [the name of the corporation] helps me. Let's not beat around the bush, if this word appears around the brand of the IT Company then the added value for this company is enormous. ... [One of the suppliers that sold us a box of software] wrote on the website that they implemented a software solution for [our corporation], and yet they sold us a product they never really implemented, they simply gave us the box. Then one of my colleagues calls and he asks, "JN, listen, they wrote here they sold you this product." I say, "yes, but it was only the product." And this is a kind of informal recommendation. "Indeed, when you can negotiate a good price for this box, then buy it, but if they want to sell this box to you at a regular price, then maybe look for some other supplier who gives you a better price, or at least has experience in implementation."

The value of formal recommendations as a means to back trustworthiness is further confirmed by providers themselves. Building-up a portfolio of good recommendations from reputable customers may even be the reason behind accepting a contract, which in financial terms may not appear profitable. From the point of view of trust, the rule of "double-check" is of interest here, as the customers do not take the reliability of presented recommendations for granted.

DISCUSSION

The field data seem to confirm and fit easily into the analytical framework of market relationships set out by Beckert (2005), where the trust-taker, not the trust-giver initiates the relationship. Indeed, the interviewees straightforwardly say that it is the task of the provider to work on trust. Before the customer enters cooperation, their trust must be won. Applying Beckert's (ibid.) terms, we would say that they must become convinced about the trustworthiness of the potential partner and, thus, they may feel certain enough to start a project together. Although the decision is made by the customer, it may be influenced by the prior actions of the provider. Then, the provider is the one who actually initializes the relationship with the partner, through undertaking actions that will make them appear trustworthy even before the actual decision is made by the customer.

In reading the interviews, it becomes clear that the decision to enter into cooperation with a particular provider must be well-justified (i.e., backed with "rational" arguments). Indeed, the decision to work with a particular supplier must be supported by specifiable arguments. In the cases I observed, there is no room for trust understood in terms of the "leap of faith" (Möllering, 2006), at the beginning of a relationship. Trust is understood only in terms of calculation (Lane, 1998)—the feeling of security that bridges uncertainty and enables active engagement into cooperation does not come from trust (i.e., "suspension" of the feeling of vulnerability) but from the understanding that being honest clearly pays off to the partner. On this ground, I would venture a claim that to be successful in building the impression of trustworthiness, the suppliers may follow the lead of customers in building up strategies applicable in the specific context of IT business. Figure 1 presents the catalogue of methods used by the customers to assess the trustworthiness of their potential business partners, as they are reflected in the field material. Accordingly, they may constitute reference points for providers that may want to implement strategies to improve the impression of trustworthiness they produce. It is evident, that in some areas, providers are aware of the way they are perceived by their customers. It is evident, for instance, when they talk about the importance of formal references or professional knowledge—how important it is to maintain the reputation of a professional in the field.

1. The process of gathering information about IT companies is partially institutionalized in the system of formal recommendations from former and present customers. Rec-

Figure 1. Methods of producing the impression of trustworthiness

Methods of producing the impression of trustworthiness	Opinions	Informal (community)
		Recommendations
	Corporate policies	
	Knowledge	Of the client (verification)
		Of the supplier (professionalism)

ommendations, however, are not reliable enough and they require further verification. So, we cannot talk about confidence in suppliers even in regard to the authenticity of recommendations—the buyers are afraid of cheating. At the same time, the customers claim they make extensive use of the opinions about IT companies within the professional community. Despite the fact that it is pervaded with gossip and unconfirmed information, according to the customers, it enables them to "feel the climate" around a particular company and it provides grounds for predicting prospects of future cooperation. Providers share the conviction about great importance of formal recommendations and they put effort to compile a good record of accomplishment of their successful jobs and best customers. However, there is a visible discrepancy regarding the weight of informal circulation of information and gossip. While the customers rely on it to quite a large extent, it seems to pass unnoticed by the providers. It is not the right place to speculate what the reasons for that are, but we may attempt a conclusion that influencing the informal opinion within the community would be a powerful tool at the hand of the provider in their effort to appear trustworthy.

2. Corporate policies actually "make the choice" for a customer and take away the burden of responsibility for the decision. On one hand, it is called "imposition," or a choice that is forced upon the customer. On the other hand, however, interviewees from the companies that use corporate procurement lists seem satisfied with the fact that they are relieved of the whole effort of decision-making and do not complain about being devoid of independence in this respect. Making choices, especially in a volatile field as IT is a risky business, and corporate policies appear to me as allowing customers to forget about this hazard. The game of trust here is played on the higher level - of headquarters of both organizations. In such a situation, the choice is predetermined to the point that any extra effort on the part of the provider to persuade the customer is of no significance.
3. Professional knowledge plays an essential role for both parties. For customers, it secures the feeling of being able to verify the product and/or service of the provider. Another issue goes beyond the scope of this chapter, whether they regard each other as IT professionals. What is important here is, indeed, the subjective confidence in their own ability of breathing down the provider's neck. For providers, appearing as knowledgeable, as an expert in the field is—from their point of view—a crucial element of self-presentation. The area of perception of both parties in terms of knowledge and being professional constitutes, I believe, a promising avenue of further exploration.

Distance toward the issue of trust expressed in the interviews seems to imply a twofold conclusion. First, it may mean that within the studied context trust is of a purely instrumental, functional value (it exists only when it clearly pays off; Coleman, 1990; Gambetta, 1998). It is desired and considered important as long as it facilitates collaboration. This understanding of the concept of trust as catalyst that smoothes cooperation has been widely acknowledge in the literature, too (Argyris, 1990; Deutsch, 1973; Smith, Carroll, & Ashford, 1995; Sheppard, 1995). Note that there was no mention of the indispensability of trust in the relationship. It is good when it is there, but if it is not, the cooperation may still last. This would confirm the theoretical considerations about the coexistence of trust and distrust (Cook, Hardin, & Levi, 2005; Lewicki, McAlliser, & Bies, 1998) as well as some insights stemming from empirical studies on trust (Huemer, 1998;

Six, 2005). Second, it may provide support for framing the cooperation by means of the dramaturgic metaphor (Goffman, 1959; Beckert, 2005). This becomes salient when we realize that both parties, and especially the provider, seem to be fully aware of "performing." The actual attitude toward the customer ("we just want their cash") is at odds with the declared will to win the trust of the customer. This realization from the empirical study may also show the promising avenue for further theoretical endeavors in the field of trust research. The benefits of the theoretical metaphor in interpretation have already been acknowledged in the literature in management and organization theory (Hatch, Kostera, & Koźmiński, 2005; Hoepfl, 1995; Mangham & Overington, 1987). Grafting it onto the research of trust may be a source of novel insights into the dynamics of trusting relationships.

REFERENCES

Altheide, D. L., & Johnson, J. M. (1994). Criteria for assessing interpretative validity in qualitive research. In N. K. Denzin, & Y. S. Lincoln (Eds.), *Hanbook of qualitative research* (pp. 485-499). Thousand Oaks: Sage.

Argyris, C. (1990). *Overcoming organizational defenses: Facilitating organizational learning.* Englewood Cliffs, NJ: Prentice Hall.

Baier, A. (1986). Trust and antitrust. *Ethics, 96*(2), 231-260.

Barber, B. (1983). *The logic and limits of trust.* New Brunswick, NJ: Rutgers University Press.

Beckert, J. (2005). *Trust and the performative construction of markets* (MPIfG Discussion Paper 05/8). Cologne: Max Planck Institute for Studies of Societies.

Berger, P. L., & Luckmann, T. (1966). *The social construction of reality.* Garden City, NY: Doubleday.

Coleman, J. C. (1990). *Foundations of social theory.* Cambridge, MA: Harvard University Press.

Cook, K., Hardin, R., & Levi M. (2005). *Cooperation without trust.* New York: Russell Sage Foundation.

Czarniawska-Joerges, B. (1992). *Exploring complex organizations: A cultural perspective.* Newbury Park, CA: Sage.

Deutsch, M. (1973). *The resolution of conflict: Constructive and destructive processes.* New Haven, CT: Yale University Press.

Glaser, B., & Strauss, A. (1957). *Discovery of grounded theory: Strategies for qualitative research.* Chicago: Aldine.

Goffman, E. (1959). *The presentation of self in everyday life*. London: Penguin.

Gambetta, D. (1988). Foreword. In D. Gambetta (Ed.). *Trust: Making and breaking cooperative relations.* Oxford: Basic Blackwell.

Hardin, R. (2004). *Distrust.* New York: Russell Sage Foundation.

Hatch, M. J., Kostera, M. I. & Koźmiński, A. K. (2005). *The three faces of leadership:*

Manager, artist, priest. Oxford: Blackwell Publishing.

Hoepfl, H. (1995). Performance and customer service: The cultivation of contempt. *Studies in Cultures, Organizations and Societies, 1*(1), 47-62.

Huemer, L. (1998). *Trust in business relations: Economic logic or social interaction?* Umea: Borea bokvorlag.

Lane, C. (1998). Introduction: Theories and issues in the study of trust. In C. Lane, & R. Bachmann (Eds.), *Trust within and between Organizations – conceptual issues and empirical applications* (pp. 1-30). Oxford: Oxford University Press.

Lane, C., & Bachmann, R. (1998). *Trust within and between organizations – conceptual issues and empirical applications*. Oxford: Oxford University Press.

Lewicki, R. and Bunker, B. (1996) Developing and maintaining trust in work relationships. In Kramer, R., & Tyler, T. (Eds.). *Trust in organizations – frontiers of theory and research* (pp. 114-139). Thousand Oaks, CA: Sage.

Lewicki, R., McAlliser, D., & Bies, R. (1998). Trust and distrust: New relationships and realities. *Academy of Management Review, 23*(3), 438-458.

Mangham, I. L., & Overington, M. A. (1987). *Organizations as theater: Social psychology and dramatic performance*. Chichester: Wiley.

Möllering, G. (2006). *Trust: Reason, routine, reflexivity*. Oxford: Elsevier.

Rosen, M. (1991) Coming to terms with the field: Understanding and doing organizational ethnography. *Journal of Management Studies, 28*(1), 1-24.

Sako, M. (1992). *Prices, quality, and trust: Inter-firm relations in Britain and Japan*. Cambridge: Cambridge University Press.

Sako, M. (1998). Does trust improve business performance? In C. Lane, & R. Bachmann (Eds.), *Trust within and between organizations: Conceptual issues and empirical applications* (pp. 88-117). Oxford: Oxford University Press.

Sheppard, B. H. (1995). Negotiating in long-term mutually interdependent relationships among relative equals. In R. J. Bies, R. J. Lewicki, & B. H. Sheppard (Eds.), *Research on negotiation in organizations* (Vol. 5, pp. 3-44), Greenwich: JAI Press.

Silverman, D. (2001). *Interpreting qualitative data*. Thousand Oaks, CA: Sage.

Six, F. (2005). *The trouble with trust – the dynamics of interpersonal trust building*. Cheltenham: Edward Elgar.

Smith, K., Carroll, S., & Ashford, S. (1995) Intra- and interorganizational cooperation: Toward a research agenda. *Academy of Management Journal, 38*, 7-23.

Tillmar, M. (2002). *Swedish Tribalism and Tanzanian Agency: Preconditions for trust and cooperation in a small-business context*. Linköping: Linköpings universitet.

Zaheer, A., McEvily, B., & Perrone, V. (1998). Does trust matter? Exploring the effects of inter-organizational and interpersonal trust on performance. *Organization Science, 9*(2), 141-159.

Zucker, L. G. (1986). Production of trust: Institutional sources of economic structure. In S. Barry & L. Cummings (Eds.), *Research in organizational behavior* (Vol. 8, pp. 53-111). Greenwich: JAI Press.

ENDNOTE

[a] I apply the masculine pronoun throughout the empirical part as it is a default pronoun in the Polish language (and it reflects the interviewees' speech).

Section III
Workplace Relations and Power

Chapter X
Social Relations and Knowledge Management Theory and Practice

Marie-Josée Legault
Téluq-UQAM[a], Canada

ABSTRACT

This chapter proposes a new hypothesis to the refusal to cooperate from qualified professionals and supports it with five arguments drawn from the fields of sociology of work and professions. The management of knowledge (KM) is based, among other things, on a system for pooling knowledge to which employees must contribute. Nevertheless, the experts of KM persistently note the relative failure of knowledge-pooling practices, particularly among the highly qualified professionals. Some experts have little to say about this issue and the scarce explanations they provide are highly unsatisfactory sociologically speaking and inspired by a folk psychology discourse. Sociology of work and professions, particularly, provide the grounds for alternative and more solid analysis of the phenomenon.

Since it is claimed that, in the so-called knowledge economy, knowledge is the most crucial economic resource in the struggle against fierce international competition, it follows that management must pay particular attention to how it manages this resource. But what does "manage" mean in this context? Does it mean promoting its development by investing in training, for example? Promoting its transfer by encouraging creativity, initiative, or teamwork, by favoring the setting up of communities of practice (if saying so makes any sense...)? Promoting the disclosure and banking of knowledge or information to make it more easily available to all, to increase access to it for the greatest number while at the same time conserving and accumulating it over time? Does it extend to appropriation by the group? Or appropriation by management? Given that these questions have not been formally answered to date, all these management attitudes are seen to varying degrees.

A so-called new discipline has sprung up since the mid-eighties, known as knowledge management (KM). KM system implementation and management specialists are producing a growing body of work, consisting primarily of handbooks and manuals. Management consultants can't seem to get enough of it and will willingly tell whoever hires them that KM is the key to the constant innovation that will help them maintain their competitive position in today's market:

A horizontal reading of these different business strategies shows, however, that this new competitiveness is based primarily on innovation. [...] The secret to an organization's success lies in its capacity to promote processes that enable interaction between different sources of individual or compartmentalized knowledge in order to generate new collective knowledge that provides a basis for diffuse innovation. 1 + 1 = 3! [...] To summarize, an organization's capacity to innovate is rooted in its capacity to transform its knowledge assets, which are organized and individualized to varying degrees, into "collective strategic intelligence." (Jacob & Pariat, 2000, p. 10-12; my emphasis) [translation]

The following chapter presents a sociologist's critical look at this body of work in which very few sociologists has so far shown little interest, whether to contribute to it or to review it. I must underline noticeable exceptions in Great Britain, that address the questions raised here without reviewing this body of work (*per se*: Robertson & O'Malley Hammersley, 2000, p. 242; Scarbrough, 1999, p. 6; Swan, Robertson, & Bresnen, 2001; Wilson, 2002). Given the popularity of this phenomenon in the world of work today, particularly since 1997 (Wilson, 2002), this lack of attention should arouse anyone's curiosity.

One of the means that the KM experts propose to manage knowledge is a system for pooling knowledge or information to which employees must contribute. I will be looking here at a phenomenon that concerns both theorists and practitioners of KM, that is, the fact that a significant proportion of workers do not contribute to these knowledge-pooling systems. KM may be implemented in companies from all sorts of industries, employing workers with very different qualifications; I have restricted my study to the case of highly qualified professionals hired by companies supplying very specialized goods and services to enterprises (B2B) using extremely sophisticated production techniques. The professionals I will be discussing are computer specialists and programmers. In Canada, they are not members of reserved-title professions, though they share many characteristics typical of professionals (university degree, shared body of specialized knowledge, high degree of qualification, close contact with customers and dedication to ensuring customer satisfaction, discretionary power[b] in the choice of means to carry out work and in the decision-making process in performing tasks, in the judgments and opinions issued about the work and, in some cases, in the ends being pursued also), but not all of them (they do not have a professional licensing body, a code of ethics, a reserved title, and self-regulation of the rules of membership: curriculum, certification and decertification of members, standards of practice, setting of rates or prices for work, penalties, peer assessment and immunity from outside control).

In this chapter, after providing a short definition of knowledge management, I will analyze and critique what KM promoters have to say about demanding that qualified professionals participate in knowledge-pooling systems, the relative lack of success of these practices and the explanations that we can deduce from the solutions they propose. I will then set out why I think these explanations are unsatisfactory in relation to what the sociology of work and the professions has learned about the appropriation of specialized knowledge. Finally, I will propose an alternative hypothesis to explain this lack of success and provide arguments to support it, ones that in my view are sufficient to submit it for examination but not to regard it as having been demonstrated.

WHAT IS KNOWLEDGE MANAGEMENT?

Several definitions of KM, that would be tedious to itemize, mirror the leading position of the information management (IM) specialists, including decision support systems (DSS), information technology (IT), information systems (IS) over the field (Robertson et al., 2000, p. 241; Swan et al., 2001, p. 10-12 & 17) and as such, those have in common to point toward:

- **A simple objective:** Promote the use of knowledge held by individuals for the purposes of innovation and/or enhancing the competitive position of companies. In a word, manage knowledge in order to create new value.
- **Means to be used:** More often than not, a combination of various information and communications technologies (ICTs) and management procedures designed to manage and provide a framework for this knowledge, from the time an idea first emerges until its "value is developed" for economic purposes, that is: formalize, collect, in other words persuade to share, validate, disseminate, transfer, store, accumulate, reuse, update, analyze, apply, etc.

Pooling Knowledge: A Necessary Precondition of KM

To achieve the objective of managing knowledge, management in organizations that subscribe to KM relies to a large extent on the practice of pooling knowledge (held or produced by their professional employees), since this knowledge "can become strategic" and "the loss of it can have serious consequences for the company" (Cotte, 1999, p. 12-13). As a result, many managers implement procedures and systems for pooling knowledge—such as forums, collaborative systems supported by intranets, or more dedicated, decentralized tools, and urge their employees to contribute to them. These are not the only means to share knowledge or information; many highly qualified professionals working in firms constantly seeking innovation do in fact share knowledge orally rather than under any recorded form (Robertson et al., 2003; Robertson et al., 2000). But the IT driven systems are the ones promoted by IM experts who, as we saw, dominate the consultancy practice in KM; now, these systems better suit the sharing of information than the sharing of knowledge of a higher level of complexity.

The importance of the knowledge of practitioners is nothing new, nor the use of that knowledge by managers in organizations. Sociologists, among others, have long studied the appropriation of workers' knowledge for the purposes of the industrialization of processes. Is the advent of KM changing this practice to the point where it makes our knowledge about the issue obsolete, or is it simply an instance of the extrapolation of known phenomena? Many arguments point in this second direction, but with subtle differences.

Pooling Knowledge: An Old and New Phenomenon

First of all, the systematic nature of efforts to make practitioners reveal their knowledge is not in itself new, as Taylorist time-and-motion studies had this same systematic characteristic (Friedmann, 1964). Nonetheless, the professionals who are the subject of KM make use of what are essentially cognitive resources and so simply observing them is not sufficient to give access to their knowledge; they must be persuaded to cooperate in order to reveal their knowledge. This raises a new problem.

Nevertheless, like the old workers' knowledge appropriation systems, typical KM knowledge-pooling systems are, more often than not, implemented in a top-down fashion and are the result of a deliberate management decision. This distinguishes them markedly from communities of practice,[c] for instance. The promoters of KM would like to promise the same performance with their systems, but such an enterprise is rather groundless (Brohm, 2005; Wenger, 1999).

More ambitious than the previous knowledge seekers, though, KM systems propose the pooling of not only formal, explicit knowledge, but also of tacit, meaning implicit, poorly defined as a "yet-to-be-codified knowledge":

Explicit knowledge refers to all knowledge that has been collected in a form that makes it easily accessible and communicable (a manual, for instance) [...] This is usually the case with methods to be followed, techniques to be used, frames of reference, policies to be implemented, legislation and regulations, etc. [...] Tacit knowledge, on the other hand, encompasses all unrecorded knowledge, usually known only by the holders of that knowledge. While by its nature more difficult to describe and archive in written form, it covers a wide range, from know-how to intuition, and includes tricks of the trade and experience acquired in human relations and other areas. Tacit knowledge is therefore generally informal, contextual, subjective and derived from experience. [...] This knowledge consists of an infinite number of firmly interlinked elements that appear to be simple, yet are incredibly complex. (Jacob et al., 2000, p. 38) [translation]

I must clarify here that I do not subscribe to a critique often addressed to the KM promoters for using the notion of tacit knowledge and, above all, their reference to Polanyi's use of the term (1958). Given the fact that tacit knowledge is "hidden from the consciousness of the knower" (Wilson, 2002, p. 38) or that it refers to preconscious operations, yet prerequisite to cognitive processes, it could not be voiced nor captured by any KM system. It is set out as an "inexpressible process [and] can only be demonstrated through our expressible knowledge and through our acts"; only the "implicit knowledge" is previously unexpressed but expressible" (Wilson, 2002, p. 39). If tacit knowledge could only be demonstrated through our acts and KM promotors' promises must thus be considered deceptive, professionals should not care about KM practices, isn't it?

In this perspective, as long as we only consider the kind of knowledge sharing that the IT tools promoted by IM experts allow, we are right to dismiss the potential of knowledge sharing. Critics agree on the fact that such a tacit knowledge, that an individual is unaware he or she has, still remains inaccessible to most KM systems; in fact, few organizations can pride themselves on having found a way to turn this knowledge into explicit knowledge. More frequently, efforts focus on transferring knowledge that people know they have and know how to explain, but have never taken the trouble to pass on, except informally, or has never formulated (though it could be so if ever necessary), that is to say, implicit or non-codified knowledge (Bukowitz & Williams, 1999, p. 4).

Alas, I'd rather stick to Polanyi's own work and distinguish knowledge of which we are explicitly aware and knowledge that is implied in this awareness:

Polanyi makes a distinction in two levels of awareness: focal awareness and subsidiary awareness: "When focusing on a whole, we are subsidirarily aware of its parts" (Polanyi, 1958, p. 57). The subsidiary awareness provides the clues that are integrated by a person into a coherent whole or entity on which one's attention can be focused. This attention is focal awareness. Polanyi relates focal awareness to explicit knowledge and subsidiary awareness to tacit knowledge. Tacit knowledge is the content of subsidiary awareness together with performative and integrative skills [...] There are two different levels of tacitness. One is the flow of impressions and traces of previous experiences [...] The other level pertains to performative and integrative skills; this (Polanyi, 1966, p. 4, s16) also describes as tacit inference. (Brohm, 2005, p. 13-14).

Of course, on a plain common sense ground, if we consider oral sharing among project teams, highly specialized knowledge is currently exchanged (Robertson et al., 2003; Robertson et al., 2000) and this is part of KM. But more, in my point of view, seen as an integrative framework as defined here (Brohm, 2005, p. 13-17), it can not be rigidly separated from explicit knowledge and can be transmitted:

As an alternative image, I constructed the stage metaphor to emphasize the dynamics and interdependence of explicit and tacit knowledge. Some of the tacit stage can become an object of focus and thereby be transformed into explicit knowledge. When the focus changes, the explicit becomes tacit once again. Obviously, my reading leads to a completely different methodological approach than the objectivistic standpoint would (which often underlies the rigid distinction between tacit and explicit knowledge. (Brohm, 2005, p. 81)[d]

In such a perspective, tacit knowledge is likely to be transferable, but maybe not in any conditions. The usefulness of the process of recording and sharing tacit knowledge for management is presented as follows:

This [...] tacit knowledge has become explicit simply by having been documented from field problems and having turned out to be effective in solving them. [...] These quick, meticulous answers save all the players [...] from having to spend their time searching or waiting, which is both tedious and often expensive. Having recourse to both an automated institutional memory (explicit knowledge) and collective work experience (tacit knowledge) enables [workers] to rapidly define the problem and give the customer a comprehensive, practical response in a very short time. (Jacob et al., 2000, p. 39-40) [translation]

There can be huge benefits for management, and often for co-workers, in having this tacit knowledge shared because it can help solve problems that lie at the heart of a company's business operations, especially for knowledge-based companies, and is often rare and highly coveted. Communities of practice have showed it (Orr, 1990). However, it is seldom highly sophisticated and rather has the information status. This information is often the most important and the most useful for experts in addressing mission-critical issues in order to satisfy specific, exacting demands from customers. At the World Bank, for instance, experts are asked to share their knowledge of the customs, constraints and problems they encounter, as well as their solutions, in the diplomatic circles of the various countries they visit in their travels. As would be expected, the resulting database is very much appreciated (Prax, 2003, p. 10-11). But does this distinguish KM from older practices? So many traditional trades could not have disappeared with the development of mass production during the industrial revolution without the assimilation of the know-how and tricks of the trade, the fruit of experience and the complex reasoning of the craftsmen.

On the other hand, the dissemination of pooled knowledge may take on a new dimension owing to the potential of ICTs for disclosing, accumulating and using this knowledge. In earlier times, knowledge could be passed on to only a limited number of people; experts who were the source of the knowledge knew fairly well to whom and to what extent they were disclosing it. Same can be said of communities of practice, but not in systems that are implemented from the top down. Current top-down KM practices are based on much more powerful technologies, ones that industrialists of the past did not have at their disposal; as a result, the impact of revealing knowledge is much quicker and more more direct and more widespread, that entails concern. Conversely, we must recall the noticeable success of intra-organizational knowledge sharing practices, management-driven, but not supported by any IT recording device that managers would have access to; they are rather oral and take place among peers (Robertson et al., 2000). More generally, those that are not management-controlled (though put forward by managers), wherein the professional groups keep a total control over the content of the exchanges, can also be successful (Robertson et al., 2003; Swan, Scarbrough, & Robertson, 2002).

KM AS A SOCIAL PHENOMENON

Hot Questions Raised by ICT Assisted Knowledge-Pooling

Implementing these practices promoted in work on KM raises a number of questions. First, what sort of knowledge can managers legally compel employees to disclose? All sorts? Are there any limits? Of what kind? Resources (knowledge) that formerly belonged to people and resulted from their training, their experience, their faculties of logical reasoning, and so on, have suddenly become *organizational resources* at the same time that ICTs—in the form of devices owned by the management—have made possible the sharing, quick circulation and storing of knowledge. In other words, from resources *within* the organization, held by the hired expert, they have become resources *belonging to* the organization, held by them. The way in which KM theorists gloss over the issues involved in this shift in ownership is quite striking:

> *The learning dynamic and the means of capitalizing on experience are what distinguish one firm from another. These types of acquired knowledge reside neither in* blueprints *nor in the memory of a given member of the organization. The* control of knowledge *is an* attribute *of the* organization *and not of the individuals composing it.* (Tarondeau, 1998, p. 29; my emphasis; see also Jacob et al., 2000, p. 10-12, cited above) [translation]

Though not pretending that we cannot find specialists' discussions on the individual/collective fundamental *locus* of knowledge from an ontological perspective (see for instance Brohm, 2005, p. 87-88; for the "individual" perspective, Argyris & Schön, 1978; Simon, 1991; for the collective perspective, Boland & Tenkasi, 1995; Cook & Yanow, 1996; Weick & Roberts, 1993; for a mixed perspective, Davenport & Prusak, 1998; Kessels, 1996; Nonaka & Takeuchi, 1995), I must insist on the fact that the legal debate *per se* is still ongoing and the issue unsolved, so the promoters can keep on saying that an organization owns the *control* of knowledge. How far do the KM promoters mean this control go? Insuring the quality of the means implemented to exploit the knowledge? Ownership? Does it simply mean that credit for this control lies with the group, with project teams, rather than with individuals? The transfer of important assets on the labor market that results from this operation raises legal questions that are treated only casually. Does management, which owns the technological system used to support KM, automatically become the owner of what is stored in it? In the works I reviewed, this legal question is not addressed. No one states, for example, that because management pays and oversees the people it hires to work for its business, the knowledge that its employees put to use for the purpose of that business is regarded as one of its resources. These underlying assumptions are never made explicit and are therefore never examined or verified. Aside from the legal aspects proper, are there no questions of legitimacy or ethics involved in the requirement to share knowledge? No one would dispute the fact that there are a number of interests at stake, so there must be such questions!

And yet, as critical as one can be toward the new KM fad and its actual innovative nature, rightly so, even if these are renamed and revamped old IM practices, these still aim at changing working practices, the way we manage work, the way people communicate at work and, ultimately, favor information sharing (Wilson, 2002; for case studies: Robertson et al., 2003; Robertson et al., 2000; Swan et al., 2002). This aim is what is at stake in this chapter, notwithstanding the delusive innovative status of the KM.

Relative Failure of Knowledge-Pooling Practices

Despite managers' obvious interest in knowledge pooling, their demands for it do not always meet with success, far from it. There are a number of barriers mentioned in the literature. Omitting those that are functional or technical in nature, I will concentrate on just one: the reluctance of highly qualified professionals to participate in knowledge pooling.

Observers invariably focus on the failure of professionals to contribute to knowledge pooling, especially with respect to so-called tacit knowledge, the passing on of which is considered to be one of the chief challenges of knowledge managers:

This is one of the major challenges to be met when implementing a formal knowledge-management system: how to persuade people, the stars, to share what has made them individually successful, in order to enrich the pooled knowledge base. A recent survey by the American Management Association (1998) of 1,051 respondents came to the same conclusion. The answer to the question "What is the biggest knowledge management problem?" that received the greatest number of responses (41%) was persuading people to share their knowledge! (Jacob et al., 2000, p. 41; see also Ballay, 1997, p. 277; Bukowitz et al., 1999, p. 162; Zack, 1999, p. 47) [translation]

This does not mean that these demands are never successful nor does that mean that these successes have never been studied (for instance, Robertson et al., 2003; Robertson et al., 2000; Swan et al., 2002). But these documented successes lay open, obviously, the flaws in the analyses of KM implementation consultants in setting out factors of success that are totally ignored in the KM guides.

Explanations of KM Implementation Experts

The authors often quoted have little to say about this issue, limiting themselves to addressing the importance, for successful KM, of fostering a "sharing culture," based on conditions expressed in pop psychology terms: the desire to pass on one's knowledge, trust, the right to make mistakes, solicitude, and mutual assistance. These are all conditions favorable to passing on tacit knowledge:

Beyond the KM roles proposed earlier, effective knowledge creation, sharing, and leveraging requires an organizational climate and reward system that values and encourages cooperation, trust, learning, and innovation and provides incentives for engaging in those knowledge-based roles, activities and processes. (Zack, 1999, p. 55; see also Jacob et al., 2000, p. 41, 43; Zara, 2004, p. 9)

So many people share this belief in the virtues of pooling knowledge, with rarely a dissenting voice, that it would be tedious to quote them all here (see, *inter alia*, Zack, 1999, p. 47, and Scarbrough, 1999, p. 6). Furthermore, writers who discuss the problem of the lack of cooperation among professionals convey the explanations given by consultants in the field. It is not hard to make out that the proposed solutions are inspired by a folk psychology explanation to the effect that professionals may be "inhibited" by a competitive, individualistic organizational culture that fosters selfishness and provokes distrust of peers, because they won't "let you make mistakes" (see also Audebert-Lasrochas, 2000, p. 234-5). This is in no way surprising, as the KM literature adopts a mostly instrumental approach (for a discussion of the relevant literature, see Brohm, 2005, p. 93-95).

The material interests of professionals are in general ignored in these explanation frameworks, at the most touched upon in some proposals, since to the conditions listed in the previous quotation, must be added the courtesy to reward "donors" by crediting them by name for their contributions:

Some surveys make no bones about it: one of the most gratifying measures [is] entering the names of "knowledge donors" into the collective bank. More than special attention showing appreciation, this gesture is highly symbolic and has far-reaching consequences. First, by signing individually, users-providers thereby assert their shared responsibility for the entire process. (Jacob et al., 2000, p. 58; see also Prax, 2003, p. 114) [translation] and Robertson et al., 2000, p. 248-250)

This solution assumes that one of the reasons behind the lack of cooperation of some professionals has to do with concerns about preserving their symbolic capital and intellectual property.

A more recent explanation—related to the "cultural" explanations, but better informed—distinguishes between the various circumstances under which professionals are asked to share and thus transcends the simple explanation of reluctance to share or selfishness by focussing on the choices offered to professionals for sharing their knowledge (Ballay, 2002, p. 165). Ballay emphasizes that experts do not refuse to collaborate, but choose where they do so; he compares experts' interest in sharing knowledge within a community of practice and their interest in sharing with coworkers, at the manager's request and within a structure imposed from the top. This analysis has the advantage of highlighting the difference between the communities of practice set up by and for peers, using their hermetic professional jargon, and the knowledge-sharing structures set up by management, which more or less oblige employees who do not *a priori* feel an affinity for one another to work together for purposes that escape them. The first transcend organizational boundaries, are informal and bring together people based on their common knowledge, without the evaluation issues that are ever-present within organizations. These, of course, are ideal types, as between these two poles we can find intermediary forms wherein professionals successfully share information and knowledge in project teams and with peers, at the managers' request, but orally and in the course of action, without keeping any written record nor putting anything at managers' disposal (Robertson et al., 2003; Robertson et al., 2000; Swan et al., 2002).

Here the source of the problem of professionals' unwillingness to cooperate is seen as being more social and less psychological, but the solutions, besides being treated very casually, are strictly folk psychology, which does not reflect the analysis of the source of the problem or contradict it even. What is worse, in the name of the undeniable stimulation that may be gained through intellectual exchanges, the threat of losing one's exclusive specialized knowledge is brushed off:

It is obvious from the incredible complexity of knowledge that any analysis that takes as a principle that an expert who communicates his knowledge will thereby lose influence and power is far too simplistic. For this is to see knowledge as a weapon that can be held and passed from hand to hand. Of course, a manufacturing secret transmitted from one expert to another in the same field can circulate in this way. But once we are talking about different types of knowledge, the exchange does not have such a mechanical effect; sometimes it is by better explaining the conditions and constituent parts of their expertise that players can highlight the complexity and importance of what they do. (Ballay, 1997, p. 277, citing Hatchuel and Weil, 1992, p. 106) [translation]

Alas, demonstrating that the threat is unfounded is one thing, if you manage to do it, but getting the professionals concerned to believe it is another—and it really is a matter of belief, and not of whether the threat is founded or not. Let me emphasize that I am not saying that there is or is not *in fact* a risk of losing major assets, but that the perception of that possibility by the professionals concerned is what motivates their actions. I cannot pronounce on whether the risk exists or not; what is worse, in my opinion, is that no one can, across the board. In the short term, the risk depends a great deal on the field of employment and a number of case studies would be required to estimate it. A medium- or long-term assessment would require a crystal ball, because these are processes that take time and are influenced by a number of forces (see Ballay, 2002, p. 182-5). But whatever the objective risk, that does not prevent the experts involved from assessing it subjectively when they decide whether to cooperate or not.

Worse, when material interests of solicited professionals are not ignored, they're denied (as are power relations; see Brohm, 2005, p. 93-95; Coutu & Willmot, 2003; LaPalombara, 2001; and Swan et al., 2002 on that topic) and even those experts who admit the existence of obstacles to knowledge pooling will denounce them as evidence of rigid corporatism in the face of the demands of sustainable development, no less (for instance, Prax, 2003, p. 1).

The steps recommended by KM promoters involve very intimate incursions into the minds and consciences of professionals involved in the process and are surprisingly normative in their statements about what management must do to enforce knowledge pooling, making short shrift of the intentions of those concerned, which are mentioned only to say that they must be "changed" if they are not "in line":

The first thing that is required is each person's commitment to cooperative behaviour. This personal commitment could take the form of signing the organization's collaboration agreement *and thereby subscribing to a* code of collaborative ethics *consistent with the agreement. Changing behaviour means obliging someone to change his or her system of values to get mind (values) in line with action (behaviour). [...] It is important to change behaviour if you want to change values, and the question of behavioural change cannot be negotiable, as it is what enables an organization to evolve.* (Zara, 2004, p. 37, 39) [translation]

Three Main Insatisfaction Sources Grounded in Sociology

The explanations of KM promoters must generally be assumed from their proposed solutions, for they are either underdeveloped or left entirely unsaid. All of them are also highly unsatisfactory sociologically speaking, for three reasons. First, materialists will find nothing convincing, because these individual behaviours, since they are fairly consistent within the environment of the knowledge economy, are beginning to look like social facts. Indeed, basic sociological principles require us to see collective effects that must be examined as such and must be linked to objective interests and strategies for action (Drazin, 1990; Swan et al., 2002). Second, the explanations are often too superficial and themselves require explanation. It goes without saying, for example, that a "climate of trust" will not hurt collaboration; but what are the conditions, the issues, the guarantees in local social relations, of such a climate? If the phenomenon is a social fact, the explanation must also be social; a strictly psychological explanation is not good enough. Even Prax's slightly more developed framework (2003, p. 110-8 and 394-9) gives only a few keys to the smooth operation of a knowledge-sharing team, but fails to attack the systematic refusal to cooperate at its source. Hardly a few of them question the basis of these refusals, but all of them recommend action, rather coercive in the case of Zara, for instance, to deal with this resistance. Last, the KM program widely promoted in consultancies so far is essentially normative, not based on a theoretical apparatus that would take into account the social relations of work (contrary to those above mentioned: Robertson et al., 2003; Robertson et al., 2000; Swan et al., 2002; Scarbrough, 1999, p. 6-9). For instance:

> *The current difficulties of many companies in implementing a sustainable KM process can be explained in part by the following reasons [...] 25% of negative leaders, opponents to any change, who struggle fiercely to maintain the status quo. Their undermining efforts will be all the more effective if they keep quiet about them. They will feign goodwill from beginning to end [...] These opponents, although in the minority, are still today the big winners and explain the very limited diffusion of collaborative intranets, for example. [...] The chances of success are therefore low today, except in some more* culturally advanced *companies.* (Zara, 2004, p. 55; my emphasis)

The explanations provided by these promoters—among other things, associated *a priori* with implicit theories—rely on the symbol of cooperation as a universal virtue, independent of social relations, and lead inevitably to a moral or "culturalist" explanation of the failure of knowledge-pooling systems.

Yet to stand up to examination, the rather unsatisfactory assumption of selfishness or mistrust should be set against the assessment of the work motivation of these professionals and take into account, where applicable, the gap separating their failure to contribute to knowledge-sharing systems and their high motivation. In a recent study of new-economy IT professionals (Chasserio & Legault, 2008; Legault, 2005), I found that they were very motivated: they put in greater effort than required, demonstrating availability, flexibility and initiative, paying special attention to costs and customers, working unlimited hours, putting their personal lives second, demonstrating productivity, cooperation, mutual assistance, etc. The assumption of the "inertia" of "unmotivated" employees would not stand up in that study population.

Loyalty has not disappeared from these technological services to business organizations, for example, but experts now have more than one loyalty: to their organizations, of course, but also to their customers, to their colleagues and to their responsibility to their boundaryless career development (see Alvesson, 2000, p. 1109-11). In other words, professionals have not stopped being loyal, but their sole loyalty is not to management; they have a number of competing or successive loyalties.

A sociological analysis of history sheds a different light on the relative failure of knowledge-pooling practices in KM. If we assume that the result of the process of having professionals pool their knowledge and make it explicit is that their managers can have access to it, even unlimited use of it, certain past experiences take on a new relevance to understanding what then becomes resistance to the appropriation by others of a hitherto hermetic knowledge that was coveted and thus negotiable. The marginalization, in many areas, of cottage-industry work methods at the time of the industrial revolution constitutes one of these useful historical reference points.

In many sectors, the cottage-industry work model was replaced by industrial work organization at the time of the industrial revolution, although the earlier form never totally disappeared. In the sociohistorical process that led to the shift from cottage-industry to mass industrial production, appropriation of the knowledge of craftspeople played a crucial role in setting up the division between the design and execution of work and the social and technical division of labor that was a precondition of industrialization.[e]

The coming together of independent craftspeople (entrepreneurs, in fact), in a collective workplace, under the legal and economic subjugation of an owner of the means of production was already a step towards the accumulation of capital, but the power derived by workers from possessing hermetic know-how essential to production acted as a brake on productivity (Clegg, 1981, p. 547). That gave rise to a major transformation of work organization, which in turn led to a major transformation of social relations: deskilling of craftspeople, incorporation of knowledge into a sociotechnical system that allows processes to be rationalized and standardized, a wage relationship instead of a trading relationship between craftspeople-entrepreneurs and their customers.

Let us be clear, however, while seeking the optimal appropriation of workers' knowledge is a condition of the capitalist mode of production, no one is claiming that knowledge has been entirely eliminated from industrial work, and the evolution of work organization cannot be summed up as the deskilling of workers. There are still traces of the crafts' or trades' organization of work in modern industry, where there is still skilled labor, in the craft guilds typical of the construction industry especially, in the professions—among independent entrepreneurs as well as in groups of professionals within large bureaucratic organizations.

A large part of the work of contemporary sociology of the professions concerns the transformations in the status of professionals that occurred as a result of the increased employment of salaried professionals in organizations, who are greater in number than self-employed professionals or those practicing in partnerships, at least in Great Britain and the United States (Evetts, 2001, p. 5), who may or may not have a professional licensing body or reserved title, working for public or private organizations (Derber, 1982, 1983; Freidson, 1994; Hirst, 1982; Hoyle & John, 1995).

Regardless of how we interpret the transformation in the status of professionals, bureaucracies that employ them seek to gain some sort of control over them. Yet the range of instruments of control of work has been refined and now includes means that vary from direct control, through bureaucratic control, to assigning "responsible autonomy" (Lincoln & Kalleberg, 1990, p. 9, citing Friedman, 1977). Professionals, as a social group, respond to this attempt to control them by reacting and resisting. They enjoy a certain power to resist efforts to control their work, however, and an entire branch of sociology of the professions has been examining this question of a new method of labor management for qualified, salaried professionals from an antideterminist perspective with regard to class theory.[f] The idea of a regulating mechanism in constant flux is more useful than that of a rigid rule structure that standardizes the social relations of work. Management controls qualified professionals less directly because it has thus far been unable to take control of their exclusive and (relatively) hermetic knowledge and eliminate it from their work process, although it can reduce it.

The issue of control is crucial to salaried professionals, who are more susceptible to pressure from outside the profession (public policies, in-house policies, public opinion) if the profession does not have its own regulating body (Hirst, 1982, p. 173). More, the professionals I have studied are hired by organizations that have a stake in a post-Fordist, liberalized economy, in post-bureaucratic organizations that are characterized by greater competition for smaller markets, by a system of capital accumulation that demands great flexibility on their part and on the part of their employees, by the tenuousness of the employment relationship, and thus, by the very great external mobility of professionals (Hargreaves & Goodson, 1996; Helsby & McCulloch, 1996; Robertson,

1996, p. 37-8). Such an economic context favours intentions of rationalization on the management side, and the resistance of professionals is merely the corollary.

One of the main instruments that enable skilled workers to resist control is the fact that they keep their knowledge quite hermetic, which guarantees them some discretion in performing their duties, since they clearly cannot avoid economic and legal subordination when they work for a salary. Developing and maintaining a hermetic language for their knowledge is the obvious way they have to protect the exclusivity of their symbolic capital and resist attempts to control their work. This means of resistance addresses a perceived threat of control from the management and has nothing to do with a refusal to share among peers, should they be outsiders; case studies in fact show an actual collaborative tendency among scientific from different fields (Robertson et al., 2003; Robertson et al., 2000; Swan et al., 2002).

This is a very inspiring notion in the process of accounting for professionals who do not contribute to KM knowledge-pooling systems, since the demand to contribute to these systems goes to the heart of their exclusive knowledge, and the hermetic language that goes with it. Interestingly, they often mention this language as an obstacle to passing on their knowledge. Indeed, one of the local issues of debate between managers implementing these systems and the experts is precisely the impossibility of translating knowledge expressed in their own jargon into the standardized format of the system, which the experts claim is too confining. For the purposes of this chapter, it is not necessary to determine whether this objection is well founded; perhaps it is only a pretext, after all, but what is interesting is that they are citing their hermetic language as a reason to refuse to share their knowledge, when this very language is a well-known means of protecting the status of professionals in hierarchies.

QUALIFIED PROFESSIONALS AND KM

Alternative Hypothesis Concerning their Lack of Cooperation

The objectives of expert systems and other knowledge-based systems challenge the notion of the exclusivity of some types of knowledge. Without claiming that they will in the short term enable professional work to be split between design and execution, they may be an important outside means of controlling work as well as significantly reducing discretionary power. For professionals who are not organized into an association, the implementation of joint resistance strategies is more difficult than for those who are, and the threat is therefore much greater.

As a result, I hypothesize that in refusing to cooperate, salaried professionals, especially those who are not represented by a professional body, are resisting what they perceive to be a threat to take over their hitherto exclusive knowledge. This they do more unofficially than officially, more surreptitiously than openly, even more unconsciously than consciously, using the means available to them individually, which are necessarily different from and more limited than those of group action, because they have no organized structure to defend their professional interests. Their resistance is based on the fear of deskilling and loss of control over how they practise their profession, related to the top-down imposition of these systems that do not ensure reciprocity. If they were organized, they could express their resistance through other (group) strategies, but the ways they choose to act are symptomatic of their isolation in a competitive market.

The fear of deskilling and loss of control is a resistance factor sorely neglected in work on KM and yet it seems to me that its potential to explain the behaviour of qualified professionals does not deserve such neglect. Let us look at why.

Importance of Assets on Job Market

The most hermetic and least codified knowledge in these highly skilled sectors is what makes these professionals valuable on the job market, whether in-house or external. Not only among the very mobile new-economy professionals, but also in large bureaucracies, the value of experts rests largely on the appreciation of the know-how that makes them necessary, that distinguishes them from their colleagues and confers upon them a certain amount of individual bargaining power, a position within the organization, and a significant asset if they should leave later (Scarbrough, 1999, p. 11-12).

Now according to KM theory, as we have seen, that knowledge which used to belong to *people* and resulted from their training, experience, powers of logical reasoning, etc., must become *organizational resources*. In terms of perceptions,[g] if the experts believe that not only their own personal assets but the very principle of the hermetic language of their profession and the existence of a territory protected by the impossibility of codifying some types of knowledge are under attack, why would they not see their professional interests threatened?

In a short online article, a specialist (Goman, 2002) gives five reasons why, in her view, experts keep their knowledge to themselves: the belief that knowledge is power, insecurity about the value of their knowledge, lack of mutual trust, fear of ridicule and lack of reciprocity. All these reasons are psychological, some overlap, and she leaves some out. In the electronic discussion that followed the article, one of the experts responded:

People don't trust management. *I am astounded that the article, "Why People Don't Tell What They Know," as well as the responses I read, leave out one of the most obvious reasons—people don't trust management with the results of the sharing! One of the goals of a knowledge sharing and gathering effort is frequently to reduce human capital (lay off people, more bluntly) and people, having seen many years of this, ask themselves, "Why should I share?" Knowledge is power and if you don't know everything I know, maybe my job is secure. The fact that the author and the respondees did not even mention this tells me that folks who are preaching about this topic are way out of touch with reality.* (Jack Hipple, Innovation-TRIZ, Inc., 8/16/2002)

From the management's point of view, why would experts rely solely on the knowledge they have already stored up individually rather than relying on that which can be found in the organization where they work, the knowledge of all their colleagues, and thus reduce the uncertainty in the task? Yet the management's plans to reduce *its* uncertainty are in competition with those of the professionals to whom knowledge pooling is nothing more than a factor increasing their own uncertainty on the job market, one that runs counter to their ongoing efforts to preserve their own value. Wilson (2002) tackles these issues and challenges the democratic utopia based on autonomy and sharing by opposing the numerous mass lay-offs in the consultancy sector to this promise. Indeed, focussing on innovation, KM is often part of a project-management approach and serves its aims; KM theorists are quickly satisfied that employee participation in decisions about what processes to adopt within the organization, work in teams with varying degrees of independence and, short, flat hierarchies are characteristics of work organization that contrast sufficiently with the Taylorist model to justify their proclaiming the progressive nature of knowledge sharing, without referring to the other conditions of this pooling of knowledge (Cotte, 1999, p. 12).

Because ICTs allow much faster circulation, storage and sharing of information, their potential for the *organization* is easy to see. Sidestepping the question of the impact on the professional categories affected, their proponents associate them with *transparency* (for the benefit of management, in fact, but at what price?) and declare them to be factors in progress. The normative content of KM handbooks promotes transparency, participation and accountability, in keeping with the new high-performance "liberal organization" model:

The ultimate aim of collective management [a precondition of KM] is to develop trust among members of the group and to foster commitment through action. This trust and commitment will emerge (1) if people are involved in making decisions that concern them, by giving their opinions and taking part in debates; (2) if, once the decision has been made, the why, the reasons for the decision, are explained, so that everyone can be sure that the decisions have been made in the interests of the organization (commitment will exist even among those whose opinion has not been followed); (3) if, once the decision has been made, the new rules are clearly set out and understood by everyone. (Zara, 2004, p. 12) [translation]

KM manuals adopt the normative version of this organizational model and associate it with great autonomy in doing the job, enhancement of skills through the dissemination of knowledge, and participation in discussions, but not decision making. Experts are asked to share their knowledge, reveal their knowledge to managers who will then decide and "explain things" to the team. Now, the normative version gives short shrift to the diagnoses of salaried professionals regarding their role in such participatory structures, their opinions of the relationship between the contribution requested and the reward offered, and their consequent interest in participating in this organizational model (see Courpasson, 2000, p. 187-211).

But therefore, it is hard to imagine, as KM experts suggest, a way of reducing management's uncertainty that would also be universally effective as a motivator, because decisions to do with work organization are made within a social relations framework and become issues at stake in the players' strategies.

The organization is only one setting where professionals feel they belong; they have other loyalties, too, including to their customers. They may well regard the know-how they use with certain customers—or in some cases have developed for a certain customer—as a secret between them and that customer, to whom they are also loyal (Robertson & Swan, 2003, p. 852). Telling the story of a given project for a given customer, for example ("feedback on the experience," one of the activities required by KM) may breach an unspoken agreement. In a context of high mobility and boundaryless careers, customers will survive the current employment relationship, and what is more, will often play a very important role in finding replacement jobs (Alvesson, 2000, p. 1109-11). And yet, replacement is an important issue as very few organizations in the sector of specialized consultancy for businesses, for instance, can pride themselves with a low turnover and a high retention rate (like the one studied by Robertson et al., 2000). Tension is inevitable between cross-organizational job markets and knowledge-sharing systems implemented separately within each organization.

Importance of Professional Discretion

The concept of professional discretion refers to the assumption that professionals have the capability, and can be given responsibility, to assess problems and their circumstances and to state their views by advising, deciding and recommending. The hermetic part of their knowledge is the best protection they have against control, guaranteeing them professional discretion founded on competency that managers and customers cannot assess unless they have the same training (Alvesson, 2001). For managers and customers, only results testify to their skills.

After pooling knowledge, professionals may fear (in the more or less long term) an increase in the number of predefined procedures for operations that until then had depended on their individual discretion (Alvesson, 2001), for is not the purpose of the operation to disclose "best practices," the new keystone of management sciences? Indeed, KM may be seen primarily as a means of shifting the weight from the first antidote to the uncertainty involved in a task (the knowledge *that a worker has already stored up)* to the second (the knowledge that *can be found* within the organization). In this regard, contemporary sociology of the professions underscores the power that specialized software gives to managers concerned with reducing the margin of discretion of salaried professionals by standardizing practices and routinizing processes (Evetts, 2001, p. 8; Freidson, 1994, p. 167). As salaried workers, they have little control over the end results of their work (*ideological autonomy,* to use Derber's term, 1982), because the manager defines them. Wrongly or rightly, these professionals may feel that they are also being asked to give up control over the organization of their work and their professional discretion over how to do their job (*technical autonomy,* Derber, 1982) as a result of knowledge pooling (Drazin, 1990; Swan et al., 2002). Conversely, that very importance of professional discretion can explain the success of knowledge pooling practices that are oral and limited to peers in a team project, as observed by Robertson et al. (2000) or, more generally, the success of those that are not imposed on professionals nor controlled by managers and wherein professional groups keep the control over the content of their exchanges (Robertson et al., 2003; Swan et al., 2002).

Legal Protection Won't do Them Much Good

If its demands to pool knowledge are successful, management will make professionals reveal knowledge that it considers implicitly to be its property, but which, up until the advent of ICTs, was largely inaccessible for purely practical reasons, because there was no technical way to store it and use it, that requires very rapid interconnexions through huge amount of words (Cotte, 1999, p. 15). Questions of intellectual property have not been raised to support what has so far remained a shift in the meaning of property rights, from rights over technical equipment to rights over what is stored using it.

In terms of labor law, in the case of salaried professionals, the concept of legal subordination of the employee to the management gives the manager the right to take all the steps that, added up, make it possible to appropriate the knowledge of employees: the right to oversee work, to give instructions concerning procedures, to set up work committees that communicate electronically, to find out what means have been implemented to reach an objective or do a job, etc. The manager has the right to watch over people at work, to question them, to look for better ways to organize the work, to change the organization of the work.

In terms of protective strategy, in the case of professionals asked to pool knowledge, intellectual property rights may serve to protect an invention, by means of patents, for example, but they will not be of much use in protecting knowledge not yet codified or materialized in the form of a product or process, or methods of doing things. Will experts be able to invoke some notion of trade secret or personal information to ward off their management's demands to share their knowledge? Self-employed professionals or consultants commissioned for their expertise on an *ad hoc* basis (expert opinion, report, study) can legitimately keep to themselves the secret of the intellectual process that leads to the final product, for their customer is only buying the final product. But the question of trade secrets is purely hypothetical in the case we are examining, having never yet been brought before the courts (at least in North America).

A fortiori, in the case of salaried professionals not organized into an association, the most clandestine strategies offer the best prospects of effectiveness without having to be justified for what they are, which is better suited to individual resistance not backed up by a collective movement.

These Professionals are not Organized

The question of the legitimacy (and not the legality) of the demand to pool knowledge is poorly dealt with in KM theory texts, either (for a discussion of that point, Brohm, 2005, p. 93-95). If it is mentioned, it is only casually, with the assumption that it goes without saying that this kind of sharing enhances the status of workers who are asked to share their knowledge, because it is supposed to be a step towards industrial democracy. The KM universe fully endorses "responsible autonomy" with regard to worker control:

Employees are our most important assets. We can't afford to lose our most important assets. We have to utilize the knowledge of every single employee. Everybody has to understand the importance of their work for us, and they have to be personally responsible to a great extent. (Andreas Rihs, one of the co-founders of *Phonak*, a leading light in KM, quoted in Von Krogh, Ichijo, & Nonaka, 2000, p. 36)

Earlier we presented the type of control of work proper to highly skilled professionals, "responsible autonomy," and the manufacturing of consent. Picking up our reasoning where we left off, let us keep in mind that, based on empirical work, "responsible autonomy," as a form of control, cannot function without clear, rational rules—consistent procedures that provide guarantees of reciprocity within which rights and obligations are shared rather than unilateral (Halaby, 1986, p. 634).

In this context, is it legitimate for management of a company involved in KM to subject its professionals to the consequences of KM in terms of the stripping of knowledge that will eventually result? In other words, if the professional perceives that he or she is losing the term "autonomy" from the method of control he or she believed in up until then, why should that person continue to be "responsible"? Here again, conversely, that very importance of professional discretion can explain the success of knowledge pooling practices that are oral and limited to peers in a team project, as observed by Robertson et al. (2000) or, more generally, the success of those that are not imposed on professionals nor controlled by managers and wherein professional groups keep the control over the content of their exchanges (Robertson et al., 2003; Swan et al., 2002).

Officially posing the question about legitimacy would require professionals to formally designate stripping and appropriation as the object of the process. But to pose the question about the ultimate aim of the management's actions, professionals asked to share their knowledge for KM purposes would have to be organized, which is not always the case (teaching professionals are one group that could ask this type of question; see Locke, 2001, p. 47-8). The question of what is at stake for professionals contributing to knowledge pooling, their individual interests in pooling knowledge in a system designed by and for the management of their companies, cannot be raised without some sort of organization. Salaried professionals who are *rather lukewarm* towards the effects of KM have no other option than underground resistance strategy.

Conclusions of Study of Communities of Practice

Experience has shown that informal communities of practice create a tremendous dynamic of innovation and integration. However, they often remain unofficial and are not always accepted by the hierarchies (Prax, 2003, p. 39). And for good reason, because they often extend beyond the boundaries of organizations and enable the dissemination of knowledge *among peers* at competing companies! Studies generally conclude that when an attempt is made to implement this type of practice from the top down, when professionals are competing with one another, whether on an in-house or external job market, sharing does not take place (Brown & Duguid, 1991). KM systems imposed from above do not have the features of the ones that experts would themselves choose spontaneously, and they are not compatible with their interests (Lesser & Storck, 2001; Merali, 2000; Wilson, 2002).

The initiative must come from below, be implemented spontaneously and remain under the control of its participants if it is to be a success. "In a very hierarchical management universe, people keep their knowledge to defend themselves" (Prax, 2003, p. 38). This KM promoter unfortunately does not explain what experts are defending themselves against—though a good analysis would take that as its starting point. For the experts are obviously not totally reluctant to share, but their attitude depends on whom they are sharing with.

There are three avenues to explore here: (1) Do those who share their knowledge in communities of practice have boundaryless careers or are they more likely to be hired in large bureaucracies where their jobs are fairly stable? (2) Does the community of practice guarantee reciprocity so that those who share their knowledge and experience are assured of benefitting equally in return in a way that will help advance their careers? (3) Are they asked to share the same type of knowledge in KM systems as they share in communities of practice? Future researches must deepen our knowledge on these issues.

CONCLUSION

At first glance, what KM promoters have to say about strategies to motivate and mobilize participants and the rewards of knowledge sharing seems naïve, in light of what is at stake for those being asked to contribute their knowledge. The promoters' literature on KM deals so unsatisfactorily with lack of cooperation that it highlights the theoretical and analytical weakness of the field as it now stands.

The inadequacy of the proposed explanations may be attributed to the fact that knowledge pooling is a practice that has given impressive results in the field of communities of practice, spontaneously implemented from the bottom up and later observed by experts or consultants, most often IM experts, who have deduced from them the ingredients for a new magic formula, without doing an in-depth analysis of the conditions of emergence (and success) of the process (on the matter of methods of dissemination of new fashions in management thinking, see Swan et al. (2001).

Yet, KM promoters and "how-to" books' authors blatantly fail to address the real issues at stake regarding the pooling of knowledge, that is, intellectual property issues, in more than one point of views. Between "managing knowledge" and "appropriating knowledge," there's a shift that hasn't been lost on experts, but nevertheless, this issue stands in a relative gap in the law as well as it's lost on any ruling body, be it professional, managerial, or else.

A study of the work of KM system promoters shows that *organizational knowledge* is defined as an *aggregate of individual knowledge*, a view that has serious practical and legal consequences. A new ethics is rapidly spreading through the literature:

Information should not be a source of power for its holder, but for the one who knows how to cre-

ate collective intelligence by sharing it. (Rochet, 2003) [translation]

Who, then, knows how to create collective intelligence? In any case, management will have extensive means at its disposal if it succeeds in obtaining access to the pooled knowledge of all its experts! However, the highly qualified professionals who are being asked to share their knowledge may perceive the initiative as threatening to their interests because of what is at stake for them. The scope of the dissemination of this shared knowledge via an automated system open to managers is an unheard-of change in practices, and no one can yet know what its effects will be.

Facing this vacuum, there are no clear guidelines that can be enforced and there is still a "free zone" regarding knowledge: managers ask for it all, while some experts share it, others keep it to themselves, though not overtly. As long as either law nor case law do not provide clear answers, and as long as professional bodies do not either provide clear guidelines (including rights and obligations), experts' propensity to share knowledge will mainly depend on their labor market situation (assuming that the more mobile they are, the less they wish to share in order to save their assets on the market), the importance they attach to professional discretion (assuming there is a threat of managers implementing standardised "good practices" based on a new *organizational knowledge)* and, finally, their control on whom are they sharing with, and the purposes of the pooling. Without any response on these, good will and fickle circumstances will bring some to share, with fickle results for the managers...

Now, what is left to be said to unfortunate knowledge managers failing in making professionals share knowledge? Sadly, I must admit that I do not believe KM promoters could do far better in giving tips to these unsuccessful managers, simply because the aim of managing knowledge is, in my point of view, another example of the overwhelming faith in management and its range, particularly in control over human beings and in the organizational capacities and reach of management. The loss of knowledge for management in a world of boundaryless careers is a simple by-product of many practices of the so-called new economy, and cannot be magically checked by the virtues of management. Far from a matter of chance, there is an important ground for knowledge-pooling to thrive in communities of practice (and far less elsewhere); cooperation and collaboration in a democratical environment is still a way of organizing, that may well have no substitute in the hierarchical environment of the firm, where workers are asked to behave like citizens while not experiencing citizenship (for an interesting discussion of this mythical faith in managerialism, see Parker, 2002, p. 1-16). In other words, sometimes, some places, management techniques are powerless. When the putting-out system was implemented in England, craftsmen did not use to refuse to share their trade secrets with peers; many of them refused to do so with factories' owners, on the grounds of their material interest, while others probably agreed to. Would *management techniques* have done better than coercition with the former? We can doubt it.

Of course, the hypothesis set out here must be subjected to empirical examination, and all I wished to do was demonstrate that it would be wise to do so; proving its validity is a task that goes well beyond the scope of this chapter. For the reasons given above, I do, however, feel that it is valid, as long as it is agreed that it does not bear on the real consequences to be expected of KM with regard to professional qualifications, but rather on these professionals' perceptions of themselves as actors who trigger action. Indeed, professionals are resting for the time being primarily on their perceptions of the risks of the operation, which are a driving force of action to no lesser extent.

REFERENCES

Allen, R. (1990). *Thinkers of our time.* Polanyi. London: Claridge Press.

Alvesson, M. (2001). Knowledge work: Ambiguity, image, and identity. *Human relations, 54*(7), 863-886.

Alvesson, M. (2000). Social identity and the problem of loyalty in knowledge-intensive companies. *Journal of Management Studies, 37*(8), 1101-1123.

Apple, M. W. (1995). *Education and power* (2nd ed.). New York: Routledge.

Argyris, C., & Schön, D. (1978). *Organizational learning: A theory of action perspective.* Reading, MA: Addison-Wesley.

Audebert-Lasrochas, P. (2000). *Les équipes intelligentes.* Paris : Éd. d'organisation.

Ballay, J. F. (2002). *Tous managers du savoir! La seule ressource qui prend de la valeur en la partageant.* Paris : Éd. d'organisation.

Ballay, J. F. (1997). *Capitaliser et transmettre les savoir-faire de l'entreprise.* Paris: Eyrolles.

Bélanger, J. (1991). La production sociale des normes de production: monographies d'entreprises en Grande-Bretagne et au Québec. *Sociologie et sociétés, 23*(2), 183-197.

Boland, R. J., & Tenkasi, R. V. (1995). Perspective making and perspective taking in communities of knowing. *Organization science, 6*(4), 350-372.

Brohm, R. (2005). *Polycentric order in organizations. A dialogue between Michael Polanyi and IT-consultants on knowledge, morality, and organization.* Erasmus Research Institute of Management (ERIM).

Brown, J. S., & Duguid, P. (1991). Organizational learning and communities of practice: Towards an unified view of working, learning, and innovation. *Organization Science, 2*(1), 40-57.

Bukowitz, W. R., & Williams, R. L. (1999). *The knowledge-management fieldbook.* London: Prentice Hall.

Chasserio, Stéphanie, & Marie-Josée Legault. (2008, in press). Strategic human resources management is irrelevant when it comes to highly skilled professionals in the Canadian new economy! *International Journal of Human Resource Management.*

Clawson, D. (1980). *Bureaucracy and the labor process: The transformation of U.S. industry, 1850-1920.* New York: Monthly Review Press.

Clegg, S. (1981). Organization and control. *Administrative Science Quarterly, 26*, 545-562.

Cook, S. D. N., & Yanow, D. (1996). Culture and organizational learning. In M. D. Cohen & L. S. Sproull (Eds.), *Organizational learning* (pp. 430-59), Thousand Oaks, CA: Sage.

Coutu, A. & Willmott, H. (2003). Re-embedding situatedness: The importance of power relation in learning theory. *Organization Science, 14*(3), 283-296.

Cotte, D. (1999). *Les enjeux du management de l'information dans les organisations: Usages, outils, techniques.* Lille, France: ADBS.

Courpasson, D. (2000). *L'action contrainte. Organisations libérales et domination.* Paris: PUF.

Davenport, T. H., & Prusak, L. (1998). *Working knowledge: How organizations manage what they know.* Boston: Harvard Business School Press.

Derber, C., & Schwartz, W. A. (1988). Les hiérarchies à l'intérieur des hiérarchies. Le pouvoir professionnel à l'œuvre. *Sociologie et Sociétés, 20*(2), 55-76.

Derber, C. (1983). Managing professionals: Ideological proletarianisation and post-industrial labor. *Theory and Society, 12*(5), 561-603.

Derber, C. (1982). *Professionals as workers: Mental labor in advanced capitalism.* Boston: GK Hall & Co.

Drazin, R. (1990). Professionals and innovation: Structural-functional vs. radical-structural perspectives. *Journal of Management Studies, 27*(3), 245-263.

Eccles, R. G. (1981). Bureaucratic vs. craft administration: The relationship of market structure to the construction firm. *Administrative Science Quarterly, 26*(3), 449-469.

Eraly, A. (1988). *La structuration de l'entreprise. La rationalité en action.* Bruxelles: Éditions de l'Université de Bruxelles.

Evetts, J. (2001). New directions in state and international professional occupations: Discretionary decision-making and acquired regulation. *Proceedings of the SASE 13th Annual Meeting on Socio-Economics Knowledge: The New Wealth of Nations*, University of Amsterdam, Pays-Bas, 28 juin - 1er juillet, retrieved from http://www.sase.org/conf2001/papers/papers.html

Friedman, A. (1977). *Industry and labor.* London: Macmillan.

Friedmann, G. (1964). *Le travail en miettes.* Paris: Gallimard.

Freidson, E. (1994). *Professionalism reborn: Theory, prophecy, and policy.* Chicago: University of Chicago Press.

Galbraith, J. R. (1973). *Designing complex organizations.* MA: Addison-Wesley.

Gelwich, R. (1977). *The way of discovery: An introduction to the thought of Michael Polanyi.* New York: Oxford University Press.

Gill, J. H. (2000). *The tacit mode: Michael Polanyi's postmodern philosophy.* Albany, NY: State University of New York Press.

Goman, C. K. (2002). *Five reasons people don't tell what they know.* DestinationKM-Viewpoint. Retrieved June 22, from DestinationKM.com

Halaby, C. N. (1986). Worker attachment and workplace authority. *American Sociological Review, 51*, 634-649.

Hargreaves, A., & Goodson, I. (1996). Chapter 1: Teachers' professional lives: Aspirations and actualities. In I. F. Goodson & A. Hargreaves (Eds.), *Teachers' professional lives* (pp. 1-27). London: Falmer Press.

Hatchuel, A., & Weil, B. (1992). *L'expert et le système. Gestion des savoirs et métamorphose des acteurs dans l'entreprise industrielle.* Paris: Economica.

Helsby, G., & McCulloch, G. (1996). Chapter 3: Teacher professionalism and curriculum control. In I. F. Goodson & Hargreaves, A. (Eds.), *Teachers' professional lives,* (pp. 56-71). London: Falmer Press.

Herman, E. S., & Chomsky, N. (2002). In Dan Frank (Ed.), *Manufacturing consent. The political economy of the Mass Media.*

Hirst, P. H. (1982). Professional authority: Its foundation and limits. *British Journal of Educational Studies, 30*(2), 172-182.

Hoyle, E., & John, P. D. (1995). *Professional knowledge and professional practice.* London: Cassell.

Jacob, R. L. & Pariat, L. (2000). *Gérer les connaissances: un défi de la nouvelle compétitivité du 21e siècle. Information, interaction, innovation,* CEFRIO. Retrieved from http://www.cefrio.qc.ca/rapports/gererconnaissance.pdf

Kessels, J. (1996). Knowledge productivity and the corporate curriculum. In J. F. Schreinemakers (Ed.), *Knowledge management: Organization, competence, and methodology* (Vol. 1, pp. 168-174). Würzburg: Ergon Verlag.

Laflamme, R. (1998). Mobilisation ou manipulation? In R. Laflamme (Ed.), *Mobilisation et efficacité au travail,* Actes du 9e congrès de *l'Association internationale de psychologie du travail de langue française* (AIPTLF), Tome 6 (p. 87-96). *Proceedings of Gestion des paradoxes dans les organisation*, Cap Rouge, Presses Interuniversitaires.

LaPalombara, J. (2001). The underestimated contributions of political science to organizational learning. In M. Dierkes, A. B. Antal, J. Child, & I. Nonaka (Eds.), *Handbook of organizational learning and knowledge*, (pp. 56-71). Oxford: Oxford University Press.

Legault, M. (1988). Le métier de chercheur scientifique en sciences sociales et la sociologie des professions. *Sociologie et sociétés, 20*(2), 163-176.

Legault, Marie-Josée. (2005). Differential gender effects of project management and management by project on skilled professionals. In Kay S. Devine, & Jean-Noël Grenier (dir.), *Reformuler les relations industrielles dans une économie de marché libérale, Actes du 41e colloque de l'Association canadienne des relations industrielles (ACRI) - Canadian Industrial Relations Association (CIRA),* (p. 105-124). Concord, Ontario: ACRI-CIRA et Captus Press.

Lesser, E. L., & Storck, J. (2001). Communities of practice and organizational performance. *IBM Systems Journal, 40*(4), 831-841.

Lincoln, J. R., & Kalleberg, A. L. (1990). *Culture, control, and commitment. Study of work organization and work attitudes in the United States and Japan.* Cambridge: Cambridge University Press.

Locke, T. (2001). Questions of professionalism: Erosion and reclamation. *CHANGE. Transformations in Education, 4*(2), 30-50.

Marglin, S. (1973). What do bosses do? The origins and functions of hierarchy in capitalist production. *Review of Radical Political Economics, 6*(2), 60-112.

Merali, Y. (2000). Individual and collective congruence in the knowledge management process. *Journal of Strategic Information Systems, 9*(2-3), 213-234.

Montgomery, D. (1979). *Worker's control in America.* Cambridge: Cambridge University Press.

Murphy, R. (1990). Proletarianization or bureaucratization: The fall of the professional? In R. Torstendahl, & M. Burrage (Eds.), *The formation of professions: Knowledge, state, and strategy* (pp. 71-96). London: Sage.

Nonaka, I., & Takeuchi, H. (1995). *The knowledge-creating company.* New York: Oxford University Press.

Orr, J. (1990). *Talking about machines: An ethnography of a Modern Job.* Ithaca, NY: Cornell University.

Parker, M. (2002). *Against management.* Cambridge: Polity Press

Polanyi, M. (1966). *The tacit dimension.* London: Routledge & Kegan Paul.

Polanyi, M. (1958). *Personal knowledge: Towards a post-critical philosophy.* Chicago, IL: University of Chicago Press.

Prax, J. Y. (2003). *Le manuel du knowledge-management.* Paris: Dunod.

Reynaud, J. D. (1991). La régulation sociale. *Revue internationale d'action communautaire, 25*(65), 121-126.

Robertson, S. L. (1996). Teachers work, restructuring, and postfordism: Constructing the new professionalism. In I. F. Goodson, & A. Hargreaves (Eds.), *Teachers' professional lives* (pp. 28-55). London: Falmer Press.

Robertson, M., & Swan, J. (2003). Control—What control? Culture and ambiguïty within a knowledge intensive firm. *Journal of Management Studies, 40*(4), 831-858.

Robertson, M., Scarbrough, H., & Swan, J. (2003). Knowledge creation in professional service firms: Institutional effects. *Organization Studies, 24*(6), 831-857.

Robertson, M., & O'Malley Hammersley, G. (2000). Knowledge management practices within a knowledge-intensive firm: The significance of the people management dimension. *Journal of European Industrial Training, 24*(2-4), 241-253.

Rochet, C. (2003). *Conduire l'action publique des objectifs aux résultats.* Paris: Éd. Village mondial.

Sanders, A. F. (1988). *Michael Polanyi's post critical epistemology. A reconstruction of some aspects of "tacit knowing."* Amsterdam: Rodopi.

Scarbrough, H. (1999). Knowledge as work: Conflicts in the management of knowledge workers. *Technology Analysis and Strategic Management, 11*(1), 5-16.

Simon, H. A. (1991). Bounded rationality and organizational learning. *Organization science, 2*, 125-134.

Stinchcombe, A. L. (1959). Bureaucratic and craft administration of production: A comparative study. *Administrative Science Quarterly*, (4), 168-187.

Stone, K. (1974). The origins of job structures in the steel industry. *The Review of Radical Political Economy, 2*(6), 113-173.

Swan, J., Scarbrough, H., & Robertson, M. (2002). The construction of "communities of practice" in the management of innovation. *Management Learning, 33*(4), 477-496.

Swan, J., Robertson, M., & Bresnen, M. (2001). Knowledge management and the colonization of knowledge. *Special Issue of the Best papers from the Critical Management Studies Conference, The Electronic Journal of Radical Organization Theory, 7*(2), (online).

Tarondeau, J. C. (1998). *Le management des savoirs.* Paris: PUF, coll. Que sais-je?

Torstendahl, R., & Burrage, M. (1990). *The formation of professions: Knowledge, state, and strategy.* London, Sage.

Thrupp, S. (1963). The guilds. *Cambridge Economical History*, (3), 230-280.

Von Krogh, G., Ichijo, K., & Nonaka, I. (2000). *Enabling knowledge creation. How to unlock the mystery of tacit knowledge and release the power of innovation.* Oxford: Oxford University Press.

Weick, K. E., & Roberts, K. H. (1993). Collective mind in organizations: Heedful interrelating on flight desks. *Administrative Science Quarterly, 38*(3), 357-381.

Wenger, E. (1999). *Communities of practice: Learning, meaning, and identity.* Cambridge University.

Wilson, T. D. (2002). The nonsense of knowledge-management. *Information research, 8*(1), paper no 144.

Zack, M. H. (1999). Managing codified knowledge. *Sloan Management Review*, 45-58.

Zara, O. (2004). *Le management de l'intelligence collective.* Retrieved from http:www.axiopole.com/intelligence_collective.php

KEY TERMS

Highly Qualified Professionals: Workers that have at least a college or university (bachelor) degree and could hold a knowledge work job.

Knowledge Worker: A term coined by Peter Drucker in 1959 (*Landmarks of Tomorrow*), is one who works primarily with information or one who develops and uses knowledge in the workplace, who works for a living at the tasks of developing or using knowledge. The knowledge worker might be someone who works at any of the tasks of planning, acquiring, searching, analyzing, organizing, storing, programming, distributing, marketing, or otherwise contributing to the transformation and commerce of information and those (often the same people) who work at using the knowledge so produced. The term includes those in the information technology fields, such as programmers, systems analysts, but also technical writers, academic professionals, researchers, lawyers, teachers, scientists. Due to the constant industrial growth in North America and globally, there is increasing need for an academically capable workforce. In direct response to this, knowledge workers are now estimated to outnumber all other workers in North America by at least a four to one margin (Haag et al., 2006, p. 4).

Pooling of Knowledge: One of the conditions of knowledge work and knowledge management, that is the sharing of the knowledge, held or produced, relevant to the task at stake by all the knowledge workers involved, be it tacit or explicit knowledge. To this effect, managers implement procedures and systems such as forums, collaborative systems supported by intranets, or more dedicated, decentralized tools. Though the phenomenon is as old as is the world, in the world of KM, IT driven systems are the ones promoted by IM experts to achieve knowledge pooling, meaning that the pooled knowledge can be saved and reused by others.

Reputational Job Market: A notion coined by Richard D. Whitley (Manchester's Business school sociologist). Mostly based on Bourdieu's materialist theory of scientific fields, Whitley's theory holds that the highly qualified professionals hold a place in their job market that is related to their social or symbolic capital, itself based on their past realizations and successes as assessed by the socially constructed hierarchy of priorities in their field.

Resistance: In the sociology of work, refers to a workers' strategy, rather individual or collective, that express disagreement over management without voicing the terms of conflict clearly or openly organising to fight against a decision, program or politics. It can be said a passive way to show disagreement, that doesn't mean an inefficient or useless strategy. It's usually set in when open opposition is impossible, inefficient, untimely or inappropriate.

Sociology of Professions: A field of sociology that studies the social organization of the professional model of work organization, its different organization models throughout history and countries, the professions' world own inner hierarchy, the different explaining theories of the phenomenon.

Social Relations at Work: Consists in the relative positions held by different social groups in the power relations at work (based on jobs, skills, authority, gender, race, age, etc.), the strategies elaborated to cope with these power relations.

ENDNOTES

[a] Author wants to thank reviewers for helpful and respectful comments; those two don't always go hand in hand.

[b] The concept of professional autonomy, which was used for many years in sociological studies of the professions, is increasingly being criticized, quite rightly, on the grounds that instances of true professional autonomy, in the strict sense of the term, and as it is defined in the weberian ideal type or model often used in work on the proletarianiza-

tion of professionals, are very rare cases. According to Freidson (1994, p. 162), these instances were limited to the medical and legal professions, in the United Kingdom and the United States, during a very short historical period and in consultancy mostly. Questions are therefore increasingly being raised about the usefulness of this concept, and that is why we will refer more to discretionary power than to autonomy (Evetts, 2001, p. 3-7).

c For the purposes of this chapter, we will distinguish communities of practice from KM systems and will define them as follows: informal, spontaneous groups, often cross-organizational, with membership being volontary, and with members being asked to participate on the basis of their competency; the chief objective of such groups is the sharing of knowledge related to a specific practice, usually professional, through social interaction. The main conditions for participation are similarity, complementarity, proximity (not necessarily physical) and reciprocity (Prax, 2003, p. 96-102).

d In taking such a standpoint, the author shares the perspective of many Polanyi's exegetes: Allen, 1990; Gelwick, 1977; Gill, 2000; Sanders, 1988. See Brohm, 2005, p. 80-81.

e Clawson (1980); Derber and Schwartz (1988); Eccles (1981); Marglin (1973); Montgomery (1979); Stinchcombe (1959); Stone (1974); Thrupp (1963).

f See, for instance, Reynaud (1991) and Bélanger (1991). For more information on the origins and development of several sociological trends on the proletarianization of salaried professionals, see Legault (1988).

g I would like to reiterate that my hypothesis does not concern the actual results to be expected from KM respecting professionals' qualifications, but the perceptions that make professionals act as they do. In other words, I am not arguing that the risk of deskilling exists objectively.

Chapter XI
"We Make Magic Here": Exploring Social and Cultural Practices Within a Global Software Organization in India

Marisa D'Mello
University of Oslo, Norway

ABSTRACT

The increased transnationalisation of work and production has re-shaped international business environments, forms of work, organizational structures, and work cultures. Employing data from an ethnographic case study from India, this chapter provides an understanding of how micro-experiences of Indian information technology (IT) or knowledge workers are intermeshed with the dynamics of a high-tech organization, also called a global software organization (GSO). It argues that while GSOs in India are workplaces that can be viewed as both models of and models for globalization processes, they are also milieus deeply imbued with personal, social, and cultural relations and processes. By delineating the various forms of culture in a GSO, the study highlights the dialectical relationship between the local and global. Further, it explicitly demonstrates they ways by which GSOs and their workers constantly construct meaning and coherence in a volatile and international business context.

INTRODUCTION

The increased transnationalization of work and production has, among other things, given birth to a multitude of transformations that have reshaped international business environments, forms of work, organizational structures, and work cultures. Theorists of globalization (e.g., Robertson, 1992) have argued that incoming global influences are encountered and selectively internalized (or "glocalized") at the local level meeting local needs. The local, as well, they contend is also a source of outgoing flows. Others (e.g., Ailon-Souday & Kunda, 2003; Beck, 2000; Giddens, 1991) refer to how the "local" is intertwined in an increasingly complex way with the global and pervades more and more our everyday life-worlds.

This chapter contributes to an understanding of global and local cultural changes through an empirically informed analysis of a high-tech organization, concerned with global software work (GSW), in India. Such high-tech firms or global software organizations (GSOs) employ knowledge workers or IT (information technology) professionals such as programers, designers, analysts, and IT managers. These employees design, develop, test, and implement software for customers around the globe. GSOs undertake such work across national boundaries through arrangements like alliances, outsourcing or subsidiaries, and use ICTs (information and communication technologies) to coordinate tasks at various stages of the software life cycle. Such work is intangible, heterogeneous, and mobile as compared to traditional service or manufacturing activities (Sahay, Nicholson, & Krishna, 2003).

An analysis of culture, in the context of the IT industry in India, is relevant and important for two major reasons. The first has to do with the global-local dialectic where events at one end of this distanciated relationship may produce divergent effects at the other (Giddens, 1991). The IT industry in India is framed within discourses of neo-liberalism and global capitalism. This context presents a challenge to inspect the interweaving of the global and cultural dynamics of the "new economy scenario" with diverse local cultures (e.g., religion, region, social) symbolically creating or modifying cultural and social spaces for IT workers within GSOs.

Secondly, GSOs, in India, face problematic issues of rapid employee turnover (or attrition) with programers remaining for two years on the average with the firm before quitting to take up more lucrative or challenging jobs (Rathi, 2003). Attrition results in much cost and energy of the organization being expended on recruiting good candidates, repeatedly building new relationships, and transferring required project and organizational knowledge with new workers. Keeping IT workers committed, productive and emotionally engaged during their work tenure, becomes a challenge for the firm working in a fiercely competitive global environment. Further, offshore software development exists in a web of highly interactive relationships (Waterson, Clegg, & Axtell, 1997), with demands particularly related to embedded knowledge, while working across national boundaries (Nicholson & Sahay, 2004). GSOs constantly strive to create corporate cultures and identification structures, that not only aim to attract and retain skilled IT professionals, but also engage their "hearts and minds" while at work (D'Mello, 2005).

This chapter thus focuses on the following key question: **How do Indian GSOs create social and cultural spaces for their employees?**

This question was analyzed through a case study done between 2002 and 2004, which drew primarily upon semi-structured in-depth interviews of 50 IT employees at different levels of a GSO (in Mumbai, India) who have undertaken software work in other countries. In addition to members, at various levels in this firm, further interviews were carried out with other consultants and opinion leaders of the industry. Participant observation and also analysis of secondary data related to company policies and other publicity material, have supplemented the interview data.

CONCEPTUAL FRAMEWORK

In this section, I outline key cultural approaches, to studying organizations and linkages between national and organizational cultures, with a focus on research in Indian organizations.

Culture and Organizations

Since the 1980s, the notion of culture, a primary and distinctive concept within Cultural Anthropology, has gained currency within corporations. Culture was seen as crucial to the shaping of values and providing meaning and purpose to daily lives of employees, as well as framing the identity, image, and reputation of the firm (Alvesson, 2002; Ashforth & Mael, 1989; Kunda, 1992). Organization studies, which has given culture a central place, is replete with diverse approaches and definitions on this construct.[a] A well-known approach is that of Schein (1984) who defined organizational culture as " ...the pattern of basic assumptions that a given group has invented, discovered, or developed in learning to cope with its problems of external adaptation and internal integration, and that have worked well enough to be considered valid and, therefore, to be taught to new members as the correct way to perceive, think, and feel in relation to those problems" (Schein, 1984, p. 3). Further, Schein posits that culture is something, which identifies and differentiates a social group, and can be managed and changed in some organizations.

Other cultural approaches to analysis of organizations that similarly borrow from anthropology and organization theory, conceive of culture in diverse ways (Smircich, 1983). Firstly, in much of cross-cultural or comparative management research, culture is conceived as an independent variable (e.g., Hofstede, 1980; Triandis, 2000). Secondly, when seen as an internal variable, culture is the social or normative glue binding an organization together and reflecting the values and beliefs shared by members symbolically through myths, rituals, legends, and stories (Deal & Kennedy, 1982; Tichy, 1982). Thirdly, as opposed to culture as a "systems metaphor," culture as a root metaphor views organizations as a subjective experience and a pattern of symbolic relationships that provide and shape meaning for its members (Smircich, 1983). Finally, a cultural framework assumes that socio-cultural characteristics of organizations reflect the wider socio-historical framework within which organizations are embedded (Smircich, 1983).

Notions of culture are often related to geographical locations. Cross-cultural researchers have, for decades, studied the relationship between national and organizational cultures and people's identities (Inkeles & Levinson, 1954; Mead, 1951). Researchers have focused on how one's national or "native" culture influences workers perceptions in the form of perceptions (Chatterjee & Cecil, 2000), values (Hampden-Turner & Trompenars, 1993), attitudes (Anastasi, 1983), and beliefs (Smith & Thomas, 1972) and how these, in turn, influence organizational culture (Hofstede, 1980). Some studies have linked "basic" Indian values and personality traits suggestive of national character with economic progress of the country (e.g., Sinha, 1988). The well-known studies of Hofstede (1980) have described national cultures on various dimensons or orientations seen as universally applicable. For example, the poles of individualistic-collectivistic orientations have been used to examine differences in cultures and, among other indices, their respective economic developments (Triandis, 1995, 2000). These studies propose that Indians are collectivists who view their self and life goals to be relational and interdependent with family and other ingroup members. Other studies, deploying different dimensions, showed that those sharing an "Indian culture" were high on context orientation (Hall, 1981), low on uncertainty avoidance, inclined towards collectivism and masculinity, high on power distance (Hofstede, 1980), and high on helping behavior (Levine, Norenzayan, & Philbrick, 2001).

While the concept of national character, values, and personality are relevant means to categorize culturally diverse behavior patterns, they have limited value in explaining any "basic" behavior and particularly in the context of India, for several reasons. Firstly with India's diversity of over a billion people with sixteen recognized languages, thousands of castes and tribes, diverse religions (Hinduism predominates), varied climate, topographies, and economic levels, there is no typical context that all Indians share. Secondly, cultural dimensions do not always match with organizational dimensions and Indians are found to be both collectivists as well as individualists and combining these modes to fit the situation (Sinha, Sinha, Verma, & Sinha, 2001; Sinha, Vohra, Singhal, Sinha, & Ushashree, 2002). Thirdly, many such concepts have been formulated in the West (Sinha & Kumar, 2004) and are often too static, objectivist, and essentialist as they carry multiple meanings across and within cultures (Tayeb, 2001). Fourthly, different structural features of organizations mirror different national values. For example, within government and traditional business organizations in India, family and caste systems have been seen to play a role (Sahay & Walsham, 1997) while private Indian businesses are seen to draw more upon professional management practices. Indian GSOs are seen to draw on family structures to refashion their internal cultures and external image and to stem attrition and enhance productivity. However, these "Indian" structures are viewed within a process of continual redefinition in response to external social and economic changes rather than as a static trait (Sahay et al., 2003).

Further, in a global ICT mediated work environment, the dynamics and issues concerning individuals and work are very distinctive and complex relating to the knowledge-intensive nature of technical language, multiple products, processes, tools and methodologies involved (Sahay et al., 2003). Coupled with different and rather complex ICTs, the task of unravelling cross-cultural issues, in terms of fixed constructs, is a very complex matter. These issues have nuanced interpretations and are better comprehended in relation to shifting business, market, social, and organizational contexts.

In this chapter, I have adopted an interpretivist cultural framework to understand how GSOs symbolically create a universe for IT employees (Avison & Myers, 1995). A cultural framework in organizations, in general, emphasizes the socio-historical context of the firm going beyond its rational and explicit goals (Pettigrew, 1979; Smircich, 1983). By focusing on the subjective interpretive aspects of organizations, a cultural framework raises questions of context and meaning, questions taken for granted as well as assumptions, and brings to the surface underlying values. Rather than something static or an entity, culture is seen as emergent, changing yet complex and as a socially constructed process (Avison & Myers, 1995). As Hannerz (1987) states: "Where there is a strain between received meanings on the one hand and personal experiences and interest on the other, and where diverse perspectives confront one another, cultures can perhaps never be worked out as stable coherent systems; they are forever cultural 'work in progress'" (pp. 550).

THE CONTEXT: THE INDIAN IT INDUSTRY AND CASE DESCRIPTION

In the late 1970s, the global software industry witnessed a trend toward outsourcing software development when it was faced with huge increases in software costs and demands for complex information systems applications as well as rapid obsolescence of IT infrastructure and a limited supply of IT workers (Bhatnagar & Madon, 1997). India became a primary destination for offshoring because of low costs and its abundant pool of English-speaking knowledgeable professionals. As a result of liberalization policies, by the Indian government during 1991-1997 and significant improvements in telecommunication infrastructure, several foreign firms entered the country in a big way. Several GSOs, which catered exclusively to the export needs of the IT industry, were founded during this period in software and hardware technology parks. In the early years, almost 75% of the export-related work was carried out at the client's location overseas while 25% was done in India. Today, these percentages are literally reversed (Heeks, 1996; Sahay et al.,

2003). The Indian outsourcing sector (including the IT as well as the business process outsourcing sector) currently employs around one million people, contributes to around 3.5% of India's total GDP and earns 13% of foreign exchange inflows (NASSCOM, 2005).[b]

The IT industry in India has two unique features. Firstly, most software firms have been founded by middle-class engineers who had previous careers as software professionals or managers in large IT companies (Upadhya, 2004), unlike many Indian firms linked to traditional business families with relatively easy access to capital. This has influenced a distinctive culture and outlook in the industry. These companies, in turn, have contributed to the creation of a professional-managerial class of workers, with high levels of formal training, occupying a unique position in the service economy as well as the social structure in India. Secondly, since the 1980s, this industry has been closely interlinked with the global economy through contractual relationships with overseas companies, foreign direct investment by multinational companies and later, in the 1990s during the IT boom, through foreign venture capital (Upadhya, 2004). In this way, the IT industry (and its workforce in India) is necessarily embedded in a global network of economic, political, social, and cultural transnational linkages.

The case (Soft Globe Ltd.),[c] a mid-size IT software applications outsourcing company established in 1980, is representative of a GSO. Founded by a team of entrepreneurs in Mumbai, it now employs 2,500 professionals worldwide. Besides three offices in Mumbai city, Soft Globe has offices in the U.S., UK, Germany, Austria, Japan, Singapore, and Malaysia with customers from primarily those countries. Assessed at Level 5, on the Capability Maturity Model[d] for its software, as well as its people processes, almost 95% of its revenues have been from repeat business. Over the years, Soft Globe Ltd. has restructured itself several times (including its vision and mission) merged functions, shut down units and opened new ones, to respond to ongoing market changes. One key feature of the company, has been the consistent leadership of the founders who encouraged place-based values of family and being Indian, while simultaneously emphasizing its global competitiveness.

CASE ANALYSIS

Organizational Culture

This illustrates a range of material, cultural, and social practices instituted by GSOs and how IT workers have responded to them.

Office Décor

The technology parks, where many IT companies in Mumbai are located, are in a campus-like environment with lots of greenery, cafeterias, and open spaces. Soft Globe office spaces are comfortably air-conditioned and aesthetically designed. Granite, marble, and glass have been generously used in the construction, both inside and outside the building. Customers from overseas locations have always been impressed with the aesthetic and spotlessly clean office interiors. In the company's reception area, large black and white photographs of smiling employees are on display with the title, "The most important thing we make here is a difference." Employee seating is organized around open-plan structures to maximize space, reduce hierarchy differences, and increase opportunities for social exchange. Project managers are usually seated in the same open office plan with their team members (their desk spaces visibly marked by extra telephones, a larger desk and most often corner seating). In several individual workstations, screensavers and images or tiny statues and deities are displayed alongside other information while the cafeteria has a prominently placed shrine on the wall with fresh flowers, coconuts and incense sticks. The

Head of the unit occupies a spacious office with a semi-transparent glass door.

Each of the Mumbai offices houses training rooms, most often a library, and convenient meeting rooms equipped with phones, computers, and white boards or projector screens. Conference rooms are equipped with state-of-the-art video-conferencing facilities. The company dress code is formal on weekdays except for Fridays when employees most often wear jeans and trainers. Cafeteria facilities, with plentiful supplies of (free) tea and coffee as well as snacks and meals, are conveniently available. For those working late, free snacks, dinner and transport facilities to the nearest railhead are provided. The canteen is an open space with flexible seating where large groups of employees are seen sitting together and sharing lunch (usually Indian preparations brought from their homes in stainless steel or plastic containers). Birthday treats are often celebrated here by teams or small groups. Those returning, from onsite assignments, have been seen to distribute chocolates from that country in the cafeteria to their group or at the workstations. Employees have been observed socializing with others at similar hierarchy levels across projects and units.

Organizational Identification Practices

Organizations are known to engage in practices aimed at socialization, or inducing commitment, to engender a strong bond between individuals and the firm. For GSOs in India, in general, building an "employee-focused culture" and even a "family culture" is seen as a means to forge this bond. Broadly, this means that IT companies actively promote informal collegial cultures with camaraderie and social events during the otherwise stressful workdays. The founders of the company evolved a set of "Soft Globe values," which have been used as a differentiator for the company's own image in the marketplace, as well as managing its internal culture. Relating to openness, respect for others, commitment to results, pride in work, and customer satisfaction were how they defined that things would be done "The Soft Globe Way." These values have been the most prominent vehicle whereby employees are socialised into, and also constantly used, to construct the ideal Soft Globe employee. Company values are displayed through internal and external company promotional material and consistently invoked by employees at all levels at gatherings, meetings and events. Employees state that these values are human values, transcending boundaries of the organization and, hence, find them also relevant for their personal lives. The audio-visual anthem of the firm (broadcast at company meetings and events) begins with the lyrics, "We make magic here." Magic is used as a metaphor to evoke wonder, delight and identification with the firm and its culture.

Yet, some company processes have collided with these values and sense of magic, particularly, during market downswings. For example, during a market downswing, a post-appraisal system was introduced where, for the first time in the company, employees were slotted within either the top 20% category, the "vital 70" (interpreted as "average" by employees) or the bottom percent, where they would be shunted out each year unless they significantly improved their performance. Employees were not directly informed of their category (with the exception of the last category). Stock options, which were earlier available to all employees based on criteria of performance and duration in the company, were now distributed only to the top 20 category. Many were annoyed and upset that this subjective process, where the assignment was based on their manager's final perception, was so "shrouded in secrecy," going against what they thought was the company's stated values of "openness" and "respect for the individual." Similarly, the post September 11th job recession in 2002-2003 also challenged the family culture notion in many Indian GSOs. At Soft Globe despite no explicit retrenchment of people, in any way, performance norms were made more stringent which may have caused some people to exit the company. An employee lamenting about the associated uncertainties said, "A company may say anything about its culture, but you are never sure of where you stand. It's all a game."

Extending the collegial relationship beyond the boundaries of the organization, ex-employees are called "alumni" as the company believes that "once an Soft Glober always a Soft Glober." Although seen as references for potential customers, as well as an attempt to evoke a feeling of continuity and remembrance, they are also encouraged to keep their ties with the company alive through invitation to company events and enrolling in the alumni corner on the Web site.

Family Feeling at Work

The company has tried to create a home-like environment in the office in various ways. The idea of "family culture" is said to be inherently appealing to the Indian psyche (Kanungo, 1995). Metaphors of family have been observed to be employed in various ways. For example, on the fact sheet of the company the CEO uses the term, family. There are various practices such as sports day (competitive sporting events for employees and their families), anniversary day (the company's "birthday" party where songs, dances, and skits are staged by employees), and funtime (a company-wide, fun plus business event held usually for three continuous days at a holiday location in India). Employees and their families are encouraged to participate and enjoy themselves while also creating "family" bonds with Soft Globe. Company instituted social practices such as table tennis matches, quiz contests and *Antakshari*[e] contests have generated enthusiastic participation from employees. In the monthly meeting for all staff in Mumbai, held close to the Independence Day of the country (August 15th), skits and events with themes from the freedom struggle movement in India are enacted by employees. "Traditional Indian dress" is usually prescribed as the dress code for that day. Statements about the the company values, its people and organizational culture are tied together in the scripts for these events. These and other such activities that draw on national sentiment have been seen to evoke a strong sense of community and a reinforcement of a shared history among employees which they have stated were kept aside in daily interactions.

These practices attempt to inscribe a strong sense of family, community, and belonging in the workplace by providing a sense of "home away from home." Employees have commented on how these events and interactions have faciliated their daily work-related interactions. However, these attempts at creating social and family-based bonding are seen to be in tension with the IT worker's constant attempts to assess his/her marketability, and viewing growth in terms of learning and "adding value" through individual work. The IT person has been seen to stay engaged to the extent that these practices match the needs of "adding value" to him or her. When they have not, the choice has often been to quit. For example, Roy, a developer who already changed jobs three times in four years, said, "See, ultimately all IT companies are almost the same. They are young, they have open cultures, and there are lots of celebrations and events in the company for the families of employees. So the main thing is to get what adds value to me. And if in the process, I can also enjoy the other things, then that is what I call a good working environment."

Cultures of Work

This illustrates some typical demands of IT outsourcing work and the contrast with Western customers work culture and attitudes to work. We note how and why employees develop and maintain social networks and their response to the impact of long hours of work on health and hobbies.

Diverse Cultures of Software Work

In the IT industry in India today, working 10-12 hours a day is normal and sometimes this goes up to 14 hours a day (even on weekends). Sometimes, employees stay overnight at the office when they have a project deadline. One major reason is the time difference between India and customer sites. Particularly, for the U.S. and Europe, employees in India have to stay back to make calls to customers in the U.S. and UK who are just beginning their work days. Another reason is the tight and often underestimated project costs and time frames in terms of "mandays"[f] where the team has to put in extra hours to meet stringent deadlines or engage in a "firefighting" mode. Soft Globe, had instituted a flexi-time policy to provide some relief to employees. There was mandated presence in the office at a committed time in the morning with seeming freedom for employees to choose the rest of their working hours (within limits) after negotiation with the manager and the project team.

Staying late in the office was necessitated by work pressure and diverse time zones. However, in spite of no work pressure, some employees were observed staying late. Part of this was related to peer pressure from their particular project team as the entire team would be sitting late. Part of this was also self-imposed, such as the desire to be "noticed by the boss" as hard working. Some employees spoke about their dilemmas of managing peer expectations with their own beliefs about work. They said that they often felt pressured to "toe the line," in terms of sitting late at work, due to their desire to be seen as "valuable by seniors" in a team-based, yet competitive environment which made them feel insecure and anxious about career opportunities. While the company has instituted a team-based appraisal system, individual performance has always been closely monitored and directly linked to monetary and non-monetary benefits and growth opportunities.

In stark contrast, to the "Indian" scenario, workers noted the difference in response to office timings and holidays while abroad. Ramesh, a project leader said:

We say unke holidays pathar ke lakeer hai *[their holidays are carved in stone]. ...this [Britain] is a very soft country. It is 5:30, I have to go home. Our people feel what is so holy about a clock ticking 5:30? I get the feeling it is a pampered country. People disappearing on holiday when they have important things. That really grates on Indians. It is such an important delivery and, look, the lady is out to Spain to spend a weekend there!*

Later in the interview, he elaborated with another example:

When a Soft Glober lands [in the UK upon arriving from India] he comes running into the office at three o'clock after landing at 12 noon on the same day. Even on a Sunday, they come in and check and see if login is working and the swipe card is working. Being there, means I'm more committed. I show it.

Some informants who had worked in the UK and USA viewed the behavior of "Westerners" as balanced and more favorable. It was often these informants who would raise the issue of estimation and lack of a work-life balance during open house sessions in the firm. On the other hand, while balancing work and home may seem a desirable goal, the constraints of the industry have been stated as an impediment. For example, a manager, in his late 30s with two children, stated very sardonically:

Clients are ever more demanding. The sales-marketing guys will put more and more pressure on developers to get things done... who cares about what the guy undergoes. Yes, we talk about it but it is more lip service. Can you visualize the CEO or Operations Head or any manager saying to a developer: "I will defend you to the customer—so go and spend time with your family?"

Social Networks

Given the strong emphasis on family in Indian society, one might ask how can the work intensity and long working hours, required in this industry, be reconciled with "Indian" values which are seen as priviledging family over work? One "way out" has been the social bonds that IT workers have been seen to create within the organization itself. Commenting on this shift, Suhasini, a senior manager, says:

Now people have more social networks within the company. We reach home normally after 8 p.m. Then, again the next day, we have to come back early. So, where is the time to socialize—and in this city commuting is so time-consuming. So, socializing, outside the office, gets limited to mainly with the family... office colleagues become much closer especially if they have families.

This was echoed by informants who said that they belonged to various networks. These networks included friends in the same company but on different projects, friends and peers from the same batch at engineering college, friends and colleagues from their ex-companies (who they frequently interacted with primarily on e-mail). Besides affiliation needs, these networks were also tapped for market information, gossip and job moves. Also, given the youthfulness of this industry, many Indian IT workers are typically younger and unmarried. Especially for male employees, the workplace becomes a site for socializing and hanging out after office hours, even without work pressure. There are frequent interpersonal interactions in teams during long working hours with team building and company structured bonding practices. The facilities provided, such as internet access, free snacks and facilities for group recreational activities including table tennis and carrom (in a physically pleasant environment) contributed to enhanced social interactions at the work site.

Consequences of Long Hours of Work

Referring to how their long working hours affected them, Sriram, a senior manager in his mid-40s, said:

*The IT person stops enjoying their life compared to the others, because of the extended working hours and odd timings... and anytime you can be called and you need to be working—it is not a regular job. Also you don't develop any hobbies. You see, life **IS** beyond the office but that realization comes to you only at forty. Till such time, you will run, busy climbing things... it is an early retirement job.*

Given the emphasis on work and health was seen to take a backseat. A newspaper article (circulated among some employees at Soft Globe and entitled "Stress Kills 6 IT Geeks" (Srinivas, 2005), reported the results of a study showing that the number of suicides, divorces, heart ailments, BP, diabetes patients, and mental depression were the highest in the software industry. The article cited lack of routine, constant deadlines, weekend working, lack of physical exercise, and new food habits in cities, such as the "pizza culture," as reasons for stress for this group. In response to this article, Arnave, a manager, wrote via e-mail:

When some incident takes place, people are stunned and shocked but life flows on. I think the community is still young and these incidents are isolated. Many people think it is a matter of individual's choice and how much risk they take. Besides, I think the need to earn more and more and pursue new lifestyles is what is driving people mad. Where do you have time to think about health and peace of mind?

Payoff for Work

While IT workers had grueling hours and constant pressures, they were rarely seen to quit working in an offshore outsourcing company and join another industry in-house IT department seen as less prestigious, less stressful and more stable. Most respondents spoke about the high salaries and the social, cultural, and professional "exposure" they enjoyed when living and working abroad. One developer talked about the freedom that the money and employment opportunities provided him. He could quit working at any time, found it easier to disagree with his boss, often ate out at restaurants and could afford a higher standard of living for his family. Frequently, newspaper articles (in Mumbai) describe the changes, in spending patterns, tilted towards branded goods and shifting social habits such as increased eating out, frequenting pubs, and an overall increase in the consumer culture and changing lifestyles in the Indian middle classes.

Cross-Cultural Working

This category illustrates contrasts observed in culturally different behaviors, social, and work practices, as noted by IT workers.

Cross-Cultural Training

Understanding the impact in international working environments, Soft Globe familiarized employees with the country where they were traveling to work in various ways. The HR Department, for example, organized programs on cross-cultural sensitivity, which were mandatory for employees travelling overseas (especially for the first time). These programs included inputs ranging from dressing styles to table manners and social etiquette in various countries. They also highlighted the importance of the participants' own cultural context, as a backdrop, seen to provide a sense of "rootedness" while urging them to be simultaneously professional in work interactions while abroad. Such training programs were usually for one day and often attended by employees a few days before the date of departure when they were also trying to complete visa and ticket formalities, work and personal matters (shopping, etc.). Participants were reminded that, while abroad, they were also "ambassadors of India." Country-specific booklets, often written by HR staff, were distributed to the employees. On the company Web site, detailed information about renting apartments, shopping, and travel tips, typical words and phrases and their meanings, as well as tips (to overcome home-sickness) for different countries, was available. Employees saw these inputs as "adding value" to him or her and acknowledging that these "soft skills" were important while overseas.

Cultural Identification Practices

When the offshore units of the various geographical territories (the USA, UK, Germany, and Asia) were started, they were expected to mirror the cultures of their overseas "parent" unit. Certain properties of the unit's culture were slowly taken on. For example, the USA unit (as well as its head) were seen to be more informal in dress code and working environment, whereas the UK unit was more formal and process driven. A hierarchy was present, among the units, mirroring the global marketplace with the USA (the most profitable unit to work in both offshore and onsite) followed by the UK, then Germany, Asia, and finally the India unit (seen as a good learning platform for exposure to the software development life cycle but relatively unglamorous as there was no scope for onsite postings). The units competed with each other in terms of revenues with "hoarding good resources" (not releasing competent IT workers to other Units) and outshining each other in their HR policies, processes and revenues. They also created t-shirts, bags, and stickers for employees and celebrated the festivals of their parent unit with much gusto. After a few years, this structure was dismantled as the competitiveness militated against a synchronized business focus, as well as being seen as mitigating the overall company culture.

Religious Festivals at Customer Sites

Soft Globe attempted to invoke cultural events such as religious festivals, to enhance business relationships with its overseas customers. An in-house writeup on these events mentioned the aims as ensuring "fun," and "sharing" as well as "total bonding." For example, at one European customer location, major Indian festivals were celebrated in a manner similar to the office celebrations in Mumbai. The office was decorated with ethnic designs, celebrations were held in the office (with explanatory notes on the ceremony) and items used and distributed to the customers. Information on the festivals, of India, was made available on the customer's intranet. Traditional sweets such as *modaks*[g] made by some of the Soft Globe employees, in their homes, were distributed at the client office site. Cricket matches with teams comprised of both Indian and customer staff, particularly at the UK location, was another popular event. HR training programs, project team managers and handbooks for overseas travel, exhorted employees to develop social relations with customers "through sharing aspects of Indian culture," which they say has often been appreciated by customers.

"Primordial" Cultures: Caste, Region, and Religion

The term primordial is used to refer to affinities or ties between those of similar kin and community within a particular locality (or "soil") with continuity over time with such ties imbued with strong affect (Weinreich, Bacova, & Rougier, 2003). In this section, I describe if and how affinities of caste, region and religion which are denoted as primordial, in the context of India, are re-produced in GSOs along with spiritual and secular practices.

Caste

Like organizations in the private sector, GSOs are seen as meritocratic organizations focusing on competence and professionalism. The empirical material has suggested that caste, as a variable in work related systems and processes, seems to have disappeared from the GSO—unlike the pre-liberalization days of licenses and quotas, where people would ply their caste-related network ties to secure employment or business licenses. As Ashish, a 32-year-old manager, (in a very decisive tone) said:

You end up reading the caste and religion from a resume and beyond that it is just a data point among others of no consequence. ...See, in this industry you are forced to hire the right kind of people. ... It just doesn't matter if [a candidate] is Hindu or Muslim or male or female ... it just does not matter.

Kinship and social networks were seen to operate at the entry level in the industry. Qualified relatives and friends could be recommended for jobs and the organization actively encouraged, as well as monetarily rewarded, employees whose proposed candidates got through the rounds of written tests and interviews and were finally recruited. If an incompetent candidate slipped easily through the selection process hoops, because of nepotism, the deficiency showed up along the work trajectory on projects where adequate competence needed to be demonstrated for continued employment.

Regional Ties

Rather than caste, affiliations based on regional belonging were observed. At Soft Globe, given that local Maharashtrians comprised the dominant community of the workforce, Marathi was typically spoken in the corridors, elevators and cafeteria. Snatches of Hinglish[h] were frequently heard in corridor and canteen conversations. It was not uncommon to see people, from the same geographical region and who spoke the same language, socializing together in the cafeteria. Andhraites from Andhra Pradesh, Malayalees from Kerala, and Tamilians from Tamil Nadu could be seen talking to each other and having lunch together even if they were not from the same team.

Religion at the Workplace

At the Mumbai location, as well as at onsite Soft Globe offices, major religious festivals (e.g., Holi, Diwali, Christmas, and Id) were celebrated officially with prescribed traditional dress codes, decorated offices, distribution of sweets, singing, or dancing events. All evoked enthusiastic responses from employees, irrespective of their religion. For example, for Christmas, the carol singers were typically from different religious groups. At many workstations, pictures or screen-savers or statues of Ganesh and other Indian deities as well as Christ were clearly visible. Some employees indicated that they felt free to express their religious beliefs in their own workspaces and that these visible forms helped them through their hectic workday. Prayers, or religious mailings, were circulated among some employees. When employees visited a religious shrine in India they brought back *prasaad* (a sweet that was offered first to the deity to be blessed) for their colleagues. Religious celebrations coexisted with celebrations of other globalized "secular" events such as Valentine's Day or Friendship Day with cake cutting practices on the company's anniversary/ birthday celebrations.

In the company, spiritual leaders (e.g., J. Krishnamurthi or Gandhi) were often referred to as role models of leadership and values in company speeches, articles on the intranet or the in-house magazine. Spiritual teachings were also invoked and linked to the firm's values or used as inspirational inputs to employees. For example, in one of the company brochures for customers, sayings from the Upanishads, the Bhagawad Gita, the way of Tao and Zen were used as similes or metaphors to illustrate the business goals of the company. Quotations of gurus or narrations from spiritual stories were noted in several articles written in the in-house magazine.

DISCUSSION

The empirical material suggests that at the physical level construction of "world class" office structures can be seen as a physical marker for Indian GSOs vying, with both national and international competitors, for a piece of the global business. Symbolically, these structures signal status, credibility and a place in the world, both externally to customers as well as internally to individuals. As awesome physical edifices, they claim a social image and position carefully constructed to dampen adverse place-related identifications (Twigger-Ross & Uzzell, 1996) commonly associated with an economically developing country.

The GSO was seen to extensively construct a cultural and social space for IT workers by drawing on a diversity of social, cultural, and material resources. The IT industry has created a new business entity in the form of a GSO. This in turn, has created a new aspirational space for many young qualified people in India who aim to get into an IT or computer science career track, and secure for themselves a bright future in this growing industry. The unanticipated burgeoning of the IT industry, the infusion of professionally managed IT companies run by Indian entrepreneurs in their early 30s and an export orientation has challenged hegemonic traditional management practices. GSOs now serve, as exemplars to other firms, as a business model that features flatter structures, quality orientation, individual incentives, an export orientation, and, often, shareholder value (Arora & Gambardella, 2005). Their cultures are generally seen as professional, youthful, open, informal, and merit based, and also modified according to market swings (D'Mello, 2005).

The case material suggests that while the GSO drew on family-related norms and values to develop their corporate culture in a conscious and and instrumental manner, these aspects of "Indian" society were not viewed as a static trait but seen within a process of continual redefinition in response to changes that necessitated an agile workplace and staff. Further, the mobility of work, workers, technologies and contexts, unique in this industry, strongly militates against fixed caste ascriptions and overly hierarchical structures. Rather than Sanskritization[i] (Srinivas, 1997) as a means to move up the caste hierarchy, individual agency and acquired competence are seen as the major means for individuals to move up a skill-based meritocratic hierarchy in a GSO. While attenuated, kinship ties and networks were also mobilized by GSOs faced with the issue of high staff turnover while seeking skilled workers who would fit into the company culture and also be motivated to join and stay longer. Such a "cultural fit" was seen as particularly important at senior levels, as managers were meant to role model the corporate culture and motivate their teams. A cultural misfit would be costly in economic terms and also adversely impact the team. If caste was a factor in hiring or promotion decisions then regular surveillance of workers, combined with stringent performance standards and mobilities across projects and teams would, sooner than later, mitigate this effect given the very nature of the software development activity. In this way, the features of global software development work have modified the hold of some cultural forces.

Regional identities and affiliation were manifested in the firm through spoken language, friendship ties, stereotypes, and social practices—reflecting the Indian social context. This was not seen by the GSO as antithetical to being global. Place-related markers of locality and community were manifested, not only physically in material artifacts (such as workstations or office decor), but also in workers' daily social lives and rituals engineered and institutionalized through organizational processes, as well as relations among employees. "Indian" values and family and cultural practices (e.g., religious festivals) were consciously co-opted and deployed by the GSO as a tool for belongingness, stemming attrition and creating boundary-spanning linkages in the service of a capitalist driven work culture demanded by software outsourcing business. Physical places concatenated with each other creating shared meanings and suggesting a porosity of boundaries. Juxtaposing evocative national sentiments closely with individual and corporate identification and IT work, in various ways, was seen to fuse (at least temporarily) national, professional, organizational and individual boundaries. The emotional charge, that this created, was reflected in the manner by which employees described and recalled these events even after leaving the organization thus suggesting that this juxtaposition was experienced as poignant and meaningful.

In these ways, "placed" meanings moved between social, cultural, and geographical as well as existential locations and re-locations of IT workers. GSOs can be seen to represent sites embodying various shifts or mobilities—of place, self, cultural, and social. Deploying Urry's (2001) metaphor of "travel" as a journey crossing process states, it can be argued that IT workers in GSOs can indeed be said to "dwell" in various mobilities" (p. 157), which are diverse yet intersecting, creating an intersecting pattern of economic, cultural and social features and experiences. Rather than the de-territorialization of culture and place (Appadurai, 1996), in the service of global capitalism, the various forms of culture suggest that GSOs (as well as individuals) construct and revise the local in the global (and the other way around) as they participate in the multiple networked regimes of offshore software development.

Further, the tandem operations, and intertwining of both global and local dynamics, suggests that, rather than a McDonaldization (Ritzer, 1998) or universal homogenization of global flows, GSOs are representative of glocalization processes (Robertson, 1992). Modern office constructions across various continents, sophisticated technical infrastructure, internationally recognized as well as standardized software development, and people management practices, are reminiscent of global features. The empirical material has illustrated how the processes of economic enterprise simultaneously intersperse the local in various forms richly meshing the universal with the particular within GSOs. Rather than a homogenous workplace, local primordial moorings have been seen to negotiate a cultural accomodation with the global in the service of global capitalism. This reinforces the view (e.g., Giddens, 1990, 1991) that globalization has multiple forms and does not create a "borderless world" (Ohmae, 1990).

This study highlights the contrasts in the theorization of place and space, the two central contours in the time-space configuration of modernity (Giddens, 1990). Place is associated with a person's sense of security, boundedness, belonging, or emotional attachment while space is associated with movement, freedom and expansiveness of possibilities (Giddens, 1990; Harvey, 1989; Tuan, 1977). Giddens (1990) claims that globalization flows create unsettling changes whereby places become increasingly "phantasmagoric." The empirical material suggests that the mobile nature of GSW certainly re-articulates place-based work, social and cultural relations within a space-based ICT mediated context. However, places are far from irrelevant and nor are they increasingly "phantasmagoric." Rather, this study suggests that GSOs and IT workers seek newness and growth while parallely searching for belongingness and rootedness.

In these ways, the GSO in India is located in a spatially specific site embodying globalization discourses, as well as receiving and transforming these discourses. Simultaneously, it draws on a diversity of social, cultural, and material resources to construct its external image and strengthen its internal culture.

CONCLUSION

In the past decade, anthropological inquiry has increasingly seen an interrogation of how the "local" is produced at the intersection of translocal, regional, as well as global cultural fields in ethnographies of local communities, identities, and spaces (Appadurai, 1996; Hannerz, 1996; Mankekar, 2002). Human geographers have also studied how the local and the global interrelate in transnational business environments (e.g., Massey, 1984; Ong, 1999). The analysis presented highlights the interconnection of both proximate (e.g., primordial) and distal (e.g., global) forms of cultures and some of the tensions therein in the context of an Indian GSO. These tensions reinforce the view (Westrup, Al Jaghoub, El Sayed, & Liu, 2002) that there are no "pristine cultures" in such firms. Secondly, it explicitly demonstrates how GSOs and their workers deploy a variety of local and global elements to construct meaning and coherence in a volatile and international business context. Finally, it reinforces the view (Giddens, 1991) that globalization is an uneven and dialectical process, with multiple or glocal forms. We can say that the workplace of GSOs is a microcosm, of globalizing as well as glocalizing processes, situated in a global network of local places with various intersecting forms of cultures which are emergent, changing and complex. Such a workplace mirrors the dialectical relations of culture, business, organization and workers responses.

REFERENCES

Ailon-Souday, G., & Kunda, G. (2003). The local selves of global workers: The social construction of national identity in the face of organizational globalization. *Organization Studies, 24*(7), 1073-1096.

Alvesson, M. (2002). *Understanding organizational culture*. London: Sage.

Anastasi, A. (1983). Evolving trait concepts. *American Psychology, 38*, 172-183.

Appadurai, A. (1996). *Modernity at large: Cultural dimensions of globalization*. Minneapolis: University of Minnesota Press.

Arora, A., & Gambardella, A. (2005). Bridging the gap: Conclusions. In A. Arora & A. Gambardella (Eds.), *From underdogs to tigers: The rise and growth of the software industry in Brazil, China, India, Ireland, and Israel* (pp. 275-302): Oxford, UK: Oxford University Press.

Ashforth, B. E., & Mael, F. (1989). Social identity theory and the organization. *Academy of Management Review, 14*(1), 20-39.

Avison, D. E., & Myers, M. D. (1995). Information systems and anthropology: An anthropological perspective on it and organizational culture. *Information Technology & People, 8*(3), 43-56.

Beck, U. (2000). *What is globalization? Translated by Patrick Camiller*. Cambridge: Polity.

Bhatnagar, S. C., & Madon, S. (1997). The Indian software industry: Moving towards maturity. *Journal of Information Technology, 12*(4), 277-288.

Chatterjee, S. R., & Cecil, A. P. (2000). Indian managers in transition: Orientation, work goals, values and ethics. *Management International Review, 40*, 81-95.

D'Mello, M. (2005). Thinking local, acting global": issues of identity and related tensions in global software organizations in India. *Electronic Journal for Information Systems in Developing Countries, 22*(2), 1-20.

Deal, T. E., & Kennedy, A. A. (1982). *Corporate cultures: The rites and rituals of corporate life*. Reading, MA: Addison-Wesley.

Giddens, A. (1990). *The consequences of modernity*. Cambridge: Polity Press.

Giddens, A. (1991). *Modernity and self-identity: Self and society in the late modern age*. Cambridge: Polity Press.

Hall, E. T. (1981). *Beyond culture*. New York: Doubleday.

Hampden-Turner, C., & Trompenars, A. (1993). *The seven cultures of capitalism*. New York: Currency-Doubleday.

Hannerz, U. (1987). The world in creolization. *Africa, LVII*, 546-549.

Hannerz, U. (1996). *Transnational connections: Culture, people, places*. London: Routledge.

Harvey, D. (1989). *The condition of postmodernity: An enquiry into the origins of cultural change*. Cambridge, MA: Blackwell.

Heeks, R. (1996). *India's software industry: State policy, liberalization, and industrial development*. New Delhi: Sage.

Hofstede, G. (1980). *Culture's consequences*. Beverly Hills, CA: Sage.

Inkeles, A., & Levinson, D. (1954). National character: The study of modal personality and sociocultural systems. In G. Lindzey (Ed.), *Handbook of social psychology* (pp. 977-1020). Cambridge: Addison-Wesley.

Kanungo, R. N. (1995). Culture and work alienation: Western models and eastern realities. In H. S. Kao, D. Sinha, & N. Sek-Hong (Eds.), *Social values and development: Asian perspectives* (pp. 233-250). New Delhi: Sage.

Kunda, G. (1992). *Engineering culture: Control and commitment in a high-tech corporation*. Philadelphia: Temple University Press.

Levine, R. V., Norenzayan, A., & Philbrick, K. (2001). Cross-cutural differences in helping strangers. *Journal of Cross-Cultural Psychology, 32*, 543-560.

Mankekar, P. (2002). "India shopping": Indian grocery stores and transnational configurations of belonging. *Ethnos, 67*(1), 75-97.

Massey, D. (1984). *Spatial divisions of labour: Social structures and the geography of production*. London: Macmillan.

Mead, M. (1951). *Soviet attitudes towards authority*. New York: W. Morrow.

NASSCOM. (2005). *Strategic review: The IT industry in India*. New Delhi: NASSCOM.

Nicholson, B., & Sahay, S. (2004). Embedded knowledge and offshore software development. *Information and Organization, 14*(4), 329-365.

Ohmae, K. (1990). *The borderless world*. New York: Harper Business.

Ong, A. (1999). *Flexible citizenship: The cultural logics of transnationality*. Durham, NC: Duke University Press.

Pettigrew, A. (1979). On studying organizational cultures. *Administrative Science Quarterly, 27*, 641-652.

Rathi, N. (2003). *Human resource challenges in the Indian software industry: An empirical study of employee turnover*. Unpublished doctoral dissertation, Indian Institute of Technology, Mumbai Mumbai.

Ritzer, G. (1998). The McDonaldization thesis: Is expansion inevitable? *International Sociology, 13*, 291-308.

Robertson, R. (1992). *Globalization: Social theory and global culture*. London: Sage.

Sahay, S., Nicholson, B., & Krishna, S. (2003). *Global IT outsourcing: Software development across borders*. Cambridge, UK: Cambridge University Press.

Sahay, S., & Walsham, G. (1997). Social structure and managerial agency in India. *Organization Studies, 18*(3), 415-444.

Schein, E. H. (1984). Coming to a new awareness of organizational culture. *Sloan Management Review, 25*(2), 3-16.

Sinha, D. (1988). Basic Indian values and behavior dispositions in the context of national development. In D. Sinha & H. S. Kao (Eds.), *Social values and development: Asian perspectives* (pp. 31-55). New Delhi: Sage.

Sinha, J. B., & Kumar, R. (2004). Methodology for understanding Indian culture. *The Copenhagen Journal of Asian Studies, 19*, 89-104.

Sinha, J. B., Vohra, N., Singhal, S., Sinha, R. B., & Ushashree, S. (2002). Normative predictions of collectivist-individualist intentions and behaviour of Indians. *International Journal of Psychology, 37*(5), 309-319.

Sinha, J. P. B., Sinha, T. N., Verma, J., & Sinha, R. B. N. (2001). Collectivism coexisting with individualism: An Indian scenario. *Asian Journal of Social Psychology, 4*, 133-145.

Smircich, L. (1983). Concepts of culture and organizational analysis. *Administrative Science Quarterly, 28*(3), 339-358.

Smith, B., & Thomas, J. (1972). Cross-cultural attitudes among managers: A case study. *Sloan Management Review, 13*, 34-51.

Srinivas, M. N. (1997). Practising social anthropology in India. *Annual Review of Anthropology, 26*, 1-24.

Srinivas, S. (2005, March 28). Stress Kills 6 IT Geeks (India). *The Asian Age*.

Tayeb, M. (2001). Conducting Research across cultures: Overcoming drawbacks and obstacles.

International Journal of Cross Cultural Management, 1, 91-108.

Tichy, N. (1982). Managing change strategically: The technical, political, and cultural keys. *Organizational Dynamics* (Autumn), 59-80.

Triandis, H. (1995). *Individualism and collectivism*. Boulder, CO: Westview.

Triandis, H. (2000). Dialectics between cultural and cross-cultural psychology. *Asian Journal of Social Psychology, 3*, 185-196.

Tuan, Y. F. (1977). *Space and place: The perspectives of experience*. MN: University of Minnesota Press.

Twigger-Ross, C. L., & Uzzell, D. L. (1996). Place and identity process. *Journal of Environmental Psychology, 16*, 205-220.

Upadhya, C. (2004). A new transnational capitalist class? Capital flows, business Nnetworks, and entrepreneurs in the Indian software industry. *Economic and Political Weekly*, 5141-5151.

Urry, J. (2001). *Sociology beyond societies: Mobilities for the twenty-first century*. London: Routledge.

Waterson, P., Clegg, C., & Axtell, C. (1997). The dynamics of work organization, knowledge, and technology during software development. *International Journal of Human-Computer Studies, 46*, 79-101.

Weinreich, P., Bacova, V., & Rougier, N. (2003). Basic primordialism in ethnic and national identity. In P. Weinreich & W. Saunderson (Eds.), *Analyzing identity: Cross-cultural, societal and clinical contexts* (pp. 113-169). London: Routledge.

Westrup, C., Al Jaghoub, S., El Sayed, & Liu, W. (2002). *Taking culture seriously: ICTs, cultures, and development*. Paper presented at the Proceedings of the 7th International Working Conference of IFIP WG 9.4 on Information and Communication Technologies and Development: New Opportunities, Perspectives and Challenges. May 29-31., Indian Institute of Management, Bangalore, India. pp. 17-28.

ENDNOTES

[a] Smircich (1983) provides a comprehensive delineation of the various ways culture is used in Anthropology as well as Organization Studies.

[b] NASSCOM, an acronym for theNational Association of Software and Services Companies is the apex body as well as the Chamber of Commerce for the IT industry in India attempts to facilitate business and trade in software and services, both within and outside the country.

[c] The name of the company and of the informants have been changed for confidentiality.

[d] Capability Maturity Models are frameworks for improving software development (as in SW-CMM) or people management processes (as in PCMM) developed by the Software Engineering Institute (SEI) in the USA (NASSCOM, 2005). Internationally recognized as standards for quality, companies use these to enhance their efficiency as well as their image in the international marketplace.

[e] *Antakshari* is a musical game played by teams based on songs from Hindi films. This game is immensely popular in India as well as with employees in this company who play it at company events or social gathering.

[f] An eight-hour unit of time, commonly used in estimations of time and cost.

[g] Steamed dessert dumplings.

[h] A hybridized language combining Hindi and English, termed "Hinglish" by the popular media (Fernandes, 2000).

i This refers to mobility processes of groups within the caste system where a group could move up the caste system by Sanskritizing themselves (i.e., by imitating the customs, rituals, and life-style of a higher caste). Over time, noble origins of the caste could be claimed (Srinivas, 1997).

Chapter XII
Outsourcing in High-Tech Corporations:
Voices of Dissent, Resistance, and Complicity in a Computer Programming Community

Erik Piñeiro
Royal Institute of Technology of Stockholm, Sweden

Peter Case
University of the West of England, UK

ABSTRACT

Management has historically sought to restrict the options for manual workers to rebel by simplifying and limiting their jobs according to Tayloristic principles. The need for their experience and knowledge has been consciously minimized, having been relocated instead to supervisors and middle managers, working routines and machines. In high-tech industries, by contrast, the workers' fundamental contribution to the enterprise is their very knowledge, offering other possibilities for rebellious activities or, at least, for rebellious plans. This chapter focuses on one of the common denominators in the exchanges among programmers, namely the concept of knowledge: how to get it, who has got it (and who hasn't), what kinds are important and its role in their conflict with management.

INTRODUCTION

Electrical, electronic, and computer engineers (IEEE-USA), a professional association with 235,000 members, starts its online introduction with the following remark: "Through years of record layoffs and bleak employment prospects for U.S. engineers…" (2006) and its 2006 President Elect, Ralph W. Wyndrum, Jr., writes in his first President's Column that the first focus of the organization is to, "ensur[e] that U.S. technology policy enhances America's future and protects American workers" (Ralph W. Wyndrum 2006). In 2004, President Elect John Steadman, testified before Congress urging it to "take a close look at overseas outsourcing to see what can be done to create and keep high-value, high-tech jobs here in the U.S." (Quan, 2003), basing his worries on the claim that "Some 160,000 U.S. IT jobs have disappeared in the past three years" (Steadman, 2004).

That programmers would need this kind of help to stop the hemorrhaging of IT jobs was unthinkable only a few years earlier, when their central role in the "new economy" made them irreplaceable high-tech professionals. Castells, for one, went as far as to suggest they were to become the elite of the labor market (Castells, 1996). Much has happened since the "new economy" collapsed and the dotcom boom crashed. The labor market has seen its share of changes, but, as William Wulf, President of the U.S.-based National Academy of Engineering says, "the rate of change in the software field is more rapid than in traditional engineering" (Costlow, 2003). One of the phenomena that has had the largest impact on the programmers' labor market is the managerial practice of outsourcing, which we introduce in the next section. Outsourcing, or the threat of it, has had several effects on the role of software engineers, the most important being probably their demotion from vital members of companies (indeed of the whole economy) to suppliers of functions; functions, moreover, that can be removed from the main body of corporate activity and which, usually on grounds of cost, placed "outside" the company, state, or continent. This move has implied a lowering of both their professional status and their salaries (Costlow, 2003).

This chapter presents a study of the responses and identity work that takes place among programmers in the face of these changes. We shall analyze discussions among programmers that deal with the subject of outsourcing from different perspectives, and show the various strategies available to them in order to maintain their identity as important and powerful members of the organization.

Various studies addressing different aspects of the programmers' identities have been conducted (Barret, 2005). Their interest, and the interest of this one too, resides in the fact that programmers have convincingly been presented as archetypical "knowledge workers" (Angell, 2000, Scarbrough, 1999). At a time when Western policy makers and company representatives express their hope in the knowledge economy as the answer to competition from low-cost markets, it is seems highly pertinent to study the different organizational processes of the "knowledge firm."

OUTSOURCING

The term outsourcing became popular in the beginning of the 90s, representing the management fashion of focusing on core activities—a fashion that is today still very much in vogue. Companies, it was said, were to concentrate on what they did best, and improve their competitive advantages through judicious elimination of secondary activities (Brown, 2005, Corbett, 2004, Power, Desouza, & Bonifazi, 2006). The activities deemed to be "outsourcable" have since expanded and now include services provided by IT-professionals (Apte et al., 1997, Kakumanu & Portanova, 2006, Kehal & Singh, 2006, Sobol & Apte, 1995). Simultaneously, the process of globalization has implied a relocation of activities in low-cost countries, generating much discussion about two neighbour concepts: off-shoring and off-shore outsourcing (Gupta & Chaudhari, 2006). Whereas "outsourcing" relates to the slimming of activities in a firm in order to concentrate on the core of its competitive advantage, "off shore outsourcing" relates to a particular kind of slimming characterized by the replacement of domestic partners with others from low-cost countries.

The related concept of "near-shore outsourcing" designates the relocation in low-cost countries that are geographically close (Mexico for the US and Eastern Europe for the EU). The fourth concept in use, "off-shoring" (and near-shoring) is applied for the relocation of activities *within the company* to low-cost countries. Even if outsourcing is often used as a general handle, it is important for our argument to distinguish between outsourcing, on the one hand, and off-shore (near-shore) outsourcing and off-shoring (near-shoring) on the other. Our argument is most clearly presented if we do not distinguish between the latter two because they both present programmers with the same threat: the relocation of their jobs abroad. It makes little difference to them whether these jobs stay within the company or not.

Off-shore outsourcing is almost always championed as a cost-reducing measure (Gupta et al., 2006), even if its detractors maintain that the costs associated with it are often underestimated (Kakumanu et al., 2006). Be that as it may, the goal of this chapter is by no means to study the profitability of this strategy but to offer an insight into a fundamental change that, following this measure, takes place in the power relations within high-tech companies. Regardless of whether or not it leads to cost savings, it would seem that at least one of its effects is undisputable: the programmers' loss of power and status. Moving IT-services outside the company (outsourcing), or away from the headquarters (off-shoring) is a clear manifestation of them *not* being considered part of the core competences. Instead of taking the long path of deskilling the IT-workforce (which is arguably what knowledge management practices seek to accomplish), management limits its power by moving it to the periphery, sending clear signals that IT-professionals are replaceable, expendable even, and thereby weakening their authority within the company. In this chapter, we shall explore how programers react to this emerging corporate agenda through the construction of alternative identities to those offered by management.

THE SLASHDOT COMMUNITY OF PROGRAMMERS

Building on earlier work (Case & Pineiro 2006), the empirical foundation of this study lies on naturalistic data derived from an online computer programmer discussion forum (www.slashdot.org). The forum initiates about 12 discussions each day, with subsequent exchanges lasting anywhere between one and three days. The discussion threads we have chosen to study had already ended by the time we drew on them for analytical purposes. We intentionally made no effort to engage in them or to influence them in any manner.

The participants, who are invariably programmers or other "techie" IT professionals, may enter the discussion as members of Slashdot, in which case their online identities are available, or they may access it anonymously, in which case they appear online under the shared name of "Anonymous Coward." Furthermore, there is nothing that prevents a participant from having several online identities. However, Slashdot is not an "identity laboratory" forum (Bruckman, 1992). Unlike other more chat-focused online "hang-outs," this is not a place to meet and get to know people. Its design does not allow for chat and picture exchange, its topics of discussion rarely touch upon personal subjects and there is a moderation system in place that has a sobering effect on the discussions. So even if there are cases of "flamebaits" and "trolls" (names given to entries whose goal is to provoke and/or mislead), the discussions can safely be assumed to present us with the participants' earnest opinions, and thus with a rich picture of what programmers think on the subject of off-shore outsourcing.

The language of Slashdot is English, and an important part of its participants are native English speakers—even if not always very careful with their mother tongue. A good deal of them, we estimate that about 85% (it is practically impossible to find out exactly how many) are either U.S. citizens or U.S. workers. So even if there are opinions from programers from other countries (particularly in those discussions that deal with, for instance, Australian issues), most of the entries originate in the U.S. It can therefore be said that the material is representative of U.S. programmers, which does not mean that it shows a homogeneous opinion on outsourcing. We have focused on negative opinions, because those are the ones which best exemplify strategies of resistance. Although they are in clear minority, however, there are also a good number of entries which side with management and express approbation and acceptance of off-shore outsourcing. The inclusion of a discussion entitled "Lowering the odds of being outsourced" is, in itself, a certain—albeit weak—manifestation of acceptance.

We have studied five different discussions, all of which dealt with issues directly related to the practice of outsourcing. These discussions include all kinds of entries, from participants nearly insulting each other through to political manifestos in favour of or against communism, capitalism, socialism and many other forms of political regime. Dealing with the subject of off-shore outsourcing, they also present outbursts of nationalism and xenophobia, and reactions against this. We touch upon these issues only tangentially, focusing instead on the question of organizational power, resistance, and identity.

Our analysis is structured around one of the discussions, "Techies asked to train foreign replacements," because due to its pungency, it offers some of the richest material. The background context to this thread is as follows: IT systems are a strategically fundamental part of the banking systems, but as early as 1998, banks are outsourcing their IT services (Ang & Straub, 1998). This is a subject certain to generate bitter comments among Slashdotters, but things get even more intense when Bank of America (BoA) presents some of its IT-workforce not only with the fact that their jobs are being outsourced but also with the threat of withholding their severance payments unless they agree to train their Indian replacements. Such a case of, according to Slashdotters, extreme lack of sensitivity from management generates a very intense debate in which participants exchange suggestions as to what those BoA programmers should do in the face of this managerial decision. These suggestions offer an insight into the programmers' perspective of: (a) the role of knowledge in power relations, (b) their own position in these relationships as holders of a specific form of knowledge, and ultimately, (c) their own role in the organization.

The chapter has been structured as follows: first we proceed to a classification of the alternative paths of action suggested by Slashdotters in reaction to BoA's policy. We shall see that they vary from quiet acceptance of the facts of life to resentful sabotage. A second analysis, that includes also material from other discussions, reorganises those suggestions around the concept of knowledge. We shall see how programmers' present knowledge as a fundamental part of their identities that differentiates them from "others," such as managers, foreign programers, and low-tech workers. Lastly, we shall see how these exchanges shed light on the processes of construction of alternative identities set in motion by power conflicts in the high-tech industries.

Apart from the BoA discussion, the following discussions have also been studied, although they do not play the same central role:

- **Replaced by Outsourcing—What's a geek to do? (2003)** (http://ask.slashdot.org/article.pl?sid=03/12/19/0456221&tid=)
    - Slashdot member SafariShane's job as a network security analyst has been outsourced and asks the community for advice: "I'd like to hear comments from folks this has happened to, and what did you do as a result?"
- **Australian workers concerned about migrants (2006)** (http://it.slashdot.org/article.pl?sid=06/01/10/0327219)

    - It would seem IT wages in Australia "not just the US" are decreasing due allegedly to immigration of IT-workers (Topsfield 2006): "Would programmers in the developed world be better off without immigration that favors IT or is there an overall benefit for the industry with skilled workers going to the developed world and thus making the industry larger?"
- **Lowering the odds of being outsourced (2006)** (http://it.slashdot.org/article.pl?sid=06/03/30/220219)
    - The discussion starts with a statement that is controversial in an open-source-friendly forum like Slashdot: "[the] best thing young IT workers can do to avoid being outsourced is beef up their management skills."

SLASHDOT ADVICE: STRATEGIES OF DISSENT AND COMPLIANCE

As previously mentioned, the most interesting aspect of the discussion "Techies asked to train foreign replacements" is that it offers participants the possibility of suggesting actions in the face of what are perceived as rather harsh managerial decisions. Usually, there is little a programmer can do in the face of imminent outsourcing, but in this case, the programmers at the BoA are asked to train their replacements. This opens up a warmly welcomed opportunity for speaking about strategies in response to management decisions. These strategies vary from compliance to illegal acts of sabotage, offering a colourful panorama of differing views about what programmers think they are entitled to do in such a situation.

Foucault (1977, 1980) contends that power embodied in and enacted through discourse is productive of particular objects, subjectivities and relations. Language is never neutral or merely descriptive, a condition that applies as much to the discourse of "outsourcing" as it does to any other. In Foucault's terms, discourses, "systematically form the object of which they speak" (1972, p. 49), and, "power produces; it produces reality; it produces domains of objects and rituals of truth" (1977, p. 194). The outsourcing discourse is productive of a new and particular set of power relations within and between organizations. Heretofore skilled IT professionals are repositioned as "redundant," "expendable," or "superfluous to needs" in the face of cheaper substitute labor now available in other parts of the world. The outsourcing discourse inevitably brings in its train a complex set of implied attitudes, purposes, dispositions and actions; in short, a set of power effects. The creation of new subjectivities and the rearrangement of power-knowledge relations is visible, we suggest, in the narratives of IT professionals within the Slashdot community. Within the online exchanges we see conflicts arising as a result of the perceived threat of outsourcing and the realization that organizations of the information age can treat programmers as expendable or substitutable in the just same way as unskilled blue workers have been under industrial capitalism. Nationalism and xenophobia get very close to the surface under such conditions. As a consequence, the exchanges we witness are often highly charged emotionally and full of invective. Respondents strike out at perceived sources of the threats involved: "foreign" workers, "unethical" capitalist corporations, "unsympathetic" and "ignorant" management, and so forth. Indeed, many of the IT professionals represented in our study seem to be acutely aware of the way in which their power positions are being redefined and demeaned. They give critical voice to what appears to be the "proletarianization" of their profession.

There is evidence of what we might call growing consciousness of a nascent class conflict of the sort that typically exercises Braverman-inspired labor process theory (Braverman, 1974). IT professionals in the Slashdot community are aware that their livelihoods are under direct threat and that a managerial class, representing the interests of capital, are responsible for the predicament in which they find themselves. Of course, issues of deskilling and "deprofessionalization" are not new to the discussion of information technology (Beirne, Ramsay, & Panteli, 1998; Kraft, 1977, 1987; Kraft & Dubnoff, 1986; Orlikowski, 1988, 1989). But whereas in the past it has been the deskilling *effects* of high technology that have been at issue, what is new with outsourcing is the extension of deskilling to the very designers and writers of software code themselves. This is precisely the situation facing a significant number of members of the Slashdot community and what interests us, as social scientific analysts, are their responses to this predicament.

For example, the scenario informing the following Slashdot exchange involves management of Bank of America (BoA) not only making a number of programmers redundant but also denying them severance pay unless they agree to train their foreign replacements. What we encounter in the ensuing discussion are a number of acts that programmers believe can be carried out by them in response to this "insult." These acts do not represent what is possible in the actual case at BoA. Few, if any, of the participants seem to have a clear view of the concrete circumstances at BoA and it would probably be incorrect to think of the suggestions as being offered mainly for the benefit of the employees being laid off. The discussion taking place at Slashdot is more directed to other members of the community. It is but one manifestation of the common and empathetic reaction to a phenomenon (outsourcing) which threatens them all. In this context, followers of the thread are invited to entertain even the most illegal and unethical of suggestions. We surmise that they are not meant to be carried out. In fact, illegal suggestions are presented, for the most part, in a humorous tone, but even if they are not offered as "real" alternatives, this is nevertheless what programmers believe to be possible. And by telling us which of their wishes they think are realizable they are offering us a key into their perspective on the nexus of power-knowledge relationships within which they find themselves positioned.

Let us start, however, close to the empirical material, with a classification of the alternatives suggested by Slashdotters. One could argue that management are paying programmers a favour when they ask them to train their replacements. It amounts to saying that they indeed have skills important to the company, which they want transferred. Programmers naturally accept this assumption, using it as the basis of many of the sabotage alternatives. The two most straightforward dissenting strategies are (1) train them badly, and (2) leave without training them—presented in the next section.

There is a presumption, in many of the Slashdot discussion threads, that replacement (outsourced) programmers will be based in the Indian subcontinent. The Slashdot narratives about Indian counterparts are, occasionally, overtly racist in tone. While we wish to distance ourselves markedly from such racist views, we feel it is important to reproduce the language in the extracts verbatim in the interests of social scientific authenticity. We trust that readers will understand and accept this decision and that—recognizing that some of the material is intrinsically offensive—absolutely no offence is intended on our part.

Train Your Replacements Badly

How do you train them badly? Participants suggest two main alternatives: (a) to lecture on technically useless subjects; and, (b) to teach trainees incorrectly. Useless lectures generally deal with material directly obtained from management (like ISO 9000 routines). Teaching those in detail will certainly make time pass without communicating to replacements anything Slashdotters consider of value. Management, however, cannot possibly oppose to the idea because, "[i]f it weren't ok, the whole ISO, SOX and ITIL charade would collapse, which would be worse for management than one poorly trained replacement" (#15509302)[a]. Here is Confused's advice:

Re: Screw 'em, Walk Out!

(Score:5, Insightful)
by Confused (34234) on Saturday June 10, @09:39AM (#15508798)
(http://slashdot.org/)
Walking out isn't a good solution, because probably you're just another cog in the big machine. If you refuse to train them, you'll just save them your severance pay without much effect.

To really spite them, you should make the effort an train your Indian replacement, in a creative way.

Step 1 is to dig all those SOX and ISO-9000 procedures, you have been forced to fill out over the years. Per definition, they describe your work in a reproducible way. These should be the base of any training. Make your replacement memorize them an apply them in harmless cases, where they don't generate too much mayhem. This alone should easily occupy about 3 months worth of training time and prove totally worthless while being the perfect employee by the book.

Step 1a, if above mentioned documentation doesn't have enough volume, take the time and have the trainee update it. This alone should cover another 6 months.

Step 2 have your trainee sign of every thing you've told him, that he learned and understood the procedure. By the end of every day, you shoudl have at least 4 signed papers by your replacement how well you've trained him. Add copies of these signed statements to your daily work log.

Step 3 stop doing any regular work - except for problems that are not easy to fix an will appear again after you've left. The trainees and the managers should get the impression your job is just easy routine. If some manageroid comes with urgent requests, let him know that you don't have time for it, as you have to train the replacement for your severance pay and the replacement isn't ready to perform this yet. If they want it done anyway, get it in written and signed on paper, best with HR also approving that you don't have to train the replacement while doing some other things. Give a reasonable time frame (e.g., Rebooting a desktop: 2 work days) [...]

A variant of this is to teach the replacements about the parts of the IT System that are either obsolete or will soon become so, leaving aside anything of importance. A more risky alternative is to teach them how a part of the program works only to modify it afterwards (without telling the replacement).

BoA "IT folk" is also advised on the option of teaching their replacements wrongly: to lie to them about what parts of the software do what. Many entries with these kind of suggestions are meant as jokes, where the character of the Indian replacement is presented as rather ignorant on programming matters. The suggestion in the title of the entry below (to type in cd /, press "enter," then type in "rm –rf *" – and press "enter" again –) is an elementary way of deleting everything in a given folder. One of the first things one learns when studying the Unix operating system (and other similar to it, such as Linux). It is highly unlikely that any programmer does not know this.

Ok Remesh, now type in cd / enter rm -rf *.......

(Score:1)
by MoronBob (574671) on Saturday June 10, @07:58PM (#15510886)
I bet it will be good training.

Leave

A very straightforward way of sabotaging the training of the future replacements is to refuse to do it and simply leave. The assumption behind the order of training the replacements is that the programmers' knowledge is important, an assumption that programmers readily accept, and that the most straightforward way of damaging the company is to deprive them of this knowledge.

There is, of course, the issue of the severance payment, which the BoA will only offer to those who train their replacements. Slashdot community "advice" on this matter appears to embrace three distinct strategies

Just Leave

This first strategy does not even mention the severance payment. Training your replacements is so far beyond what you as a programmer with professional integrity can possibly accept that there are no mitigating considerations. Leave and never look back. Or, as some point out, expect a "callback" and get ready for retaliation:

Reality of Outsourcing
(Score:0)
by Anonymous Coward on Saturday June 10, @04:34PM (#15510319)
Having worked in IT for several investment banks, here are a few first hand observations on outsourcing.

[...] 5. Be prepared for a callback. If they outsource your role, and then discover the muppet from the services company they chose, or even the grad fresh from uni in India with zero experience does not have the same level or skills and experience, they may call you back offering to rehire you. If they do this, calmly report you will come back on a contract basis, and name an astronomical rate. You know they are over a barrel, they know they are over a barrel. Enjoy!

6. Never, ever, train your replacement. Perhaps you would also like to go and park a big ol' kiss on the backside of the greasy senior mismanager who authorised you losing your job in the first place. Most jobs take time not because of the technology, but because of the process. the local or company-wide processes, the red tape, the electronic paperwork. Let the new guy or girl struggle. Instead, use your time looking for the next role.

Leave Making Sure you Have Money in the Bank

Everyone should save a little bit of their money every month to be able to do this. This suggestion triggered a series of responses around the subject of whether it is or is not possible to save this money. Some programmers complained that this is impossible if you have to support a "mortgage and a family," implying that they were being exploited by companies, and more specifically, management.

Leave and Sue BoA for your Severance Pay

One discussant alleged that what BoA are doing is illegal. There ensued a discussion on whether or not it is illegal to require employees to train replacements under these conditions; a discussion that appeared to have no clear end. Once again, the subject was raised of the pros and cons of a less "raw" capitalism, with a more restrictive law regarding contract-termination.

Train the Replacements Well, Things Will Take Care of Themselves

This suggestion is founded in the seemingly common idea that the programmers' knowledge about the system is not as easily transferable as management would like to believe. IT systems are too complex to be explained to novices (foreign programmers are treated like novices) so there is no point in getting excited about conscious acts of sabotage. Doing what management wants you to do, that is, training the new replacements, is a perfectly plausible alternative: the time allotted to this is too short, they simply will not learn what they need to know, regardless of how hard you try. So by all means, train them, give it your best shot.

Re: Screw 'em, Walk Out!
(Score:1, Interesting)
by Anonymous Coward on Saturday June 10, @11:58AM (#15509296)
-To really spite them, you should make the effort an train your indian replacement, in a creative way.-
I don't agree. I advise making your very best effort.
Twenty years ago, something like his happened to me. I was the engineer in charge of manufacturing systems test for a packet switched network when the headquarters decided to move manufacturing to Southern California and shut us down. The test guys wanted to sabotage everything and asked me to help. I told them I had a better idea. Let's do the very best we can knowing that they will foul it up anyway.
You know what? I was right. They fouled it up despite all our efforts to help them. Frankly, I think it was more satisfying that way. Even when I found out years later it was an article of faith down there that I had sabotaged everything and that's why they had so much trouble.

Regardless of Anything Else, Take Your Money to Another Bank

"Vote with your dollars," put your money in a Credit Union, or anywhere else. This suggestion appeared in quite a number of entries, and triggered discussions about where to put the money instead. There were also a number of commentaries along the lines of "Bank of America" changing its name to "Bank of India."

Unite!

A few participants mentioned the idea of building a programmers' union so as to be able to put pressure on companies such as Bank of America, and eliminating the worst of managerial practices, such as forcing programmers to train their own replacements. These contributions are interesting for the fact that they give voice to a clear sense of proletariat class consciousness and the need to form some kind of global labor movement (reminiscent of the nineteenth and early twentieth century Marxist International movement). However, suggestions regarding the need for collective action met in cases with a convinced opposition based on the idea of workers' unions becoming, with time, centres of power that ignored their members' needs and interests.

Re: bunch of things
(Score:3, Insightful)
by smchris (464899) on Saturday June 10, @11:16AM (#15509127)
[...] IF programmers had been unionized like commercial pilots, I think we can see there would probably have been outsourcing eventually anyway. But you might have been able to strike, get union relief, and threaten BoA with a world of trouble. And perhaps have worked for a few more years.
As it is, I see three options:
[...] 3) As others have suggested, you can do everything you can think of to monkey wrench the transition. I suggest small group meetings off-site, off-hours to coordinate. Document. At the appropriate time publish and teach others. Get BoA's name everywhere they don't want it. 'BoA The Movie." There are small voices to publicize this outsourcing like the AirAmerica and Jones

Networks and plenty of web sites you can glean from portals like Buzzflash.com.
And become honestly close to each other in the process. You'll need the networking.

world-wide worker union needed
(Score:0)
by polar red (215081) on Saturday June 10, @12:17PM (#15509354)
There's a world-wide workers union needed, with as most important objective : world-wide minimum wages. This is absolutely necessary, because employers already work globally, and by pitching different countries(the workers of those countries) against each other, they can exploit maximally leaing to 21th century feudalism! This is why nationalism is the new opiate for the masses, keep 'em stupid !

Accept it

A few participants were of the opinion that outsourcing is part of the game, and that there is nothing inherently bad about it and that it should be accepted. It may not be a good management practice but it does exist, it is not going to go away and, even if it hurts, it does not break against decency. However, this position is so extreme in the Slashdot context that whenever it appears, one suspects that the author might be trying to pull out a "troll": an entry that argues for an opinion very much against the community's with the hope of setting in motion a wave of outraged answers. This tactic has to be employed subtly, otherwise it risks being moderated out as flamebait.

Sabotage the System

Quite a number of participants recommended sabotage, based on their knowledge of the system, as a viable alternative:

Re: In a capitalist economy, stuff like this happen
(Score:2)
by dgatwood (11270) on Saturday June 10, @02:13PM (#15509858)
Three words: temporary root passwords.
'Nuff said.

Root passwords are passwords that allow you to access everything in a system (from the root). dgatwood is suggesting that BoA IT-workers should create a series of "backdoors" into the system so that, one would imagine, they can access it later on from outside the company for malicious purposes.

KNOWLEDGE AND POWER IN HIGH-TECH ENVIRONMENTS

In this section we shall consider the role of knowledge in the programmers' perspective on the relationships of power in the organization. We seek to address the question of what kind of power programmers think that their knowledge offers. Programmers seem to have different views: from "very little" to "a lot," the latter end of the spectrum having a far larger representation in the discussion. However, what is interesting to us here is not so much the quantitative proportions of views as an analysis of the different ways in which knowledge and power are connected in the discussion narratives.

Now, in the main, the overall message in the discussions is that programmers are not replaceable or expendable. This claim is simply a way of expressing their views about their role in the organization and about the power they actually hold, even if management does not perceive or recognize that power. It is also a strategy for constructing an alternative identity to that offered by outsourcing-happy managers, but more about this in a later section. Let us now see how this claim of being irreplaceable is argued for.

Not Argued, Just Stated

This message is sometimes presented on its own, as if it were plain to everyone why programmers should not be replaced:

From Australian IT Workers...
Re: Oh geez....
(Score:1)
by sgt_doom (655561) on Wednesday January 11, @02:17PM (#14447246)
No, they didn't take our jobs - our jobs were handed over to them by very shortsighted, unimaginative corps-heads who think everyone is an easily replaceable module - which has been the corporate America credo for the past thirty to forty years. [...]

It is undoubtedly telling that a number of participants would state that they are irreplaceable without even trying to explain why. Their identity is strongly based on the idea of being fundamental parts of the organization. It is however more interesting to study the arguments presented as the basis for this statement.

Irreplaceable on a Large Scale

For instance, the "humanitarian" view that people should be valued as human beings and, hence, not laid off simply for reasons of economic expediency seems—perhaps surprisingly—not to be expressed by Slashdotters. On the other hand, there are frequent nationalistic comments regarding the native government's responsibility of making sure jobs "stay in the country." The reasons behind this argument as presented by Slashdotters are based on the idea that it is of national strategic importance that the skills associated with those jobs do not disappear from the country. Two underlying assumptions seem to lie at the heart of this: (a) that IT-skills are a fundamental strategic asset; and (b) that IT-skills are not easy to obtain. Participants do not spell these assumptions out, at least not in a structured manner, but they are implied in the ways they speak about outsourcing. These assumptions naturally reinforce their identity as vital members, not only to the firm, but even of society; so sure are they about this that it is not unusual to find comments that seem to revel in the "inevitable" effects of outsourcing:

Remember this . . .
(Score:1)
by Ph33r th3 g(O)at (592622) on Saturday June 10, @11:29AM (#15509166)
. . . next time you're reading about the "tech shortage" in the United States.

More Fuel for the Fire
(Score:3, Interesting)
by blueZhift (652272) on Saturday June 10, @08:27AM (#15508570)
(http://bluezhift.proliphus.com/ | Last Journal: Tuesday August 08, @10:52AM)
This is just more fuel for the fire. Next there will be an article about some CEO complaining about how there aren't enough skilled IT workers in the U.S. and how college students are not entering the field. I just don't understand how U.S. companies can continue to build up so much ill-will (or bad karma if you will) with practices that BoA [Bank of America] at least acknowledges are offensive and yet continue said practices. A big price is going to be paid for these betrayals someday, a very big price.

Irreplaceable in the Company

By and large the most frequent argument for the claim of being irreplaceable—whenever it is explained by Slashdotters—is based on the assumed centrality of their knowledge and its role in the company. For instance, according to a number of entries, it is true that native programmers cost more but this is because they know more and hence can deliver better quality. This kind of argument is a central, if contested, dimension of the discussion, with narratives contrasting various positions. For example, some Slashdotters argue that native programmers cost more because they know more while others maintain they cost more for other reasons. Similarly, some contend that native programmers know more and deliver higher quality outputs while others espouse the countervailing view that not all foreigners are poor programmers.

Due to their personal experiences, or to prejudice, they seem to assume that services that cost less are of inferior quality. This, applied to the problem at hand, results in many voices suggesting that outsourced IT services will never offer the same quality as those obtained from the native workforce.

From Techies asked...
Re: Time to change banks...
(Score:5, Informative)
by Zemran (3101) on Saturday June 10, @08:32AM (#15508585)
(http://www.geocities.com/zemran | Last Journal: Friday November 07, @07:07AM)
Lots of British companies that outsourced to India 5-10 years ago found that it cost more overall because the quality suffered and so much time was spent with managers flying backwards and forwards. Many of those companies returned to the UK because it is cheaper to pay for something to be done properly in the first place that to get a cheap job done that needs twice as much spent on fixing it.

Why exactly outsourcing does not offer the same quality is not always argued for in the entries (as above). But whenever it is, we can find two key ideas: one is based on knowledge (native programmers are more knowledgeable) and the other in general prejudices. The general prejudices have to do with the perceived degree of foreigners' loyalty to the company and on their level of trustworthiness. There are a number of entries that are outright xenophobic, most of them are contested, both by foreign and native participants (or, more precisely, by participants who claim to be foreigners and native).

KNOWLEDGE AS IDENTITY OPPOSITION

We see from our data set that knowledge plays an essential role in the construction of programmers' identities. The question of whether or not foreign programmers have that knowledge is contested but what is never questioned is the necessity for a programmer to possess that knowledge. This having/not having knowledge is articulated around three sets of opposition concerning: managers, foreign programmers, and low-tech workers.

Managers

Very little positive is said about managers in Slashdot discussions in general, and even less when the subject at hand is outsourcing. Basically, managers are placed on the opposite side of programmers in the Slashdot narratives: everything that managers are said to be can safely be interpreted as what programmers are not, and vice versa. Managers are definitively "other," taking the role of the scapegoat in terms of Slashdot identity formation (Burke, 1969) and possessing two demeaning characteristics in the eyes of techies: managers are, above all else, incompetent and greedy.

Incompetent

One of the basic reasons to dislike managers is that they are incapable of doing anything vaguely productive. Their decisions are based on ignorance of what an IT system actually is. Such alleged ignorance extends not to the programming side of things but also to more "managerial" aspects such as the roots of bugs and why low-quality software requires far more long term maintenance than well constructed programmes (and hence incurs further costs for the company). Moreover, managers are oblivious to the principles that result in high-quality software, and, more generally, to what decisions are in the best interests of the company. For example, the general view (though it is not without its opponents) among participants is that the increasingly frequent appearance of "Another silly outsourcerer..." (#15508465) is a yet further confirmation of managers' lack of understanding of the basics of their business:

From Australian IT Workers...
(Score:2)
by BVis (267028) on Tuesday January 10, @08:20PM (#14441818)
[...] What I'm saying is that someone who you pay half of what you pay someone else is going to be perceived by the short-sighted useless douchebags in middle management to be the better value, no matter how craptacular their skillset is.[...]

From Techies asked...

Re: In a capitalist economy, stuff like this happen
(Score:5, Insightful)
by AB3A (192265) on Saturday June 10, @10:09AM (#15508894)
(http://slashdot.org/~AB3A/journal | Last Journal: Wednesday August 02, @11:52AM)
It may not be wrong. But I don't think legality is the point.
Over the longer term, what have these idiotic managers done? They've disconnected themselves from the very communities they serve. Sure, it's cheaper to have a bunch of lower paid foreigners do this work. But do they really understand what they're doing and why? Will they know when they've run in to a problem? Will they be able to reason their way through regulations and laws they had nothing to do with? [...]
Frankly, I see this as a huge disconnect between management and the techies who actually make things go. If anyone here owns this stock, I recommend they sell within the next year or so.[...]

Maybe it's time for the technocratic war to begin.
(Score:5, Interesting)
by Anonymous Coward on Friday December 19, @12:14PM (#7764834)
The managers and CEOs of this country have no idea about how to make router connection or how to correct a line of code in their payroll systems.

I'm on call 24x7x365 while the CEO sleeps.

The none technical types need to understand where info power resides.

In contrast to alleged managerial ignorance, programmers construe themselves as the ones who comprehend the complexities of software development and, crucially, those who actually write software, in other words, the value adding members of the organization. They, the workers, know what must be done and, unlike managers, they undertake productive work. This is a fundamental dimension of their identity as programmers.

The general view is that outsourcing is bad, for a number of reasons, but a number of participants actually defended this managerial practice, explaining that it made economic sense to move development to zones of the world with cheaper labor. Even if these entries showed more sympathy for the managers' position, there was little in them about managers' knowledge and skills. The decision to outsource was in all cases presented as an obvious one possessing a compelling logic, as follows: cheaper labor, hence reduced costs, hence move. The decision that requires thought and that would manifest both intelligence and courage, according to Slashdotters, is that of *not* outsourcing. The opposition between managers and "IT folk" is therefore upheld with respect to the knowledge/productivity axis even when outsourcing is not directly criticized.

Greedy

Slashdot was originally a forum for open source programmers and even if it now gathers programmers of all kinds, the concept of profit has retained some of its former "bad karma." Whereas making money in itself is not negative, there are limits that should not be transgressed:

Profit?
(Score:2)
by CaptDeuce (84529) on Saturday June 10, @09:06AM (#15508690)
(Last Journal: Thursday June 08, @08:07PM)
From TFA:
BofA estimates that outsourcing has allowed the bank to save about $100 million over the past five years.

From Fortune 500 [cnn.com]:
** Profits in 2005: $16.465 billion*
** Growth: 16.4% over 2004*
** 1995-2005 annual growth rate: 8.5%*
So... that profit figure was damaged considerably by the $10 million or so in IT expenses. Which means—using a very rough calculation—profits rose from $16.464 to $16.465, a whopping %.009 (a thousandth of one percent) by eliminating those jobs!
It's amazing BofA showed any growth over the past five years what with carrying the dead weight of 500 IT people!

Arithmetic
(Score:1)
by blue.strider (737082) on Saturday June 10, @11:25AM (#15509153)
*Savings from f***ing up workers:*
100 million over 5 years = 20 million / year.
Income of (only one! of) the blood suckers:
BoA CEO pay (in 2003) = 20 million.

Class war at its best.[...]

This aspect of the programmers' identity, opposed to the "greed-is-good weasels" (#15509184) of management, is however not as prominent as the knowledge/productivity one. There are several entries that touch upon the subject of profit making but the culprit is more often ignorance about the importance of long-term sustainable strategies rather than greed.

Foreign Programmers

There is much debate in the discussions about foreign programmers, and they are not always presented in a favourable light. However, in spite of certain participants' racism and xenophobia it is interesting to observe how Slashdotters seem often to return to fundamental considerations of knowledge and skill in promoting or preserving their sense of identity.

Those who are most insistent about the unsuitability of foreign programmers do so mainly by denouncing their lack of knowledge and skills around two areas: lack of understanding of the national culture (including language and laws) and lack of programming skills. Very occasionally they are also presented as untrustworthy. These accusations are contested, and a number of entries are devoted to expressing admiration or at least respect for the skills of specific foreign programmers "met [...] over the years" (#15508616), but no participant questions the idea that advanced technical knowledge is necessary to be a good programmer. The issue is whether foreign programmers possess this knowledge. Some Slashdot participants claim that they do; others that they do not. No-one suggests, however, that the level of knowledge required to provide IT-services may have declined (due, for instance, to advances in technology). If anything, it would seem that it comprises not only technological skills but also social and language skills beyond the reach of foreigners.

Low-Tech Workers

Low-tech workers do not appear at any length in the discussions we have studied. When they do, however, they constitute yet another kind of "other." However, unlike the previous two categories of other—managers and foreign workers—this one is not perceived as a threat. Low-tech workers are not going to take their jobs, neither are they going to make the outsourcing decisions. Hence, we find them presented in, at worst, a rather neutral manner, and, indeed, often rather sympathetically. After all, they are facing similar problems. The main difference between blue and white collar workers, according to Slashdotters, lies in the level of knowledge necessary to do their job. Whereas technical know-how is a central element in the programmers' view of themselves, participants do not seem to consider it an important attribute of the low-tech blue collar work:

Well the thing is
(Score:2)
by Sycraft-fu (314770) on Saturday June 10, @01:47PM (#15509762)
With manufacturing, it doesn't matter where it happens or who does it. So long as the workers can do the job right, nothing else matters, just ship the goods to where they need to go. Not so with IT. There are distinct disadvantages to IT outsourcing.[...]
Low-skill manufacturing is always going to be a lowest-bidder kind of situation, but that's not the case with everything. Even higher skill manufacturing isn't that way. Notice that Intel doesn't fab all there chips in China, in fact they don't fab any there. Their fabs are in places like Arizona, California, Israel, and Ireland. Same with AMD (Germany, Texas, Japan, etc). For various reasons, outsourcing to a lowest-bidder kind of country doesn't work out for them, despite it being a manufacturing job in the end.

In the face of outsourcing, it is difficult to maintain that IT workers have a special kind of knowledge that makes them irreplaceable. Some, as we have seen, blame outsourcing on the ignorance about the importance of US IT workers' knowledge and predict dire consequences to the native companies; others, such as SlashSquatch, go to the other extreme, equating programmers, and blue-collar workers:

Software Engineers NEED a Professional Society
(Score:1)
by SlashSquatch (928150) on Saturday June 10, @10:39AM (#15508989)
(http://cryptostenchies.com/)

The professionalism is gone from software engineering.

It grew from computer science with an influx of new systems and new employees in the 90's. Professionalism went out the window, long hair was grown. For 100 crappy, expensive projects, one survived. A negative image was burnt on people's minds.

Next, the programmer started to become a blue collar worker. They do more for less pay, punch a clock and no one lets them make any business decisions.

Now comes the next backlash, outsourcing. Very few company officers come from IT. Administration is in the dark about IT decisions. They rely on marketing information both coming and going to make decisions. IT gets pissed at installing yet another crap package and stupid patch. It's an US v. THEM situation.

Every software developer that runs around saying "I don't care about the EVIL business side, I just want to program" this is your reward. Complete impotence.

The solution: Software Engineer Professional society. One very effective Professional Society is the Society of Actuaries. They make GREAT money, business decisions and all they have to do is apply mathematics.

dyslexics of America, Untie!

It is interesting to note that the equation does not build on a lack of special knowledge on the part of the IT workers, the "problem" is rather that this knowledge is not valued any longer. SlashSquatch's solution, thus, is to force the firms to value it through the establishment of a Software Engineer Professional Society: such an institution should be capable of securing jobs and salaries when even actuaries' application of mathematics (allegedly trivial in comparison with the skills required of IT workers) can be made valuable. But not everyone likes the idea of unions, as we saw in the previous section, reasoning that if you are being outsourced that is due to your lack of programming skills, which means you are not a worthy programmer:

Re:And then get arrested, convicted...
(Score:0)
by Anonymous Coward on Friday December 19, @06:07PM (#7769188)
-that's why you should organize before you need to... A union may have been able to stop this.-

Yeah - I want to sit on my butt all day long and get paid big bucks for doing mediocre work. Unions do have their place in the world. But just because _YOU_ have lousy skills and fear for your job doesn't mean that the rest of us feel the same. Unions seek to make reward based not on skill, but on length of union membership on who you know and on who's apple you're polishing

CONCLUSION

The discussions presented here, arguably, do not constitute cases of resistance per se but rather of dissent. The voices of dissent are not directed toward the managers, at least not directly, but to other peers. As Collinson (2003) argues, dissent in this case is not so much about sabotage as about the construction and performance of identities, expressing empathy and solidarity. In some cases this is very obvious, since what is suggested is that "complying with managers" is what should be done, things "will f@ck up anyhow." For some of the participants, it is not so much doing something that counts but assuring one another that they form part of the community of the initiated; they "belong" to the community and know what will happen. They are the one's with the knowledge to run BoA, the others are not only 'greedy ignorant idiots' but, above all, they are outsiders.

As we have attempted to demonstrate above, the identities constructed and performed in the entries do not present a monolithic set. Instead, there are differences on at least two levels. Firstly, there are differing, and at times opposing, views on the subject of outsourcing. The kinds of dilemmas and questions debated in the narratives are numerous. Is outsourcing good, bad, or simply a neutral inevitability? Is it the result of greed or the unfolding of inexorable capitalist imperatives? Is it devastating for US programmers or is it the only way to maintain a healthy economy? Is it anti-nationalistic or a necessary step toward the improvement of living conditions in other countries? Secondly, as we have seen, there are differing and also at times opposing opinions as to how to deal with outsourcing, and as to what programmers can do about it.

We opened the chapter by indicating that some social observers conclude that we are currently undergoing an epochal shift away from the socio-economic dynamics of industrial capitalism and witnessing the emergence of an informational global social order. Castells (1996), for example, argues enthusiastically that recent ICT developments, constituting a new technological paradigm, have ushered in a new era of "informationalism" and "networking" that is distinct from capitalist or statist industrialism. By contrast, we suggest that a more even minded and measured consideration of the social and historical conditions in play is required. More compelling, we contend, are arguments that understand the distinctive social and material technologies of "post-industrial," "late," or "disorganized" capitalism as being genealogically related and continuous with an industrial (or statist) legacy (Harvey, 1989, Lash & Urry, 1987). Castells, we suggest, is inclined to overstate the case for paradigm shift with the resulting danger that historical continuities with respect to labor relations or the geographical redistribution of production under late capitalism are at risk of being overlooked. Reflecting on the employment relationships explicitly articulated in the Slashdot exchanges previously presented, for example, one might reasonably conclude that certain asymmetries characteristic of *industrial* capitalism persist within *informational* capitalism. As we have seen, Slashdotters respond vociferously to the threats that outsourcing poses to their livelihoods and professional status. Indeed, the narratives testify overtly to an antagonistic relationship between employers and employees which are reminiscent of, if not isomorphic with, the kinds of labor unrest that characterised industrial capitalism in the nineteenth and twentieth centuries. Many of the remarks made in Slashdot entries appear to be consistent with the forms of dialectical conflict present in Marxist critiques of industrial labor relations. The very language of Slashdotters concerning the formation of collective bodies to protect their professional interests, or, indeed, the speculative creation of an "international" IT resistance movement in the face of outsourcing are entirely consistent with the sentiments of *The Communist Manifesto*, for example (Marx & Engels, 1996/1848).

This will be unsurprising to those familiar with certain stands of critical management studies that recognize the continuity between industrial and informational capitalism. Willmott (1995, p. 95-6), for example, argues that the employment relationship within contemporary capitalist organizations (informational or otherwise) is such that, all other things being equal: (1) investors will attempt at all times to ensure that managers direct and control organizations in ways that are deemed to serve investors' interests (i.e., the generation of profits that, in turn, translate into dividends); (2) employees will endeavor wherever possible to improve their pay, job security and conditions of employment, and; (3) Managers (as materially and symbolically privileged employees) will endeavor to protect their employment status. Many members of the Slashdot community, responding to the perceived threat of outsourcing, seem acutely aware of these dynamics of employment relations and give voice to what looks remarkably like a new iteration of industrial class consciousness. Whether dissent of this sort will eventually give rise to organized resistance on the part of IT professionals working under conditions of contemporary capitalism remains, of course, a moot point. The current political climate and labor law frameworks in western economies will make such orchestrated resistance very difficult, added to which divisions within the Slashdot community itself and scepticism, from some quarters, regarding the value of unionization cast serious doubt over the possibility of effective collective action. We hope, nonetheless, that reporting on the voices of dissent from within this community sheds fresh and interesting light, from the perspective of the workers affected, on how the management of outsourcing in high-tech environments is being perceived and responded to.

REFERENCES

Ang, S., & Straub, D.W. (1998). Production and transaction economies and IS outsourcing: A study of the U.S. banking industry. *MIS Quarterly, 22*(2), 535-552.

Apte, U. M., Sobol, M. G., Hanaoka, S., Shimada, T., Saarinen, T., Salmela, T., & Vepsalainen, A. P. J. (1997). IS outsourcing practices in the USA, Japan, and Finland: A comparative study. *Journal of Information Technology, 12*(4), 289-304.

Angell, I. (2000). *The new barbarian manifesto: How to survive in the information age.* London: Kogan Page.

Barret, R. (2005). *Management, labor process, and software development.* London: Routledge.

Beirne, M., Ramsay, H., & Panteli, A. (1998). Developments in computing work: Control and contradiction in the software labor process. In P. Thompson & C. Warhurst (Eds.), *Workplaces of the future* (pp. 142-62). London: Macmillan.

Braverman, H. (1974). *Labour and monopoly capital.* London: Monthly Review Press.

Brown, D. (2005). *The black book of outsourcing.* Chichester: Wiley.

Bruckman, A. (1992). Identity workshop: Emergent social and psychological phenomena in text-based virtual reality. Retrieved November 27, 2006, from ftp://ftp.cc.gatech.edu/pub/people/asb/papers/identity-workshop.rtf

Burke, K. A. (1969). *A rhetoric of motives.* London: University of California Press.

Collinson, D. L. (2003). Selves at work. *Organization, 10*(3), 527-547.

Case, P., & Pineiro, E. (2006). Aesthetics, performativity, and resistance in the narratives of a computer programming community. *Human Relations, 59*(6), 753-782.

Castells, M. (1996). *The rise of the network society.* Oxford: Blackwell.

Corbett, M. F. (2004). *The outsourcing revolution: Why it makes sense and how to do it right.* Chicago: Dearborn.

Costlow, T. (2003). Globalization drives changes in software careers. *IEEE Software, 20*(6), 12-16.

IEEE-USA. (2006). *About us.* Retrieved November 27, 2006 from http://www.ieeeusa.org/about/default.asp

Gupta, S., & Chaudhari, N. S. (2006). Information technology offshore outsourcing: A perspective of advanced countries. In H. S. Kehal & V. P. Singh (Eds.), *Outsourcing and offshoring in the 21st century: A socio-economic perspective* (pp. 122-139). Hershey, PA: Idea Group Publishing.

Harvey, D. (1989). *The condition of postmodernity. An enquiry into the origins of cultural change.* Oxford: Blackwell.

Kakumanu, P., & Portanova, A. (2006). Outsourcing: Its benefits, drawbacks, and other related issues. *Journal of American Academy* of *Business, 9*(2), 1-7.

Kehal, H. S., & Singh, V. P. (Eds.). (2006). *Outsourcing and offshoring in the 21st Century: A socio-economic perspective.* Hershey, PA: Idea Group Publishing.

Kraft, P. (1987). Computers and the automation of work. In R. Kraut (Ed.), *Technology and the transformation of white collar work* (pp. 91-111). NJ: Lawrence Erlbaum.

Kraft, P. (1977). *Programmers and managers: The routinisation of computer programming in the United States.* New York: Springer Verlag.

Kraft, P., & Dubnoff, S. (1986). Job content, fragmentation, and control in computer software work. *Industrial Relations, 25*(2), 184-196.

Lash, S., & Urry, J. (1987). *The end of organised capitalism.* Oxford: Polity

Marx, K., & Engels, F. (1996/1848). *The communist manifesto.* London: Junius.

Orlikowski, W. (1989). Software work. *Industrial Relations, 25*(2), 184-96.

Orlikowski, W. (1988). The data processing occupation: Professionalisation or proletarianisation. *Research in the Sociology of Work, 4*, 95-124.

Power, M., Desouza, K. C., & Bonifazi, C. (2006). *The outsourcing handbook: How to implement a successful outsourcing process.* London: Kogan Page.

Quan, M. (2003). *IEEE USA presses Congress on visa curbs.* CommsDesign. Retrieved November 27, 2006, from http://www.commsdesign.com/news/market_news/OEG20030326S0029

Ralph, W. & Wyndrum, J. (2006). *Welcome to the new year.* IEEE-USA, President's Column. Retrieved November 26, 2006, from http://www.ieeeusa.org/communications/presidentscolumn/Wyndrum/jan06.html

Scarbrough, H. (1999). Knowledge as work: Conflicts in the management of knowledge workers. *Technology Analysis and Strategic Management, 11*(1), 5-16.

Sobol, M., & Apte, U. (1995). Domestic and global outsourcing practices of America's most effective IS Users. *Journal of Information Technology, 10*(4), 269-81.

Steadman, J. (2004). *Career trends.* IEEE-USA. Retrieved November 27, 2006, from http://www.ieeeusa.org/communications/Steadman/steadmannov04.html

Topsfield, J. (2006). Migrants blamed for IT jobs cut. *The Age.* Retrieved November 27, 2006, from http://www.theage.com.au/articles/2006/01/09/1136771500496.html?from=top5

Willmott, H. (1995). The odd couple? Re-engineering business processes; managing human relations. *Technology, Work, and Employment, 10*(2), 89-98.

ENDNOTE

[a] All the data extracts from *Slashdot* are reproduced verbatim, complete with spelling, grammatical and typographical errors. The # numbers in parentheses are used to code *Slashdot* entries and may be used to locate them in the on-line archive.

Chapter XIII
Power and Ethics in IS Evaluation

José-Rodrigo Córdoba
University of Hull, UK

Wendy Robson
University of Hull, UK

ABSTRACT

In this chapter, we widen the perspective on power as an issue in the practice of information systems (IS) evaluation. Power is frequently seen as a resource that some people have and some do not. Therefore it is primarily discussed in the context of IS development methodologies and little is said about power in evaluation. Based on Michel Foucault's ideas, we discuss the role(s) of power and ethics strategies in the evaluation of information systems plans. Reflections from practice of an evaluation at Javeriana University in Colombia led us to identify some salient features of ethics and power in IS evaluation, which we could use to inform further IS evaluation exercises. The experience also led us to acknowledge two layers of power that influenced the evaluation: An institutional level and an "emerging" local level. At the time of the evaluation, these layers exhibited their own mechanisms by which information systems projects were defined, justified, implemented, and evaluated at the organization. These layers also influenced the ethics of evaluators. Recognising their existence could influence the strategies that evaluators define to deal with issues that emerge in the process of information systems development and implementation.

INTRODUCTION

Information systems (IS) evaluation has been a popular and long-standing research topic for many years (Smithson & Tsiavos, 2004). Through this process, the value of information systems investments and plans are assessed prior to the investments themselves (*exante* evaluation) or after their implementation (*expost*) (McAulay, Keval, & Doherty, 2002). There are different perspectives about what, how, and why evaluations need to be carried out. For instance, those individuals responsible for investments need to identify benefits and tradeoffs that are to be made. At the same time, those individuals being affected by the investments need to see how they can benefit (and hopefully not being harmed) themselves from the process (Serafeimidis & Smithson, 2003). With a variety of perspectives emerging in IS evaluation, and participation becoming an important element of it (Remenyi & Sherwood-Smith, 1999), it is worth asking how organizations are dealing with IS evaluation, and what could be done to improve it.

In this chapter, we focus on the relationships between power, ethics, and IS evaluation as a way of advancing evaluation practice. We detail a relational view of the concept of power by exploring how power relations shape—and are shaped by—individual and group ethical behavior, and how this affects evaluation of information systems investments. We use Michel Foucault's ideas on power and ethics. These ideas allow us to draw some elements for analysis and reflection in IS evaluation. Using these elements, we describe and reflect on a practical exercise of evaluating an information systems plan at Javeriana University in Colombia. The insights gained from this experience lead us to suggest some insights that could be used to recognise the practical nature of power and ethics in evaluation processes in organizations. As Foucault, we see power as an analytical property attributed to human relations by which actions influence other actions (Foucault, 1984a) and this includes ethical behavior.

Our aim to use power as an aide is not to declare that power needs to be considered as an inherent feature of IS evaluation. Instead, we would like to use the notion of power to generate insights as to how evaluation is influenced by issues that are not traditionally considered in the realm of information systems. About ethics, we hold the view that ethics defines "acceptable" or "ethically appropriate" ways of relation between individuals, themselves and others (Foucault, 1984b).

The chapter is structured as follows. We begin by briefly situating IS evaluation within the context of organizational power (often called "politics"). We then asses how power has been dealt with in the IS literature. This leads us to argue that the existence of a relational view of power can help us enrich our understandings of evaluation. We present some of Foucault's ideas on power and ethics that we consider useful to inform the practice of information systems evaluation, and draw three (3) elements for analysis and reflection. Using these elements, we then present our reflections on a practical exercise of evaluation of an information systems plan at Javeriana University. Our discussion leads us to identify two layers of power, which at Javeriana shaped the definition, implementation, and evaluation of information systems at the institution. We conclude the chapter by suggesting how practitioners could improve their evaluations by being aware of power relations and ethics.

POWER IN INFORMATION SYSTEMS EVALUATION

We begin by situating power within an organizational context in which it has been often understood as "politics" (Checkland & Scholes, 1990; Markus, 2002; Symons & Walsham, 1988). This view coincides with that of an organization as a set of "coalitions," alliances, and struggles between individuals and groups, in which some people try to privilege their interests at the expense of other people, often succeeding in their effort (Morgan, 1997). The view of power as politics seems to impact understandings of IS evaluation, where this process is seen not as "the result of a single group of stakeholders, but rather of the complex interplay of various actors, both human and non-human" (Smithson et al., 2004, p. 207).

What this "political" view of power and IS evaluation could mean for evaluation activities is that the knowledge used for evaluation as well as the knowledge produced (i.e., in relation to the "value" of investments) is somehow influenced by people's interests. That knowledge could be for instance on which evaluation methods are regarded as suitable during an IS evaluation.

Serafeimidis et al. for instance (2003) report on an IS evaluation case in which these methods were mostly financial in their nature. By privileging the use of such methods, Serafeimidis et al. (2003) report that those in financial positions were able to scrutinise, determine and validate the outcomes of the evaluation itself. They attribute this situation to organizational power. Unfortunately, they do not provide a clear description as to how this power originates, operates, and changes throughout an evaluation process. They simply call power an "external variable" that needs to be identified and managed. What is also missing is a deeper understanding of how evaluators not only interpret evaluation as a political process (Walsham, 1993), but influence evaluation itself with their decisions and actions.

Table 1 contains a summary of four (4) different notions of power that we relate to IS evaluation as a process. These notions are drawn from existing classifications in the IS literature (Dhillon, 2004; Horton, 2000; Jasperson et al., 2002) and elsewhere (Lukes, 1974; Oliga, 1996). As seen in the table, in IS evaluation we can associate power with tangible or distinguishable resources (i.e., information), skills, or authority, which some people have and use to control others (Bariff & Galbraith, 1978; Horton, 2000) in the evaluation process. Power can be also associated with institutional structures that influence and are influenced by IS evaluation. The use of such structures in IS evaluation can contribute to reinforce, perpetuate, or resist existing organizational hierarchies and "games" (Bloomfield & Coombs, 1992; Dhillon, 2004; Markus, 2002). Power can also be seen as the *influence* that any action of particular individuals have in the behavior of others (Handy, 1976; Walsham, 1993; Walsham & Waema, 1994). In IS evaluation this includes for instance, the influence that evaluation experts could have over systems users (Horton, 2000; Serafeimidis et al., 2003); the political skills used to convey certain meanings (Checkland, 1981; Checkland et al., 1990); or the style that managers have to define, implement, and evaluate IS plans (Walsham et al., 1994).

The previous three views about power show individual notions, as if power had different but not intersecting manifestations. In IS practice, it has been acknowledged that explicit exercise of power can contribute to systems implementation (Markus, 2002; Serafeimidis et al., 2003; Walsham et al., 1994). However, this does not fully consider the often indistinguishable, unintended, contradictory, and complex consequences of power in IS/IT implementations in a context of intervention (Jasperson et al., 2002; Robey & Boudreau, 1999). We see these first three views as incomplete as they rely on the explicit identification of the source of power (i.e., allocation and control of resources, facilities, or influences by someone). Implicit or "hidden" forms of power, for example forms of behavior not explicitly having power attributes

Table 1. Power in orientations for IS evaluation

Power as	Manifestations
Resources (Bariff et al., 1978)	Authority, skills, information, use of technology.
Capacity (Markus, 2002)	Structures that facilitate (or inhibit) communication.
Influence (Checkland et al., 1990; Handy, 1976; Walsham et al., 1994)	Expertise and styles used to facilitate (or inhibit) knowledge exchange and change.
Relational	In the relations between people (M. Foucault, 1977b), as a backdrop (Horton, 2000) and in the conditions that make evaluation (im)possible.

or intentions or not generating direct impact on projects are not contemplated (Horton, 2000; Robey et al., 1999). Viewing power as something that resides in commodities, authority, political relationships, or structures does not fully explain why functioning information systems may be seen as failures by their users, whilst non-functional examples may be articulated as successes. Neither does it explain why information systems developments or failures could be seen by some individuals as the by-product of struggles between groups of people who frequently have different interests (Jackson, 1992; Walsham, 1993, 1999; Walsham et al., 1994).

To this limitation, there is an emerging interest by IS practitioners in understanding how it is that power can be identified in the *relationships* between systems users and other stakeholders, and how power can be an influence in information systems implementation (Bloomfield et al., 1992; Damsgaard & Scheepers, 1999; Dhillon, 2004; Doherty & King, 2001; Doolin, 2004; Dooling, 2004). The notion of power as something that resides in the relations between people (see Table 1) has been put forward, and could include different manifestations of power as exhibited in such relations. Power could be seen as a "property" of the relationships between individuals which is reinforced, modified or resisted with the implementation of information systems (Bloomfield et al., 1992; Dhillon, 2004; Dooling, 2004; Introna, 1997) Nevertheless, we argue that this notion is incomplete. Little has been said about extending and using it in information systems evaluation. We also need to account for the relationships that evaluators establish and develop through the evaluation process.

In providing a relational understanding of power, researchers have used the ideas of the French philosopher and historian Michel Foucault. His ideas for instance have been used to interpret the dynamics of implementation and use of information technologies in organizations (Ball & Wilson, 2000; Ballantine, Levy, Munro, & Powell, 2003; Bloomfield et al., 1992; Dooling, 2004; Introna, 1997; Knights & Sturdy, 1990; Zuboff, 1988). The use of Foucault's ideas has emphasized exploration of how technology affects worker's autonomy and how behavior is shaped by specific technologies (i.e., surveillance or performance based). We concur with the general view of these authors that information systems practice could be better seen as continuous process of negotiation of power relations between individuals and groups (McLaughlin, Rosen, Skinner, & Webster, 1999; Smithson et al., 2004). However, we argue that little has been said about how power influences participation of individuals in evaluation (Gregory, 2000), and how individuals (including evaluators) make use of power according to their own (ethical) views in order to operate within the dynamics of IS evaluation activities. To address these issues and explain the relational nature of power, we now turn our attention to Foucault's ideas on power and ethics.

FOUCAULT ON POWER AND ETHICS

It is difficult to provide a summary of Foucault's work on the history of Western civilization without missing or misunderstanding important issues from his original works or further interpretations. Foucault has provided an interesting insight for the problem of the *human subject*, be it individual or collective. For Foucault, the main question in modern society is how human beings are constituted as subjects (Foucault, 1982a, 1982b). In his work, the meaning of "subject" is twofold: "someone subject to someone else by control and dependence, and tied to his own identity by a conscience or self-knowledge" (Foucault, 1982a, p. 212). Both meanings in the above definition suggest a form of *power*, which subjugates and makes subject to.

In his historical analysis of the different forms of constitution of subjects, Foucault (Foucault,

1977a, 1980a, 1984b, 1985, 1989, 1994c) employs the concept of power as an aide, which enables him to describe how these forms have emerged and evolved and how knowledge is produced to "normalize" (regulate) individuals' actions (Foucault, 1982a). In other words, how power has constituted different forms of subjectivity for individuals. For Foucault power is constituted by the relations between subjects and at the same time, it constitutes those relations. Power is seen as *power strategies* that individuals use or follow to *influence* the actions of others (Foucault, 1994a, 1994b). Power could be defined as a "total structure of actions brought to bear upon possible actions: in that it incites, it induces, it seduces, it makes easier or difficult" (Foucault, 1982a, p. 220).

In Foucault's work, power could be seen as a kind of matrix of influences of actions over other actions at a particular moment in time (Rabinow, 1984). Power adopts many forms, some of them associated to economic systems, institutions or individual rights (Foucault, 1980b). Power is dynamic, elusive; it operates at different levels, and targets individuals as well as collectives. Power influences, seduces; it makes difficult or easier to do something (Foucault, 1977b). Because of their uncertainty, their mobile and transient nature, power relations can reinforce, cancel or resist each other via multiple connections that find their manifestation in the individuals (Foucault, 1977b).

From the above, three important features of power in Foucault's work can be highlighted which could influence our understandings of IS evaluation. We use some of Foucault's original ideas to support our reflections, as well as some other authors that have used Foucault's work. The first one is that power operates across groups (i.e., departments, organizations, groups of individuals), and its use could produce intentional and unintentional consequences. With this, Foucault aims at avoiding any analysis of power that sees it as a "negative" or constraining force to be denounced, resisted, or imposed on others. Analysis of power relations could identify some existing sources of power (i.e., authority, structure, influence), but this identification should not only aim to show certain effects, but also contrary ones (Robey et al., 1999). For IS evaluation, this would mean that power should be studied in their constraints, but also in their possibilities for action. Moreover, and as Foucault argues, today's resistance in a situation could become tomorrow's power in another (Foucault, 1982a). Awareness on the existence on power also means flexibility to incorporate different manifestations in the analysis; to adapt this analysis to power's changing circumstances, and to be able to identify continuously new power formations, and how these influence relations between individuals.

Second, any action in relation to power (including resistance) operates within power itself (Foucault, 1982a, 1984a). Foucault does not see power where there is no possibility to resist it. Otherwise individuals will not have freedom to act in relation to power (this state of affairs he calls domination). Within power, individuals do not operate externally to it (Foucault, 1977b). However, there is space for individuals to *manoeuvre* within the possibilities and constraints that they could see at a particular time (Foucault, 1984c). To the existence of power in relations, it is still possible to identify some space for individual or collective action. For IS evaluation, the prevalence of some manifestations of power (i.e., as authority, resources, influence) can also means that some action can be exerted with them, and that other power sources can be generated.

Thirdly, and in relation to the first two, power is simply an aid that could illuminate the dangers that power or resistance to it could have for individuals, but it is up to individuals to make decisions about this (Foucault, 1982b). With this, Foucault is suggesting that we should reflect on how to make use of power available according to our purposes. In his analyses of sexuality, Foucault introduces *ethics* as a space for reflection:

Morality [ethics] also refers to the real *behavior of individuals to the rules and values that are recommended to them...the manner in which they respect or disregard a set of values... (p. 25)...those intentional and voluntary actions by which men not only set themselves rules of conduct, but also seek to transform themselves, to change themselves in their singular being.* (Foucault, 1984b, p. 10) [brackets and emphasis added]

To the existence of power relations (by identifying and analysing them), we can use our ethics, or the set of practices we use to relate to ourselves and others (Foucault, 1977b, 1982b, 1984b), in order to define appropriate courses of action. However, ethical reflection with "Foucaultvian" lenses does not presuppose a set of norms to be followed. Rather, it means that we need to take into account which ways of being ethical or "being seen" as ethical (Hodgson, 2000) can be available within power relations, and how these could contribute to inhibit or enable positive change. For example, the tensions between the implementation of common technological software platforms against moving forward via individualization (Bloomfield et al., 1992) could be used as starting points to identify strategies for ethical action. Another example is any moral or code-based obligations (Gotterbarn, 1997) and constraints that affect individuals and therefore their deployment or use of systems. In short, the use of power could be made positively productive as at the same time ethically relevant for those taking part or conducting IS evaluation.

ELEMENTS OF ANALYSIS TO INFORMATION SYSTEMS EVALUATION

Having highlighted the previous ideas from Foucault, we would like to present elements of analysis which we think can provide insights into the practice of information systems evaluation.

First, IS evaluation can be seen as a process of *power development*. As such, it affects and is affected continuously by existing forms of power. An evaluation can reinforce, resist, or generate different possibilities to power relations. It can for instance reinforce existing practices (i.e., financial) (Atkinson, 1998; Kendall & Wickham, 1999; Serafeimidis et al., 2003), and define particular ways of valuing information systems plans and projects (Serafeimidis & Smithson, 1999), in other words of "normalizing" evaluation practices. Analysis of power requires identifying relations to which IS evaluation taps into, relations that could generate normalizations for individuals, and their possibilities to act within such relations.

An adequate starting point for identifying such power relations is situations where there are conflicts or tensions about different forms of individualization (Foucault, 1982a). These possibilities for instance lead groups of individuals to claim that new practices are being generated to improve individual freedom in a situation, whereas for others the same possibilities undermines their freedom, normalizing their behavior (Bloomfield et al., 1992; Doolin, 2004; Robey et al., 1999). In these situations, power might be developing by cancellation, reinforcement, or generation of new relations, and with these new influences in individuals (Foucault, 1977b).

The identification of power relations and their development will also require IS evaluators to identify roles that they assume within the network of such relations. In this network, and at different points in time, evaluators can find themselves being for instance accepted, rejected, ignored or "led somewhere else" by evaluation participants. It is important therefore to identify roles and relations that influence evaluation, and reflect on the effects that taking part in such relations is having for the evaluation process. As power is dynamic, evaluators' activities will take them to different directions; new roles can be shaped up, and new power relations can be generated which will influence the process and outcomes of evaluation activities.

Second, any attempt to improve implementation of systems by IS evaluation needs to consider *possibilities and constraints to act within power.* This power may be about possible actions to develop, possible relations to form (i.e., alliances, conversations, communications, ruptures, etc). This identification would require evaluators to map the terrain of possibilities and constraints for action in a given situation (Brocklesby & Cummings, 1996), to show how these have formed and which effects are they having in individuals. "Rolling out" of this terrain needs to leave some space open for a variety of interpretations and critical reflection on what can be done within power (Taket, 1994). The aim is to incorporate evaluators' own situation in relation to other people, and to provide insights about what is possible and available to do in the evaluation context. With this critical reflection, evaluators can acknowledge their *inevitable* participation within power relations. By doing so, they can then proceed to identify which relations they are (and can be) using, resisting, or creating to generate positive evaluation changes.

Third, the structuring of power analysis should lead IS evaluators to consider their own *ethics*, in other words IS evaluation influences the ethics of evaluators, and at the same time is influenced by it. If this is the case, evaluators should strive to work to avoid *either situations of domination*, where we do not have freedom to change existing power relations, or of *isolation* and closure to new possibilities for action (Foucault, 1984a, 1994b). These possibilities can lead to closure of debate in IS evaluation, or to isolation of the evaluation process from the wider organizational context. It could be for instance that too much emphasis on financially driven IS evaluation is leading to "standardization" but to difficulties in organizational innovation. Alternatively, it could be the case that following ethical standards of conduct is generating lack of flexibility in addressing new issues. In both cases, the imbalance restricts the ethical stance of individual or collective freedom.

However, there is the possibility of enhancing reflection and *self-reflection* between people (including evaluators) about their own situation in evaluation, how knowledge is constituted or could be redefined, and how they could act according to what they consider is right or wrong. This element of reflection could constitute a normative (ethical) stance that evaluators should consider when exploring the purposes and process of evaluation.

With these three elements of reflection in mind, we present an illustrative case of information systems evaluation at Javeriana University in Colombia, in which our main purpose was to identify and reflect on the relevance of power and ethics for IS evaluation. Therefore, during evaluation activities, we followed them but not in very rigorous or strict ways. Rather we tried to identify some of these elements' salient features, and see how we as evaluators used them in practice. Our presentation therefore contains some descriptive and reflective (*e.g., expost*) accounts of what we found and what we did, not necessarily what we should be finding or doing. In this regard, we try to follow our perspective on Foucault's ideas about describing ourselves as a work of art, in continuous progress, by which we possibly try by describing what we did, with the purpose of "no longer being, doing or thinking what we are, do, or think" (Foucault, 1984c, p. 46). Although it is not our entire purpose to draw managerial implications, we somehow hope that our description will provide some insights to IS evaluators.

AN ILLUSTRATIVE CASE: INFORMATION SYSTEMS EVALUATION AT JAVERIANA UNIVERSITY

We now present and reflect on a brief case study of an IS evaluation exercise that one of us (José) undertook at Javeriana University in Colombia in 2003. The word "brief" is important, as we took an opportunity during a two-week visit to Colombia to get an update of how an information systems plan developed at that institution in 1999 (Córdoba & Midgley, 2006) was being developed. From our perspective, we were carrying out some IS evaluation activities. By finding out if the IS plan carried in 1999 had brought some degree of value to the institution, we could also identify the relevance of power and ethics for evaluation.

Before the visit of one of us (José) to the institution, we re-established contact with some of the stakeholders which had been involved in the 1999 plan, and explained that we wanted to identify and assess any impact of this project in the institution. The 1999 plan had generated a number of suggestions which stakeholders (mainly within the institution) were encouraged to take forward and implement. The main suggestions included:

1. Improve the coordination between IS projects and the evaluation of the social impacts that information systems were to have.
2. Develop an information system to provide flexibility and support in the delivery of lectures and in the definition of research projects in conjunction with the industry.
3. Encourage members of the institution to develop a culture of solidarity in which there could be tolerance and respect for diversity.
4. Establish mechanisms for the evaluation of impacts of information systems projects, including their consequences for different groups of stakeholders.

At the time of being presented, these suggestions were received with enthusiasm by senior research managers. There was also the feeling that a participative process had been creating expectations amongst the people that could not be satisfied, and that the suggestions could be going against established plans. Those who took part were instinctively recognising the potential tensions between local departments and the wider institutional environment.

The narrative that follows will relate the unfolding of the evaluation to the three elements previously highlighted. As said before, our purpose was to see which features of these elements could be salient in evaluation activities, and reflect on the activities and decisions made by evaluators (i.e., José). Considering this, we contacted stakeholders who had been involved in the 1999 project. We were able to gain access to five of them (there were initially around 40), and we sent them a semi-structured questionnaire with the purpose of finding out how far the above suggestions from 1999 had been taken forward. In our contacts with the institution, we were also allowed to get access to an office in the computer science-engineering department where the 1999 project was based, and this helped us to have informal conversations with people who had participated previously. To support our reflection about the evaluator's role and decisions, José kept a diary of the activities that he conducted during his visit to the institution. Using this and the above sources described, we now present the features of IS evaluation as (1) a process of power development; (2) offering possibilities but also constraints for action; and (3) influencing the ethics of evaluators.

Evaluation as Power Development

Prior to arrival in Colombia, we contacted a number of stakeholders who had taken part in the project in 1999. Considering the time that had elapsed between the original project and this one, an effective strategy to help establishing contact again was for José to introduce himself as a UK-based researcher. Those who did not initially agree to take part in the interviews were happy though to offer some informal conversations. Moreover, José was offered space in the computer science department and an office from a member of staff who was in maternity leave was allocated to him.

It appeared as if the role of "foreign" researcher was being conceived of as a possibility to exert some individual freedom by the evaluator, and it enabled José to start his evaluation activities, and establish a number of contacts. Moreover, it called the attention of some other staff members for whom a "foreign expert" could be seen as a valuable source of *knowledge*. At the computer science department, people recognized this role and began conversations with José to enquire about possibilities of studying or doing research abroad. Other people asked him about how universities were developing their electronic information services. Although the role was not leading José to seek an involvement with "all stakeholders involved and affected" by a previous IS plan (Remenyi et al., 1999), it was enabling him to enter into a set of relations, and he was becoming influenced by (as well as influencing) other people.

José took informal conversations as opportunities to exchange his views on various subjects, but also to enquire about the project of 1999, and what new initiatives had been undertaken. Some conversations did not progress as intended. For instance, the institution was engaged into a process of international accreditation, and José offered his help as a facilitator in an incoming planning session but the offer was not taken forward. At a more institutional level, a set of relations was focused on the practice of accreditation, to which José's evaluation did not offer too much. Although it can be said that this conversation could have been taken forward to explore wider power relations, these did not either seem to impact or being impacted by evaluation activities.

This and other conversations though, prompted José to identify the temporary existence of two different layers of power relations. One that operated at a centralized (i.e., institutional) level, and another which operated at departmental (i.e., local) level. Later on, José was also able to reinforce *and operate within* some of these departmental layer relations by organising and taking part in an evaluation workshop with staff members from the library whose purpose was to learn how to evaluate library services at the institution. These aspects will be elaborated in more detail later in the chapter.

As an evaluator who had initially assumed the role of "foreign expert," José had been prompted into an intended direction: That of enquiring about the value of the 1999 plan. However, he was also prompted into other directions. Some of these were rejected whilst some others which in principle were not intended could be pursued within the existing network of relations at the institution. This highlights a potential influence of power as a notion in the development of evaluation activities. Existing relations could have been mapped more thoroughly, and their possibilities and constraints for action assessed in relation to the purpose of enquiring about the value of previous plans. These relations were having impacts into the purpose(s), activities, and role(s) of the evaluator.

Possibilities and Constraints for Action Within Power

In the role of a foreign expert and evaluator, and among the people to be interviewed José had access to talk to a senior research manager to discuss the progress of the recommendations made in 1999. Initially, this manager was interested in learning about José's recent research experience in the UK. When presented with the recommendations of 1999, it was noted by José in a rather constraining way that these recommendations did not consider the institution's pre-existing project approval scheme. This assessment and approval scheme stated a sequence:

- Recommendations should be justified by departments.
- Then taken to the corresponding research committees and approved by faculty directors.

- After this, senior management would allocate resources for their implementation (which in some cases required further analysis).
- Monitoring of the progress of projects should be made by those responsible for their development.

The senior manager's judgement was that any proposal for change needed to be framed within a particular discourse about what constituted a good research methodology. In addition, for this senior manager, the suggestions made in 1999 *lacked credibility*, as they had been the result of using a qualitative (not quantitative) methodology. Not enough people had been included for their definition (although around 40 people were interviewed and took part in the project). For this manager, to properly justify any recommendation, a quantitative survey should be done, and of a "representative" sample of the population at the university. Recommendations should be then passed to higher levels for further review.

This example is chosen to demonstrate the relation between suggestions, their discourse for justification and the perceptions of the adequacy of the methodology used to define them. In this instance, we think that we identified the existence of constraints and possibilities for action in power relations. In this case, power was being manifested in relations that control the project approval rhetoric, the definition of acceptable ways of defining knowledge (i.e., methods and size of the sample) and the knowledge itself within the institution (what is acceptable as improvement and what is not). At that particular moment in time at the institution, this form of power was involving senior managers, project managers and officers, or anyone else offering suggestions for the use of information systems at the institution.

The previous can be seen as instance of the issue, noted earlier, where explicit power is associated with processes to standardise the ways of behaving, acting or thinking of people into "normal" or "good." In getting suggestions and project initiatives approved, it could be said that there was a set of practices (schemes), which needed to be used in order to produce adequate or appropriate evaluation (e.g., ethically acceptable and supported). These practices seemed to be embraced in a type of **"institutional layer"** of knowledge and power about information systems projects or initiatives. In this sense, the idea of power is being used in a way consistent with Foucault: Actions influencing other actions, projects and decisions about them influencing others, and ways of behaving deemed necessary to be able to relate to others. Knowledge involves knowledge of how to justify projects according to what is being regarded as acceptable to do: Knowledge of methods used, schemes for approvals, reasons for the need of a project, meetings in which it is advisable to present ideas, communications to generate, methods to be used, ways forward, etc. The practices discussed were complemented by some initiatives that were generated by a "proper" process of justification (as previously described) or the direct definition of projects by senior managers. One of these projects was the implementation of a corporate information system to support academic registration. This project had been organised under the direct supervision of senior management, and had been allocated management resources, which gave it continuous recognition and support within the institution. This speaks about the nature of the justification for a project and the relationships that exist between typically used justification methods and the institutional layer reinforcing existing power-ethics structures. Justification methods are a key power strategy. In the case of the senior research manager, particular forms of justification were used to profile projects, to channel activities and resources, to reinforce his existing plans.

The operation of this institutional layer has contributed to the implementation of some projects and has affected the success of others. If appropriate procedures were followed to obtain knowledge (i.e., by using quantitative methods), and to present it in the form of findings, there could be a chance that this knowledge entered into relations of approval and therefore it could be validated and implemented. However, if circumstances did not allow reinforcing of the relations that enabled projects to be "delivered," there could be time delays or even rejections.

In another interview, and by asking about the development of project initiatives at the institution since 1999, José identified one example of this within another (more departmentally localized) project. This project had aimed to bring the use of a new (virtual reality) technology to the university. According to the interviewee (a research coordinator), the development of this project had gone through different stages, also characterised by the change in project leader role several times. The project has been continuously re-started and one of its main activities is the gathering of information about the existing use of this technology in departments and faculties. The articulated reasoning behind this is that with this identification, a "best practice" is selected and implemented across different areas.

From our perspective, this was a way of saying that the project was not ready to fit within the discourse of acceptable institutional projects, possibly because it was challenging the power that existed within department, or because existing relations at the institutional layer did not see it fit for them. Unfortunately, this particular project still needed to discover that. Despite having knowledge about the virtual use of technology, those responsible for this project had to accommodate their views to what they were required and allowed to do with it. The turnover in leadership can be partly explained partly as a manifestation of behavior failing to comply with institutional forms.

The characterization of this "institutional layer" of power does not mean that its different expressions are homogeneous. In fact, at the time of conducting the evaluation, there were important differences. For example, the approval process required seemed to be mainly directed at projects suggested by department and faculties, not to those "corporate" projects (those being defined by senior managers). Moreover, "best practice" projects could be organised in different ways and have different resources allocated to them. This also accords with the dynamic nature of power relations in which different practices could be at a particular time resisting, supporting, defying or undermining each other (Foucault, 1977b).

Power Influencing the Ethics of Evaluators

From the additional interviews, informal conversations, and invitations to run other (not originally planned) evaluation activities, an unexpected evaluation picture emerged. In it there was, parallel to the institutional layer, a **"local layer"** of IS projects. That local layer was of projects developed to address the specific IS needs that certain departments and areas had. The projects developed their own justification that was accepted at the level that provided them with resources for their completion (i.e., faculty or department). Those responsible for these local projects conducted those projects in ways so as not to get the attention of the senior managers. Therefore their scope, and the relations developed between those involved, did not transcend the boundaries of an area or department even though their projects could have been adapted to satisfy the organizational layer's approval schemes. It seemed a rational and adequate response to avoid the need to re-cast their justification and organization and the possibility of delay or cancellation if forced into being regarded as institutional projects. Thus, they *avoided* the potential for tension with the institutional layer.

Two examples of "local layer" projects explain further the nature of the tension with the institutional layer. The first was an application to support the monitoring of progress of students. The second was a project to define possible roles for technology in education. Each project is in a different department. In the case of the first project local need came about because of large numbers of students, it was led by the computer science engineering department team who were driven by technical credibility. In the case of the second project, local need was identified by the "open education unit" who saw the necessity to integrate technology with the educational process. Both projects were developed despite the apparent lack of power to command resources. The way people assessed the relevance and progress of these projects was embedded into each department's style and did not echo the institutional approval schemes and methodologies. The *behavior* of those involved in projects was not explicitly "seen" as a threat to institutional activities.

From the previous, it can be argued that an additional set or layer of power knowledge and ethics relations was operating at Javeriana. This layer would have been in conflict with the formal and institutional structure of relations if openly declared. The differences between the projects demonstrate the potential but not certainty for the local layer to be in tension with the institutional layer. The second project was producing conceptual definitions placing an obligation to the institutional layer to select particular projects. Additionally, there were demands to change the course of some current institutional projects. However, the institutional layer had its own justification mechanisms that would not necessarily promote the same suggestions. Furthermore, the institutional layer had ways of issuing policies and definitions that had some conflict with the recommendations made by the second project. Part of the tension was the lack of institutional definitions for the use of technology.

An additional difference can be noted between these two projects. The first lay within a high-tech environment (the computer science engineering department); the second in an environment characterised by low to medium-tech (the open learning section). The institutional layer associated itself with a technology attitude closer to the second than the first. The first project had the ability to be positioned beyond the technological "comfort zone" of the institutional layer whilst the second project had no such possibility. This is an instance of the third element of our earlier analysis of Foucault's ideas as they can apply to IS evaluation. The participants in the first project behavior was one in which they used their high-tech situation in playing out the type of subjects they wished to be.

Ethics Influencing Evaluation

The previous reflections show that at the time of the evaluation, there was interplay between what we call local and institutional power layers. These were offering both possibilities for action under certain constraints. The institutional layer was offering possibilities for action (which had to be approved first, but which could have institutional impact). The local layer was offering possibilities to do projects with limited resources, and without calling the attention of senior figures. In each of these, evaluation could have been developed further by using existing power possibilities and constraints, and in our case, the evaluators' ethics seem to have played an important role in deciding what to do.

In the interview with the library director, it emerged that she felt that the project of 1999 had not been properly followed up, and that it lacked any practical application for the library. She complained that if any evaluation activity was to be done, it needed to be relevant for those involved. By feeling the need to "make it up" to the library director as part of what he considered ethical to do, and considering the importance of producing relevant knowledge for evaluation, José took this as an opportunity. He suggested to the director to organise and facilitate a participative workshop with library staff and users. The agreed purpose was to identify opportunities to improve the delivery of library services. The director accepted the idea and a dozen people took part in the workshop.

The format consisted of two activities: First to identify the type of services that could be improved. Secondly, people could define indicators that could be used to monitor the improvements of services. Throughout the workshop, discussions were strongly influenced by the director's views on which services and indicators could be implemented. These views supported existing institutional policies, and it seemed as if the workshop had become *dominated* by them as if there was no other alternative.

To this situation and driven by a desire to empower users, José used his role as facilitator to ask participants about how they thought users could be consulted on the opportunity of services, and from this point onwards he became an advocator of users. This led the group to define a survey with questions to be asked to users, some of these questions being accepted by the director.

The process and outcomes of this exercise can be seen as an example of a strategy to use existing power (in the form of an activity and a role) by a stakeholder (José) to develop some form of action in evaluation, and this involved inviting others to engage in this action. The workshop was framed within what institutionally was useful to do and possible to achieve, given the situation and what people at Javeriana considered possible to do. The workshop was serving two purposes at the same time. First, it was "reinforcing" the influences that the library director, subordinates and users had on each other; these groups also were exerting influence over José to obtain his knowledge. Second, the evaluator was modifying existing relations by introducing new knowledge, but also was introducing his own ethical concern to empower users. The net result was a combination of influences and a shift in the role of the evaluator (from evaluation "foreign" expert to users' advocate).

CONCLUDING REMARKS

This chapter has illustrated how the ideas of Foucault on power and ethics can help in evaluating information systems plans. We have developed a view of evaluation as a process of *power development*. As such, evaluation can be framed within a set of practices and justifications that render particular types of knowledge as *true* and accepted or ethically appropriate. This view was described in terms of two further elements for analysis in evaluation: *Evaluation as offering possibilities and constraints for action, and influencing the ethics of evaluators*. By reflecting on these elements in the evaluation exercise at Javeriana University, we have identified a variety of sets of relations by which the evaluation was possible, and generated useful knowledge in various forms (some intended, some unintended). The relations which we were able to identify were intertwined in a complex structure, in which actions were related to others. Such relations offered opportunities, but also limitations.

A particular role for evaluation (we called it that of a "foreign expert") led José (the evaluator) to engage in certain conversations to exchange knowledge, and to decide which actions were possible and valuable to undertake during evaluation. The role assumed seemed appropriate to establish relations with those in formal positions of power, and collaborate with their purposes in the institution of improving the delivery of information services and their competitiveness. It was also useful to reinforce the power of some organizational units (i.e., the library), and to help the evaluator to reflect on his own ethics.

At the time of the evaluation exercise, it was possible to identify an "apparatus of power" that operated at two levels: institutional and local. Each level had their own way of justifying initiatives, putting them in place and implementing them. Both layers influenced individual behavior about what was acceptable to do or to be. Implicitly, the institutional layer intended to shape the practice of the local layer. The local layer might have not intended to impact on the institutional layer but generated some unexpected opportunities for its change. Initially, one of the evaluators (José) had possibilities to engage in both layers, something that he could do partly because of his role as an outsider to the dynamics of the institution. However, as the evaluation unfolded it became difficult to balance the variety of expectations. This became more evident when doing the evaluation workshop at the library, which suited José's purposes as an evaluator to produce "relevant" knowledge, empower users and become an ethically responsible evaluator. It appeared as if moving through power relations in IS evaluation required also a *shift* in the roles of the evaluator, and this shift was motivated by but also influencing the evaluator's own ethics.

Our interpretation of events explores one perspective, and by taking this, we do not intend to invalidate other's understanding of how Javeriana develop their information systems. The elements of analysis and our description of the exercise did not allow us assess in more depth the implications of power layers in the ethics of individuals. We need further research on how this could be developed. However, we hope that the ideas presented in this chapter could help practitioners to deal with the complexities encountered in evaluation, and to be aware of the dynamic nature of relations between people that influences successful deployment of information systems projects in organizations.

Overall, the use of Foucault's power-ethics concepts creates the conditions for reflection about evaluation of information systems based on the potential opportunities to deliver knowledge that could be made acceptable (e.g., "good") among those involved in it. These situations of power-ethics happen anyway. We can see power only as an issue of resource allocation and control or we can be aware of its role in shaping evaluation and the *relations* within. In short, IS evaluators who adopt even all of the first three perspectives on power (summarised in table 1) will risk having an incomplete understanding of the dynamics of a situation. Foucault's relational lens onto power improves on those because it embraces the other three and allows us to appreciate their relationships. A further reflection on power from the ideas presented in this chapter might also lead practitioners and participants in evaluation to reflect on the purposes of the process in relation to their own purposes as ethical subjects, and act accordingly. This does not only mean that power can be only resisted, but instead *used strategically* to create opportunities for innovation and co-ordination of actions. Those who are evaluating and being evaluated could develop their own agendas according to their ethics while still being in evaluation, as a way of operating within the constraints and possibilities for action offered by existing power relations.

ACKNOWLEDGMENT

The authors would like to thank people at Javeriana University who were involved in the practical exercise of evaluation that we report in this paper. In particular, to the staff in the Computer Science and Systems Engineering Department. We would also like to thank the Economic and Social Research Council (ESRC) of the UK who funded the visit to Colombia.

REFERENCES

Atkinson, D. (1998). The production of the pupil as a subject within the art curriculum. *Journal of Curriculum Studies, 30*(1), 27-42.

Ball, K., & Wilson, D. (2000). Power, control, and computer-based performance monitoring: A subjectivist approach to repertoires and resistance. *Organization Studies, 21*(3), 539-565.

Ballantine, J., Levy, M., Munro, I., & Powell, P. (2003). An ethical perspective on information systems evaluation. *International Journal of Agile Management Systems, 2*(3), 233-241.

Bariff, M., & Galbraith, J. R. (1978). Intraorganizational power considerations for designing information systems. *Accounting, Organizations, and Society, 3*(1), 15-27.

Bloomfield, B., & Coombs, R. (1992). Information technology, control, and power: The centralization and decentralization debate revisited. *Journal of Management Studies, 29*(4), 459-484.

Brocklesby, J., & Cummings, S. (1996). Foucault plays Habermas: An alternative philosophical underpinning for critical systems thinking. *Journal of the Operational Research Society, 47*(6), 741-754.

Checkland, P. (1981). *Systems thinking, systems practice*. London: John Wiley and Sons.

Checkland, P., & Scholes, P. (1990). *Soft systems methodology in action*. Chichester: John Wiley and Sons.

Córdoba, J. R., & Midgley, G. (2006). Broadening the boundaries: An application of critical systems thinking to IS planning in Colombia. *Journal of the Operational Research Society, 57*(9), 1064-1080.

Damsgaard, J., & Scheepers, R. (1999). Power, influence, and intranet implementation: A safari of South African organizations. *Information Technology and People, 12*(4), 333.

Dhillon, G. (2004). Dimensions of power and IS implementation. *Information & Management, 41*, 635-644.

Doherty, N., & King, M. (2001). An investigation of the factors affecting the successful treatment of organizational issues in systems development. *European Journal of Information Systems, 10*, 147-160.

Doolin, B. (2004). Power and resistance in the implementation of a medical management information system. *Information Systems Journal, 14*(4), 343-362.

Dooling, B. (2004). Power and resistance in the implementation of a medical management information system. *Information Systems Journal, 14*(4), 343-362.

Foucault, M. (1994a). *La Hermeneutica del Sujeto (Hermeneutics of the Subject)* (F. Alvarez-Uría, Trans.). Madrid: Ediciones de la Piqueta.

Foucault, M. (1994b). La práctica del cuidado de uno mismo como práctica de la libertad. In F. Alvarez-Uría (Ed.), *La Hermeneutica del Sujeto (Hermeneutics of the Subject)* (pp. 106-142). Madrid: Ediciones de la Piqueta.

Foucault, M. (1994c). Technologies of the self. In P. Rabinow (Ed.), *Michel Foucault: Ethics, subjectivity and truth.* (pp. 223-251). London: Penguin.

Foucault, M. (1989). *The archaeology of knowledge*. London: New York: Routledge.

Foucault, M. (1985). *The history of sexuality volume three: The care of the self.* London: Penguin.

Foucault, M. (1984a). The ethics of the concern of the self as a practice of freedom (R. e. a. Hurley, Trans.). In P. Rabinow (Ed.), *Michel Foucault: Ethics subjectivity and truth: Essential works of Foucault 1954-1984* (pp. 281-301). London: Penguin.

Foucault, M. (1984b). *The history of sexuality volume two: The use of pleasure*. London: Penguin.

Foucault, M. (1984c). What is enlightenment? (C. Porter, Trans.). In P. Rabinow (Ed.), *The Foucault reader: An introduction to Foucault's thought* (pp. 32-50). London: Penguin.

Foucault, M. (1982a). Afterword: The subject and power. In H. Dreyfus & P. Rabinow (Eds.), *Michel Foucault: Beyond structuralism and hermeneutics* (pp. 208-226). Brighton: The Harvester Press.

Foucault, M. (1982b). On the genealogy of ethics: An overview of work in progress. In P. Rabinow (Ed.), *The Foucault reader: An introduction to Foucault's thought* (pp. 340-372). London: Penguin.

Foucault, M. (1980a). Truth and power. In P. Rabinow (Ed.), *The Foucault reader: An introduction to Foucault's thought* (pp. 51-75). London: Penguin.

Foucault, M. (1980b). Two lectures. In C. Gordon (Ed.), *Power/knowledge: Selected interviews and other writings Michel Foucault* (pp. 78-108). New York: Harvester Wheatsheaf.

Foucault, M. (1977a). *Discipline and punish: The birth of the prison*. London: Allen Lane.

Foucault, M. (1977b). *The history of sexuality volume one: The will to knowledge* (Vol. 1). London: Penguin.

Gotterbarn, D. (1997). Software engineering: A new professionalism. In C. Myers, T. Hall, & D. Pitt (Eds.), *The responsible software engineer: Selected readings in IT professionalism* (pp. 21-31). London: Springer.

Gregory, A. (2000). Problematizing participation: A critical review of approaches to participation in evaluation theory. *Evaluation, 6*(2), 179-199.

Handy, C. (1976). *Understanding organizations*. Aylesbury: Penguin.

Hodgson, D. (2000). *Discourse, discipline and the subject: A Foucauldian analysis of the UK financial services industry*. Aldershot: Ashgate.

Horton, K. S. (2000). *The exercise of power and information systems strategy: The need for a new perspective*. Paper presented at the 8th European Conference on Information Systems (ECIS), Vienna.

Introna, L. D. (1997). *Management, information and power: A narrative of the involved manager*. Basingstoke: Macmillan.

Jackson, M. C. (1992). An integrated programme for critical thinking in information systems research. *Information Systems Journal, 2*, 83-95.

Jasperson, J. S., Carte, T., Saunders, C. S., Butler, B. S., Croes, H. J. P., & Zheng, W. (2002). Power and information technology research: A metatriangulation review. *MIS Quarterly, 26*(4), 397-459.

Kendall, G., & Wickham, G. (1999). *Using Foucault's methods*. London: Sage.

Knights, D., & Sturdy, A. (1990). New technology and the self-disciplined worker in the insurance industry. In I. Varcoe, M. McNeil, & S. Yearley (Eds.), *Deciphering science and technology* (pp. 126-154). London: Macmillan.

Lukes, S. (1974). *Power: A radical view*. London: Macmillan.

Markus, M. L. (2002). Power, politics, and MIS implementation. In M. Myers & D. Avison (Eds.), *Qualitative research in information systems*. London: Sage.

McAulay, L., Keval, N., & Doherty, N. (2002). The stakeholder dimension in information systems evaluation. *Journal of Information Technology, 17*(4), 241-255.

McLaughlin, J., Rosen, P., Skinner, D., & Webster, A. (1999). *Valuing technology: Organizations, culture, and change*. London: Routledge.

Morgan, G. (1997). *Images of organization*. Thousand Oaks, CA: Sage.

Oliga, J. (1996). *Power, ideology, and control*. New York: Plenum.

Rabinow, P. (1984). Introduction. In P. Rabinow (Ed.), *The Foucault reader: An introduction to Foucault's thought* (pp. 3-29). London: Penguin.

Remenyi, D., & Sherwood-Smith, M. (1999). Maximise information systems value by continuous participative evaluation. *Logistics Information Management, 12*(1/2), 145-156.

Robey, D., & Boudreau, M. (1999). Accounting for the contradictory organizational consequences of information technology: Theoretical directions and methodological implications. *Information Systems Research, 10*(2), 167-185.

Serafeimidis, V., & Smithson, S. (2003). Information systems evaluation as an organizational institution experience from a case study. *Information Systems Journal, 13*, 251-274.

Serafeimidis, V., & Smithson, S. (1999). Rethinking the approaches to information systems evaluation. *Logistics Information Management, 12*(1-2), 94-107.

Smithson, S., & Tsiavos, P. (2004). Re-constructing information systems evaluation. In C. Avgerou, C. Ciborra, & F. Land (Eds.), *The social study of information and communication technology: Innovation, actors and contexts* (pp. 207-230). Oxford: Oxford University Press.

Symons, V., & Walsham, G. (1988). The evaluation of information systems: A critique. *Journal of Applied Systems Analysis, 15*, 119-132.

Taket, A. (1994). Undercover agency? Ethics, responsibility, and the practice of OR. *Journal of the Operational Research Society, 45*(2), 123-132.

Walsham, G. (1999). Interpretive evaluation design for information systems. In L. Willcocks & S. Lester (Eds.), *Beyond the IT productivity paradox* (pp. 363-380). Chichester: John Wiley and Sons.

Walsham, G. (1993). *Interpreting information systems in organizations*. Chichester: John Wiley and Sons.

Walsham, G., & Waema, T. (1994). Information systems strategy and implementation: A case study of a building society. *ACM Transactions on Information Systems, 12*(2), 150-173.

Zuboff, S. (1988). *In the age of the smart machine: The future of work and power*. Oxford: Heinemann Professional.

Chapter XIV
Critical Insights into NHS Information Systems Deployment

Rajneesh Chowdhury
CHR Global Consulting Services, India

Alan Nobbs
National Health Service, UK

ABSTRACT

This chapter discusses a systems methodology called strategic assumption surfacing and testing (SAST) that was used to understand the design and deployment of information systems in the healthcare context. It is based on the experiences of conducting SAST with a group of healthcare professionals, working in the National Health Service (NHS) in England. This application of SAST in the NHS setting highlighted deep politico-cultural concerns in the organizational setting, and it helped towards the conception of a normative inclusive approach for health informatics design and deployment. This approach introduces the understanding that the development of information systems in healthcare is a complex agenda, the success of which demands the active involvement of all stakeholders through all the key stages of the process. Critical perspectives on SAST have also been considered and the assumptions fostered towards arriving at the conclusions, have been highlighted.

INTRODUCTION

The responsibility of provision of healthcare services in England rests with the National Health Service (NHS). The NHS is the largest employer in Europe with an annual budget set to exceed £92 billion in 2007/08 (Department of Health, 2005). In October 2002, the NHS launched the National Programme for Information Technology (NPfIT) with the objective to create an integrated healthcare information system (IS) supported by information technology (IT). NPfIT has an anticipated investment of over £6.2 billion (Health Informatics Community, 2004). However, there has been tremendous scepticism amongst key stakeholders regarding the success of NPfIT due to its lack of consultation with end users. The fundamentally top-down policy-led approach to the design and implementation of NPfIT and its lack of adherence to effective IS project principles have come under much spotlight as contributing to the feared failure of the project (Ballard, 2006).

This chapter argues that effective deployment of healthcare IS can be achieved by considering the interaction between a diverse range of factors within the organizational setting. This kind of an approach is inherent in the systems philosophy of management thought. Systems thinking has influenced a range of methodologies and techniques that facilitate stakeholder participation, boundary critique, and inclusive decision-making. Boundary critique (Midgley, Munlo, & Brown, 1998) is the idea that one's understanding of the world is bounded by the position and worldview they occupy. Therefore, the more one's boundaries are critiqued, the more informed and inclusive do understanding and perspective become. Strategic assumption surfacing and testing (SAST) is one such methodology that has been discussed in this chapter. Experiences from a SAST exercise with a group of NHS professionals and its resultant normative approach to health informatics has been illustrated.

HEALTH INFORMATICS AND THE SYSTEMS APPROACH

The NHS model of healthcare information system is epitomized in its NPfIT project, which in turn makes the promise to re-create the NHS as a high-tech environment. NPfIT aspires to deliver an integrated healthcare system for the NHS in England with its core in effective IS. The main elements of NPfIT are as follows:

- **NHS Care Record Service:** This is the central database of patients that will be available to authorized clinicians in the country, whenever and wherever required.
- **Choose and Book:** This is the electronic booking service whereby general practitioners (GPs) and other primary care staff are able to make hospital bookings at the convenience of their patients' date, time, and place.
- **Electronic Transfer of Prescriptions:** This service seeks to electronically link up the prescribers and dispensers of medicines in England. The objective is the connection between all GPs, community pharmacies and other dispensers.
- **Picture Archiving and Communication Service:** This enables medical images like X-Rays and scans to be stored electronically that can be viewed by clinicians in their video screens or computers. This is expected to eradicate hard films for recording radiographic medical images.
- **NHSmail:** This is the national e-mail and directory service that will be provided free of charge to NHS staff.
- **NHS Network (N3) Broadband:** This is the fast and reliable broadband service that will support the whole system. It is claimed to deliver the robust demands that will be made by the new system to deliver all the above services.

The previous developments, if implemented as planned, are set to pose the NHS as a truly high-tech organization with state-of-the-art IS support.

NPfIT and Related Challenges

Despite record levels of investments in NPfIT, there is considerable scepticism within the NHS and beyond that the project is heading towards failure. The Institute of Public Policy Research warned "that the program could be undermined by a failure to consult properly with medical professionals, a dearth of IT skills within the healthcare service and poor understanding of exactly what the health benefits are supposed to be" (Sherriff, 2004). Research and expert opinion suggest that core activities like clinical engagement, staff training, and system compatibility may have been compromised in the design and deployment of NPfIT. Gillies (2000) similarly notes that the primary reason why IT has not been a success in the NHS is that "IT has been technology driven not information driven" (p. 16). There are various factors resulting in this situation. Important ones among these may include the sheer size of the NHS, which posits tremendous challenges, management-clinician conflict where both perceive each other's roles as differing in common grounds, and underestimation on the complexity that IS projects have to deal with. For success in a project like NPfIT, clinical engagement and user commitment are paramount, along with the appreciation that different factors do not operate in isolation, but in interaction with one another. This kind of an understanding is the focus of the systems approach, which is discussed in more detail below.

The Systems Approach

Systems thinking is the philosophy in management thought that encourages holistic understanding. It supports the idea that organizations are constituted with elements that are in interaction with one another, and this interaction gives rise to the character and nature of the system (Jackson, 2003). Hence, emphasis is shifted from individual elements to the interrelationships between elements. Systems thinking encourages boundaries to be approached with criticality, and that boundaries can always adapt as a result of its interaction with its environment. Hence, a systems approach argues for the "sweeping-in" of immediate and non-immediate factors that influence the behavior of complex systems (Churchman, 1968). It therefore lends an inclusive and participatory perspective in the decision making process. As Hammond (2002) illustrates in Table 1, a systems approach has considerable implications for organizational planning and decision-making – those that are indicated in the left hand column; these ideas are in contrast to the ones in the right hand column:

Systems thinking can be of tremendous value in the design and deployment of IS. This is pertinent to the implications of considering the human, technological, and contextual factors in the conception and design of organizational IS. A systems approach can enable the understanding that IS of the present day are dominantly technology enabled human activity systems within specific contexts. Xu comments: "Systems science has been considered the basis of information systems. A wealth of research in information systems in the framework of systems science has produced an astonishing array of theoretical results and empirical insights, and a large suite of tools and methods" (Xu, 2000, p. 105).

The systems approach has influenced a wide array of methodologies and techniques to foster stakeholder involvement and facilitate participa-

Table 1. Social implications of systems theory: Contrasting views (Hammond, 2002, p. 431)

Participatory decision-making processes	Hierarchical decision-making processes
Self-organization	Externally imposed order and control
Free will, creativity, spontaneity	Determinism
Democracy	Technocracy

tive decision making in complex situations. One of such methodologies is SAST. SAST has been discussed below in theory and practice in the context of IS design for the NHS. SAST has been selected as an appropriate methodology in the current context as it facilitates decision making when there are two groups with distinctly opposing viewpoints. In this case, these are the management-led approach and the service provider-led approach in healthcare IS. This has been discussed in more detail below.

STRATEGIC ASSUMPTION SURFACING AND TESTING (SAST)

SAST is a methodology that was developed by Mason and Mitroff (1981) to enable managers to deal with complex situations in modern organizations. Mason et al. prefer to call complex problem situations "wicked problems" where issues are multidimensional, interconnected, and uncertain. Wicked problems that arise in these situations have social, political and organizational ramifications. Their understanding was governed by the idea that "in tackling wicked problems, problem structuring assumes greater importance than problem-solving using conventional techniques" (Jackson, 2003, p. 137). This leads to the understanding that unless the formulation and structuring of problems are addressed effectively in the beginning, we may end up tackling the wrong problems. SAST is therefore designed to formulate and explicate assumptions that people harbor in organizations.

SAST has been greatly informed by the following ideas of Rosenhead (1987), as described by Jackson (2003):

- A satisficing rather than optimizing rationale
- An acceptance of conflict over goals
- Different objectives measured in their own terms
- The employment of transparent methods that clarify conflict and facilitate negotiation
- The use of analysis to support judgement with no aspiration to replace it
- The treatment of human elements as active subjects
- Problem formulation on the basis of a bottom-up process
- Decision taken as far down the hierarchy as there is expertise to resolve them
- The acceptance of uncertainty as an inherent characteristic of the future and a consequent emphasis on keeping options open (p. 138).

Influenced by the philosophy of Churchman (1968)[a], Mason et al. embarked on a systems project that would accept the existence of a variety of worldviews, or *Weltanschauungen*, as an unavoidable prospect and embrace divergent subjectivity as a strength. Further, all worldviews are restrictive and a holistic perspective can only be achieved by synthesis of a variety of worldviews. A systems mindset would encourage one to question and formulate ones own assumptions and worldviews, and critically debate the same with opposing assumptions and worldviews (Churchman, 1968; Mason et al., 1981). Borrowing from Hegel, SAST is driven by the understanding that in any organization there would be a dominant set of worldviews--*thesis*, an opposing set of worldviews, *antithesis*, and there is always a possibility for the opposing worldviews to enter a state of constructive debate, and arrive at a higher level of understanding, *synthesis* (Jackson, 2003). This is however a never ending process, and the synthesis would always give rise to opposing set of beliefs. What is important for Mason et al. is that the worldviews and beliefs are derived from deep rooted assumptions that people hold in their minds. Management decisions are in turn dependent on these assumptions and beliefs. However, an effective organization is one that is able to formulate these assumptions and counter assumptions amongst its members and learn how it can behave differently from the knowledge that emerge. As Jackson notes:

An organization only really begins to learn when its most cherished assumptions are challenged by counterassumptions. Assumptions underpinning existing policies and procedures should therefore be unearthed and alternatives put forward based on counterassumptions. (p. 141).

Constructive criticism and investigative debate is central to the previous philosophy. This philosophy is essential for "wicked problems" in complex organizations where not only there are a variety of opposing assumptions and beliefs, but also a tendency to subjugate the assumptions of the weak and the underdogs. SAST has therefore been designed to be participative, adversarial, integrative, and managerial mind supporting (Jackson, 2003). Decision-making process ought to involve different stakeholders with different assumptions and different ideas about how problems should be addressed. Hence, the situation should be adversarial, apart from being participative. Further, there ought to be the opportunity to bring together divergent views to a higher level of integrative understanding from which decision makers can gain deeper insights into wicked problems. Hence, this methodology has the potential to make a real contribution in the practical and operational level. The methodology of SAST follows four stages (discussed in detail in the next section): group formation, assumption surfacing, dialectic debate, and synthesis. The following is an account of SAST in action in the high-tech NHS environment.

A SAST exercise was conducted with a group of NHS professionals in June 2006. The purpose was to examine if the dominant and opposing viewpoints in the context of IS in the NHS can be brought together in synthesis. This was intended to inform the design of a route-map for IS in the context of UK public sector healthcare. This exercise was supported and funded by the NHS North and East Yorkshire and Northern Lincolnshire Network of Cardiac Care (NEYNLMCN).[b] The following discussion follows the methodology of SAST and the insights it generated in the exercise.

Group Formation

Group formation is the first stage where participants are divided into two distinct groups. The effort should be to "maximize convergence of perspectives" (Jackson, 2003) within each group and "maximize divergence of perspectives" between the groups. The result is two groups of opposing viewpoints with each group consisting of relatively like minded people.

There were eight participants in the exercise: one consultant clinician, one general practice manager, two nurses, one information support officer, one clerical staff, one service improvement facilitator, and one service improvement manager. The conflicting idea that was prevalent in the group was the design and deployment strategy of NPfIT and how a new health informatics strategy could be conceived and implemented. Certain participants believed that the prevalent top-down approach to the current NHS IS strategy was working and making progress. This was the dominant perspective overtly cherished by the NHS management. They believed that the new NPfIT system could be used for improvement in patient care and for the monitoring of clinicians' performance. They were of the opinion that there had been considerable consultation with frontline service providers before NPfIT was implemented. This represented the viewpoint of the management-led approach to health informatics. At the same time, there were other participants who believed that there had not been appropriate consultation before NPfIT was implemented. This was the opposing viewpoint. They believed that the NHS was wasting its resources in delivering functions that are not required by patients and the public. They were of the opinion that NPfIT was in a way imposed upon them by the management. This represented the service provider-led approach to healthcare information systems. They advocated that system specification and definition ought to come from frontline service providers like the clinicians, and that administrative staff should also have a say in the system as their role would radically change with the deployment of the new systems.

Considering the overt difference of opinion between the participants, they were divided into the following two groups, based on the approach they favored:

- **Group-1 (management-led approach)**
  - General Practice Manager
  - Service Improvement Facilitator
  - Service Improvement Manager
- **Group-2 (service provider-led approach)**
  - Consultant clinician
  - Nurse 1
  - Nurse 2
  - Clerical member of staff
  - Information support officer

The result was two groups where difference of opinion was maximized between the groups. However, as the following stages will show, this methodology facilitated the opposing groups to structure assumptions, many of which were quite similar instead of being opposing.

Assumption Surfacing

This is the second stage where the aim is to formulate and express key assumptions that members in the groups harbor. As Jackson (2003) notes, this should be done in a "supportive environment," where the aim should be as "imaginative and creative" as possible.

This stage may be facilitated with three methods: stakeholder analysis, assumption specification, and assumption rating. Groups can be asked to conduct a stakeholder analysis to identify who they think the relevant stakeholders in a particular project are, in terms of who the affected parties would be as a result of the implementation of a particular project. Groups may then be asked to specify their assumptions for each of their stakeholders. These specifications should be related to how these would influence the success of the project under consideration. Groups may then be asked to rate their assumptions in a chart rating them against two criteria: importance and certainty. A spokesperson from each group then has to make a presentation about the assumptions.

To facilitate this stage, both the groups were asked to state who they thought the stakeholders were in the implementation of IS in the NHS. The groups brainstormed their views in different rooms and agreed to a list of stakeholders. Following were the results:

- **Group-1 (stakeholders)**
  - Clinical professionals
  - Finance department of NHS
  - General public
  - Government, including the Department of Health (DoH)[c]
  - Healthcare IT Leads
  - IT Industry
  - Managers
  - Media
  - Patients
  - Regional and local NHS organizations
- **Group-2 (stakeholders)**
  - Administrative staff
  - Clinicians
  - General public
  - Government including the DoH and the Treasury[d]
  - Industry (Pharmaceutical and medical equipment companies)
  - NHS Management
  - Patients

- o Private healthcare providers
- o Suppliers (IT services including software and hardware and telecommunications)
- o Support staff (e.g. IT staff)
- o Universities (Research and epidemiology)[e]

The groups were then taken forward to the method of assumption specification, where they had to state what their assumptions were. These assumptions were thought to effect the success of NHS IS strategies, with specific consideration to NPfIT. The responses were as follows:

- **Group-1 (assumptions)**
  1. IT industry has vested interest in personal gains.
  2. The media wants to portray a negative image and always wants to highlight problems.
  3. The media should be more positively engaged by NPfIT.
  4. The general public have a one-sided view, as portrayed by the media.
  5. The general public has a lack of confidence in NPfIT.
  6. The IT industry has a conflict of interest.
  7. The general public has a lack of understanding of the aims of NPfIT.
  8. The general public believes that money should rather be spent on healthcare and on professionals, than on IT.
  9. The government expects too much too soon from a complicated project.
  10. The government has an unrealistic timeframe for delivery of the project.
  11. The government is politically driven and does things that are locally irrelevant.
  12. The healthcare sector has a lack of expertise and lack adequate IT staff to deliver the project.
  13. The finance department underestimated resources needed for the national and local delivery of the project.
  14. Clinicians believe that they have not been consulted.
  15. Clinicians have an unrealistic expectation of participation.
  16. Clinicians have a fear of their IT skills.
  17. Patients are mostly not interested in getting involved.
  18. There is lack of training capacity to ensure skills for delivery.
- **Group-2 (assumptions)**
  1. NPfIT will go over cost. It is a white elephant.
  2. Administrative staff will be resistant to change.
  3. Administrative staff has no time to work with the new systems.
  4. Clinicians are not computer literate.
  5. Clinicians are time limited to work with the new systems.
  6. Patients are not computer literate.
  7. Clinicians are conservative to accept change.
  8. The government is control freak. It wants to control professionals with the information from NPfIT.
  9. Clinicians are sceptical about success of NPfIT.
  10. Patients want local treatment.
  11. Private healthcare services are only interested in profit.
  12. Clinicians are sceptical about patient confidentiality in the new system.
  13. NHS managers need numbers.
  14. Suppliers see NPfIT as an opportunity for profit.
  15. Administrative staff will find it difficult to use the new system.
  16. The treasury wants to keep costs under control.

17. Patients lack knowledge about the system.
18. The government wants to impose solutions all the time.
19. Universities need to do more research for information and funding.
20. Private healthcare services are always after more NHS work.
21. The government will blame someone else when the system does not work.

The groups then rated their assumptions in a chart against the axes of certainty and importance. A spokesperson from each group then presented their stakeholders, assumptions, and ratings to the other group. At this stage, it was interesting to note that although both the groups were supposed to be opposing in their viewpoints, there were many issues, which were common to both the groups. In addition to this, there were few elements, which both the groups seem to support as the root cause of many of the challenges faced by NPfIT. This is elaborated in more detail in the discussion below focusing on the presentation by each group.

Group-1 Assumptions

Group-1, which favored a management-led approach in healthcare information systems, believed that the IT industry had a vested interest in the implementation of NPfIT. The group however felt that this was a "gross assumption" in their part and they were not certain about it. The group felt that the media had a big role in portraying the NPfIT negatively. The media always highlighted problems rather than adopting a balanced position. This led to increasing scepticism of the project amongst clinicians and the general public. It was the view of group-1 that the media ought to be more "positively engaged" by NPfIT and the wider NHS as it is a "very important" stakeholder for the long term success of the project. This is because the media shapes public opinion. They also assumed that the general public had a fairly one-sided view about what NPfIT was, influenced purely on what they heard in the media. The public also suffered from a general lack of understanding about the aims and objectives of NPfIT. The public would rather be interested to see money being spent on healthcare *per se* and healthcare professionals, rather than support systems like healthcare IS.

The group thought that the government was to blame for making the project too complicated. This is because the government wanted too much sophistication in too little time. Therefore, the government had a "fairly unrealistic" time frame for delivery of the project. Moreover, it was the view of the group that the project was politically driven rather than being locally relevant to patients and clinicians. Members of group-1 also felt that there was a lack of expertise of IT skills in healthcare to realistically deliver the project. They felt that the finance for delivery of the project was also being underestimated. However, they were not certain about this as this may be the result of media reporting, and the way viewers, including the group itself, were picking up messages from the media. Regarding clinical consultation, the

Chart 1. Assumption rating by group-1

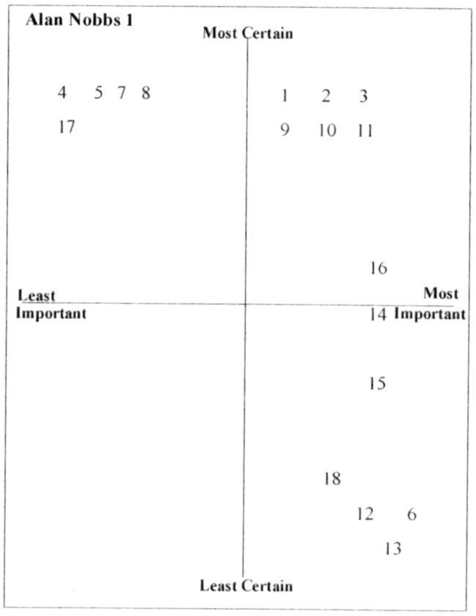

group felt that even it may have been misled by the DoH with the idea that appropriate consultation had taken place. The group was not sure whether or not clinicians were consulted appropriately. At the same time, they also felt that clinicians had an unrealistic expectation about what participation and consultation was supposed to mean, as consultation with every clinician is unrealistic. Fear of IT and skills deficit amongst clinicians was also featured as assumptions that came in the way of the success of NPfIT. Related to this, the group also featured that there was a general lack of appropriate IT training across all levels in the NHS.

Group-2 Assumptions

Group-2 favored a service provider led-approach to health informatics. Members of this group were certain that inspite of phenomenal investments in NPfIT, it was not yielding any of the anticipated benefits. However, at the same time, they also believed that administrative staff, patients, and clinicians may not be sufficiently IT literate to work with the new system. This may also have led to resistance to the change process, creating more challenges for the project. They also believed that patients lacked appropriate knowledge about the system and there was no appropriate initiative made to educate the public about the new system. The group highlighted that there was no realistic planning for training and development of clinical staff.

Group-2 was also critical of some of the objectives of NPfIT. For instance, one of the hallmarks of NPfIT is patient choice, but the group was confident that patients wanted local treatment. This discards one of the most important features of NPfIT itself. Members of this group also had grave concerns about patient confidentiality in the new system. They believed that the new system has been deployed without much consideration of security of access to patient details, which puts confidential patient information at risk.

This group also assumed that clinicians were conservative of change and not receptive to the new IS. This was creating grave challenges for the successful implementation of NPfIT. It was the view of this group that inspite of knowing about these challenges, the DoH did not take any specific measures to address them. The DoH rather proceeded with its own plan of deploying a system that would enable itself to have more control over the clinicians and management processes.

The previous insights interestingly shifted the blame from NHS managers to the government itself. This brought group-2 closer with group-1 in some of their viewpoints. The unintended consequence of these insights was that the groups already started to sympathize with each others' position within the organization.

Dialectic Debate

This is the third stage where both the groups are asked to debate the assumptions and viewpoints of each other. Whilst facilitating this stage, consideration should be paid to the following points (Jackson, 2003):

Chart 2. Assumption rating by group-2

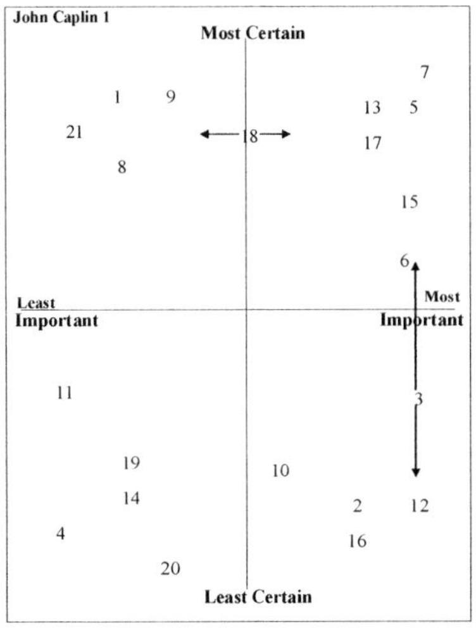

- How are the assumptions of the group different?
- Which stakeholders feature most strongly in giving rise to the significant assumptions made by each group?
- Do groups rate assumptions differently?
- What assumptions of other groups does each group find the most troubling with respect to its own proposals? (p. 144)

After a certain period of time for which the debate has proceeded, groups can be offered an opportunity to modify its assumptions. This is called "assumption modification." However, as the following narration of this particular exercise will show, participants from both the groups were already beginning to see common grounds even before the following stage of anticipated consensus.

Group-1, that represented the management-led approach to healthcare information systems, was of the opinion that much of the resistance to NPfIT from clinicians was a generation issue. Members of this group felt that there was no problem with the younger clinicians accepting the new systems and they are more adept in using IT. They therefore felt that probably the problem would solve itself over time when the younger generation of clinicians would replace the older generation. This was however, taken with much contempt by group-2, who advocated that the main issue with NPfIT was its lack of consultation. This was immediately refuted by group-1 who was of the opinion that clinicians have never recognized their initiatives in helping them with service improvement.

As the debate progressed, the groups also began to see some common grounds. For instance, one of the key members of group-1 expressed scepticism of the DoH actually carrying out robust consultation with clinicians and patients about its IS strategies. They thought that they themselves might have been misguided by the DoH. In this regard, group-2 added that no one had actually ever approached them and asked what they really wanted. They felt that management would always make decisions about NHS reorganization or implementation of new strategies in complete isolation from clinicians. Identifying themselves closer to group-2, group-1 felt that there was always talk of a patient-led NHS, but the NHS never asked patients before it formulated its policies. Most of the consultation process in the NHS were actually "information giving" sessions, rather than consultation sessions, in which patients and the public are just informed about what the NHS was going to implement. When policies fail, management would try and backtrack the consultation process with patients and the public. This insight from group-1 immediately reflected a disparity between the NHS managers and the government, represented by the DoH. This disparity was more pertinent in discussions surrounding the choose-and-book feature of NPfIT, which offers choice of five treatment sites for patients in England. Group-1 felt that this was the "most ridiculous" feature as both patients and clinicians prefer local treatment. This view was immediately accepted by group-2. However, the managers felt that they had to work towards supporting Choose-and-Book as they had to deliver what they had been paid to deliver. Failure to do so would probably see their funding withdrawn. Hence, group-1 was surfacing frustration with their own position in the sense that they were implementing a strategy, some of its features they themselves did not support.

Both groups-1 and -2 expressed concern about the performance of the private companies that were contracted to deploy NPfIT. It was highlighted that there was considerable delay in the deployment of specific features of NPfIT, and some of the features that had been deployed were not functioning as expected, or not functioning at all. At this stage, it may be easy to blame the contracted companies for the fiasco. However, the participants suggested that there was a lack of realism that existed not only in the side of the NHS, but also in the side of the suppliers regarding what could be delivered and in what time scale the same could be delivered. However, at this stage, group-1 was again radical to question whether the contracted companies were failing to deliver due to their incompetence, or because the NHS did not clarify its specifications in the first place. One member of group-1 said: "If we can't tell them exactly what we want and stick to it, they are bound to go wrong."

However, it was stated by both the groups that people's expectations and demands change from time to time, and change is unavoidable. There was some agreement at this stage that IS should have built-in flexibility so that it can be adaptive to changing requirements. Group-1 believed that this can be possible only through true "partnership working" between management, clinicians and the public right from the beginning. Partnership working is about involvement of key stakeholders in the whole process of design and deployment of strategies. Certain comments from group-1 indicated that NHS management itself was operating under the pressures of DoH. This issue becomes more explicit in the discussions that followed.

After a lapse of time, the groups were asked if they would like to change ratings of any of their assumptions or the assumptions themselves. Participants returned to their own groups and spent sometime discussing how their assumptions have been informed and influenced after the debate and discussion. Both the groups decided to change how they rated certain assumptions initially. The following charts show the ratings that were changed.

The groups then discussed their assumptions again and started to consider how the present unwelcome situation in health informatics in the NHS could be addressed. This led to an interest to envisage a normative approach for health informatics that would be able to involve stakeholders in partnership, with the element of learning built into the model so that system flexibility and change is not seen as a challenge, but as an opportunity

Chart 3. Changed assumption rating by group-1

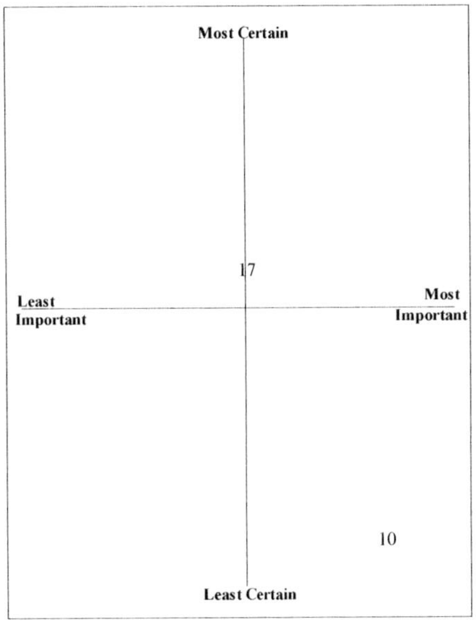

Chart 4. Changed assumption rating by group-2

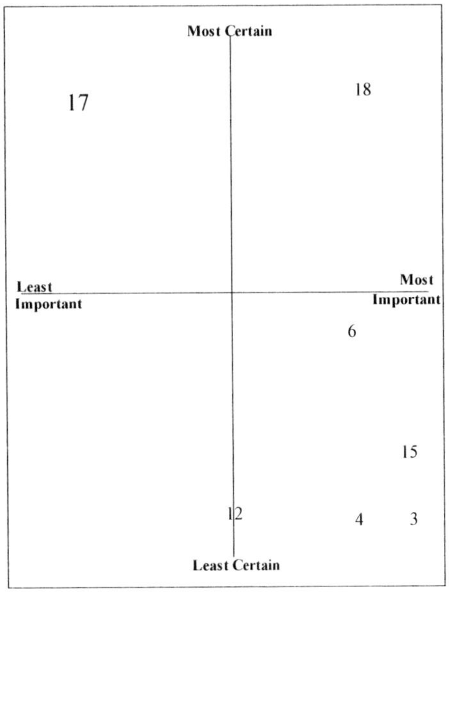

for the system to evolve. Working toward this approach was the final stage of this exercise.

Synthesis

This is the final stage of SAST where the previous stages are expected to lead to a synthesis of views. This will be the result of modification of assumptions, negotiation, and accommodation of viewpoints. Synthesis is expected to result in a reformed strategy for the organization to adopt. However, if the groups fail to arrive at a synthesis of views, the problematic assumptions and conflicting viewpoints should be taken up for further research and consideration.

In the exercise under consideration, the final stage was concentrated on both groups working towards overcoming their differences and envisaging a normative approach for public sector health informatics. This stage facilitated the groups to be critical of their own boundaries and perspectives. It implied the groups "sweeping in" the viewpoints of the opposing group and conceives a more inclusive approach towards healthcare IS. All insights in the discussion to follow have been the result of a synthesis of ideas of both the groups, and represent an approach conceived by the participants supported by the facilitator.

This approach to health informatics in the NHS ought to follow the following stages.

Needs Assessment

The conception of ideas for new IS emerges only from the need of such ideas to improve the operational situation. In the healthcare sector needs may be realized at the level of healthcare service delivery for the improvement of services and enhanced effectiveness in care delivery. If systems are introduced without any need these can be a dangerous toll on organizational resources and efforts. Once needs are realized, they have to be assessed to see if there is actually a requirement to introduce new IS or can those needs be met with existing facilities. Once needs are assessed and there is an agreement that new IS is a requirement, should service providers and management conceive of the new technology. If new systems are introduced without any need, they may come as a management-led initiative to cater to management needs. This can lead to members of an organization feeling imposed by new systems and not being committed to what has been introduced. Therefore, the conception of any new system is the realization and assessment of needs by members of the organization.

Stakeholder Analysis

Once management is convinced that there is a need for the introduction of new IS, there ought to be an understanding of who the stakeholders are in the new project. A thorough stakeholder analysis should be initiated to identify all those people who would be involved in the project or would be affected by it. Therefore, it goes beyond just the people who would use the system. A set of primary and secondary stakeholders were complied. This list was by no means meant to be exhaustive. For a health informatics project in the NHS, stakeholders would include what is detailed in Figure 1.

In the previous map, stakeholders are categorized into primary and secondary stakeholders. It was the view of the groups that other successful systems should be noted as secondary stakeholders. These are systems from where learning can be gained for the development of the new IS. This is because designing a system from scratch may lead to lack of realism in what can be achieved and how the system can be designed. This may further lead to lack of direction posing considerable challenges in the deployment of the system. Other successful systems would be specifically from the healthcare sector, but not confined geographically to the area for which the system is designed for. Insights may be drawn from successful systems in other countries. The guiding

Figure 1. Stakeholder map for information systems in the NHS

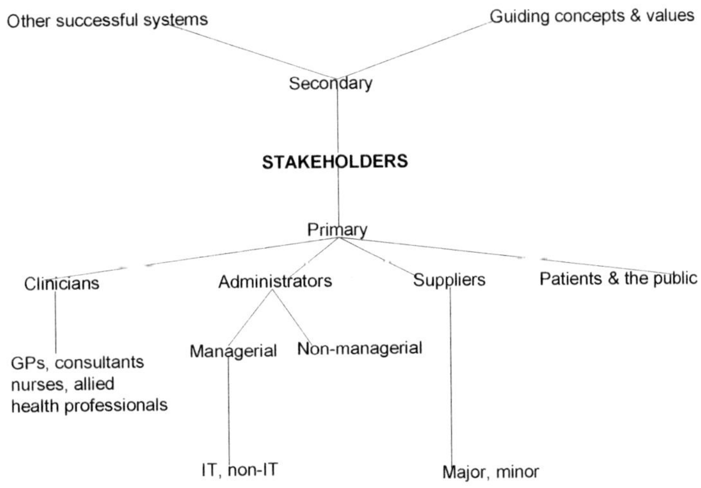

concepts and values that influence the new system and that which would be influenced by the new system have also been considered as stakeholder values. Primary stakeholders include clinicians, administrators, suppliers and patients and the public, who would be immediately effected by the new system (See Figure 1).

There should be a prime initiative to understand how the stakeholders should be involved in partnership right from the beginning. This should move beyond just information giving about the project, and there ought to be true consultation and partnership working from the beginning.

System Specification

Needs assessment and stakeholder analysis may lead to the specification of what is actually desired from the system. This involves understanding of what the vision of the system is, what the vision of the organization is, and what can be realistically delivered. In a healthcare organization the vision may be to improve the quality of care; this will be supported by a whole gamut of technological activities and features. However, everything may not be possible to be delivered in a realistic sense. This was feared to have happened with NPfIT, in which the planners demanded too much within a limited time span. System specification should therefore be realistic and should be aligned with the organizational culture (discussed in more depth in the next stage). In another sense, this may also be called the stage of feasibility study to examine whether the whole idea is feasible or not. At this stage there may also be the requirement to go back to the first stage of needs assessment if there arise any scepticism about the success of the specified system.

It has to be realized that IS is integral to the process of healthcare service delivery. Therefore it has to be noted that the specified system is not treated as just one more piece of IT, but that which supports effective healthcare delivery underpinned by a systemic approach.

Context Analysis

Every organization is different. Design and implementation of any IS has to be suited to its context including the core business, the people and its culture.

The core business of the NHS include collating and dealing with confidential patient information. This entails that the nature of information that would be dealt with by the system is highly sensitive and confidential. This is a significant and decisive factor in the design of new systems. Any IS should regard confidentiality of patient information and safety of patients as its top priority. For instance, the NHS Care Record Service, in NPfIT, would record information for the whole population of England registered in the system. This sort of information is unlike of anything that can be found in other businesses. Security services and access criteria to manage and maintain this information is specific to the healthcare context. If the system has to be developed effectively, it has to take the specific nature of this information into consideration.

In terms of people, the NHS employs highly qualified professionals. It has to be noted that these professionals cannot be forced by management to adopt an IS that the management wants. Any IS would have to prove its potential functionality and benefits to the professionals and patients. This is best achieved when professionals are directive of the design of the system. This takes us back to the first point of needs assessment, the primary impetus for the development of new IS. Due to the nature of work, it is the professionals who would normally recognize the need of new IS. Therefore, the participation and active involvement of professionals is paramount.

Organizational culture is important because any change not respecting culture may face stiff opposition to the change process. Organizational culture will include existing levels of receptivity of IT amongst staff, and also myths and stories associated with the same. The NHS finds itself in a unique position within healthcare as well, due to its sheer size and disparate organization. Myths and stories include comments such as one that was featured in the exercise by group-2: "we have not seen a single government IT project in a large scale succeed." This is an example of cultural scepticism of IS projects in the NHS. These feelings and opinions should be taken into consideration in the design of new IS. Scepticism ought to be understood and addressed in a manner that is culturally sensitive and appropriate to the organization.

Application of a generic IS model for the NHS will not be adequate, but any approach needs to be firmly based in the context itself including core business, people, and culture.

Risk Analysis

System specification may be followed by understanding the nature of risks the IS may face. Risks may be both technical and human. Following are some of the technical risks that may surface:

- Failure of the systems to be delivered on time
- Failure of the technology to deliver what has been aspired to be delivered
- Failure of suppliers to deliver what they have been contracted for
- The system failing to cope with changes in requirements and project specification
- Confidentiality of patient information breaking down
- Rapid change in technology rendering the original systems to be obsolete

Inspite of involving stakeholders appropriately, and considering the socio-cultural factors of new IS, there may still be human challenges in the way of implementing successful systems. Following are some of the human risks that may come with an IS project:

- Staff unwilling to use the new system
- Staff incapable to use the new system
- The system clashing with the organizational culture

The analysis of potential risks may even require designers to go back to the first stage of needs assessment and follow up the rest of the stages. Risk analysis is a crucial stage and if this is not undertaken in a detailed and critical manner, there will always be the fear that in spite of undertaking the rest of the stages effectively, the system may still fail.

Development & Implementation

The previous stages are expected to provide a robust background for the development and implementation of the desired IS. This is more the technical aspect of IS. The challenges here are to select the appropriate contractors and suppliers who will be able to technically deliver what they have been asked to. Success for the technical teams can be achieved with their work in partnership with the rest of the stakeholders.

It may not be appropriate to regard the implemented system as the final solution. Human expectation and system specifications are deemed to change. IS ought to be developed in such a way that changes do not threaten the existing system, but aids in the evolution of existing systems. This is what brings us to the overarching idea of cogenerative learning, discussed below.

Cogenerative Learning

The idea of cogenerative learning is borrowed from Elden and Leven (1991) who talk of the term in the context of action research. They are of the opinion that cogenerative learning is the process where the power relations between the researcher and the research participants dilute, due to the active process the researcher involves the participants in. They note:

The insiders are not simply sources of data or sanctioners of studies and reports but actively help create and codetermine in every phase of the research process--especially in creating new meaning. They are not merely consulted in each phase of knowledge production; they participate as cocreators. *We call this empowering participation.* (Elden et al., 1991, p. 133)

In the context of IS implementation, system designers, managers and planners ought to work cogeneratively with system users, "the insiders," to encounter challenges, learn from pitfalls and cherish knowledge. This has to be achieved in partnership and not in isolation between stakeholders. Constant learning ought to be integrated into IS so that the system is able to evolve from one stage to another and technology is designed to be adaptive. Cogenerative learning may be treated as the overarching philosophy. This lends the iterative angle to the IS design approach.

The previous approach has been illustrated in Figure 2.

The previous model indicates a normative approach for healthcare IS, as informed by the SAST exercise. The insights that have been discussed in the above paragraphs are pertinent in the context of UK public sector healthcare. These insights were conceived by professionals working in a wide range of roles within the NHS. Hence, these insights and the resultant normative model for healthcare IS are invaluable for health informatics designers and planners in the context under consideration.

WEAKNESSES OF THE APPROACH

The previous discussion attempts to present SAST as an effective approach for the enablement of organizational decision making. However, taking a critical perspective, there may arise several aspects where the effectiveness of SAST can be called into question. This section highlights some of these critical aspects.

First, the methodology starts with the identification of participants into two diametrically opposing groups. However, it is not always necessary that even though participants are divided into two opposing groups, they will need to have

Figure 2. Normative approach to healthcare IS

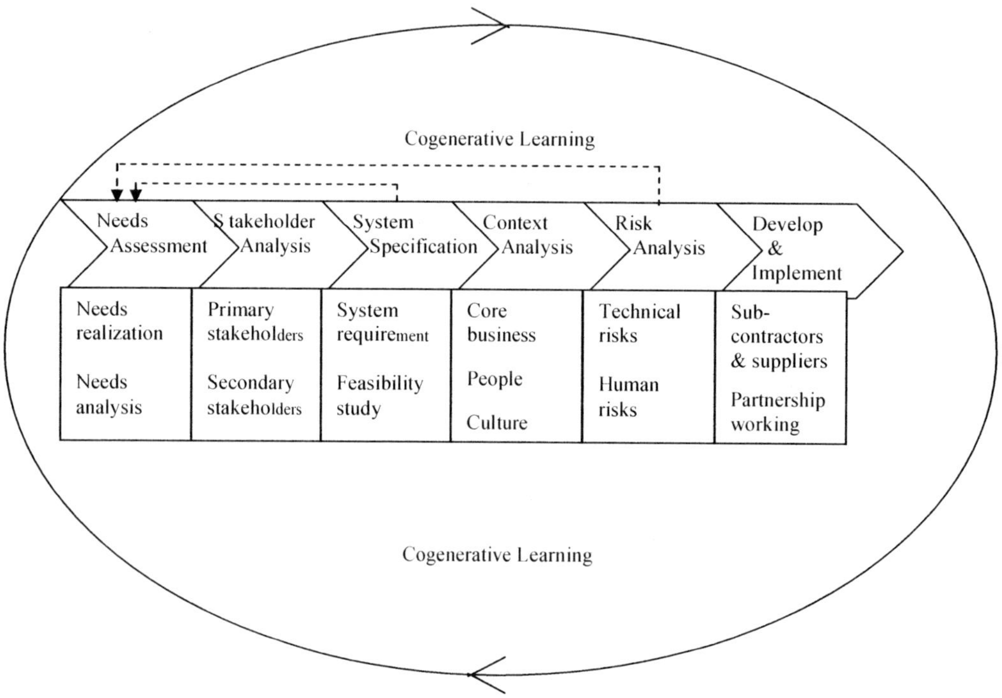

radically opposing viewpoints. As this exercise shows, although there were two opposing groups, there were numerous instances where members of one group closely identified with the viewpoints from the 'opposing' group. The whole idea of maximising divergence of perspectives between the groups may be context dependent. Hence, although SAST aspires to unearth assumptions, it itself starts with an assumption that viewpoints of members from the two groups will always be radically opposing. This introduces a fundamental critique to the whole methodology.

Second, SAST has been portrayed above as an "ideal type" approach for management decision making, which has been aptly applicable in the context of healthcare IS. However, when it comes for the actual implementation of the approach in real life, SAST is best only as an "ideal type" approach. It has the limitation of optimistic simplification. Jackson (2000) notes that the methodology assumes that if the people's attitudes change, so will the social system. This indicates to the methodology's simplistic assumption that social systems are readily adaptable to human systems. However, given the tremendous complexity that gets incorporated into social systems once alterations and new structures are introduced, it may not be very simplistic to alter the same social systems. In the "ideal typical" approach previously taken for the formulation of a health informatics model, SAST has not allowed these challenges to be surfaced.

Third, the SAST approach does not pay due attention to the influence of power relations in the environment in which the exercise takes place. It presumes that in the dialectical debate stage every participant in the teams will present arguments and defend their own stands equally competently to arrive at an understanding. However, if participants come from a different set of hierarchy in the organization, certain members may be reserved about their participation. In the current case study, debating on the situation and arriving at a consensus inspite of radically opposing viewpoints was certainly possible. However, under what conditions this was made possible is also to be scrutinized. This stems from the feedback that was received after the exercise was conducted. One of the participants said that her inputs and vocal arguments were possible because she already knew the rest of the participants very well and felt comfortable to share her opinion; however if the group were different, her involvement could have been very different. A similar feedback was shared by another participant as well. Hence, the whole dimension of power and repression in the overtly consensual atmosphere needs to be considered. This may introduce a Foucauldian dimension of power into the methodology, which SAST very clearly does not consider.

Fourth, arrival at consensus in this particular exercise should not give the impression that consensus is a definite outcome of SAST. There may be cases where no consensus is arrived even after extensive deliberation, and that differences widen instead. This can happen due to a variety of reason including lack of enthusiasm of participants, power relations or simply because the methodology is not situated for that particular context. DP Dash (Dash, date not available) cites his experience of an instance where he applied SAST with two opposing groups, but could not reach a consensus. As he notes "Despite attempts at assumption negotiation and modification, it proved impossible to arrive at any overall synthesis during the final stage of the methodological process" (Dash, date not available). However, the whole idea of consensus is overplayed in the methodology and it gives an impression that a single or repeated deliberation of SAST will certainly lead to a consensus of opinion. Therefore, the idea of consensus at the end of SAST should be regarded with some criticality and it should be recognized that consensus is context and situation dependent.

CONCLUSION

Application of SAST in the context under consideration yielded interesting conclusions for IS design and deployment in the high-tech NHS environment, and for the methodology itself. These conclusions have been summarized in the next three levels.

Methodological

- The application of SAST was found to be a successful intervention strategy in the conflict situation of IS development in healthcare. SAST not only helped bringing the two opposing groups to a level of agreement, but it also identified the common grounds that both the groups had, but did not receive an opportunity to articulate prior to the exercise. Therefore, methodologically, SAST was found to be a useful tool in context under consideration. The systems approach was able to effectively consider a range of socio-technical factors in the design and deployment of healthcare IS.
- It was understood from the feedback that the effectiveness of SAST can be context and situation dependent. The impression that the methodology offers regarding the universal arrival at consensus at the end should be regarded with certain scepticism and consensus and for that matter the progress of the methodology is dependent on the kind of participants, the agenda, and the power relations between participants.

Organizational

- There was a new light on the often perceived conflict between management and clinicians in the NHS. It was understood that management itself had to operate within the pressures of the government. It was agreed in the exercise that many of the strategies adopted with respect to IS were not management led, but government led. Management itself did not agree with some of these strategies, but had to support these as they had no choice. This insight shed new light on how decisions were made in the high-tech NHS environment. This also carries great ramifications on organizational development factors in the context under consideration.

- It was the view of the groups that some of the decisions taken by the DoH were meant to meet political ends, rather than for the best interest of the NHS. As the groups suggested, NHS management itself may have been misled in many instances by the DoH with regard to the inclusiveness and robustness of planning for IS strategies. This portrays the NHS as organizationally fragmented and politically secretive.

Operational

- Healthcare IS is best supported by robust needs analysis, stakeholder analysis, feasibility study (system specification), context analysis, and risk analysis, before systems are designed and deployed.
- IS ought to be designed not as fixed solutions, but as systems that are capable to change and transform. This is specifically pertinent for healthcare due to its multi-dimensional and complex nature. Healthcare IS should be designed as complex adaptive systems, which is able to adapt to complexity and change in the wider environment. This character should be built into healthcare IS projects.

The previous conclusions carry pertinent implications for the design and deployment of healthcare IS. Introduction of new IS in a high-tech environment is not straight forward, but it has to be sensitive to a diverse range of factors. If this is neglected, the chances for success of the new IS may be compromised. A systems approach is vital in cases where issues are divergent and complexity is a rule rather than an exception. Application of systems methodologies like the one discussed in this chapter can be of immense benefit in understanding these issues and enabling variety to be addressed with variety for greater success in complex projects. It may be appropriate to conclude with the following quotation:

In information systems development, it is necessary to systematically view information systems as a socio-technical system and develop information systems using a wide spectrum of technologies, which is superimposed over the complex socio-technical interactions. Systems science provides the basis for taking such a broader view of information systems (Checkland, 1981, 1988; McLeod, 1995, Sommerville, 1996 in Xu, 2000, p. 106).

ACKNOWLEDGMENT

The authors would like to thank Dr Amanda Gregory from the University of Hull Business School, UK, for reviewing this chapter and for her valuable comments.

REFERENCES

Ballard, M. (2006). *NHS IT probe useless.* Retrieved August 20, 2006, from http://www.theregister.co.uk/2006/03/24/nao_npfit_too_late/

Churchman, C. W. (1968). *The systems approach.* New York: Dell Publishing.

Dash, D. P. (Date not available). *Strategic assumption surfacing and testing (SAST).* Retrieved on March 16, 2007, from http://www1.ximb.ac.in/users/fac/dpdash/dpdash.nsf/pages/CP_SAST

Department of Health. (2005). *Hewitt announces action to turnaround NHS finances.*

Retrieved July 21, 2006, from http://www.dh.gov.uk/PublicationsAndStatistics/PressReleases/PressReleasesNotices/fs/en?CONTENT_ID=4127292&chk=HDOR9C

Elden, M., & Leven, M. (1991). Cogenerative learning: Bringing participation into action research. In W. F. White (Ed.), *Participatory action research* (pp.127-142). Newbury Park, NJ: Sage Publications.

Gillies, A. (2000). *Information and IT for primary care: Everything you need to know but are afraid to ask*. Oxon: Radcliffe Medical Press.

Hammond, D. (2002). Exploring the genealogy of systems thinking. *Systems Research and Behavioural Science*, *5*(19), 429-439.

Health Informatics Community. (2004). NPfIT says vast unforeseen expenditure claims are "complete nonsense." Retrieved July 15, 2006, from http://www.informatics.nhs.uk/cgibin/item.cgi?id=922&d=11&h=24&f=46&dateformat=%25o-%25b-%25Y

Jackson, M. (2003). *Systems thinking: Creative holism for managers*. Wiley: Chichester.

Jackson, M. (2000). *Systems approaches to management*. Wiley: Chichester.

Mason, R. O., & Mitroff, I. I. (1981). *Challenging strategic planning assumptions*. Chichester: John Wiley & Sons.

Midgley, G., Munlo, I., & Brown, M. (1998). The theory and practice of boundary critique: Developing housing services for older people. *Journal of the Operational Research Society*, *49*(5), 467-478.

Rosenhead, J. (1987). From management science to workers' science. In M. C. Jackson, & P. Keys (Eds.), *New directions in management science* (pp. 109-131). Aldershot: Gower.

Sherriff, L. (2004). *What's the point of NHS IT? No one knows*. Retrieved January 15, 2006, from http://www.theregister.co.uk/2004/07/28/nhs_it_bens/

Xu, L. D. (2000). The contribution of systems science to information science research. *Systems Research and Behavioural Science*, *2*(17), 105-116.

ENDNOTES

[a] Chuchman's (1968) fundamental idea is that every worldview is terribly restricted and to make sensible decisions for lasting beneficial effects, one needs to "sweep in" a variety of opposing perspectives and worldviews. He advocated that whilst it is not possible to know everything, it is indeed beneficial to understand the implications of the lack of comprehensive knowledge. Therefore, self criticality in theory and practice was central to the work of Churchman and his philosophy.

[b] NEYNLMCN is responsible for design and deployment of service improvement strategies for cardiac care in the North and East Yorkshire and Northern Lincolnshire region of England. This is headed by Alan Nobbs with a core team of four staff and representation from a range of service providing professionals.

[c] In the UK, the responsibility of the provision of health and social care welfare is under the Department of Health (DoH) and Social Security. DoH was formed in 1966 as a result of a merger between the Ministry of Health and the Ministry of Social Security. DoH is answerable to Parliament for the strategic control and direction of the NHS and social services. DoH is therefore the apex government body for healthcare matters in the UK (Source: Department of Health).

[d] The Treasury is the United Kingdom's economics and finance ministry. It is re-

sponsible for formulating and implementing the Government's financial and economic policy (Source: HM Treasury).

- Epidemiology is the scientific study of factors affecting the health and illness of individuals and populations, and serves as the foundation and logic of interventions made in the interest of public health and preventive medicine (Source: Wikipedia online dictionary).

Chapter XV
Managerial Image, Social Capital, and Risk in a Czech Engineering Enterprise

Ben Passmore
University System of Maryland, USA

ABSTRACT

High tech Czech enterprises have faced a transformation, which is shared with similar enterprises throughout the developed world. The value of manufacturing has declined while the value of highly skilled and educated workers has grown. This has placed new demands on workers and managers. Previous research indicates that Czechs harbor considerable doubt about the honesty of their political and economic system. One response to this perceived dishonesty has been disengagement from their work and public life. This behavior has proven problematic for the managers who have aggressively pursued strategies to control perception of risk, create trust, and build engagement in work. Based on ethnographic fieldwork at a Brno engineering firm, this chapter will analyze these strategies and their results, and suggest how the understandings developed in this case might be more broadly applied.

(The boss) is something between friend and father. You have no worries to come to him and tell what problem you have, but, on the other hand, your friend will tell you 'you can do that' (or) 'you can't do that.' But you are waiting not only for decisions but for real help. (High-level Manager at Alstom Brno)

The problem facing managers in high-skill manufacturing industries in the Czech Republic has been one common to manufacturers in the developed world generally. It has been a transition not, as one may suppose, from Communism to Capitalism. This initial transition although difficult was completed for many industries with a few years after 1989 (see Clark & Soulsby 1999;

Možny, Musil, Mareš, Řezniček, & Sirovatka, 1995, Soulsby, 2002). The far more difficult change has instead been from Fordist nationally-oriented production to globally oriented flexible production/accumulation (see Bell, 1973, Harvey, 1989). This change left Czech industry to confront the major problem of how to transform the day to day and interpersonal way industrial work functioned in a fashion that would allow the workers and managers within that industry to deliver on new goals. This article will examine this problem on a micro-scale through the ethnography of a company known today as Alstom Brno s.r.o., a single high-tech engineering firm located in the Southern Moravian city of Brno. As such it will focus on the creation of new relationships between workers and with their work in a political economy, which has radically redefined these relationships. Three particular features of the transition will be highlighted in this chapter. They are the rise of human capital as a key component of enterprise success, the rise of social capital as the main avenue for maintaining engagement of those possessing that human capital, and the rise of a discourse of risk communication centered on corruption and morality.

In the early 1990s, Czech industry was well provisioned with educated and skilled workers, particularly in highly technical areas such as engineering, who were working in relatively modern production facilities. By the 2000s, it was the skills of these workers, which had become the most attractive aspect of many Czech businesses, particularly in technical fields. However, the transition to capitalism required new demands to be made upon workers. Firms, founded on production of highly complex mechanical and electronic goods, found the emphasis shift away from production to services. The shift devalued in relative terms the production capital embodied in the physical plant of these enterprises and substantially raised the value of the human capital embodied in professional and technical workers. Thus, management was forced to cope with the difficult task of developing methods to insure the engagement of workers in the new tasks, which are valuable because of the human capital of those working on them. It is this shift that has proven to be a particular challenge in the Czech context.

Previous research indicates that Czechs harbor considerable doubt about the honesty of their political and economic system, whether that system is the late Communist one or the new Capitalist one (Eyal 2003, Eyal, Szelenyi, & Townsley, 1998). This is expressed at the level of the individual enterprise in very low levels of trust in management and perception of very high levels of economic risk to the enterprise and individual. One response to this perceived dishonesty has been disengagement, to the extent possible, from their work and public life. This behavior has proven particularly problematic for the managers in high-technology private enterprises which have come to trade heavily on employee skills. The level of technological and problem-solving innovation demanded in these settings makes this disengagement a particular challenge to managers. Indeed in these circumstances where trust is in short supply, risk aversion is likely to typify behavior (Gambetta, 1988). This favors simple or normative participation where simple task completion is the extent of employee participation (Clark et al., 1999). This accurately describes the Czech context, especially in the early to mid-nineties, and indeed this level of participation by workers was the noted preference of both managers and workers at that time (Musil 1995). Overcommitment to the company was perceived as holding little reward and considerable risk in terms of time and effort. This left managers with a quandary, how can one establish legitimate leadership and dramatically alter the way work is done when the workforce does not trust you and refuses to go beyond the absolute minimum required? The general answer reached by management has been a combination of enticements, manipulation of perceived external risks to alter the equation in favor of engagement, and an attempt to utilize managerial social capital to reduce the perceived risks within the enterprise.

The research presented in this article is the result of participant observation, formal, and informal interviews, surveys, and documentary research gathered during more than a year of ethnographic fieldwork. Most activities took place at the primary company under discussion and most respondents were interviewed at their workplace. Additional work done with other companies is included as relevant to the discussion. Surveys included a consensus survey to determine the general agreement with statements made during interviewing and a company-administered satisfaction survey provided by Alstom. Documentary research included review and analysis of materials from the Czech Statistical Office (*Český statistický úřad*), professional organizations, and the individual enterprises. The majority of the data and themes presented in this article are derived from the formal interviews. The characteristics of the participants in those interviews are summarized in Table 1.

The analysis of the dynamics of social capital emerged through a series of interviews and the course of participant observation with the employees and managers of these two enterprises. It resulted in two analyses, descriptive analysis of the results of specific inquiries, and a qualitative data analysis of those themes that emerged from these responses and participants' broader perspectives and conclusions. This second analysis was accomplished through immersion in the data set, which produced inductively identified themes (Lincoln & Guba, 1985).

A brief note is necessary regarding the use of data from Alstom Brno. This company is of such prominence in the region that attempts to make anonymous the company's identity would inevitably fail. For this reason and to protect the confidentiality of statements made in interviews, minor details regarding identities of informants and specific information about the company's operations have been omitted or altered throughout this chapter.

This chapter will analyze the strategies adopted by managers, the most important of which has been the manipulation of the image of "the manager" and his/her role, and the impact these have had on the maintenance of social capital (trust) and perception of risk within the organization. To accomplish this, this paper will examine a successful engineering and manufacturing firm in the Southern Moravian city of Brno during the period preceding E.U. accession to accomplish three objectives. It will analyze the perception and communication of risk within the company as it bears on the issue of worker engagement. It will

Table 1. Interviewees characteristics by age group, gender, and work level

	Young (under 30 yrs)		Early Career (Age 30-44 yrs)		Late Career (Age 45-65 yrs)		Pensioner (over 65 yrs)		Total
	Male	**Female**	**Male**	**Female**	**Male**	**Female**	**Male**	**Female**	**All**
Top management	0	0	0	0	3	0	1	0	**4**
Middle management	0	0	1	0	4	2	0	0	**7**
Professional	1	0	3	1	1	2	0	0	**8**
Highly skilled worker/ Low level supervisor	0	0	4	0	2	3	0	0	**9**
Skilled Worker	3	1	4	5	2	8	0	1	**23**
Total	**4**	**1**	**12**	**6**	**12**	**15**	**1**	**1**	**52**

document the attempts by managers to remake their image with workers. Finally, it will present a theoretical framework to consider the interplay of risk and the social capital held by managers in that enterprise.

CONTEXT

Alstom Brno, originally PBS, is one of the oldest manufacturers in the Czech Republic and indeed in Central Europe generally. It was originally founded just outside of Brno in 1814, and has operated at its present site for more than a century. It is a company of such importance to the region that it is consistently mentioned by name in any history of the area or city. It is currently a producer of energy related heavy manufactured goods, particularly mid-sized steam turbines, but has produced a variety of complex mechanical equipment in the past, including textile manufacturing equipment and steam engines. PBS began the 1990s by ceasing to exist as the state company it had been since 1946, and re-opening one day later as a newly "privatized" company (really a company in an official position to become a fully private company). The company was initially quite successful as the new markets came into play. It was able to successfully continue to maintain clients and sales within Czechoslovakian markets (and later the Czech and Slovak markets). However, these markets were eventually exhausted as the needs for these large, expensive, long-lived products were met. The lack of developed foreign distribution networks and contacts placed the company in a position which made passing into foreign ownership the most effective solution to the its financial problems. It also generated immediate problems of legitimacy for the Czech management, problems, which will be discussed at length in the strategies section below. It entered into a joint venture in the early nineties and eventually became a wholly owned part of its venture partner. In the late nineties, the company passed into the hands of another multi-national company. In fact though, this description glosses over a complicated process wherein with each purchase or change in status elements of the company continued operating in the earlier form. Thus, as of the end of 2003, PBS continued to exist as a company (though not on its original location and after multiple legal incarnations) while three separate elements of the original company are operating as separate companies on the "campus" which once housed the main workshops of the company. Additionally, groups of technicians and engineers have formed companies, which compete for maintenance and other contracts on machinery they built while with PBS or its successors. (See Figure 1)

This study focuses primarily on Alstom Brno (with a few forays into the other companies), which operate the boiler and turbine works and which employed, with few exceptions, former PBS employees. The management had not changed significantly at Alstom Brno during the ten years preceding this study. However, the majority ownership of the company had passed into the hands of foreign owners by 1997. The company, although now foreign owned, continues to have all of its major management positions filled by Czechs. Most of these managers had been with the company 15 years or longer. Managers proudly stated that this has allowed the company to continue to maintain its "Czech character" even within a multinational corporation. This "Czech character" proved to be an important part of carefully protected image of managers at the company.

Beginning in the early and mid-nineties, Czech industry (and by extension society through the working lives of its citizens) began fundamentally reordering itself. This reordering has been characterized by a series of changes, for purposes of this discussion the most important of which include: the shift from industrial production to service work, changes in the occupational environment including labor force reduction and labor market segmentation, and enhanced competition and opportunity.

Figure 1. Fission of PBS Companies, 1990-2003

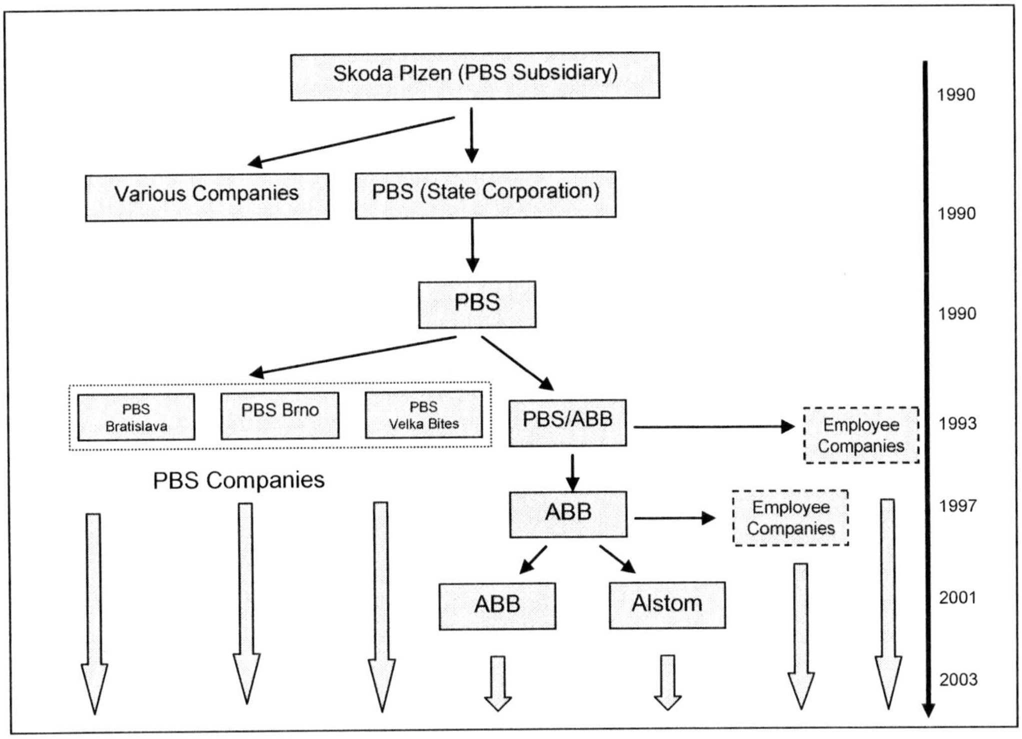

The change from an industrial production base to a service base is evident within the manufacturing enterprise examined in this study. Alstom Brno's managers explained that the workforce they employ today is two-thirds white collar or professional and only about one-third blue collar. This almost exactly reverses the proportions present in 1989. The changes are in large part due to a shift from manufacturing with little or no service after the "sale," to a staff that deals as much or more with high-quality design work and customer service/maintenance contracts as it does with the production of products. The skilled services of the Czech engineers and professionals are the product many managers and professionals at the company wish to sell. This is particularly true now as the Czech wages rise, and the ability to compete with the low wages of manufacturers farther east declines. This shift is not to new areas but a refocusing on skilled services and away from manufacturing capacity. The importance of accessing the companies' human capital (in the form of the employees' skills and experience) has become the key to this new focus.

At Alstom Brno (then ABB/PBS), this change is best understood through the first major R&D task in which the Czech engineering firm participated. In 1996, the Czech engineers joined multi-national teams, which were developing a new generation of modular steam turbines. A portion of this work included the development of common design tools, which would allow the development of further projects by teams scattered across multiple countries. It was this experience which validated the ability of the Czechs to effectively deliver technical work at the highest level and created the relationships which allowed them to participate in this work. It was also during this period in which the company grappled most visibly with the need to develop high levels of English language skills in its workforce, with as much as 60% of the company's workforce in English classes during the late 1990s. This perceived necessity to speak English led to significant conflict and eventually the exit of a number of the technical staff. By the close of the process the engineering workforce had been retooled to be effective within the structure of a multi-national corporation. It was perhaps not coincidental that the company became full foreign owned in 1997 following this period (Alstom, 2001).

The employment environment has also shifted dramatically. One the most striking changes to workforces during the nineties were the dramatic change in the size and nature of workforces at individual enterprises. Alstom Brno reduced its staff significantly. A workforce over 10,000 in the late eighties has been reduced to 1,700. In addition to reductions, the segmentation of the workforce into primary and secondary markets has been recognized as a long-term trend by management, if not always in clear terms (Piore, 1973). Alstom has successfully begun using immigrant Ukrainian, Russian, and Polish labor in its manufacturing positions, eliminating the lowest skill Czech workers almost altogether. Simultaneously, the remaining Czech employees are expected to possess a broad range of up-to-date skills, and to learn to speak English fluently.

Finally, large companies have splintered into many smaller enterprises. This process is described above for PBS, and has dramatically increased competition both for customers and for the most highly skilled employees. In some cases, groups of these employees have left companies en masse in order to join other organizations or start their own companies.

The result of each of these trends has contributed to a series of changes. First, far fewer but more highly skilled employees is the standard. Within individual enterprises, managers and workers experience a far greater degree of employment insecurity than during the communist era. At the same time, and seemingly paradoxically, the employment security of these high skill employees across the entire employment sector is dramatically improved and these individuals enjoy considerable autonomy and mobility between jobs. It would be fair to say that this last point is not always fully evident to workers.

This poses a significant challenge for managers who must maximize the productivity of these very employees while wielding far less effective measures to coerce compliance from them. The key to understanding how they have effectively established a new legitimacy for their management and managed (relatively) radical change lies in two concepts. First, that a series of risk points exist in relationship of employees, managers and the broader economic environment. The consequences of failure to avoid these risks are considerable, job loss, low salary and ultimately the demise of the organization. Secondly, that these risks if they can be convincingly located outside the company (and the character of the managers) in the perception of workers can provide an important source of social capital for management.

A FOLK MODEL OF RISK TO INDUSTRY

What is it that motivates workers? The managers at Alstom and at a number of other enterprises in the Czech Republic had a simple answer. One manager put it succinctly:

Above all it is financial reward of course. If the person has a lot of money, he or she is satisfied, this is clear. Besides (money), it is what surrounds you, such as work environment and then the team he or she works (with). But money--it is, unfortunately, the first priority.

Unfortunately, it is in the area of financial remuneration that managers and workers agree that Alstom has fallen most notably short. Yet satisfaction, as measured by satisfaction surveys, has not fallen but instead risen significantly since 1997 even as satisfaction with wages remains low (see Table 2).

In the first two cases, the head of human resources at Alstom interpreted these changes as successes of the Czech management within the now multinational organization at building the kind of systematic engagement, which had brought the company renewed success. She interpreted the last item as natural given the tough times in the country, and indicated that this does not affect the other indicators because the workers understand the current economic circumstances of the company. This now-shared understanding of what the "facts" of the economic situation has been a major part of Alstom's success at gaining engagement through its entire workforce. It also reflects the successful placement of a risk (failure to share the company's wealth) to management authority outside the immediate purview of the managers. This effectively preserves or even enhances the trust noted in the first questions.

What are those "facts" as perceived by the managers and workers at Alstom? They are, in a word, dire, and managers have worked to convince workers and themselves that the economic facts support sacrifice (most notably financial) on everyone's part. The perception is that the economy is horrible, unemployment is high and "becoming redundant" is a constant danger. The EU and other globalizing influences are going to force these Czech companies through another difficult transition, and the only realistic response is for everyone to suffer together. These opinions were reflected in virtually every interview. The following quotations are representative:

Generally I perceive (life in the Czech republic) as better - besides free borders and freedom of traveling you see more respect. What scares me is the unemployment, to be made redundant, the issue of job security. (Mid-level Manager)

It is the effect of the market economy and unemployment is a global problem. It is the part of the system and we have to accept it as it is. Job security was the only certainty we had in the past. (Technical Office Worker)

The degree to which all of these "facts" are objectively correct and the amount of suffering management and workers (especially those of different ages) undergo is debatable, but it is made clear that an outside threat, often expressed simply as "globalization," exists and that the only real option is to follow the lead of the management. This external threat is presented as largely beyond the control of any individual, but it might possibly be influenced by the skilled planning and direction of managers. This leaves workers with the quandary of whether they should trust these skills in their managers, and indeed more broadly with whether these managers can be trusted at all.

Unfortunately, the threats to the success of the company exist not just from the outside but

Table 2. Selected results of worker satisfaction surveys

Question	% responding "Yes" or "Most Likely Yes"	
	2001	**1997**
Do you trust the company's management?	65%	34%
Would you recommend to a friend to seek a job at Alstom?	75%	39%
Are you satisfied that your salary is fair and reasonable for the type of work you perform?	26%	21%

from the widespread perception that damaging corruption is common throughout society. This is not in itself surprising, post-communist states including the Czech Republic have been plagued by endemic corruption. However, this situation is particularly likely to draw accusation of corruption specifically because the outside threat is largely beyond the control of individuals. These risks contrast sharply with the risks of the earlier era, which could be managed through the political acumen of the company's leadership. They are new and not entirely concrete or explicable to the average (even well educated) worker. Ericson and Doyle writing on risk and morality capture this quandary:

In order to sustain 'the logic in which someone must be held to blame for any event that threatens an individual's 'quality of life,' risk must be construed as a product of human agency and therefore controllable through attributions of responsibility and processes of accountability. (2003, p. 7)

This suggests that the "outside" risk has been effectively established in the minds of workers, but that it is not possible to effectively locate all risk on the outside of the company. The distant and nebulous risks inherent in post-Fordist flexible production create a circumstance in which those who claim some minor ability to affect them are likely to receive disproportionate blame when those dangers are encountered. At Alstom, the mechanisms of informal accountability were built largely around the need to avoid even the suggestion of corruption, and on the ability of the management to successfully communicate their incorruptibility to their fellow workers.

CORRUPTION AND SOCIAL CAPITAL

There are some groups of people that have been here for a long time and because they have been here for a long time and know plenty of people ...(managers) overlook them(when there are problems). The managers think they know the problems of these people because they have known these people for so long... And when you report on him you are told "Ignore him." I'd think they would welcome it, (but) when I stressed this, I was told "Take care of what you are supposed to." rather than being told "Yes, you are right." It is rather the opposite of (what they told you). (Alstom Technical Line Worker)

Here you get to the important positions through the so-called "Let's be friends system", through inviting important people out for a beer and hoping the person will help me to be promoted. (Retired Manager)

The collapse of communism in Central and Eastern Europe and subsequent struggles to establish civil society with a functional economic environment led to a revival in interest in the ideas of social capital. Nichols (1996) writing about Russia, but equally applicable to the Czech Republic, stated directly, "the hallmark of the Soviet system was the purposeful destruction of what is now understood as "social capital" (p. 631). This had been done specifically to paralyze non-governmental social action. Other argued that the presence of so-called "non-communitarian" social capital (growing from black markets and personal connections) presented a substantial roadblock, in the form of cronyism and other corruption, to successful re-development (Hayri & McDermott, 1998; Kolankiewicz, 1996; Stark, 1997). The basic thesis of this scholarship has been that post-communist nations suffered from a dearth of formal social capital (expressed as public trust in institutions, solidarity with legal organizations or formalized networks and norms of public behavior) instead they were characterized by a surplus of informal social capital (expressed as insider relationships, informal and closed networks, and norms which supported self interest ahead of community). It is the operation of these informal networks, which forms the basis for most day-to-day "corruption" in these settings.

Ethnographic and other sociological work in worksites throughout the region has tended to confirm this thesis and add an additional layer in its examination of what has happened during the post-communist period. Grabher and Stark (1997) have argued that these "legacies" have interfered with the development of institutionalized formal social capital but that it also "preserves diversity for future recombinant strategies" (p. 22). Trust, generally, has figured prominently in some of the sociological literature (Buechler & Buechler, 2003; Dunn, 2004; Koch & Thomas, 1997; Kuczi & Mako, 1997; Torsello, 2002). This research has suggested that "trust" has often been built on the existing informal networks, particularly kin networks, but that these networks have been challenged and eroded during the post-communist period. In this vein a useful operational definition of social capital might follow from Portes and Sensenbrenner:

> *...those expectations for action within a collectivity that affect the economic goals and goal seeking behavior of its members, even if these expectations are not oriented toward the economic sphere.* (1993, p. 1323)

Within the organization and networks then the same behaviors, which are often described as corrupt are the keys to flexible responses within the economic field. Although the focus varies, these informal relationships have undergone a process of redefinition as individuals have tried to preserve their valued ties even as the social world has been fundamentally altered. This process has often leaned heavily upon ideas, which seem to have little to do with economic activity in any classic sense.

Similarly, there was broad agreement among informants in this study that society and the basic nature of the workplace had been transformed in the years since 1989. New criteria of what matters and what various types of relationships and behavior mean had come into existence during the intervening years. A process of legitimation, defined here as the process by which a social group establishes, "...new meanings that serve to integrate the meanings already attached to disparate institutional processes" (Berger & Luckman 1985, p. 91), had taken place. Many work tasks, for example welding, had not changed dramatically but the entire context in which they occurred had. At the time of its inception, little opposition existed to the idea of private ownership of industry in the Czech Republic (Mareš, 1995) so it was less an issue of legitimizing the system as a whole as legitimizing the specific relationships with the factory setting. However, neither the loss of the old system's legitimacy nor the growing legitimacy of the new system of private ownership guaranteed or mechanistically determined the specifics of a functioning private workplace. Workers, managers, and owners of companies found that they were dealing very directly with what the new basis for legitimation of their actions would be. This led them to negotiate the new arrangements not on the strength of existing means or formal systems, but on the basis of long-standing personal relationships within the enterprise. This has created the irony of social capital as the source of both risk (personal connections equaling corruption) and of risk reduction (trust in managers allowing enterprise reorganization). Managers have sought to utilize their personal relationships to facilitate the latter while manipulating their image to avoid the appearance of the former.

AVOIDING CORRUPTION AND MAINTAINING MANAGERIAL IMAGE

Information Symmetry and Internal Transparency

One of the key components for avoiding the appearance of corruption was the creation of the perception of information symmetry between workers and managers. The major avenue for achieving this has been management's use of "internal transparency" regarding plans and organization. If information symmetry is the desired goal, with each group knowing what the other is doing, then internal transparency is the method they use to achieve it. This transparency largely concerns the public (i.e. within the company as a whole) presentation of management plans, sales issues, project descriptions and the like in order to avoid surprises and establish that the workers understand the importance of various plans. Following the initial administration of a job satisfaction survey in 1997, the company set out to deliver the message that it was and would remain a Czech company despite the presence of multi-national ownership. The management sought to reassure the employees that the success of the company would remain in the hands of those who worked there and they worked to make it clear what the company's plans were. In addition, managers in individual departments have worked and continue to work to detail the issues surrounding each contract to insure a knowledgeable workforce.

The non-managers, whether line workers or highly skilled professionals, all expressed a desire for this information. They noted that this openness is said to be in contrast to both the communist past and the behavior of early transition companies headed by "dishonest" or "rough" people. They noted that in the past companies had routinely gone bankrupt and yet the owners and managers had somehow managed to profit. They suspected that this was due to a combination of personal dishonesty but also deception about the state of the companies. They expressed a determination that this type of deception should not occur within their company and for this reason a desire to have an open flow of this type of information at all times. To insure this happened all those interviewed, from general directors to line workers, expressed a profound interest in information concerning the honesty of management and workers. In this sentiment, one can see three important elements. First, it articulates (not unjustified) perception of the role of morally compromised individuals in putting their enterprises at risk. Secondly, it points clearly to the necessity for managers to express in a convincing manner openness and commitment to transparency as a key element in maintaining trust. Finally, it points up the emphasis on group solidarity of all levels of management and workers in the attempt to build close workgroups and companies.

Culturally Appropriate Tropes

The perception that corruption is the source of much of the controllable risk to enterprises creates a contradiction for managers. Through their attempts to establish information symmetry, they hope to establish their fundamental honesty and thus utilize the value of the social capital, which they hold within the company through long-standing relationships with employees. Unfortunately, this informal utilization of relationships is often considered the defining characteristic of corruption. This is in part countered by raising perceived risks from outside to improve the bounded solidarity of the group. Keeping that risk level high enough can be used to excuse some minor (or not so minor) violations of the now "shared beliefs" of proper moral action. The other method used is the establish managers as insiders within the group of employees, a process accomplished when managers called on a series of culturally specific tropes for action and rectitude. They attempted to establish their image as being, if not above reproach, deeply trustworthy through enacting these deeply understood cultural forms. The following outlines the two key areas: "family" and "the Czech as European."

"Family"

One of the first things most people working at these companies will tell anyone visiting Alstom is that the company has deep roots in Brno and that they are part of the "family." This metaphoric family and the uses to which it is put to are key elements in the organization of engagement. Workgroups were also described in family terms. A "good worker" was described by some as "one of us", someone who could work in a team, or even simply as "family."

Family entered the discussion through the history of the organization. Prior to 1990, it had not been uncommon to see family members of two or even three generations working for the company. As one employee described it:

There was some tradition, not only in technicians but for workers. If father works here, his son ... (In the past) your daughter or your son worked in (the) same company, not always the same group or same department but in the same company. And a lot of people started and still continue here.

The post-communist era has challenged these ideas in a number of ways. The number of people whose children are following them into these companies is in sharp decline. The economic changes of the era have so altered the circumstances, it is argued, as to break what were once highly desirable patterns of recruitment. One sales director commented on whether multigenerational patterns are persisting:

No. I am sure. No, because it was done because of the situation in the socialist period. There wasn't difference if you worked in PBS or Kralopolska (or wherever) ... You were a welder and you received money... But now where you have a higher salary, you go... (Before 1989) there were few differences, it wasn't important for people to change company. Why? ... And now it is totally different, if one company goes for bankruptcy, there is high salary, low salary, and there is immigration (and) it is very, very high.

Additionally, the current workgroups are now made up of workers who have been with the companies for significant periods of time. Because the focus of human resources management has been almost entirely on reducing staff size, one manager in this area estimating 90% of her time is spent dismissing and only 10% hiring employees, few new workers are forced to integrate into the workgroups. The result is only a handful of the most motivated younger workers are in the groups and they expressed perhaps the highest levels of solidarity with these groups, which it was so difficult to join. The result is smaller, more tightly knit workgroups who have to rely heavily on one another.

These changes have driven engagement in ways that are perhaps unexpected and even counterintuitive. The atomized workgroup structure has encouraged tighter and more exclusive bonds as these groups have come increasingly to represent the entire universe of work relations. These people become the only people one works with and those one depends on to meet the much higher workload of the present era. For all of these reasons, the workgroups, despite reduced levels of family, or outside of work ties, are extremely cohesive units. Much of the cohesion is based around rhetorical or metaphoric family relations in the workplace.

The destruction of existing family networks in the workplace, which held social capital during the communist-era is paralleled with a process wherein the attempt has been made to transfer some of the value of those networks, at least rhetorically to another system. Family, always an important locus of trust, has become the gold standard of trust made even more valuable in the workplace by its relative scarcity in the workplace. Anything leaders can do to effectively appropriate these images for their own use is of tremendous value. Thus, actual family is in decline but the discourse regarding family is of increasing importance.

Nationality and Europe

If family underpins the ideal nature and qualities one expects from trust relationships, then history may define the discussion of the meaning of the current broad political-economic system and the value of this organization within it. Within the category of history, two intersecting narratives inform this discussion: the meaning of Europe and the meaning of being Czech. Each has been of major importance across a broad range of issues in Czech society in the post-communist era and each has translated itself into meaning at work.

The importance of being European, being like Europeans, or preparing to rejoin Europe and the broader international community (in acceding to the E.U.) was a remarkably common theme during interviews. This was not surprising given that the referendum to accede to the E.U was only a few months away at the time of the research, but it also reflected a broader theme regularly articulated in many sectors of Czech society.

This theme was raised in several ways. It was used rhetorically to make the point that the company's standards were up to any in the world. The management aggressively promotes the company as a European and international one. The Human Resources manager spoke with deep pride over the success of a local engineer who has had significant success in a position within the multinational in Australia. She felt it demonstrated the strength of the company and Czech Republic. This promotion of the connections to other elements of the multinational, along with the heavy emphasis on English language training and actual placement of Czech workers in foreign assignments make up the main elements of the company's effort to create itself as an "international" company.

The use of English has developed as a potent symbol for cosmopolitan Europeaness. In theory, the local employee might work with the non-Czechs during foreign assignments, during international meetings, or if an employee of another subsidiary of the multinational came on assignment to Brno. In reality, only the first two have occurred with any regularity. The last has not occurred regularly because of the lower wages available in the Czech Republic and the weak appeal of living in Brno, and each instance where it has occurred is an event worthy of mention.

Finally, the successes of the company both in competing in foreign markets and of acting as an equal company with the others in the multinational are heavily promoted inside the company. The company newsletter specifically documents this and disseminates it to the staff. This international quality will, they hope, make the transition into the EU less a shock and more of an acknowledgement of already existing relationships.

This last point is the key to the value of "European-ness." It is fundamentally an acknowledgement of the relationship that has existed but has been neglected. The impact of this valued status in the workplace is that it shapes the behavior and attitudes on the workfloor as workers and managers strive to be European enough in attitude, skill, and organization.

"Czechness" is similarly a state of being presumably shared by Czechs as shared through a set of self-referential stereotypes (see Holy, 1996). These characterizations are impossible to avoid in discussions of any of the major issues of the day, and, in this instance, they are also often used in contrast with the imagined "European." Quality and high levels of skill are two things, it was regularly said, which the Czech worker and a Czech company can bring to the marketplace. Perhaps most importantly, Czechs are widely believed to possess a quality of adaptability which defines both the individual and Czech companies. This quality is described in Czech as *zlaté české ručičky* (Golden Czech Hands). A general definition provided for this trait was:

'Zlate Ceske Rucicky is literally Golden Czech Hands, which implies that Czechs are very handy, dexterous, skillful, able to fix or adapt things in such a way that it serves a special purpose.

In the post-communist era, this trait has repeatedly been cited as the single most important Czech trait. An example is an April 2003 survey by a leading Czech daily, found it to "still" be the trait that most defined Czech character (Machacek, 2003, p. A6). Even when not mentioned by name it is a common theme:

We (Czechs) are adaptable... We do this by not becoming desperate and focusing on keeping what you have got.

This adaptability is not entirely a simple thing, however. It is argued that although it has allowed Czechs to survive great difficulties, it has also created in them a certain moral ambiguity. In the consensus survey, the question: "An honest person can seek "loopholes" to circumvent laws and regulations they do not agree with" had the second lowest level of agreement (only "The majority of Czechs are honest people" had a lower agreement level). Discussing this question, the seeking of ways around the rules was seen as a typically Czech response. Indeed, one respondent that this behavior was a reflection on the much admired trait of adaptability. This rule avoidance was often greeted with a shrug and the comment "what do you expect." Even the widely reviled practice of "tunneling", or asset stripping, which had cost average Czechs millions of crowns was viewed as the not unexpected result of the adaptability of Czechs. Thus, one of the same traits that makes Czechs (and their enterprises) so good for Europe are also among those which have given them such a bad reputation.

These discourses of what it means to be Czech and how this Czech will fit into the new Europe and world are not limited to the workplace, but are particularly potent in that locale because (along with travel) it is the main venue through which these Czechs interact with Europe and the world. The value structure reconfigured from this discourse suggests both the nature of the new system they have joined, which skills they should cultivate and whether long valued traits will remain important.

The interplay of "Czechness" and "Europeaness" yield a series of results which draw a sharp picture of the difficult negotiation in which managers find themselves involved. Managers seek to portray themselves as Czechs to urge workers to work harder and to be ever more adaptable. This common heritage, along with the family discourse previously discussed, allows them to access existing relationships and establishes expectations of performance. The position of managers as the most integrated with Europe gives them the power to motivate as well as inoculating them against accusations of corruption. The appeal of these managers is considerably enhanced because it is they who can help the professional worker become more "European" through access to valued foreign internships, assignments and other connections. It is however, the inoculation of managers from corruption accusations where the greatest value of this discourse lies for the management. When managers choose or are forced to take relatively harsh measures to improve performance, such as lay-offs or lack of salary increase, it is because of the uncontrollable (globalization) risks associated with integration with Europe rather than the controllable (corruption) associated with the Czech Republic.

CONCLUSION

BP: ...Is there something that tells you that an employee is a problem?

B: Yes... Come to find out shortly that results don't come. I mean especially in my job (which is) to give money and tasks to do something, to solve some problems responsibly. Then simply their output doesn't come, that's the first signal. I am not talking about the quality of the result or response, but the first signal is that (the work) doesn't come at all. It happens very very often that you need to come to this guy, or girl, and talk again and again... "Please do this, it is quite important"... If you need to talk like this with the same person several times, (it) is not good. It is a problem for me... we can solve the problem in minutes if they say, "Okay, it is not my job", "I don't like to do this." That is fair, but to say, "okay, I'll do it" and results don't come, that's a problem. That I would say is the first signal to me... (as opposed to) the quality of the output, but that's always the point of view because it can only be perspective. So that's not so important. (Mr. B. and Author)

From this case study, one may take away several points regarding the management of high-tech industries. The first concerns the primary challenge to management of established high-tech industry in the Czech Republic and to high tech industry more generally. This challenge is not so much quality or skill related, but instead motivational and inter-personal. At Alstom, the tasks associated with work may be more advanced or complex than they were a few years earlier but they have followed in a clear arc from past to present. The relationships, of worker to manager and of company to client, have on the other hand been radically and completely transformed. This case of change is admittedly an extreme one where one advanced industrial economic system has superseded another, but it is different more in scale and rapidity of change than in the type of change. It is the shift in what is of value in the operation from "production capital" (for example, the physical means of production) to the human capital (for example, the skills of engineers and other technical staff), which has by necessity altered the relationship between worker and manager. This same shift has been taking place in other industrial settings since the late 1970s and is arguably accelerating with the advent of the so-called innovation economy created by ever more flexible industrial production. The changing job market, wherein a worker is both free to leave but also free to be dismissed, has simultaneously strengthened and weakened the position of managers at Alstom. They face problems familiar to many managers in skill-intensive industries. How do you identify and retain the best employees who are now aggressively courted by competitors? How do you insure that your staff is really doing their best work? How, in short, do you keep a group of workers fully and routinely engaged with their work when they harbor the significant, and not wholly unjustified, belief that a full commitment to the work may be riskier than a more limited involvement.

In order to answer these and similar questions, managers at Alstom (and other companies) have had to ask another question: what is it that motivates people to work? As previously discussed, the easy answer is salary and other benefits, but this answer has proven to be of limited utility as the company has confronted the limitations of its ability to raise these levels to the desired points. Instead, they have had to consider those things, which are both valued and feared by employees, essentially understand the workers perception of risk in these new circumstances. They have sought to gain a measured gamble by the workers that managers can lead them to long-term prosperity if they will accept the program of change and behavior which the management has created. This program has included the building of new skills, such as facility with the English language, but far more importantly an acceptance of the need for serious engagement in the work at all levels.

In order to effectively utilize these personal relationships to maintain this engagement, managers are essentially engaged in a protracted act of communicating the level of risk to the company and what they are doing to alter it in a favorable fashion. They attempt to hew closely to certain culturally appropriate tropes to portray themselves as being of a particular character. Their actions form a kind of narrative about risk and success. When they utilize the longstanding relationships within the company, they rely that it will be through the positive narrative of success that their actions will be interpreted. This essentially insulates their behavior from criticism in much the same way that a trusted public institution might act in another circumstance. That institution might do things which observers do not agree with, but given sufficient social capital they avoid serious criticism. More generally, there is considerable flexibility as to which tropes are used and the resulting narratives. Indeed, Dunn (2004) found the "family" metaphor used by workers to effectively generate precisely the opposite result of that seen here, one of greater dependence and reduced engagement.

The managers of Alstom have succeeded in creating change through the promulgation of a specific narrative of risk. Risk from outside in the form of globalization and risk from inside in the form of corruption. They have offered themselves as a trusted group with a program, which will allow the enterprise to reach goals which satisfy both managers and workers. Social capital expressed as trust in management is not therefore by itself a motivator; it is instead a collateral hedge against risk. The desire to maintain it acts as a constraint of actions through expectation which indicates to all participants that only a relatively limited range of actions are likely to occur. Managers act out their roles in such a way that their behavior, which might be considered detrimental in some instances is accepted as necessary. This complex of behaviors and narratives forms a key element in gaining and maintaining the engagement need to pursue the work of this high-tech enterprise.

REFERENCES

Alstom Power s.r.o. (2002). *100 years of steam turbines in Brno*. Brno, Czech Republic: Alstom Power.

Bell, D. (1973). *The coming of post-industrial society*. New York: Basic Books.

Berger, P., & Luckman, P. (1966). *The social construction of reality: A treatise in the sociology of knowledge*. Garden City, NY: Doubleday.

Buechler, H., & Buechler, J. M. (2003). *Contesting agriculture: Cooperativism and privatization in the New Eastern Germany*. SUNY Series in the Anthropology of Work. New York: State University of New York Press.

Clark, E., & Soulsby, A. (1999). *Organizational change in Post-Communist Europe: Management and transformation in the Czech Republic*. London: Routledge

Dunn, E. C. (2004). *Privatizing Poland: Baby food, big business, and the remaking of labor*. Ithaca, NY: Cornell University Press.

Ericson, R., & Doyle, A. (2003). *Risk and morality*. Toronto: University of Toronto Press

Eyal, G. (2003). *The origins of post-communist elites*. Minneapolis: University of Minnesota Press.

Eyal, G., Szelenyi, I., & Townsley, E. (1998). *Making capitalism without capitalists: The new ruling elites in Eastern Europe*. London: Verso.

Gambetta, D. G. (1988). Can we trust trust? In D. G. Gambetta (Ed.), *Trust: Making and breaking cooperative relations* (pp. 213-237). New York: Basil Blackwell.

Grabher, G., & Stark, D. (1997). Organizing diversity: Evolutionary theory, network analysis, and post-socialism. In G. Grabher & D. Stark (Eds.), *Restructuring networks in post-socialism: Legacies, linkages, and localities* (pp. 1-32). Oxford: Oxford University Press.

Harvey, D. (1989). *The condition of post-modernity*. Oxford: Blackwell

Hayri, A., & McDermott, G. A. (1998). The network properties of corporate governance and industrial restructuring: A post-socialist lesson. *Industrial and Corporate Change, 7*(1), 153-193.

Holy, L. (1996). *The little Czech and the great Czech Nation: National identity and post communist social transformation*. Cambridge: Cambridge University Press.

Koch, T., & Thomas, M. (1997). The social and cultural embeddedness of entrepreneurs in Eastern Germany. In G. Grabher & D. Stark (Eds.), *Restructuring networks in post-socialism: Legacies, linkages, and localities* (pp. 242-262). Oxford: Oxford University Press.

Kolankiewicz, G. (1996). Social capital and social change. *British Journal of Sociology, 47*(3), 427-441.

Kuczi, T., & Mako, C. (1997). Toward industrial districts? Small-firm networking in Hungary. In G. Grabher & D. Stark (Eds.), *Restructuring networks in post-socialism: Legacies, linkages, and localities* (pp. 176-189). Oxford: Oxford University Press.

Lincoln, Y. S., & Guba, E.G. (1985). *Naturalistic inquiry*. Beverly Hills, CA: Sage

Machacek, D. (2003). Stale uznavame 'Zlaté české ručičký'. *Mlada Fronta Dnes*, April 17, 2003.

Mareš, P. (1995). Legitimacy. In I. Možny, L. Musil, P. Mareš, I Řezniček, & T. Sirovatka (Eds.), *Social consequences of change in ownership: Two case studies in industrial enterprises in the Czech Republic– Spring 1993* (pp. 53-70). Brno, Czech Republic: Masarykova Univerzita.

Možny, I., Musil, L., Mareš, P., Řezniček, I., & Sirovatka T. (1995). *Social consequences of change in ownership: Two case studies in industrial enterprises in the Czech Republic– Spring 1993*. Brno, Czech Republic: Masarykova Univerzita.

Musil, L. (1995). Participation. In I. Možny, L. Musil, P. Mareš, I Řezniček, & T. Sirovatka (Eds.), *Social consequences of change in ownership: Two case studies in industrial enterprises in the Czech Republic– Spring 1993* (pp. 26-52). Brno, Czech Republic: Masarykova Univerzita.

Nichols, T. (1996). Russian Democracy and Social Capital. *Social Science Information, 35*(4), 629-642.

Piore, M. (1973). Notes for a theory of labor market stratification. In M. Reich, R. Edwards, & D. Gordon (Eds.), *Labor market segmentation* (pp. 125-150). Lexington, MA: Heath and Company.

Portes, A., & Sensenbrenner, J. (1993). Embeddedness and immigration: Notes on the social determinants of economic action. *American Journal of Sociology, 98*(6), 1320-50.

Soulsby, A. (2002). The impact of societal transformation on Czech managers: a study of post-communist careers. In M. Kelemen & M. Kostera (Eds.), *Critical management research in Eastern Europe* (pp. 164-186). London: Palgrave.

Stark, D. (1997). Recombinant property in East European Capitalism. In G. Grabher & D. Stark (Eds.), *Restructuring networks in post-socialism: Legacies, linkages, and localities* (pp. 35-69). Oxford: Oxford University Press.

Strašil, K. (1994). *180 Let Prvni Brnenska Strojirna (180 Years of PBS)*. Brno, Czech Republic: Moravska Typografie

Torsello, D. (2002). Managing instability: Trust, social relations, and the strategic use of ideas and practices in a southern Slovakian Village. In D. Torsello & M. Pappova (Eds.), *Social networks in movement: Time, interaction, and interethnic spaces in Central Eastern Europe*. Nostra Tempora 8, Forum Minority Research Institute. Samorin, Slovakia.

Section IV
Self Management

Chapter XVI
Self-Entrepreneurial Careers: Current Management Practices in Swiss ICT Work

Elisabeth K Kelan
London Business School, UK

ABSTRACT

With the changes in the new economy, careers are said to be in flux. Although employees were for a long time at the mercy of organizations to make a career, this has supposedly changed in recent times as now careers are often said to be in the hands of individuals. This trend is called self-entrepreneurial careers. In this chapter, management practices of careers in information communication technology (ICT) work in Switzerland are analyzed and it is explored in how far ICT careers follow a self-entrepreneurial career pattern. It is shown that there are divergent career patterns in Swiss ICT work. The career patterns in two ICT companies are analyzed and it is highlighted that in one company the concept of increasing responsibility was predominant while in the other one network like careers were more common. It is argued that both companies reflect and support the idea of self-entrepreneurial careers.

INTRODUCTION

The world of work is undergoing fundamental changes and careers have not been left unaffected by this (e.g., Arthur & Rousseau, 1996; Castells, 1996). While traditional organizational hierarchies were characterized by delayed gratification and steady career progression, in the new economy careers appear increasingly as in employees' hands. This means that instead of relying on climbing organizational career ladders, new workers are encouraged to manage their own careers (Handy, 1994; Peters, 2001; Pinchot, 1985; Pink, 2001). Instead of assuming that a company will ensure employment, individuals are taking charge of their careers. In doing so, they deploy strategies that encourage them to think of themselves as a product or a brand (Peters, 2001). People have to sell themselves on the marketplace and have to make sure that the skills they have are valued by people who may hire them.

This self-entrepreneurial trend is, not surprisingly, most pronounced in entrepreneurial and freelance forms of employment. However, companies are also striving to introduce more entrepreneurialism to manage their resources more efficiently. In times when employment for life is no longer a given, individuals increasingly adopt self-entrepreneurial management practices to develop their careers. Enterprising careers are therefore no longer confined to entrepreneurs and freelancers but are for more and more people a common way of planning a career. Enterprising careers have often been attributed to environments quick in adopting new trends of working. Information communication technology (ICT) work is a prominent example. The article should contribute to the understanding of how entrepreneurial careers are managed in ICT work in Switzerland.

This chapter is structured as follows. First, I draw on sociological theories of the changing workplace to show how sociologists have made sense of the changing workplace. Second, I explain the methodology, which was used to collect the material on which the study is based. In the third part, I outline and contrast the different career models that were employed in two companies in Switzerland to manage ICT work and highlight which differences and similarities in managing ICT careers there were. Finally, I offer some conclusions on entrepreneurial careers in ICT work.

SELF-ENTREPRENEURIAL CAREERS

The mid-1990s saw a radical shift in economic production, a shift often compared to the introduction of steam machinery, which changed the economic mode of production in the eighteenth and nineteenth centuries. Like the steam engine, ICTs transform the way in which surplus value is created. These new technologies are the central pivot of the changing economy because surplus value is created through the application of knowledge based on information. Technologies speed up this process and simultaneously are a result of these knowledge-generating processes (Castells, 1996, 2000). This means that these new technologies both shape and are shaped by society (MacKenzie & Wajcman, 1999). The new economic formation is referred to as the information society, knowledge society, or new economy.

These changes in the economic mode of production alter current workplace organizations and give rise to new management forms. Instead of the muscle power, which shifted heavy industrial goods, the shared brainpower of individuals and institutions now fuel the economic engine. The means of production are no longer heavy machinery in which an industrialist has to invest, but less capital-intensive goods such as computers or brainpower. Work in the new economy no longer has to be done in a sweatshop factory, but can be done either in the home as in pre-industrial times or in state-of–the-art offices, where free gourmet food, a gym, childcare, laundry, and dry cleaning services on-site render the office into a place where one wants to live (Elgin, 2005). These work relations sound on first hearing much more empowering than previous work relations.

However, the new managerial forms emerging with these transformed work environments are infused with individualization and flexibility. While work relations used to be fairly standardized for most people, today's working practices result from individual negotiations and are accordingly much more flexible. While previously employers bought the labor time of a worker/employee for a fixed sum of hours, today's work is organized based on targets to be met. How people organize their work is up to themselves as long as the final result fulfils the goals set (Pongratz & Voß, 2003). Moreover, jobs for life and the idea of delayed gratification seem to become more the exception than the rule. This means that workers are increasingly required to be flexible in line with demands of market cycles and to find work from a variety of sources.

These changes, together with the alleged decline of other socially structuring parameters like gender, class, and race, lead Beck (2000) to state that individuals now have the chance of being author of their own biography, although this involves risk at the same time. Sennett (1998) in particular points to the risk attached to such flexibility. Being flexible means that the individual time horizon shrinks to the short term and this short term mentality results in people having difficulties in creating consistent narratives of their selves. This results, Sennett argues, in a "corrosion of character." In this light, du Gay (1996) argues that people are encouraged to be entrepreneurs of the self and to see their own work identity as a project. However if work is not going to plan then it is also the individual who shoulders the risk. There seem to be new costs and risks attached to laboring under these changed circumstances. Power now seems to reside in individuals to shape their own career but at the same time, this new rhetoric may disguise much of the power framework through which workers sell their labor.

Given these drastic changes in how labor is managed and how power relations are shifting, it is worthwhile exploring in more detail how social theorists have made sense of these changes at work. Beck uses the term reflexive modernization to denote the transition from first to second modernity (Beck, 2000). While first modernity "was characterized by collective lifestyles, full employment, the national and welfare state and exploitation of nature," the central features of the second modernity are "the ecological crisis, the decline of paid labor, individualization, globalization, and the gender revolution" (Beck, 2000, p. 18). Reflexive modernization refers firstly to the idea that the new society is a reflexive one and secondly to the idea of a society that reacts reflex-like to changes (Beck, 2000, p. 21). This results in de-traditionalization, whereby traditions become negotiable and decidable on an individual basis. Therefore individualization and risk increase. Traditional "markers" of identity like occupation, class, and gender, which have determined the life-course of individuals in the past, are losing their importance (Beck, 2002; Beck & Beck-Gernsheim, 2002b). It is now up to the individual to shape his or her own life.

The choosing, deciding, shaping human being who aspires to be the author of his or her own life, the creator of an individual identity, is the central character of our time. It is the fundamental cause behind changes in the family and the global gender revolution in relation to work and politics. (Beck, 2002, p. 22f)

Under reflexive modernity institutions are tailored to self-organized and individualized biographies, resulting in "institutionalized individualism" (Beck & Beck-Gernsheim, 2002a) or "collectivised individualism" (Beck, 2002). An individual biography is not only an "elective biography," "reflexive biography," or "do-it-yourself biography" (Beck et al., 2002b) but also always a "risk biography" (Beck, 2002). Beck (2000) describes how people are encouraged to seem themselves as companies adopting a Me & Co mindset, in which they see themselves as a small company that has to be managed. The term risk biography highlights the fact that having a choice is also a burden, as the individual is personally liable for wrong decisions and consequent failures (Beck, 2002; Beck et al., 2002b). Beck stresses that "(a)ctive management (and that does seem the right word) is necessary for the conduct of life in a context of conflicting demands and space of global uncertainty" (Beck, 2002, p. 26). Biographies now have to be managed and become a task and an accomplishment at once. This means that failing at work is often equated with failing as an individual. It also means that individuals are "condemned" to decide and to carry the burden of their decisions. Constructing an appropriate self becomes an ongoing task and Beck highlights how risks have shifted to the individual.

Building on Beck's work but adding more detail on how working relations are changing, Pongratz et al. (2003) argue that the new economic order sees the rise of the entrepreloyee or "Arbeitskraftunternehmer." The work relations of the entrepreloyee are characterized by three elements. First, people are more and more self-responsible, giving rise to new forms of surveillance. The new form of surveillance no longer comes from the management but increasingly from workers themselves. The new mantra of the self-managed workforce is *"(i)t doesn't matter how you manage your job and what you do in detail, the main thing is you achieve at least the goals set"* (Pongratz et al., 2003, p. 244, italics in original). A second element of the entrepreloyee is the "economization" of work. Efficiency is no longer something measured by management; instead individuals have to make sure that their are efficient. Only if they are able to produce an output as efficiently as possible they can successfully sell it on the market. A third element is that the private life is increasingly managed like an organization. This is facilitated by new ICTs, through which the worker can work anywhere. However, this introduces an expectation that workers will work everywhere, so that private life has increasingly to be organized and coordinated with the demands of work (Pongratz et al., 2003; Voß, 1999; Voß & Pongratz, 1998, 2000). This means that individualized, frenetically planned life-styles and biographies are becoming the norm (Voß, 1999).

This marks a shift of power relations form organizations to individuals. However, these new forms of work are seen as more efficient than previous ways of organising work. Now it is the individual who engages in a new form of self-exploitation. Individualization of work stands in this understanding for the market orientation of individual lives. In order to be successful, individual lives have to be geared towards the market and entreployees have to internalize market relations. This involves a subjectification of work (Voß et al., 1998) through which the individual is responsible for their own success (Voß & Pongratz, 2000, p. 239). The new working arrangements with its power relations are internalized and lead to new narratives about a market- and success-oriented self.

The changes previously described are related to changed career models. In recent years, careers have moved away from bureaucratic hierarchies to being flat and flexible. With these changes, the career structure itself has changed from being linear to being more network-like. Bureaucratic careers are built on the assumptions of gradual career progress, climbing a career ladder with small pay rises. This gradual progression is no longer assumed in flat and flexible careers, where work is characterized by project-based orientation, portfolio employability and irregular pay (e.g., Barley & Kunda, 2004; Blair, 2001; Jones, 1996; Pratt, 2000; Ursell, 2000). The new independents seem to call the shots in the new economic formation (Leadbeater & Oakley, 1999). It appears to be the case that the new model for careers is one in which individuals are in charge of building their own careers. Instead of relying on an organization, people are now self-responsible for shaping their careers. It can be expected that these new forms of careers are much more self-entrepreneurial. In these new employment relationships, power has shifted to the individual and this raises the question of how people experience the new employment relations and make sense of the changes in power relations.

High-end ICT workplaces are a particularly good example for self-entrepreneurial careers for the following reasons. ICT work makes use of the new organizational forms, such as team work and project work, which are said to make up the new workplace experience. There are rarely clocking-in machines in ICT firms and the individual workers are responsible for getting results regardless of how long it takes. This is in line with the theory of the entreployee. ICT workers have to self-manage their time and meet often tight deadlines. The general imperative seems to be that ICT workers are flexible and "fancy-free" (Manske, 2003) and can work whenever and however. Secondly, one can expect that ICT workers are among the individualized elites that can author their own lives and feel self-responsible for their decisions. While ICT workers were once highly sought after, global outsourcing and the bursting of the dot.com bubble have increased job insecurity in recent years. ICT companies once offered careers for life but with the economic changes ICT work is now becoming precarious. This potentially creates more exposure to risk, as the reflexive modernization theorists point out. Now workers are supposed to be in charge of shaping their own careers, becoming entrepreneurs of the self.

Although it is commonly assumed that these tendencies can be observed to different degrees in different localities, there is a lack of research that provides in-depth analyses of career perceptions in different contexts. In this paper the focus is on how ICT workers in Switzerland describe their careers. Although Switzerland is not a technology hot spot like Silicon Valley, Switzerland has a reputation for technical excellence through institutions like the world famous Swiss Federal Institute of Technology and also of service excellence. ICT companies are attracted to Switzerland and particularly Zürich because of an highly educated workforce, a high work ethic among employees and also a "business-friendly" employment law which means that employees can be laid off more easily than in other European countries (NZZ, 2004). If the new economy changes the global organization of the economy, then these tendencies must be observable in different locations. This article thus considers the role these self-entrepreneurial tendencies play in contemporary ICT work in Switzerland.

METHODOLOGY

In order to study the extent to which careers in ICT work in Switzerland are self-entrepreneurial and how workers experience these new work relations, the chapter draws on theories and methods in discourse analysis as developed by Potter and Wetherell (Potter & Wetherell, 1987; Wetherell & Potter, 1988). A discourse analysis is well suited to exploring the language people use to make sense of their selves and their careers. Discourse is here understood as a social practice, constructed and constructive of social reality and as such functional, occasioned and rhetorically organized (Gill, 2000; Potter et al., 1987; Tonkiss, 1998). The aim of a discourse analysis is to show which interpretative repertoires and subject positions people use (Billig et al., 1988; Davies & Harré, 1990; Edley, 2001; Potter, 1996; Potter et al., 1987; Wetherell et al., 1988). A discourse analysis is a useful instrument to explore how people make sense of their realities and as such particularly suited to exploring how ICT workers conceptualize their careers.

The material presented in this article originates from an organizational ethnography conducted in two companies in Zürich, Switzerland. The larger company, referred to as Bluetech, is a local subsidiary of a multinational ICT company selling hardware, software, and ICT services; the smaller company, referred to as Redtech, employs around 50 people and develops business software. The research is based on in-depth interviews with and observations of 16 men and 10 women who worked in technical positions in both companies. The interviewees were at the time of the interviews in 2003 and 2004 between 24 and 54 years old. The interviews were conducted in Swiss High German, a dialect of High German, and lasted around 60 minutes. In the interviews I asked questions around the topics work, biography and careers, technology, gender and future plans. The interviews were fully transcribed and the observations noted in field notes. The interviews and field notes were coded into three main areas, which were work, biography, and gender and then sub-coded into various topic areas assisted by qualitative research software[a]. For the transcription a simplified notation system was used to present the talk[b] and the interview extracts are my own translation from Swiss High-German. The names for the companies and the interviewees are pseudonyms and quotes are followed by information in parentheses about which company people worked for and how old they were.

DIVERGENT CAREER MODELS IN ICT WORK IN SWITZERLAND

As outlined in the literature review, careers are changing from being bureaucratic to being more network-like. Although in this study Bluetech careers share more similarities with bureaucratic careers and Redtech careers with flat and flexible careers, they do not comply with these career types in the purest form. After a process of delayering, Bluetech's career ladder steps have been significantly reduced and Redtech employees do not carry the full burden of self-employment and the portfolio employment lifestyle. In the following section, I explore how employees in both companies described their career trajectories.

Increasing Responsibility

At Bluetech ICT workers have the choice between two career models: they can stay in a professional track continuing to work closely with technology or they can move into a managerial function. The two career tracks are organized into different bands and with each higher band leadership and skills demanded increase. Management ensures through annual performance evaluations not only that employees' performance is satisfactory but also that they progress through Bluetech's career system. Most people in this study were in the professional track focusing on technology but a move into the managerial track was possible at any time. Women in particularly reported that there was a strong wish on the part of the management to have more women in managerial and leadership roles. Much of the talk about career development at Bluetech was geared towards accounting for why employees wanted to stay in the technical area.

Tim (54, Bluetech) joined Bluetech after an apprenticeship as electronic mechanic and working for various other companies. He remained at Bluetech in the technical stream and he accounts for this as follows:

Tim: I prefer this (the technical side), that is certain. I'm rather the type who solves and analyses technical problems, rather than leading people or more administrative tasks. I don't like that very much.

Tim points to his personal preference for technical problems as reason for staying in the technical track. A management career denotes for him mainly managing people and administration, areas which he does not want to move into. The pleasure people gain from technology is not only important in selecting an ICT career but also for remaining in it.

If one stays in the technical area, then one gets increasing responsibility with the length of service. Many people talked to me about the different functions they have fulfilled in the course of their ICT careers. An example is Steven (30, Bluetech):

Steven: I started in 199X as systems technician for large frame machines. Er, I did that for two years and then I had the feeling I needed a new challenge and the opportunity came up to enter a project team and to become project leader later on.

After two years at Bluetech, Steven says he looked for a new "challenge." The language of challenge was regularly used and reflects the official career development policy of Bluetech where new challenges feature prominently. Bluetech is proud of giving its employees the chance of self-development. The new challenge Steven took on was to join a project team first as a team member. We see again the notion of rising responsibility slipping in as he soon moved up the hierarchical career ladder by becoming a project leader. There was a strong discourse on this rising responsibility and moving up the career ladder at Bluetech and people construct themselves as improving their career position with each rung of the career ladder they climb.

However, the hierarchical career ladder was not emphasized in all accounts. For instance, Quinta (42, Bluetech) did not talk much about her own career development within ICTs but in a questionnaire distributed before the interview she indicated that she worked first as a systems engineer, then as a product manager and consequently moved into sales. She also mentioned in an informal interview that her recent career change from sales into people manager for ICT workers was motivated by her own life situation as she has two children and a position in sales did not allow her the flexibility she needed to care for them. When talking to me informally she constructed Bluetech as giving her the opportunity to adjust her working hours and times to her own needs, using both her family responsibilities and Bluetech's flexibility to account for her career changes.

It was very interesting to find that remaining flexible was a common theme in many interviews. Taking personal responsibility for remaining flexible was a common theme, used, for instance, by Steven (30, Bluetech) when we were talking about his future and how important learning is:

Steven: What is new today is old tomorrow and that requires that one trains oneself further and further.

He draws on a repertoire that constructs times as fast changing in ICTs and the consequence for him is that he constantly has to learn new things to stay on the ball. It is up to the individual to make sure that one gets the appropriate training. This repertoire suggests that it is no longer a company who wants to have well trained employees but it is oneself who has to find ways to remain up-to-date. Other people talked in parallel ways about how they are able to keep their value on the market through learning new skills and keeping abreast of current developments in the field.

Robert (46, Bluetech) particularly referred to how the new work relationships are not only empowering but also more precarious. He explained that one has to train oneself further to ensure that one always finds a job. If one's own training is outdated he believes it is difficult to find a position. As long as one has trained oneself and has the newest skills it seems much easier to find new work. He conveys the image that stagnation is not tolerated as it exposes oneself to failure. One's own market value, which can be enhanced through learning new things, is one of the central features in the new employment contract. Instead of necessarily relying on a company to provide long-term employment, one always has to be ready to find work elsewhere. This staying up to date is particularly relevant in times when employment security is no longer a given and Robert said that after Bluetech made many people redundant his attitude towards work changed. He now only assumes that he will be employed for at least three months, which is the company's notice period, but he does not expect to be employed for a longer period of time. This clearly shows that a new power regime appears to be operating in ICT work. ICT workers have to rely on themselves to have a career and need to make sure that they are good enough to find work elsewhere. Instead of seeing Bluetech a secure option, people have to come to terms with that their employment may not be forever. It is no longer a company, which ensures a career; rather, the career appears now to be in the hands of individuals with all the pressures and burdens but potentially also options that brings. At the same time individuals face market forces and are hired and fired based on market cycles. They thus have to ensure that they remain employable.

Most people shared the view that Bluetech is not the central component in their career development but they are self-responsible for their careers. Bluetech can function as a facilitator for careers, however, by providing regular career consultations, performance evaluations and ensuring that employees have enough resources to develop their careers further. The idea that careers are chains of increasing responsibility is a strong feature of how people account for their careers at Bluetech. This increasing responsibility can be realized either in a managerial stream or in a technical stream through different positions and titles. It became clear that older forms of more traditional careers continue to exist even though companies are now less hierarchical. However even in these cases it was clearly observable that people see themselves as in charge of their careers. The language of delayed gratification that was previously dominant is now being replaced by a discourse of increasing responsibility. This increasing responsibility is a responsibility for others but also a responsibility for the self. Now individuals have to ensure that they retain their market value and responsibility. This marks an important shift from managing people in ICT organizations by promising them a career for life to managing them by giving them responsibilities for their own careers.

Positioning Oneself

In contrast to Bluetech, Redtech is far less hierarchical and people working at Redtech are either among the large group of programmers or in the small management team. Sometimes programmers can take on leadership roles when they become team leader for a project, but this is only a temporary status as this function rotates. Therefore it is not necessarily seen as an additional hierarchy but it is seen as a way to showcase one's skills in ways which other employers may appreciate. As such it is a way to promote one's skills to other people who may want to "purchase" these skills. In this environment the notion of career takes a new form and is more like positioning oneself as a node in a network. As Redtech careers are different from classic careers a lot of discursive work was done by Redtech employees to justify why they have decided against a more formal career in a large organization and prefer where they are.

This special way of organising at Redtech is expressed by Pascal (33, Redtech) when he talks about the difference between his former employer, a large bank, and Redtech.

Pascal: Er, yes, well, Redtech (is) a relatively small company, you know everyone, er, hierarchies are in a sense not very strong, clearly we have the management and the project leader, which is officially not a seperate hierarchical level, but (she/he) has to make decisions and has to lead the team. That is for me the major difference, the missing hierarchies. (...) That is the big difference. It is all much more direct than at (bank name), one builds strongly on personal responsibilities, people take the competencies[c] they need, they want and well, you just take them in advance (before one does something) and you are not punished for what you have done but that is what is expected.

One of the advantages of Redtech is for Pascal the lack of formal hierarchies. Firstly, Pascal acknowledges that there are differences in the formal hierarchy between programmers and management and he also adds that project leaders have special responsibilities which gives them power over others although this is not a formal hierarchical level. Pascal constructs it as if one has to ask permission in a larger organization, while at Redtech it is expected that one just does things as this competence or authority is assumed. Through this account, Redtech is constructed as a company structured only marginally by hierarchies and in which people are credited with the authority to do whatever they feel is necessary to get the job done.

Although there are few official hierarchies, there are ways of differentiation leading to an informal internal hierarchy. A first element is how long one has worked in ICTs and for Redtech, which also influences the pay level. A second element is being project leader. As Pascal alluded to, project leaders, especially in larger projects, have special authority in that they are able to allocate people to tasks, mediate the contact between the team and the customers and also pay bonuses. Günther (32, Redtech) for instance mentioned that he took on a project leader role as a career move to position himself for future employers.

A third way in which an internal hierarchy was constructed was through becoming technologically recognized. Some people occupy "guru" positions through having the reputation of being excellent with technology. A fourth way to position oneself within the company is by participating in particular projects. Creating technology for a well-known company generates prestige for those who develop that technology. The importance of the projects one worked on can be found on the company website: most employees have personal homepages where they introduce themselves and a large part of these websites is filled with information on which projects and in which functions they have worked. The projects thus are a way to showcase people's skills and experience.

Interestingly, in the interviews most people at Redtech defended against not working for larger companies where they could have had more classic careers, linking this to their own career path in ICTs. A first repertoire regularly cited was the long working hours needed to climb the career ladder in larger companies. Danielle (36, Redtech) answered the following when I asked her about what role climbing a career ladder has for her.

Danielle: I do not look too much at the career ladder. I have seen it and I know how that works. That would simply mean to work from 7 (a.m.) to 11 (p.m.) without break and when I look at these people they have all somehow a destroyed social life. That is not my goal; I always try to have some balance. I could not imagine my self just doing the job either, working for 8 hours and going home. I sometimes take something home, to read something or so. I try to take a middle path. And that is hard to reconcile with a classical career. Well, I need to be happy and I do not want to sell myself.

Danielle denies wanting a classic career and refers to her own experience of working in a large bank. She rejects a classic career because this would mean having no private life and this is not attractive to her. This comment could be construed as Danielle wanting an easy unstressful job. To counteract this portrayal she says that she is not a person who just works for eight hours and then goes home, as she also does reading at home. She sees this "middle path" between working long hours and doing just the required hours as not sitting well with a classic career. What she constructs here is a healthy work-life balance incompatible with a classic career and she therefore prefers for the moment to work in a company where networks rather than hierarchies dominate the working life.

People who decided not to have a classic career in this study justified their decision also with the meetings required by them. At Redtech I found that people disliked meetings and the general view was that people preferred doing things rather than just sitting in meetings talking. This became evident through comments in the team meetings I observed in which the common theme was that meetings only create talk and little action and thus did not get any work done. The general impression voiced at Redtech was that in larger companies regular attendance of meetings was a necessary part of career progression. In contrast, people at

Redtech did not have to attend too many meetings. Many people also voiced discomfort with the internal politics encountered in meetings. Pascal (33, Redtech), for instance, uses office politics as a reason he does not want to have a classic career.

Pascal: I think that one has to do (-) many things, simply due to career (...) and I don't want this.
Elisabeth: What kind of things?
Pascal: To accept political decisions, subordinate myself (-) make decisions which don't make sense and to try to please certain people, simply so that (-) one is considered by them when the next promotions are due. I don't want this.

What emerges from this justification is that a classic career requires certain politics and Pascal does not want to play this game just to make a career. This sounds similar to Danielle's use of "selling oneself" discussed earlier as it would be expected that one subsumes the personality under a fake personality to make a career. Redtech workers commonly used such a strategy to construct themselves as individuals with their own identity that can be distorted in making a career. This uniqueness is almost worshipped at Redtech in that everybody is said to be equal and has the authority to do what he or she thinks is required by the job. This is much more in line with the spirit of working in a start-up atmosphere. I found an expression Esther (33, Redtech) used very insightful in this respect. Esther talks here about the advantages of Redtech.

Esther: I think it is a bit like having set up a business collectively[4]*. We share the risk and we share the profit. (...) One has the feeling of being a part of it and not simply of being a slave, who is kept as cheaply as possible, but you really belong to the company.*

The expression of "collectively setting up a business" is what many people were expressing implicitly in their accounts. Esther feels "a part of" the company. The notion of the "slave" sounds exploitative and also like having no individual identity, like being an anonymous person who labors for the well-being of the company but never gets the fruits of this hard labor. At Redtech the labor one does reflects in company results and employees get bonuses for their performance. Employees referred to how they feel valued, trusted and accepted at Redtech. Having the feeling of freedom, earning well but having not too many risks is what attracts people to Redtech and for most people the decision to work for Redtech was a decision for its working climate and against larger companies and classic careers. The belief in "collective individuality" offered by Redtech is one of the most common features of how people describe their career paths within ICTs and this collective individuality gives them the opportunity to become entrepreneurs of the self. This reflects in that it was not uncommon to talk about one's own skills as a core competence which is management jargon (cf. Prahalad & Hamel, 1994). Usually, core competencies relate to the fact that organizations have to focus on what is unique to them and develop this further, but in this context ICT workers also talked about or alluded to core competencies. This may be because this language is used by human resource professionals but it shows that ICT workers perceive themselves as operating like small companies and their skills become thought about as market like features which they are able to sell.

This leads to what Redtech is doing to encourage these self-management practices. Workers perceived it as an unwritten yet well understood rule that career progression within Redtech is rather an exception than a rule. This is clearly communicated when people are hired but people are nevertheless attracted to Redtech as it provides an environment relatively free of hierarchies and politicking for the sole goal of having a classical career. Redtech is also a company that prides itself of offering interesting jobs through which employees can develop. Employees appreciate that Redtech is providing the tools such as a well stocked library and subscription to the main magazines in the area which facilitates learning as well as the option to gain managerial experience as a project leader without making these things compulsory to what it means to have a successful career.

In the absence of much hierarchy, individuals find ways of differentiating themselves and this analysis has outlined the elements on which people at Redtech draw to do this. Much of the justification for having chosen a rather non-traditional career was linked to the chances and opportunities attached to this career path. Although people had a clear idea about the chances this career has for them, they were also aware that they had to keep their skills up to date to be able to operate like entrepreneurs of the self. This was indeed one of the similarities between Redtech and Bluetech. The careers at Redtech and Bluetech appear as diametrically opposed in that at Bluetech a system of increasing responsibility was used, which meant that with each career ladder step one received more responsibilities. At Redtech the advantages of working for Redtech in contrast to working for a larger company were singled out to account for employees' own career choices. Thus it became clear how diverse career paths within ICTs could be and this adds to the understanding of the shapes careers take in the new economy. However, in both organizations it was common to refer to one's own skills and the importance of keeping one's own skills up to date to make sure that people retain their market value. People thought about themselves as small companies, which had to manage the labor resources efficiently to stay in business. This is what it means to have a self-entrepreneurial career.

DISCUSSION

The previous discussion has shown that there are distinct differences in how careers at Redtech and Bluetech are managed. While Redtech has done away with hierarchies, at least to a large extent, rising through the ranks with more responsibility is the important characteristic of careers at Bluetech. Management structures support these career patterns but what is surprising is in how many areas the discussions converge and little difference is detectable. In both companies elements like loyalty to a company and a job for life were seen as outdated concepts. Instead what companies tried to offer is in line with what Kanter (1995) describes as making employers attractive in these times of change: offering interesting and challenging jobs and the opportunity for people to develop. All that companies do to enhance this new form of labor is to provide the resources and tools for people to become an entrepreneur of the self (Chiapello & Fairclough, 2002), such as the opportunity to learn and interesting projects to enhance once employability. In both companies, people had developed mechanisms to deal with the changing nature of the workplace. The responses were strikingly similar: to become an entrepreneur of the self and to have a self-entrepreneurial career. People presented themselves as self-responsible for their careers and the concepts of career developed were thus very individualistic.

Even though the companies were very different and various ways of constructing self-entrepreneurial careers were possible, many of the discourses were rather similar in both companies. Both companies studied developed different ways to embrace self-entrepreneurial careers. At the same time, the discourses people developed were rather similar in many respects. One of these elements is certainly the high level of individualism that shines through these accounts. It is no longer the company or management that is seen as responsible for a career but individuals are. Any notions of collectivity or that many people may have similar experiences were generally ignored and it was rather presented as if careers were made by individuals themselves. This individualism was seen as a normal part of ICT work and this is just how career happens. That this creates new burdens for the individual was acknowledged but not seen as a problem of power and influence but just as how business operates. The discourse of individualism and self-entrepreneurial career thus crowds any other understandings out.

The material originates from discussions with and observations of ICT workers in Switzerland and one can expect that the discourses these people mobilized are very specific. However, in looking at other literature on changes in career patterns as well as ICT work, Switzerland does not appear to be that special. The pressures to create a life and a career in times when career are eroding that for instance Beck (2000) theorizes are very visible in the accounts ICT workers produced. ICT workers presented their careers as individualistic endeavours for which they rather than a company are ultimately self-responsible. The marketization of career concepts that others have referred to (Pongratz et al., 2003) was also an issue that became very visible during the fieldwork in the ICT sector in Switzerland. People constructed their careers as being the result of an individualistic effort rather than of a company and the language they used reflects a strong market orientation.

The changing nature of careers in technology-related work has mainly been observed in various other parts of the world. The precariousness of ICT work in Ireland and its consequences for career and identity construction have been shown (Ó'Riain, 2000, 2002). Gill (2002) highlights the persistence of gender inequality in project-based and meritocratic new media work in the United Kingdom. The flexible working conditions under which webworkers in Germany craft a career are well documented (Manske, 2003). The impact on individuals of flexiblization in careers in IT and new media work in the United States of America has also been explored in recent research (Barley et al., 2004; Batt, Christopherson, Rightor, & Van Jaarsveld, 2001). Although the circumstances in these countries are different from the Swiss context, there seem to be many commonalities in how careers in technology are constructed. ICT work in Switzerland is equally under pressure as jobs for life can no longer be taken for granted, and a meritocratic job market leads people to see themselves as responsible for their success. ICT workers in Switzerland thus have to be flexible in how they approach their careers, as the sector is increasingly becoming individualized and self-entrepreneurial. This chapter adds a Swiss perspective to previous studies.

One could argue that the new work relations embodied here through self-entrepreneurial careers are freeing individuals from the Weberian "iron cage" of bureaucracies. In the account collected from ICT workers in Switzerland it became clear that many people saw these new work relations providing them with new freedoms to self-manage their career. No longer having to rely on a company to have a career was widely seen as liberating. The downside of these self-entrepreneurial careers is that insecurity and risk also shifted to workers. It became evident in the interviews that people could no longer expect a stable and long-term work relationship but rather assumed that their employment was on a short term basis and that they would have to move on if their current employer no longer needed their services. In this light it was essential to keep their own knowledge up to date (Kotamraju, 2002). These new perceptions of risk, insecurity and the death of a long term perspective is not only an element of ICT work but can also be observed in other areas of work (Beck, 2000; Blair, 2001; Sennett, 1998, 2000, 2006; Ursell, 2000; Wittel, 2001). It shows how the new power relations in self-entrepreneurial work flow and that these new work relationships are not only empowering but also disenfranchising.

CONCLUSION

Changes in management practices of careers in the new economy have been at the center of this chapter. The way careers are changing was characterized by careers becoming more self-entrepreneurial. This means that individuals feel increasingly in charge of shaping their careers and organising them as if people were small firms. The article reviewed how social theorists like Beck and Pongratz and Voss made sense of the changes in the workplace and it was argued that ICT work in Switzerland is a particularly good example for studying self-entrepreneurial careers. The article has shown that two different ideas dominated how careers were conceptualized in the two ICT companies studied. While the idea of increasing responsibility was a strong feature of people's talk at Bluetech, this increasing responsibility was seen as self-responsibility. Individuals have to ensure that they maintain market value to remain employable. At Redtech, in absence of much hierarchy, individuals carved out certain areas of skills they used to market themselves. These skills made up a portfolio to be used when seeking new jobs. It became evident that many people at Redtech saw themselves as operating like small companies.

Although formal career patterns at Bluetech and Redtech were very different, it was staggering how similar the talk about keeping up skills and seeing themselves as responsible for the own career was. People at Redtech and Bluetech did not rely on a company to give them the skills they needed but saw it as their own responsibility to keep skills up to date. Their own skills were seen as the single most important shaping factor of their career and these skills were their own responsibility. Thereby people conceptualized themselves like small companies or brands and their self interest was to keep these small companies and brands up to date as otherwise they may have to go out of business. The discourses people in ICT work in Switzerland adopted reflect very clearly the Me & Co mentality Beck (2000) talks about as well as the entreployee mentality Pongratz et al. (2003) theorize. The findings are also in line with other research on the ICT industry in other areas of the world. The management practices in the two companies studied reflect this mentality that the workers are in charge of managing their careers and people adopted these assumptions when talking about their careers, but the article also points to that these changes may not free individuals and may create new burdens. This shows that the tendency to see careers as a self-entrepreneurial endeavour is strong in contemporary ICT work and shapes management practices in the ICT sector.

ACKNOWLEDGMENT

I would like to thank Rosalind Gill and the anonymous reviewer for their comments on earlier versions of this chapter. Thanks also to Rachel Dunkley Jones for proofreading the text. I would like to acknowledge the support I received from the Graduate College "Knowledge Society and Gender Relations" at the University of Zurich and the German Educational Exchange Service (DAAD) to conduct the research on which the chapter is based.

REFERENCES

Arthur, M. B., & Rousseau, D. M. (1996). Introduction: The boundaryless career as a new employment principle. In M. B. Arthur & D. M. Rousseau (Eds.), *The boundaryless career: A new employment principle for a new organizational era* (pp. 3-20). Oxford: Oxford University Press.

Barley, S. R., & Kunda, G. (2004). *Gurus, hired guns, and warm bodies: Itinerant experts in a knowledge economy.* Princeton, NJ: Princeton University Press.

Batt, R., Christopherson, S., Rightor, N., & Van Jaarsveld, D. (2001). *Networking. Work patterns and workforce policies for the new media industries.* Washington, DC: Economic Policy Institute.

Beck, U. (2002). A life of one's own in a runaway world. In U. Beck & E. Beck-Gernsheim (Eds.), *Individualization: Institutionalized individualism and its social and political consequences* (pp. 22-29). London: Sage.

Beck, U. (2000). *The brave new world of work* (P. Camiller, Trans.). Cambridge: Polity.

Beck, U., & Beck-Gernsheim, E. (2002a). Authors' preface. In U. Beck & E. Beck-Gernsheim (Eds.), *Individualization: Institutionalized individualism and its social and political consequences* (pp. xx-xxv). London: Sage.

Beck, U., & Beck-Gernsheim, E. (2002b). Losing the traditional. In U. Beck & E. Beck-Gernsheim (Eds.), *Individualization: Institutionalized individualism and its social and political consequences* (pp. 1-21). London: Sage.

Billig, M., Condor, S., Edwards, D., Gane, M., Middleton, D., & Radley, A. (1988). *Ideological dilemmas: A social psychology of everyday thinking*. London: Sage.

Blair, H. (2001). You're only as good as your last job: The labor process and the market in the British film industry. *Work, Employment and Society, 15*(1), 149-169.

Castells, M. (2000). Materials for an exploratory theory of the network society. *The British Journal of Sociology, 51*(1), 5-24.

Castells, M. (1996). *The rise of the network society*. Cambridge, MA: Blackwell.

Chiapello, E., & Fairclough, N. L. (2002). Understanding the new management ideology: A transdisciplinary contribution from critical discourse analysis and new sociology of capitalism. *Discourse and Society, 13*(2), 185-208.

Davies, B., & Harré, R. (1990). Positioning: The discursive production of Selves. *Journal of Theory of Social Behaviour, 20*, 43-65.

Du Gay, P. (1996). *Consumption and identity at work*. London: Sage.

Edley, N. (2001). Analysing masculinity: Interpretative repertoires, ideological dilemmas and subject positions. In M. Wetherell, S. Taylor, & S. Yates (Eds.), *Discourse as data: A guide for analysis* (pp. 189-228). London: Sage.

Elgin, B. (2005). Revenge of the nerds: Again. *Business Week, July 28, 2005.*

Gill, R. (2000). Discourse analysis. In M. Bauer & G. Gaskell (Eds.), *Qualitative researching with text, image and sound: A practical handbook for social research* (pp. 172-190). London: Sage.

Gill, R. (2002). Cool, creative, and egalitarian? Exploring Gender in Project-based New Media Work in Europe. *Information, Communication, and Society, 5*(1), 70-89.

Handy, C. B. (1994). *The empty raincoat: Making sense of the future*. London: Hutchinson/Arrow.

Heritage, J. (1984). *Garfinkel and ethnomethodology*. Cambridge: Polity Press.

Jones, C. (1996). Career in project networks: The case of the film industry. In M. B. Arthur & D. M. Rousseau (Eds.), *The boundaryless career: A new employment principle for a new organizational era* (pp. 58-75). Oxford: Oxford University Press.

Kanter, R. M. (1995). *World class: Thriving locally in the global economy*. London: Simon & Schuster.

Kotamraju, N. P. (2002). Keeping up: Web design skill and the reinvented worker. *Information, Communication, & Society, 5*(1), 1-26.

Leadbeater, C., & Oakley, K. (1999). *The independents: Britain's new cultural entrepreneurs*. London: Demos.

MacKenzie, D., & Wajcman, J. (1999). *The social shaping of technology*. Buckingham: Open University Press.

Manske, A. (2003). Arbeits- und Lebensarrangements in der Multimediabranche unter Vermarktlichungsdruck - Rationalisierungspotenzial für den Markterfolg. In E. Kuhlmann & S. Betzelt (Eds.), *Geschlechterverhältnisse im Dienstleistungssektor - Dynamiken, Differenzierungen und neue Horizonte* (pp. 133-146). Baden-Baden: Nomos.

NZZ. (2004). Auftrieb für Zürich - Positive Reaktionen auf den Standortentscheid von Google. *Neue Zuericher Zeitung, 29 January 2004.*

Ó'Riain, S. (2000). Net-working for a living: Irish software developers in the global workplace. In M. Burawoy, J. A. Blum, S. George, Z. Gille, T. Gowan, L. Haney, M. Klawitter, S. H. Lopez, S. Ó'Riain, & M. Thayer (Eds.), *Global ethnography: Forces, connections, and imaginations in a postmodern world.* Berkeley: University of California Press.

Ó'Riain, S. (2002). High-tech communities: Better work or just more work? *Contexts, Fall/Winter 2002.*

Peters, T. (2001). *The brand you 50.* New York: Alfred A. Knopf.

Pinchot, G. (1985). *Intrapreneuring: Why you don't have to leave the corporation to become an entrepreneur.* New York: Harper Collins.

Pink, D. H. (2001). *Free agent nation: How America's new independent workers are transforming the way we live.* New York: Warner Books.

Pongratz, H. J., & Voß, G. G. (2003). From employee to "entreployee": Towards a "self-entrepreneurial" work force? *Concepts and Transformation, 8*(3), 239-254.

Potter, J. (1996). Discourse analysis and constructionist approaches: Theoretical background. In J. T. E. Richardson (Ed.), *Handbook of qualitative research methods for psychology and the social sciences* (pp. 125-140). Leicester: BPS.

Potter, J., & Wetherell, M. (1987). *Discourse and social psychology: Beyond attitudes and behaviour.* London: Sage.

Prahalad, C. K., & Hamel, G. (1994). *Competing for the future.* Boston: Harvard Business School Press.

Pratt, A. C. (2000). New media, the new economy, and new spaces. *Geoforum, 31*, 425-436.

Sennett, R. (1998). *The corrosion of character: The personal consequences of work in the new capitalism.* New York: Norton.

Sennett, R. (2000). The new political economy and its culture. *The Hedgehog Review, Spring* (2000), (Online).

Sennett, R. (2006). *The culture of the new capitalism.* London: Yale University Press.

Tonkiss, F. (1998). Analysing discourse. In C. Seale (Ed.), *Researching society and culture* (pp. 245-260). London: Sage.

Ursell, G. (2000). Television production: Issues of exploitation, commodification, and subjectivity in UK television labor markets. *Media, Culture, & Society 2000, 22*(6), 805-825.

Voß, G. G. (1999). Menschen als Unternehmer Ihrer Selbst. Der Arbeitskraftunternehmer - Ein neuer Typus von Arbeitskraft und seine Konsequenzen. *Leben im Transit. Neue Arbeitsgesellschaft und die Suche nach dem Beruf, Evangelische Akademie Tutzing, 15th-17th January, 1999.*

Voß, G. G., & Pongratz, H. J. (1998). Der Arbeitskraftunternehmer: Eine neue Grundform der Ware Arbeit? *Kölner Zeitschrift für Soziologie und Sozialpsychologie, 30*(1), 131-158.

Voß, G. G., & Pongratz, H. J. (2000). Vom Arbeitnehmer zum Arbeitskraftunternehmer - Zur Entgrenzung der Ware Arbeit. In H. Minssen (Ed.), *Begrenzte Entgrenzungen: Wandlungen von Organization und Arbeit* (pp. 225-247). Berlin: Edition Sigma.

Wetherell, M., & Potter, J. (1988). Discourse analysis and the identification of interpretative repertoires. In C. Antaki (Ed.), *Analysing everyday explanation: A case book of methods* (pp. 168-183). London: Sage.

Wittel, A. (2001). Toward a network sociality. *Theory, Culture, & Society, 18*(6), 51-76.

ENDNOTES

[a] TAMS Analyzer by Matthew Weinstein.

[b] A simplified version of the Jefferson transcription system (cf. Heritage, 1984) was used: (-) indicates a short notable pause, (0.9) an exactly timed longer pause (here 9 seconds), (inaud) is inaudible, (text) indicates a transcriber clarification on unclear parts of tape or additional clarification such as gender of German word, (...) material deliberately omitted and HEHE stands for laughter.

[c] Competencies refer in German very strongly to the formal right and of authority to do something.

[d] Esther used the expression 'sich kollektiv selbstständig machen', which means to set up a business and to become an entrepreneur not individually but with a group of people.

Chapter XVII
Reflections on Organizing and Managing in Self-Managed Knowledge-Work Teams: A Constructionist Turn

James J. Keenan
Fairfield University, USA

ABSTRACT

This chapter employs social construction, communication, and leadership perspectives as framing for a retrospective analysis of the construction and management of two highly syntalic, cohesive self-managed knowledge-work teams in a service organization operating across culturo-graphic boundaries. The retrospection focuses on the social-narrative construction and development of the self-managed teams and teamwork in the context of fast changing conditions in the business environment. Grounded in the constructionist epistemology, the self-managed teamwork is re-examined from an updated symbolic convergence perspective. The bona fide teams' stages of development and progress toward convergence and coalition are described. The high cohesiveness and syntality of the teams are re-examined as mindful and heedful interrelating in the light of constructionist theory. The chapter posits communication, especially talk, as critical to constructing and organizing the sets of interacts, roles, and behaviors that are involved in self-managing and self-leading teams.

INTRODUCTION

This chapter is a reflection from the constructionist perspective on two groups of experienced knowledge-work managers who worked together as *bona fide* teams that were self-managed to accomplish personal goals, which they created in common and toward which they had never before conspired. The two teams were successful in self-managing their tasks and relationships and their teamwork was marked with high cohesiveness and syntality.

It is not easy to be succinct about teams in large part because there are so many variations of small, purposive, non-hierarchical human work groups. To cite a few of the many types of groups, there are formal and informal, short-term and long-term groups, social and work, support, small and large, face-to-face, and virtual. There are project teams with the ability to organize their own work and there are the prefabricated rigidly structured service teams apparent in the fast-food establishments dotting byways and business landscapes. Teamwork varies among industries, firms, locations, and levels of managing and leading. Teamwork varies with the characteristics and transactions among actors and actants: people, technologies, ambient physical conditions, organizations, nations, and cultures.

One variant of work-team form and process is the self-managed team, which has seen increasing use by large organizations (e.g., Kirkman, Gibson, & Shapiro, 2001; Kirkman & Rosen, 1999; Kirkman & Shapiro, 1997, 2001; Lawler, Mohrmann, & Ledford, 1995; Wellins, et al. 1990). In addition to charting the spreading use of self-managed work teams, research focused on the characteristics of such teams and their advantages and disadvantages has been growing (e.g., Chaston, 1998; Gibson & Tesone 2001; Kirkman et al., 1999; Moravec, Johannessen, & Hjelmas, 1997; Stewart, & Manz, 1995; Wegner, Erber, & Raymond, 1991; Zarraga & Bonache, 2005).

Conceptualizing and performing self-managed teaming and working are particularly salient needs in the context of fast-changing organizational forms and work technologies. Acceleration in the scale and scope of such changes earmarks much contemporary human work and is centrally a process of increasing, spreading, and speeding social interaction among multiple constituencies including different groups of producers, investors, consumers, and regulators (Bijker, 1995). Complex human performances and organizations constitute, and are constituted by, technological change. Self-managed teams are such constituent processes (Murray & Moses, 2005). At the same time, technologies are constituents of teams.

In the abstract, all teamwork involves communicatively constituted task-work and relationship-work. This chapter employs a constructionist perspective in reflecting on the communicative *constitution* and the multidirectional, multidimensional *managing* and *languaging* of two self-managed teams. Self-managing teams are defined here "as small numbers of people with complementary skills who are committed to a common purpose, have a defined set of performance goals, and execute an approach for which they hold themselves accountable" (Katzenbach & Smith, 1993). Self-managing teams share leadership, decide working procedures, and schedule and conduct their own meetings. Self-managing teams develop relatively unique attributes such as group culture, image, vision, cognitions, emotions, syntality, synergy, and cohesiveness (Harris & Sherblom, 2005, p. 157). Self-managed teams are co-emergent with the on-going cognitive and emotional experiences and the behavioral routines in individual worklives. Some such experiences become transformed and institutionalized in organizational action taking (Crossen, Lane, & White, 1999; DiMaggio & Powell, 1983). Reframed in the constructionist epistemology, teams are viewed here as systems of interlocked sensemaking behaviors (e.g., Krippendorf, 1971). Teams are sets of interacts that are assembled into processes that constitute organizations (Weick, 1979). Teams are socio-cognitive processes that emerge in communicating and through the agency of language.

PURPOSE

The purpose of this chapter is to reflect on and conceptualize from constructionist communication perspectives the emergence and characteristics of two self-managed knowledge-work teams who completed graduate-level academic researches successfully. The two self-managed teams were *bona fide* work teams whose members were also full-time managers in a service organization operating across culturo-graphic boundaries in the context of fast changing conditions in technology and the organization's business environment.

The chapter is developed in three sections. The first section contains an Introduction, a description of the intellectual approach to the chapter and a statement of the chapter's Purpose. The second section describes the self-managing *bona fide* teams and the data upon which the chapter's reflections are based. The third section presents, from a communication-constructionist perspective, retrospections, and interpretations concerning six aspects of the self-managing teams. The aspects include: team construction, convergence and communion, collective mind, self-managing roles and behaviors, talk technology, and cohesiveness as heedful technology. The third section also contains conclusions concerning self-managing knowledge-work teams when viewed from perspectives based in communication-constructionist epistemology

INTELLECTUAL APPROACH

This chapter employs social construction, communication, and leadership perspectives as framing for a retrospective analysis of the communicative construction and management of two highly syntalic and cohesive self-managed teams in a service organization operating across culturo-graphic boundaries. The analysis focuses on the social-narrative construction and development of individual and group identities and the team's work in the context of fast changing conditions in the business environment. Grounded in the constructionist epistemology, the data are interpreted from a symbolic convergence perspective that seeks to understand the coalescence of the participants' motivations and intellectual resources in strategic relations and activities. Descriptions of the *bona fide* team's stages of development and progress toward convergence and coalition are charted and the key characteristics of the team as a highly syntalic convergent group are described. This chapter posits communication, especially language, as critical to discursively constructing and organizing the sets of interacts that constitute the self-managing, self-leading processes, and the on-going organizing of teams. Talk is viewed as the essential technology for team managing and leading. The chapter conceptualizes multidirectional and multidimensional talk in the emergence and practice of self-managing and self-leading teamwork.

THE *BONA FIDE* TEAMS AND THEIR WORK

Two groups of adult full-time managers in an international technology-enabled service organization operated as successful self-managed teams. For all of the managers in the two teams their teamwork was in addition to and contemporaneous with their full-time individual company managerial work assignments in organizing and providing cross-cultural, high technology worldwide relocation services. Such "company work" involved the team members in highly pressured, scripted relations with local internal company employees as well as with distant, virtual employees and contractors.

Team Members: The "teamwork," which is the focus of this chapter, was carried out by two groups of managers who elected to self-manage as two teams for four months. The team memberships did not overlap and the two teams worked contemporaneously toward the same objective: the completion of the members' individual graduate thesis reports. Membership in either of the teams was voluntary and did not allow interchange or duplicate membership. The purpose of each team was to help its members mobilize, develop, and complete their individual final research projects required for a graduate degree in organizational communication. One team included 12 females and 5 males between the ages 34 to 43. The average age of members in this team was 36 years. The members of this team were managers with an average tenure of 10 years. A second team included 12 females and 4 males between the ages 29 to 44 years. The average age of the members in the second team was 39 years. The managers in this team had an average tenure of 12 years with the company.

Teamwork: Each team was a *bona fide* cohort of individuals matriculating in and completing the work requirements of a *bona fide* company-sponsored advanced degree program of learning. The cohorts were stable groups of managers who had taken all their academic coursework together during two years. The graduate degree-granting program and the managers' company required the conduct of advanced research and the submission of a graduate-level research thesis. At the time of this study's focal performance, that is, the cohorts' acting as self-managed teams, cohort members were starting to work individually and to concentrate their efforts on individually carrying out and completing their Master's research and Thesis. In their organizing into two self-managing teams, the cohort members saw a way to help each other work through the tasks involved in writing about their individual research activities. Each cohort opted for self-managing its teamwork without outside managing. Acting as self-managing teams was proposed by the managers and not by the company or the university. The teamwork proposal was, however, approved with encouragement to be "self-directed learning groups." Each cohort formed, organized, and carried out all details of its functioning as self-managed and was not managed from outside the team. All members of the two cohorts participated in one of the self-managed teams. All but one team member developed and completed on-time approved thesis research projects.

The members of both teams considered their teams and teamwork highly successful and many have maintained personal relations with other team members since the completion of their teamwork. Judging from outside the group, the two advisors also considered the teamwork as highly successful. In particular, they were impressed with the high degree of diligence, cohesiveness, and syntality among the members of both self-managed teams.

The retrospective reported here was carried out approximately one year after the work of the two self-managed teams was completed. Figure 1 recapitulates what had happened in the earlier situation. The retrospective is focused on gaining additional insights into the teams' characteristics and performance in the light of communicational-constructionist and sensemaking perspectives.

The original empirical data were collected through non-participant observation by two faculty research advisors members and brief individual status progress reports provided every two weeks by all members of both teams. After all of their personal research products were completed and approved, all team members reported on their individual and team work in personal, individual interviews conducted by one of the two faculty research advisors. Interviews were semistructured, narrative and inquisitive about topics such as: which team members did what at the beginning and at later stages in the member's team; leadership behaviors (types, by whom, shared) during the early and later life stages of the team; number, topics and locations of meetings; communication behaviors of team members over the course of the team's work; emotional ups and downs of "storming, norming, and conflicting"; crisis points including midlife crises in the group's usefulness to individuals; and ratings by the interview respondents of the effectiveness of the team for its members.

Figure 1. The teamwork situation: Constituent common tasks in self-managed teamwork

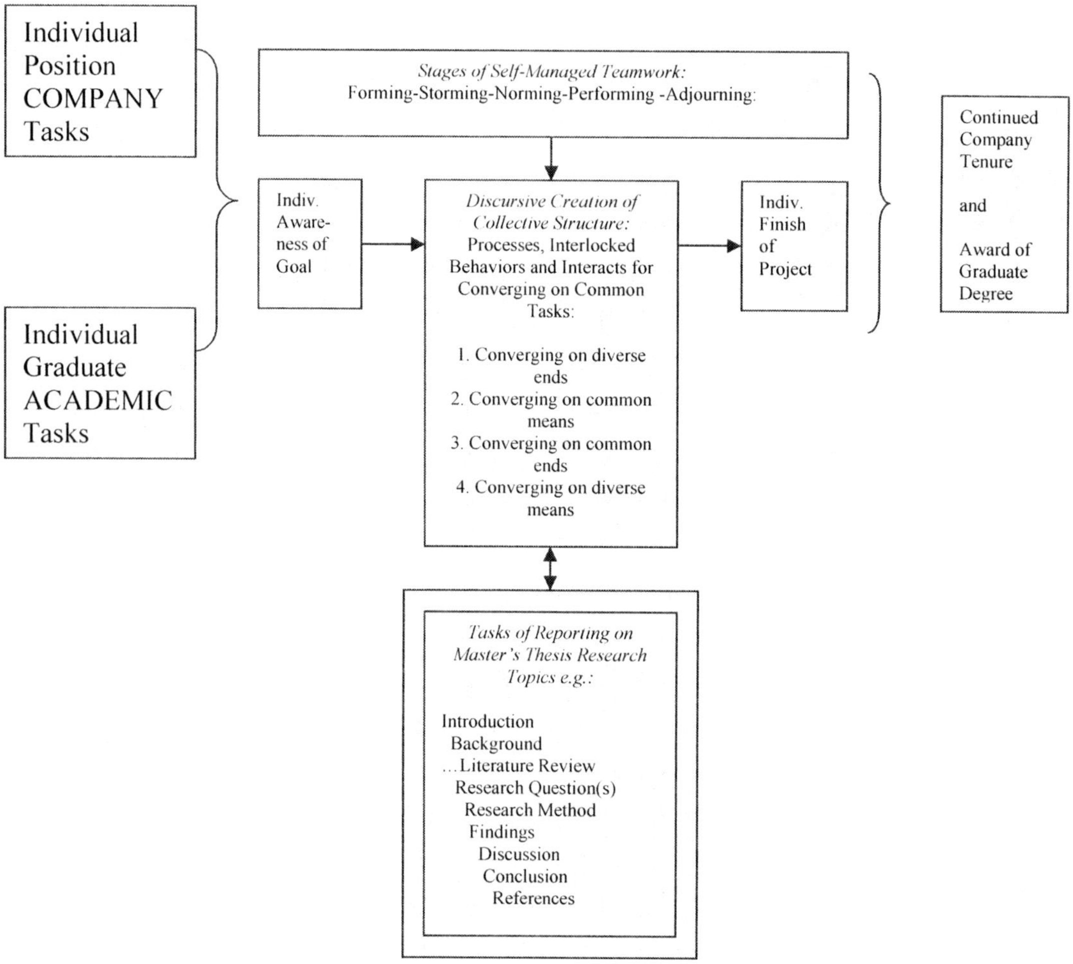

REFLECTIONS: CONSTRUCTIONIST PERSPECTIVES ON SELF-MANAGED KNOWLEDGE-WORK TEAMS

The apparent successes of two self-managed knowledge-work teams and, in particular, their cohesiveness, and syntality, was the impetus for a retrospective exploration of the self-managed teams' common reality construction and processes. Several research questions were prompted by the retrospective review. Explanations principally from the constructionist perspective and from communication and management studies were found to be salient. The questions included: How can self-managed teams be defined from a communication perspective? How are self-managed knowledge-teams communicationally constituted? How do such teams develop? How do self-managed teams sensemake, coalesce, and converge? How are self-managed teams managed? What roles and skills are involved in self-managing teamwork? What are the features of the talk-technology essential to self-managing? This retrospective is an attempt to capture constructionist theory and perspectives in response to such questions.

Reflecting on Team Construction

Faced with the necessity of planning, carrying out and delivering an approvable product/outcome (graduate-level thesis document), and encouraged by the course faculty at the outset of their development as teams, to act as self-managed learning groups, the manager-students were facing what they said was a formidable task. They decided to form two teams of learners who could assist each other. Organizing, they said, was a way of moving forward on the important work ahead of them. Organizing offered ways of benefiting each other and the possibility of continuing such benefits even though their projects were individual and different.

The self-managed teams started their life cycles and began developing through stages, which Tuckman and Jensen (1977) have labeled: forming, storming, norming, performing, and adjourning. In the constructionist perspective, with the managers' preliminary convergence on interest and means, they had started a four-step process of group development. Teaming had begun with "convergence on means." Having first converged on shared ideas of how a structure can form (i.e., on means), the managers then activated a repetitive series of interlocked behaviors to form a collective structure (Allport, 1962, p. 17; Weick (1979, p. 90). Weick (1979, p. 88) points out that collective structure is "accomplished by processes. Processes contain individual behaviors that are interlocked among two or more people." Their interlockings are interacts that are assembled into processes that constitute the team's collective organizational structure (Allport, 1962, p. 17; Weick, 1979, p. 89). Working together in teams offers the team members the possibility of repetitively helping one another, satisfying needs, and self-expression (Brockett, 1975; Weick, 1979, p. 90). With the recognition that each had something to contribute, even if individuals do not agree on goals or ends, team members were converging on issues of means rather than on issues of ends. The formation of collective structure had started with the prioritizing of means convergence (Allport, 1962, p. 17). Subsequent stages would follow (see Figure 2) and proceed through diverse ends to common means and common ends. Concerning the reversal of the familiar postulation that diverse ends are followed by common means, Weick (1979, p. 92) and constructionist epistemology assert that "perhaps the most important consequence...is that it preserves the constructionist crucial point that people *create* structure." Members of the new teams were discursively creating collective structure and eventually common means and common ends.

Reflecting on Convergence and Communion

The high cohesiveness and, in particular, the apparent convergence of intellectual and emotional awarenesses in each team reflect and argue for the constitutive character of communication. Symbolic convergence theory (Bormann, 1985) is an attempt to analyze the communicative constitution of collective action and thus is useful in these retrospections about the teams. The theory offers insights about team members' sharing symbolic reality through discursive processes in which collective consciousness emerges. In the SCT perspective, the teams can be understood as coalitions that can be variously characterized, for example as: including individuals who communicated with one another about issues and actions; as deliberatively constructed groups; as groups

Figure 2. Allport's model of team development (Weick 1979, p. 91)

outside the formal structure of the (company) organization, as lacking formal internal structure; as groups in which members know who is and is not a member; and as coalescing around common objectives (Stevenson, Pearce, & Porter, 1995).

SCT posits that group members dramatize reality and share both fantasy and logic and that the sharing of consciousness and group awareness begins with a few interacting individuals dramatizing the past and setting the stage for others to take up the drama and share it with still others. Collective consciousness emerges discursively and when dramas have strongly shared meaning, generality, and time-space depth (Olufowote, 2006, p. 459). Shared group consciousness includes its constitutive emotions, motives, and meanings. Applied to reflecting on the teams, SCT underscores narrativity as a central communication tool in socially constructing group awareness. (The narrativity constituting the teams can be understood additionally through the lenses of other behavioral research traditions (e.g., rhetoric, semiotics, psychology, sociology, cultural anthropology, information science, and critical studies).

Olufowote (2006, p. 464-466) sees important links between symbolic convergence theory and Weick's model of sensemaking. Intrinsic to both perspectives is concern for actor's creating reality, structuring the unknown, managing equivocality and uncertainty, and organizing. "Uncertainty triggers sensemaking when actors are faced with unpredictable, unanticipated, or unwanted futures" (Weick, 1995). Weick's model sees dramatizing as "essentially a sensemaking activity for understanding and artistically organizing experience (equivocal past, uncertain future). Actors, such as the teams, shared narratives (stories, conversations and other verbal and non-verbal constructions) and proselytized fantasy to reach a collective understanding and organization of lived experience." (Olufowote, 2006, p. 465).

Particularly relevant to understanding the cohesiveness and syntality in the teams is Bormann's explanation that "symbolic convergence creates, maintains, and allows people to achieve empathetic *communion* (emphasis mine) as well as meeting of the minds" (Bormann, 1983, p. 102).

Reflecting on Collective Mind

Another retrospective on the teams, and, in particular, on the high cohesiveness and syntality of the teams' members suggests that each team functioned as what might be called "collective mind" (Ryle, 1949) and as "heedful interrelating" (Weick, 2001). The supposition is examined in a brief reflection here.

Research on the construct of group or collective mind is relatively little and mostly concerned with shared individual mental processes and products, essentially focusing on cognitive interdependence and memory processes. Sandelands & Stablein (1987, p. 139-141) argue that organizations can think and are mental entities. Hutchins (1991) underscored organizations as distributed information processing systems. Connectionist studies of organizations and groups have focused on neural-like networks (Quinlan, 1991, p. 41); neuron-like units (Churchland, 1992, p. 32; and brain-style processing (Rumelhart, 1992, p. 69).

Weick (2001, p. 264), following the social constructionist perspective and avoiding phrases such as group mind in favor of "collective mind," asserts that collective mind inheres in the pattern of interrelated social activities among many people. "Collective," unlike "group" or "organization," refers to individuals who act like they are a group. "People act heedfully when they act more or less carefully, critically, consistently, purposefully, attentively, studiously, vigilantly, conscientiously, pertinaciously--qualities of mind attached as adverbs to the performance. When heed declines, performances are said to be heedless, careless, unmindful, thoughtless, unconcerned, and indifferent. Heedful interrelating is collective mind (Weick, 2001, p. 264).

Collective mind is embodied in the interrelating of social activities. Collective mind is developed more or less fully depending on the amount of heedfulness with which the interrelating is done. What accounts for variations in heed (Weick, 2001, p. 268)? Weick finds the answer in Mead's insight that mind is "the importation of social process" (Mead, 1934, p. 186).

In a way that provides additional insight into the cohesiveness and other characteristics of the two teams in this retrospective, Weick (2001, p. 266, 267) describes collective mind further. Dispositions toward heed are expressed in actions that constitute interrelating. "A heedful contribution enacts collective mind. The collective mind is located in the process of interrelating."

Reflecting on Self-*Managing* Roles and Behaviors

Reflecting on the teams as processes of interlocked social interacts provided support for and suggests the salience of an array of research perspectives and findings concerning managing and leading. While this retrospective is not focused on managing and leading *per se*, the observer cannot easily ignore the relevance of theory and practice on these topics.

Members of the two teams took responsibility for the work processes, that is, the interlocked behaviors of the collective structure. Together, the members shared in the management and leadership of the teams. As members of the collective structure, they shared space, time, and energy, but they did not necessarily, and perhaps only partially, if at all, share visions, aspirations, intentions or all managerial roles and behaviors (Weick, 1979, p. 91). In research specifically on virtual teams, Carte, Chidambaram, and Becker (2006, p. 323) found that members of high performing self-managed teams shared responsibility for leading more than did members of lower performing teams. They concentrated their leadership behavior on performance and production and shared leadership behaviors focused on housekeeping and keeping track of group work. The researchers also found that the teams' behaviors emerged strongly during the first half of the groups' life span, stayed throughout the life of the groups, and steadily dissipated over time.

It should be noted that all of the original members of the two teams stayed in their teams and participated in the interlocked behaviors that formed the collective structure. The constructionist postulations concerning interlocked behaviors in the creation of teams and other collective structures implicitly allow for the possibility that individuals will not invest all their behavior in one team or consistently in one managerial behavior. Over the life of the teams' collective structures, individual members may invest relational behavior only partially or sporadically in a team or person, or task or relationship. Any member may, of course, discontinue being involved in the team's interacts, partially or wholly, before the common goal is achieved.

In this constructionist retrospective of the characteristics of the self-managed teamwork, the members of each team can be viewed as involved in complicated mixes of roles and behaviors that are used to influence the regular patterns of interlocked behaviors in the teams' collective structures. These interlockings are essentially communicational. The realities of team members and their teamwork are discursively constructed. In this epistemologic, managing, and leading are discursive and constructive processes. Discursively constructive processes are constituents of managing and leading.

Recent research confirms that members of self-managed work teams take responsibility for the quality of the work processes and products as well as share in the management and/or leadership functions of their teams (O'Connell, Doverspike, & Cober, 2002). In the absence of assigned managers/leaders inside or outside the team, the emergent task managing, and/or leading behaviors of team members are important. What can be retrospectively deduced from the narratives of team members concerning their participation in the managing and/or leading of their team in the absence of assigned leader/managers inside or outside the two teams?

In retrospect, the team members illustrated their self-managing teamwork as having been varied in many ways. In both teams, managing/leading was performed by more than one individual. At times, multiple individuals acted managerially. Different members performed the same function at different times. The team members talked about their team's self-managing as best characterized as "task leadership" (using Bormann's 1990 term) exercised by team members in either a delegated, co-leader, or peer arrangement (Brown & Gioia, 2002) and as proceeding from task to task in the team's plan for helping members. Members of both teams commented that they thought their team's members to be very involved and participating in self-managed processes.

Each team's self-managing was characterized by team members as generally responding to the perceived needs of the teams at the moment. In the judgment of the members of both teams, team performance was successful and enjoyed because the teamwork also met their individual needs. The members said that their teamwork was seen as urgent for everyone and yet was considerate of each other's feelings and ideas. Their teamwork was marked by caring and high mutual trust among team members. Recent research on leading and managing in self-managed teams has found that such characteristics correlated positively and highly with successful team performance. A finding worth noting in this retrospective is that such research also emphasizes that "self-managing" does not mean "non-managing." To the contrary, self-managing involves complex mixtures of managerial roles and behaviors that emerge from the team-member's ambitions and from among the responses of the team's members to their ongoing environment changes (e.g., Erez, Lepine, & Elms, 2002).

Research by Carte et al. (2006), Denison, Hooijberg, and Quinn (1995), and Quinn, (1984) suggests a framework for identifying and organizing managing and leading behaviors. Their leadership rubric will be used here for a retrospection of managerial behaviors that were mentioned in the team members' narratives about their teamwork. Such managerial behaviors would appear to have been enacted by various team members at various times in the self-managing of the teamwork in retrospect here. The leadership rubric asserts that self-managing the two teams consisted of eight different managerial roles: *innovator, broker, producer, director, coordinator, monitor, facilitator*, and *mentor*. In the constructionist perspective, these roles are discursively developed and enacted. The managerial behaviors involved in each role both influence, and are influenced by the ongoing processes and interacts of the teamwork. The roles focus sensemaking and sensetaking by the team members. The managerial roles problematize the situation as a whole and/or aspects of it. In describing their task-work, team members referred to and illustrated managerial roles and role-behaviors variously employed in their self-managing their teamwork. For example, they talked about their coping with "swerves" as a group. Swerves were perceived as large, often dramatic, changes in direction or effort that affected intra- or extra-team work. This situation, the members reported, called for their discussing and changing, as necessary, some aspect of their teamwork in order to meet the external pressures or realize an opportunity. The "swerve" very often required creating some new intra-team capacity to implement the change required. In the perspective of the leadership rubric, the situation called for, and the team carried out, the *innovator and/or broker* role behaviors for understanding and meeting the actions called for by requirements from outside the team. Some "swerves" originated from within the teams. That is, a perceived need and/or opportunity suggested to the team members that changes in their team's focus, processes and interacts were needed or could be very useful in their teamwork. In responding to the need and/or opportunity for changing intra-team task-work, the members of the team needed to meet and discuss the matter. In the perspective and terminology of the leadership rubric, the managerial behaviors and skills of the *facilitator* and *mentor* roles for developing people and processes capable of the change were required.

In each team, members also stressed the team's need to "make steady progress" toward the team's goal of helping every member, as needed, in carrying out and completing individual thesis research and writing. What was required was for the two teams to develop and follow agreed upon schedules of agenda and responsibilities for completing what needed to be done. Self-managing had to be authoritatively agenda setting and insistent on clear definition of team and individual goals and the detailed components of project deliverables. In terms of the leadership rubric, the two teams needed the managerial behaviors of *producer* and *director* roles. This was a need during the life of the teamwork and for both teams. In addition to the managerial behaviors associated with the *producer* and *director* roles, the self-managing of each team also persistently called for the skills and behaviors of one or more team members, not necessarily the same individuals at all times, who could enact, again in the terminology of the leadership rubric, the roles of *coordinator* and *monitor*. For accomplishing group and individual objectives, the team members said that managerial skills for maintaining "forward motion" in completing agenda were, together with the *producer* and the *director* roles, most important because the behaviors were required at all times and stages in their teamwork.

In their narratives and comments about the different ways in which teamwork was managed without external direction, the roles of *monitor* and *coordinator* were illustrated. The roles were needed to assure the execution of plans, schedules, and outcomes of the everyday teamwork. Progress toward the accomplishment of common ends and means as well as the projects of individual members needed monitoring and coordinating.

Producing and *directing* role behaviors were also illustrated in the narratives of the team members. The self-management of the two teams was focused on assuring the conformity and delivery of the products required by the company and university. The deliverables included milestone reports and the submission of approvable documentation. Producing and directing included coordination with external stakeholders concerning overall requirements and results. Coordination in the behaviors associated with other managerial roles was needed to assure delivery-as-promised of end-items from all members.

In the constructionist perspective, the self-managing roles and behaviors were sets of interacts that were assembled discursively into processes that constituted each team. In retrospect, and with the application of a leadership rubric, the team members may be viewed as having been involved in eight sets of managerial behaviors the goal of which was self-directed managing as teams to succeed in meeting the requirements of company-sponsored graduate-level education, individual research and writing of a graduate thesis document.

Using Quinn's framework (1984, 1996), the role behaviors and types of managerial concerns related to each role enacted by the teams can be summarized as follows. Self-managing that was focused on encouraging and facilitating change in teams' means and/or ends involved *innovator, broker* and alliance-building role behaviors, interacts and processes. Self-managing that was focused on encouraging and facilitating *extra-team relationships* (i.e., with stakeholders outside the team) involved *producer* and *director* role behaviors, interacts, and processes. Self-managing focused on encouraging and facilitating the stability of team processes and routines involved *monitor* and *coordinator* role behaviors, interacts, and processes. Self-managing focused on encouraging and facilitating *intra-team relationships* (i.e., with members and stakeholders in the team) involved *facilitator* and *mentor* role behaviors, interacts, and processes (Aggestam, M., & Keenan J., 2005, p.306).

Reflecting on Talk-Technology

Reflecting on the teams also underscored communicating in various forms and, in particular, talk as fundamental technology for the teams' self-developing and self-managing.

It is perhaps everyday experience that what people say (more precisely, who says what, to whom, how, when, why, and with what effect) is or can be important in social and societal relations. Constructionist theory relies heavily on the importance of social interaction, conversation, and, in particular, language not only in reflecting meanings but in constructing and organizing social realities. The self-managing teams can be conceptualized as discursively-constituted and as discursively constituting the selves and other realities of the members.

Social constructionism (e.g., Berger & Luckmann, 1967; Gergen, 1999) assumes that multiple realities are created in social action through the organizing properties of communicating and particularly through the agency of language. In this view, the self-managing teams might be viewed as existing in and as a variety of communication-based constructions involving individuals in conversations. Following the phenomenological tradition, constructionism regards the realities of the self-managed teamwork as subjective and intersubjective. Weick (1979) has argued that social construction regards collective structures such as teams as cognitive patterns imposed on a multitude of fluid and chaotic phenomena.

The constructionist-interpretive perspective argues that actors such as the team members create and enact the realities in which they live and work (Berger et al., 1967; Czarniawska, 1997; Weick 1979). Both the present and the future are predicated on creating a "material and symbolic record" (Silverman, 1970; Smircich & Stubbart, 1985, p. 726). The teams created frames of reference or cognitive consensualities (Gioia & Sims, 1986). The teams themselves were created, negotiated, and re-created through social relations (Bettenhausen & Murnighan, 1985; Burrell & Morgan, 1979; Griffith, Devadas, & Neale, 2003; Prahalad & Bettis, 1986).

The teams were socially constructed and language-based (e.g., Berger et al., 1967; Mitroff 1979; Smirich et al., 1985). The team members' communication about their teamwork and about their individual projects involved a multiplicity of intersubjective relations, conversations and discourses that reflected their pluri-vocal meanings.

Reflecting on the teams, it is clear that the self-managing, communicating, and importance or outcome of the teamwork was *multidirectional* and *multidimensional*. Teamwork was self-managed directionally: directed toward what was for each of the teams' members dramatic newness (the completed graduate theses and course work) and, to a lesser degree, to accomplishing incremental matriculation advancement. Teamwork and self-managing was also directed toward relating to the member's stakeholders inside and outside the company (e.g., managers and co-workers inside the company; university and family outside the team members).

Team members reported in their interviews that their teamwork was "intensely communicational." In both giving information and receiving information, "people were talking with everyone all the time" about their research and progress on their theses. Teamwork, they said, involved self-managing and "managing meant communicating with many people all the time!" Self-managing increased the amount of communicational activities of many team members, especially those who were directing, producing, and/or coordinating tasks. In each team, there was a need to rotate or share the work and members were generally willing to do so but some team members tried to avoid teamwork largely because of heavy company-work or home-life demands at the time.

After the success of the teamwork and the adjournment of the teams, members were interviewed and asked to retrospect and reconstruct, in terms of the leadership rubric, the communicational activities that were involved in self-managing. Looking back and with the advantage of hindsight, they agreed that there were indeed managerial roles and activities required of everyone in the teams. Such managing was very similar to the managing and leading they were experiencing in their company work. To the interviewees the team was like a company and needed to be run as a company replete with stakeholders inside and outside the company. At any time, the interviewees continued, there were needs for all of the types of managing (listed in the leadership rubric). Reflecting on their own company, the managers noted that the managerial roles and activities, even the names given to them in their company, became institutionalized over time. This insight was the stimulus for much discussion. Several of the ex-team members commented that they feared that, in the course of institutionalizing, the team members, and the company lost the new thinking and refreshed creativity that can emerge in sharing and rotating the responsibility and activities for managing work. They noted that self-managed teamwork makes owners and executives out of all members and the structure of the teamwork makes everyone a manager of both work and people.

Different managerial roles made different demands for communicational activities. The communicational activities of self-managing were both shared and rotated over the course of the two teams' work. For example, they reported that when the need was discussed for important change or redirection, or improvement in the work of one or several individuals in the teams timely *Innovator* and/or *Broker* roles were enacted. The role behaviors involved suggesting improvement in existing research ideas and practices, work schedule, meeting and exceeding requirements or promises of quality or quantity of thesis work. One or several team members suggested, as a result of looking externally beyond the team, consideration and use of state-of-the-art approaches, technology and leading edge practices.

Similarly, the *producer-director* roles were evidenced and were emphasised as important to the continuing tracking and understanding of the "deliverable" research report and the dictates of timely and standardized individual products. Both individuals and groups within the teams needed to assure their compliance with external requirements and deliverables. Self-managing, the teams' producer-director roles included maintaining regular contacts, as necessary, with university, company, research associates and research subjects and their meeting all requirements for academic and professional standards.

Self-managing by the teams involved one or several members to fill the roles of *coordinator-monitor*. This involved keeping track of team meetings, facilities, work schedules, research data, assuring group discussions of research ideas, progress, problems, and reporting. The roles also included maintaining and, as necessary, assuring the security and confidentiality of research-relevant records, references, books, and reports.

There was, the team members reported, an urgent need, especially in both teams' mid-course, for members to encourage and communicate with each other concerning progress and locomotion and task and relationships in working together. There was the need for *facilitator* and *mentor* role behaviors, for example: meeting with individuals involved in differences over schedules, work habits, and research approaches, encouraging team and personal creativity, and facilitating the development of individual and team capacities for innovation.

Reflecting on Team Cohesiveness as Heedful Technology

This reflection was prompted by the obvious syntality of the teams and the cohesiveness of the members in each team. The reflection reported here was and continues to be an opportunity to think about the cohesiveness of the team members and, indeed, cohesiveness as itself a technology. The cohesiveness and syntality in the teams was all at once the product and producer, as well as, the process of the members' careful, rather than careless, interacts and heedful interrelating. Cohesiveness and syntality involved team members' minding to connect with their team colleagues. Syntalic teamwork is collective mind that is caring and heedful. Cohesive and syntalic mind is action that connects members and develops group intimacy and interrelating jointly with the groups' maturing (e.g., Tuckman's, 1977 forming, storming, norming, and performing) (Weick, 2001, p. 275) and, as indicated in the post-teamwork interviews with the team members, even after the adjournment of the teams. After their teamwork had resulted in success, the team members reported continuing to feel happy about the heedful performance of their colleagues as individuals and the teams.

Upon reflection, the cohesiveness and syntality of the self-managed teams can also be viewed technologically, that is, as logics about means-ends relations. The heedful interrelations of cohesiveness and syntality are means of successful teamworking. Highly cohesive and syntalic teams are heedful. Heed is a disposition to act with attentiveness, alertness, and care. Heedful teamwork involves a communion of minds, that is, feeling, thinking, and willing to interrelate (Weick, 2001, p. 275) Paraphrasing Berniker (1987, p. 10), in the constructionist perspective, "cohesiveness as technology refers to a logic of awarenesses about the means by which we work on the world, our arts and our methods."

Technology is awareness that is communicatively constructed and organized, and that can be studied, codified, and taught to others through discourse, talk, and language. The cohesiveness of the members of the teams emerged in sensemaking and in making heedful contributions in a joint situation of interrelations among activities, which Asch (1952, p. 252) referred to as a system) that enacted collective mind. Cohesiveness was a technology embodied in the processes of interrelating (Weick, 2001, p. 267) as part of a technical system, a combination of methods used to produce some valued outcome. The valued outcome was the successful accomplishment of all requirements. To all appearances and from their self-reports, the members of the teams attained their goal with high cohesiveness evidenced in the patterns of their interrelated activities, that is, the style and strength with which their ties were tied together. The styles of the self-managed teams were highly heedful and their ties were tightly bound. Highly heedful and with their contributions tightly tied, the members worked to benefit each other and the whole.

CONCLUSION

The apparent successes of two self-managed knowledge work teams and, in particular, their cohesiveness and syntality were the impetus for a retrospective exploration of the developing and managing involved in the teams' work. Several questions were prompted by the retrospective review of the communicative construction and management of the two highly cohesive self-managed teams. Explanations principally from the constructionist perspective and from communication and management studies were found to be salient. The questions included: How can self-managed teams be defined from a communication perspective? How are self-managed knowledge teams communicationally constituted? How do such teams develop? How do self-managed teams sensemake, coalesce, and converge? How are self-managed teams managed? What roles and skills are involved in self-managing teamwork? This chapter is principally an attempt to capture constructionist perspectives in response to such questions.

Research has increased concerning teams as concomitants of globalized, international, and intercultural organizational forms and relations. Teams, especially self-managed and virtual teams of knowledge workers, have been problematized as key features of the fast changing intellectual and social capital of organizations (Aggestam & Keenan, 2000, 2001b; Keenan, 1999, 2000). Teamwork is technology for understanding and managing technological change. Technological change, especially systemic and rapid change, is a trigger for teamwork (Weick, 2001, p. 261). Framed in the constructionist perspective, teams have been viewed here as systems of interlocked sensemaking behaviors and processes that emerge in communicating through the agency of language. Weick (1979) has described teams as sets of interacts that are assembled into processes that constitute organizations. Communicative teamwork is constitutive of managing technology.

Reflecting on the teams draws attention to understanding the variety of technologies in which the teams' performances were involved cognitively and emotionally. They were involved in several communicationally created technological constructions emergent in their social interactions. Adopting "techno-logic" as essentially their learned understandings concerning means-to-ends relations, the team members may be viewed as individual psychophysical technologies working in a variety of social technologies (other groups, task forces, work departments and indeed the whole company for technologically developing a product which itself can be seen as a technology, that is, a means to an end, in this case the completed and approved thesis document. Reminiscent of Hulin and Roznowski's (1985, p. 47) view, the team members were technological, that is, combining the physical with their intellects and knowledge processes to transform something into a form useful by others. They were contributing to technology as a body of knowledge about the means by which we work on the world, our arts, and our methods. They were contributing to technology as a body of knowledge that can be studied, codified, and taught to others (Berniker, 1987, p. 10). Their interacts as team constructed and embodied technology. The team members were experiencing technology as ongoing, communicatively constituted awareness and at times presenting conditions for arousal and emotional experiences that were, paradoxically, both energizing as well as debilitating, urging them onto fight (continue on) or flight (give up trying). In the work of the teams, members were constructing and enacting new technology as "equivogue" (Weick, 2001, p. 162).

In retrospect, the two teams were technologies for constructing, sharing, and enacting understandings of their members' about their individual research projects. Shared understandings emerged in the interacts and served as glue for the teams' self-managing and for exploring the members' taken-for-granted understandings, their "local theories" (Elden, 1983), their cognitive maps (Lowestedt, 1993; Weick, 2001, p. 261). The teams were technologies for trans-porting, trans-forming, and trans-acting individual worldviews and mental models as well. These explorations and exchanges (e.g., about the requirements for their individual research projects) guided the team members in the settings in which they found themselves and helped steer their behaviors. Their understandings were constructed through sensemaking processes in which their worlds emerged as sensible and understandable. Their sensemaking processes were individual but they took place in a social context (i.e., the perceptions and the knowledge shared among fellow team members) (Weick, 2001, p. 261).

REFERENCES

Aggestam, M., & Keenan, J. (2005). *Narrating organizational knowledge: A communicationist turn*. Paper delivered at the Irish Academy of Management meetings, Galway, Ireland, September 2005.

Aggestam, M., & Keenan, J. (2001a). *A constitutive meta-analysis of communication capital in organisations undergoing dramatic change: A case study.* Paper presented at International Communication Association, 24-28 May, 2001, Washington, DC.

Aggestam, M., & Keenan, J. (2001b). Corporate governance and organizational knowledge capital: Some conceptualisations. *Corporate Governance Journal*, *9*(4), (October), 259-275.

Aggestam, M., & Keenan, J. (2000). *Communication capital in organisation: Managerial mindsets and activities during dramatic change.* Paper presented at International Communication Association, 1-5 June 2000, Acapulco, Mexico.

Allport, F. H. (1962). A structuronomic conception of behavior: Individual and collective. *Journal of Abnormal and Social Psychology, 64*, 3-30.

Asch, S. E. (1952). *Social psychology*. Englewood Cliffs, NJ: Prentice-Hall.

Avolio, B. J. (1999). *Full leadership development: Building the vital forces in organizations*. Thousand Oaks, CA: Sage.

Bass, B. M. (1990). *Bass and Stogdill's handbook of leadership*. New York: Free Press.

Berger, P., & Luckmann T. (1966). *The social construction of reality*. New York: Doubleday.

Berniker E. (1987). *Understanding technical systems*. Paper presented at Symposium on Management Training Programs: Implications of New Technologies, Geneva, Switzerland.

Bettenhausen, K., & Murnighan, J. K. (1985). The emergence of norms in competitive decision-making groups. *Administrative Science Quarterly, 30*, 350-372.

Bijker, W. (1995). *Of bicycles, bakelite, and bulbs: Toward a theory of sociotechnical change*. Cambridge, MA: MIT Press.

Bormann, E. G. (1985). Symbolic convergence theory: A communication formulation. *Journal of Communication, 35*, 128-138.

Bormann, E. G. (1983). Symbolic convergence: Organizational communication and culture. In L. L. Putnam & M. E. Pacanowsky (Eds.), *Communication and organizations*. Beverly Hills: Sage.

Brockett, C. (1975). Toward a clarification of the need hierarchy theory: Some extensions of Maslow's conceptualization. *Interpersonal Development, 6*, 77-90

Brown, M. E., & Gioia, D. A. (2002). Making things click: Distributive leadership in an online division of an offline organization. *Leadership Quarterly, 13*(4), 397-419.

Burrell, G., & Morgan, G. (1979). *Sociological paradigms and organizational analysis*. London: Heinemann.

Carte, T. A., Chidambaram, L., & Becker, A. (2006). Emergent leadership in self-managed-virtual teams. *Group Decision and Negotiation, 15*, 324-343.

Chaston, I. (1998). Self-managed teams: Assessing the benefits for small service-sector firms. *British Journal of Management, 9*, 1-12.

Churchland, P. M. (1992). A deeper unity: Some Feyerabendian themes in neurocomputational form. In S. Davis (Ed.), *Connectionism: Theory and practice* (pp. 30-50). New York: Oxford University Press.

Crossen, M., Lane, H. W., & White, R. E. (1999). An organization learning framework: From intuition to institution. *Academy of Management Review, 24*(3), 522-537.

Czarniawska, B (1997). *Narrating the organisation: Dramas of institutional identity.* Chicago: University of Chicago Press.

Denison, D. R., Hooijberg, R., & Quinn, R. E. (1995). Paradox and performance: Toward a theory of behavioral complexity in managerial leadership. *Organization Science, 6*, 524-540.

DiMaggio, P. W., & Powell, W. W. (1983). The iron cage revisited: Institutional isomorphism and collective rationality in organizational fields. *American Sociological Review, 48*, 147-160.

Elden, M. (1983). Democritization and participative research in developing local theory. *Journal of Occupational Behaviour, 4*, 21-33.

Erez, A., Lepine, J., & Elms, H. (2002). Effects of rotated leadership and peer evaluation on the functioning and effectiveness of self-managed teams: A quasi experiment. *Personnel Psychology, 55*, 932-947.

Gergen, K. J. (1999). *An invitation to social construction.* London: Sage.

Gibson, J. W., & Tesone, D. V. (2001). Management fads: Emergence, evolution, and implications for managers. *Academy of Management Executive, 15*(4), 122-133.

Gioia, D. A., & Sims, H. P. (1986). Introduction: Social cognition in organizations. In H. P. Sims, Jr., D. A. Sims, et al. (Eds.), *The thinking organization.* San Francisco: Jossey-Bass Publishers.

Griffith, T. L., Devadas, R., & Neale, M. A. (2003). Virtualness and knowledge in teams: Managing the love triangle of organizations, individuals, and information technology. *MIS Quarterly, 27*, 265-287.

Harris, T. E., & Sherblom, J. C. (2005). *Small group and team communication.* Boston: Pearson.

Hulin, C. L., & Roznowski, M. (1985). Organizational technologies: Effects on organizations' characteristics and individual's responses" In L. L. Cummins & B. M. Staw (Eds.), *Research in organizational behavior.* Greenwich, CT: JAI Press.

Katzenbach, J., & Smith, D. (1993). *The wisdom of teams.* Boston: Harvard Business School Press.

Keenan, J. (1999). *Systems of intellectual capital in collectivities from work organizations to human settlements: some conceptualizations.* Paper presented at Human-Computer Interface International Conference, Munich, Germany.

Keenan, J. (2000). Organizational knowledge capital for informatic systems. In G. Bradley (Ed.), *Humans on the net: Information and communication technology, work organization, and human beings.* Stockholm, Sweden: Prevent.

Kirkman, B. L., & Shapiro, D. L. (2001). The impact of cultural values on job satisfaction and organizational commitment in self-managing work teams: The mediating role of employee resistance. *Academy of Management Journal, 44*(3), 557-569.

Kirkman, B. L., Gibson, C. B., & Shapiro, D. L. (2001). "Exporting" teams: Enhancing the implementation and effectiveness of work teams in global affiliates. *Organizational Dynamics, 30*(1), 12-29.

Kirkman, B. L., & Rosen, B. (1999). Beyond self-management: Antecedents and consequences of team empowerment. *Academy of Management Journal, 42*(1), 58-74.

Krippendorf, K. (1971). Communication and the genesis of structure. *General Systems, 20*, 15-35.

Lawler, E. E., Mohrmann, S. A., & Ledford, G. (1995). *Creating high performance organizations: Practices and results of employee involvement and total quality management, practice, and results in Fortune 1000 companies.* San Francisco: Jossey-Bass.

Lowestedt, J. (1993). Organizing frameworks in emerging organizations: A cognitive approach to the analysis of change. *Human Relations, 46*(4), 501-526.

Mead, G. H. (1934). *Mind, self, and society.* Chicago: University of Chicago Press.

Moravec, M., Johannessen, O. J., & Hjelmas, T. A. (1997). Thumbs up for self-managed teams. *Management Review, 86*(7), 42-47.

Murray, P., & Moses, M. (2005). The centrality of teams in the organizational learning process. *Management Decision, 43*(9), 1186-1202.

O'Connell, M. S., Doverspike, D., & Cober, A. B. (2002). Leadership and semi-autonomous work team performance: A field study. *Group and Organization Management, 27*(1), 50-65.

Olufowote, J. O. (2006). Rousing and redirecting a sleeping giant. *Management Communication Quarterly, 19*, 431-492.

Prahalad, C. K., & Bettis, R. A. (1986). The dominant logic: A new linkage between diversity and performance. *Strategic Management Journal, 7*, 485-501.

Quinlan, P. (1991). Connectionism and psychology. Chicago: University of Chicago Press.

Quinn, R., Faerman, S., Thompson, M., & McGrath, M. (1996*). Becoming a master manager.* New York: John Wiley & Sons.

Quinn, R. E. (1984). Applying the competing values approach to leadership: Toward an integrative model. In J. G. Hunt, R. Stewart, C. Schriesham, & D. Hosking (Eds.), *Managers and leaders: An international perspective.* New York: Pergamon.

Rumelhart, D. E. (1992). Towards a microstructural account of human reason. In S. Davis (Ed.), *Connectionism: Theory and practice* (pp. 69-83). New York: Oxford University Press.

Ryle, G. (1949). *The concept of mind.* Chicago: University of Chicago Press.

Sandelands, L. E., & Stablein, R. E. (1987). The concept of organization mind. In S. Bacharach, & DiTomaso (Eds.), *Research in the sociology of organizations* (Vol. 5, pp. 131-161). Greenwich, CT: JAI Press.

Silverman, D. (1970). *The theory of organizations.* London: Heinemann.

Smircich, L., & Stubbart, C. (1985). Strategic management in an enacted world. *Academy of Management Review, 10*(4), 724-736.

Stevenson, W. B., Pearce, J. L., & Porter, L. W. (1995). The concept of "coalition" in organizational theory and research. *Academy of Management Review, 10*, 256-258.

Stewart, G. L., & Manz, C. C. (1995). Leadership for self-managing work teams: A typology and integrative model. *Human Relations, 48*(7), 747-770.

Tuckman, B., & Jensen, M. (1977). Stages of small group development. *Group and Organizational Studies, 2*, 419-427.

Wegner, D. M., Erber, R., & Raymond, E. (1991). Transactive memory in close relationships. *Journal of Personality and Social Psychology, 61*, 923-929.

Weick, K. (2001). *Making sense of the organization.* Malden, MA: Blackwell Publishing.

Weick, K. (1995). *Sensemaking in organization.* Thousand Oaks, CA: Sage.

Weick, K. (1979). *The social psychology of organizing*. New York: McGraw-Hill.

Wellins, R., Byham, S., & Wilson, C. (1991). *Empowered teams: Creating self-directed work groups that improve quality, productivity, and participation*. San Francisco: Jossey-Bass.

Zarraga, C., & Bonache, J. (2005). The impact of team atmosphere on knowledge outcomes in self-managed teams. *Organization Studies, 26*(5), 661-681.

Chapter XVIII
The Entrepreneurial Constitution of High–Tech Work Environments

Maria Aggestam
Lund University, Sweden

The information highway is a tool of the individual.
(Gates, 1995, pp. 166–167)

ABSTRACT

In recent years, business practice has shown an interest in virtuality and virtual organizations as one of the responses to global-organizational development and company restructuring. High-tech environments demand continued innovations. The rational, authoritarian, bureaucratic organizations characterizing industrial and production-oriented companies are being rapidly replaced by the new, boundaryless, flexible, and high-technology-driven environments. Entrepreneurs respond to global opportunities, find partners, and build cross-border communication networks to link corporate processes and create empowered workplaces. A distinctive feature of these processes is the reliance on free-floating cybertrust where each partner is involved in decision-making processes and in building capabilities in continuously changing international markets. This chapter reflects on how contemporary entrepreneurs benefit from the high technology, virtuality, and cyber-culture of the modern organization. The findings draw attention to the technology-driven opportunism within virtual media and the role of entrepreneur within human–computer interaction. The chapter seeks to demystify views about organizations as virtual domains.

THE VIRTUAL ORGANIZATION

The global economy has created the virtual organization as a new form of organizing and entrepreneuring (Aggestam, 2003). Metaphors such as "borderless world" and "global village" extol interdependencies among countries and organizations employing business environments facilitated by high-technological improvements, which provide easier cross-border communications. The pursuit of a new entrepreneurial approach in the global world has made the existing bureaucratic and hierarchical structures of businesses outmoded, and there is a need to adapt to the constant emergence of new information and communication technologies. Progress in the new global economy is marked by confrontation with rapid technological development and change processes and emphasizes the abilities of actors to take advantage of entrepreneurial opportunities.

The term "virtual" originates from the Latin word *virtus* and concerns "proficiency, manliness" (Scholz, 1997). But for all that the term carries connotations of unreality, a virtual resource is very much real. Virtuality is the general term for discerning the unreal in terms of a familiar physical (real) resource or space, that is, a "real society or real person" with other virtual landscapes that do not exist within the same physical reality (Jordan, 1999).

The term "virtual organization" refers to a new and evolving organizational form where the boundaries are not easily detected and where the founding entrepreneurs are concerned with business efficiency and effectiveness. The virtual organization has many different facets and depictions. The notion of virtuality was first used within information and communication technology where the issue of virtual memory—memory created by software—became crucial. The fascination with the virtual organization has been sustained over two decades and started when Moshowitz (1986) used the term for the first time. Since then, the virtual organization has been referred to as the "virtual company" (Goldman & Nagel, 1993), "virtual factory" (Upton & McAfee, 1996), "virtual enterprise" (Hardwick, Spooner, Rando, & Morris, 1996), "virtual workplace" (Pruit & Barett, 1991), and "virtual Web" (Goldman, Nagel, & Preiss, 1995).

The virtual organization provides a new means and new resources for entrepreneurial activities with its relatively low cost, ease of use, and broad diffusion. It also helps to overcome barriers in interorganizational communications among firms. One example of the virtual organization is the temporary cooperative model of the "virtual corporation," that is, a model derived and used in virtual Web communication that exists only for a certain project or task and is time-limited. Byrne (1993) pointed to technology as playing a central role in the development of the virtual corporation. Teams of people in different companies work together concurrently rather than sequentially via computer networks in real time. An example of such virtual cooperation might be international accounting and auditing companies. Virtual corporations tend to exercise great flexibility of action and provide close collaborations among partners concerning specific professional and business tasks. The partners are focused on a limited set of topics and problems and their solutions. Their knowledge is "action-centric," that is, focused on relating cooperatively.

There has been a long debate about what constitutes a virtual organization. Some writers, such as Gerlach and Hamilton (2000), have argued that the virtual organization is based on cyber-technology that enables working practices, sharing desks, groupware, telecommunicating, and the empowerment of virtual teams and individuals.

In the virtual organization, communicating partners become part of a system of, for example, shifting responsibilities for work-practices, clarifying fuzzy distinctions between suppliers and customers, shifting lines of authority, and synthesizing information and communication processes, rather than dealing with material structures. Virtual organization processes are flexible, highly adaptable, and exploit internal and external resources that are largely non-physical assets. One consistent pattern appears to be that once the product has been developed, virtual organizations focus on reproducing and imitating their successes. Entrepreneurial effort creates a product or process, which becomes a subject of re-production and thereby generates the next stage of the agenda: concerns about problems of survival and profitability for the new creation.

The literature also suggests that virtual organizations tend to exist within cyberspace as a way of gathering, processing, and evaluating information. This has implications for the entrepreneur who can use the information and communication technology and the connectivity of virtual cooperation to uncover and follow up opportunities. The entrepreneur often appears to envision future possibilities that others fail to recognize using information and communication technology to interact with both physical and virtual marketplaces. Kirzner (1997) has developed a theory of entrepreneurial alertness, which asserts that entrepreneurs employ information differently than non-entrepreneurs for sensing opportunities. Furthermore, entrepreneurs coordinate scarce resources that can be built, promoted, and communicated in cyberspace. This enables entrepreneurs to find and link resources discovered and investigated through virtuality. These discoveries often occur at unconventional times and places. For the entrepreneur, virtuality enables boundaryless scanning for opportunities and a useful stream of information flow. Rushkoff (1994), described virtuality as bounded territory by digital information and called it cyberspace—and called the unbounded territory cyberia. (p. 2-3). This line of thought extends the traditional view of entrepreneurs by seeing them as cyborgs that combine high technology, space, and their own bodies/mindsets in captivating new configurations. To address these issues, below I include a discursive perspective on cyber-culture and the borderless world as background for entreprenuring within the virtual organization following by understanding high-tech entrepreneurs as cyborgs and a main resource in virtual business world.

CYBER-CULTURE AND THE BORDERLESS WORLD

There has recently been an upsurge of interest among practitioners and academics in new ways of entrepreneuring. These new ways embody the latest developments in information technology and include the virtual organization and "cyberspace," a term used for the first time in Gibson's novel, *Neuromancer* (1984). Cyberspace can be described as "virtual lands, with virtual lives and virtual societies, because these lives and societies do not exist with the same physical reality that 'real' societies do. With the emergence of cyberspace, the virtual becomes counterposed to the real" (Jordan, 1999, p. 1). In this way the Gibsonian concept of cyberspace began to transmute into a tangible reality and to find its place within social and cultural analysis as a significant characteristic of the forward-looking world. Another significant term is "cybernetics," coined by Wiener (1948) to describe a science connecting statistical mechanics with communication and control theory. Cybernetics was originally created on an interdisciplinary basis connecting patterns from biology and technology to create machines capable of imitating human mental activities. For Wiener, cybernetics embodied human mind and body that were used to communicate and control technology and automatic machines. In other words, cybernetics embodied the technology for communication with and control of automatic machines. This indicated the importance of information, messages, and feedback that accommodated both communication and control within the body of humans and machines. It was one of the first acknowledgements of machines as communicating organisms with the capacity to link with each other. As a cybernetician, Wiener used the model of an organism from biology, which he connected to engineering, thus effectively transforming it into hardware and technology-related problems.

Haraway (1991, p. 150), refers to the cyborg as "a cybernetic organism, a hybrid of machine and organism, a creature of social reality as well as a creature of fiction." Cyborgs also have various symbolic meanings. Dick[a] (1968, p. 211) locates cyborgs in the borderless place between mechanized humans and humanized machines, making it difficult to tell them apart. *Webster's New World College Dictionary* defines "cyborg" as a "hypothetical human being modified for life in a hostile or alien environment by substitution of artificial organs and other body parts" (1996, p. 343). From this perspective, many of us are already cyborgs, with technical devices such as pacemakers implanted in our bodies.[b]

Cybernetics lacks the usual understandings of location and time. Sotto (1997) has pointed out that in cyberspace time is not subjective time but refers to speed of information transfer, and space does not involve location in usual terms but the flow of electrons that are invisible. Information and communication capabilities connected to "non-penetrative cyborg technologies" (Clark, 2003) are ways of becoming competitively knowledgeable and smart. These technologies are ways of keeping confidential matters from falling into hostile hands. Cyber-culture is characterized by a focus on the future and on the newest technologies. Some recent technologies in cyber-culture include developments such as mobile phones, interactive computer software, smart-chipped automobiles, PDAs, and Web bots. Clark (2003) points out that our brains are cybernetic, adapting, and capable of learning in new ways depending on the perceived demands of the environment. In this sense, one might say that entrepreneurs are individuals who cybernetically adapt to environmental changes and find undiscovered opportunities. The notion of cyborg technology for Clark is a combination of, or interaction between, technology and people, particularly communication technology. For other scholars, cyborgs represent a combination of the biological and technological and also the human and non-human. In this sense, the idea of cyborgs breaches boundaries, breaks rules and challenges prejudices by confronting assumptions and terms.

By integrating new techniques into our environment, we develop new avenues for using them, which contribute to a large range of new business venture possibilities. The view that emerges here is that understanding cyber-culture is largely a matter of keeping up to date with developments in the newest consumer technologies, often called "boys toys" or "cyborgs for boys." In cyber-culture, entrepreneurs use electronic devices to extend their mental capabilities and to enhance their "practical intellect" (Göranzon, 1991), which has market implications for work performance and our future work life.

ENTREPRENEURS AS A RESOURCE IN THE VIRTUAL ORGANIZATION

There is a long-established view that information and communication technologies create flexibility and efficiency and are driving forces behind the unconventional exploitation of resources and development in business organizations. Entrepreneurs use technological devices as highways for the virtual organization of both resources and capabilities of human agency. The virtual organization becomes an entrepreneurial work life, which, unlike the traditional organization, provides conditions for the exchange of goods and services along with virtual value-chains in an electronic and cyber-world. In this sense, entrepreneurial efforts towards new venture creation might be called "virtucommuting" (Sherman & Judkin, 1992, p. 151). Virtucommuting could be a part of entrepreneuring and a crucial distinguishing characteristic of entrepreneurs in high-tech firms. It occurs when actors combine, exchange, enact, and create new resources by utilizing other resources in virtual organizations. Entrepreneuring that merges entrepreneurs with resources and actions in multiple ways and operations constitutes the strength of virtual organizations. Entrepreneuring is a powerful activity allowing technology, technological devices, and entrepreneurial actors to master the tasks with which they are confronted. The virtual organization can be regarded as a significant personal resource that entrepreneurs use in business venturing. Burt (1992) points to the entrepreneur's capabilities as essentially building, maintaining, and coupling connections that provide opportunities and profits to the organization. He highlights the importance of distance between two or more business players. The entrepreneur uses market niches in order to influence the relationship between two or more actors for his own advantage, thus allowing one person to take the opportunity to be "tertius" (an individual who benefits from the disunity of others). This person is an entrepreneur in the literal sense, one who generates profit from being between two or more partners (Franke, 1999). In this sense, social skills that manipulate communication among actors can be seen as the core resource of some entrepreneurs. Given that entrepreneurial activity as manipulative, the "tertius" concept is useful in defining a successful entrepreneur. Pihkala, Varamaki, and Vesalainen (1999) highlighted informal communication in the process of venture creation as being a resource critical to the company's development, while Chell (2000) stressed the importance of identifying and enacting all kinds of relationships, even virtual, in order to survive.

ENTREPRENEURING THE VIRTUAL ORGANIZATION

The emergence of the new mode of entrepreneuring through computers and high-tech opportunities has been one of the most interesting topics in entrepreneurship scholarship over the last few years. Long before the concept of the virtual organization came into common use, the role of computers in the life of the entrepreneur was apparent but not much researched. In other words, entrepreneurs also adjust to the imperatives of virtual technology. The progress in advancing our knowledge about the role of opportunistic entrepreneuring within the virtual organization and high-tech technology has many aspects, some of which are discussed here. The virtual organization and its attributes are one of the current strengths that impact business performance and can be conceptualized as the endogenous power of entrepreneuring. In the business environment, computer technology is integrated with this power and with company structures to enhance the ability to entrepreneur. Linking entrepreneuring to computers, and thereby to a certain form of virtual organization, the communication, memory, mathematical abilities and extrasensory capabilities of the entrepreneur are improved and contribute to making the entrepreneur intellectually adept. On the one hand, new technologies often create opportunities while, on the other, posing threats to continuing entrepreneuring. The entrepreneur typically uses a multiplicity of possible ways of dealing with the uncertain situation (Burns & Stalker, 1961) created by new technology. One feature of this multiplicity is the spatial ingredient in entrepreneuring. Space is an important dimension in the interactions among small firms and is intrinsic to their business, social, and institutional mutual aid (Johannisson, 1990). This is the case, for example, in Gnösjö, Sweden, and in Silicon Valley in the USA. Spatial entrepreneuring is defined here as the process of making connections and communicating among physically co-located individuals and companies. Spatial entrepreneuring is always in the picture, affecting the situation more or less. In some cases, the entrepreneuring has been artificially induced by the design and location of relatively proximate facilities and activities (such as in Sweden and USA). In other cases, space is a dimension that is naturally part of the entrepreneuring process but can be regarded as relatively "natural" and can be proximate or distal (Birely, Cromie, & Myers, 1991). Whether the entrepreneurial newness is artificially induced or natural, it is clear that new developments in the contemporary dynamic and globalized world involve virtual environments that are dependent on information technology. In this view, entrepreneuring within virtual environments is the temporary or stable linking of individual actors and companies through information technology, and the keeping in reserve of other potential partners, which enables rapid exploitation and enactment of fast-changing opportunities (Barnatt, 1995; Byrne, 1993,). The virtual organization engages entrepreneurs in particular issues but lacks a stable or enduring structure. The environment in which entrepreneuring takes place is characterized by time pressure, ambiguity, uncertainty, and often incomplete information. In the light of these features, virtual organizations provide special conditions for entrepreneuring which include special rules for formality, informality, efficiency, flexibility, team-work, competitive activities and opportunistic behaviour.

In other words, a virtual organization can be seen as a temporary setting (e.g., time, place and process) for entrepreneuring. This temporary setting affords the possibility of connecting through information technology all or some of the principals actors—idea and venture creators, suppliers, buyers, and customers—that are related to specific opportunities in fast-changing markets.

THE ENTREPRENEUR AS A NET-BROKER

It has been argued that the use of virtuality and virtual organization is systemically linked to complex sets of markets and power relations. These processes inform the actions and, in part, constitute the identity of entrepreneurs. The identity of entrepreneurs refers to their actions and is contingent on their fulfilling expectations regarding entrepreneurial capabilities through demonstrating the effectiveness of their practices as goal-oriented activities. With the continuing concern about high technology, virtual organizations have been actively creating new industries based on breakthroughs in net-brokering. These industries involved technical entrepreneurs, here called net-brokers, who possess distinct characteristics. These net-brokers have the significant ability of using telecommuting and interacting with others in order to create new business ventures. The traditional and new channels of communication are being augmented by skilful net-brokers in the pursuit of perceived business opportunity. Franke (1999) cogently pointed out that virtual organization requires a different organizing and managing approach to accomplish creative goals. Net-brokering is a response to that requirement. Broadly, a net-broker is a person who is motivated entrepreneurially and who uses the characteristics and powers of net-working people and companies toward entrepreneurial ends. Net-brokering is a personalized process in which a constellation of happenings creates new business opportunities. For example, the net-broker could organize the team or net-working actors and companies for a short period of time in furtherance of network progress. Generally, the entrepreneur as a net-broker is self-taught, self-disciplined, and self-organized, and is self-empowered to fulfil personal goals. There is a long-established view that entrepreneurs differ significantly from managers (Allinson, Chell, & Hayes, 2000), particularly in the technical sectors. Roberts (1989) found that technically-oriented entrepreneurs are more cognitive, alert, and intuitive than salaried managers. This view is supported by Franke (1999) who understands the entrepreneur as a net-broker who is highly cognitive and motivated and who has the ability to capitalize on untapped market niches hidden within virtual connections. Clearly, the entrepreneur can be seen as an accomplished net-broker.

Studies have indicated the importance of social capital involved in net-brokering. They attach special significance to the net-broker's social assets such as trust, gratitude, friendship, and kinship, various obligations and companionship that assist their function in securing and arranging access to resources (Starr & MacMillan, 1990). Studies in this area are, however, few in number and have not addressed the many activities and characteristics of net-brokering in high-tech virtual environments. Miles and Snow (1986) characterized the role of the broker in organizations. Brokers act as facilitators, analysts and catalysts, and create the partnership processes that are an essential part of entrepreneuring. Entrepreneurs as net-brokers focus upon organizing partnerships, communicating ideas, and maintaining relationships predictive of identifying new business opportunities. These activities call for holistic and relational thinking, in other words, creativity. Allinson et al. (2000) and Minzberg (1976) link these activities to the right hemisphere of the brain. Entrepreneurs as net-brokers also use the left hemisphere, which controls language capabilities. In practical terms, entrepreneurs use both the right and left hemispheres in organizing virtual teams in the work environment, connecting groups, designing and marketing products, and using information technology to approach their markets in the global village. It was indicated by Hatch (1995) that net-brokering entrepreneurs, using virtual organization increase the number and quality of relevant and involved actors. A more detailed description of entrepreneurs as net-brokers in virtual organizations is provided by Kanet and Faisst (1997). They envision the virtual workplace as including the entrepreneur as a net-broker, acting as information broker and initiator of various activities. As such, the net-broker is the foremost dynamic driving force in the virtual organization. The goal of net-brokering entrepreneuring is recognizing market opportunities and the focal challenge is to combine and connect capabilities of various professional knowledges and knowing together. Entrepreneuring processes as net-brokering within virtual organizations are also an extension of the execution of the operations and activities of the company. Net-brokering brings with it the power, for example, to resolve controversy among partners and to represent the company in the larger forum of customers. Franke (1999) emphasized additional roles for the entrepreneurial net-broker in virtual organizations. The net-broker initiates and prepares the virtual Web, maintains the Web, deskills and re-skills virtual Web interactions, and takes the initiative for entrepreneuring within the virtual environment. The view that emerges here is that the virtual organization sets the stage for entrepreneuring and for realignment of bureaucratical and hierarchical orders through redefining and redistributing qualifications, capabilities, and responsibilities.

THE NEW WAVE OF NET-RELATED ENTREPRENEURS

Among the latest internet developments is the appearance of a new wave of net-related entrepreneurs. They have anticipated and benefited from some of the profound shifts in organizational performance that the borderless world and global village are bringing about, yet they do not make the sort of provocative or visionary claims that characterized the first generation of "dotcom entrepreneurs." The new entrepreneurs exhibit a mix of pragmatism and opportunism in building new ventures and then selling them quickly. Viewed as sources of innovation, these entrepreneurs have been much criticized, especially those whose principal motivation was to set up companies that could be 'flipped' for a profit. The complaint is that they have been driven by selfish, pecuniary ideals and even, startling as it may seem, contributed to the decline of national competitiveness (Florida & Kenney, 1990). Critical views suggest that new companies created through virtual channels often fragment industries and fail to become major competitors themselves. Keeley and Tabrizi (1995) studied a sample of 49 companies from nine industries to understand the long-term impact of technological entrepreneuring on the industries. They found few cases in which the start-up might be judged as competing with or damaging a former employer. In many cases, both the start-ups and the former employers became leading firms in their industries. Overall, the benefits from start-ups were found to be considerable while the negative effects were minor. Other scholars have pointed to start-ups of high-tech organizations and their difficulties in retaining key people when forced to compete with other companies and start-ups. For example, Florida et al. (1990), in the context of the situation in Silicon Valley, noted high employee turnover as a result of the entrepreneurial failures. Both Ferguson (1988) and Reich (1987) noted some concomitants of start-up failures, for example, fragmentation of industries, inefficient production, extreme competition, and production of faulty products inherited from key technologists in established industries.

The new wave of net-related entrepreneurs is a very interesting phenomenon. Future studies will need to develop a multidisciplinary understanding of the variety of negative and positive outcomes of this striking development in globalized high-technology environments.

THE CASE OF YOUTUBE[C]

The case of YouTube illustrates also the nature of entrepreneuring and the characteristics of high-tech entrepreneurs in the very fast-changing and competitive environment. YouTube is a recent example of a virtual organization within the emerging market of free video distribution that exemplifies high technology and serves as an illustration of both new wave and net-brokering entrepreneurs. Net-brokering entrepreneurs can be understood as cyborgs that possess the ability to combine technology, space, and investment models with their mindsets in new business configurations. The company stands at the forefront of a cultural shift within traditional media companies towards digital media. Digital media and high-tech companies operate in markets characterized by strong and dynamic competition and high growth potential. High-tech growth potential implies an important unmet need for newness in the borderless virtual Web environment that also provides access to the active customers who are constantly looking for technical novelties. YouTube could be referred to as a high potential firm, where the outcome is rapid in developments and in profits.

YouTube is a trendy, free video-sharing Website that allows users to upload, view, and share video clips. The wide variety of content includes movies, TV programmes and music videos, as well as amateur-made content such as video blogging. The history of the company is relatively short, embracing 18 months from foundation to sale. YouTube represents a company with the largest financial reward made in the shortest time. In that short time, it exploited its intangible, knowledge-based information-rich assets for maximum financial gain. YouTube was a business angel-founded (See: Paul, Whittam, & Wyper, 2007) company in February 2005 from employees of PayPal in San Bruno, California. Business-angels look for the most tempestuous business opportunities to invest in and tend to focus on the personal factors and attributes of entrepreneurs. They are willing to take massive risks, because the rewards can be commensurate.

The domain name, YouTube.com, was registered at the beginning of 2005 and the company was developed in-depth over the following months. At the end of 2005, the venture capital and investment company Sequoia Capital provided $3.5 million as seed money and then at the beginning of 2006 poured an additional $8 million into the company. The company employed 67 persons in 2006 and was one of the fastest-growing Web sites on the World Wide Web, with about 20 million hits each month (which were 44% women, of all ages and the rest males aged 12 and 17 years old), according to Nielsen/NetRatings (http://www.usatoday.com, July 16, 2006). According to a July 2006 survey conducted by Alexa.com, every day 100 million clips were downloaded on YouTube, and an additional 65,000 new videos uploaded (Alexa.com., Amazon.com, July 26, 2006). In reaching development decisions a key issue has been to balance risks and profits. The business model of YouTube was advertising-based, a model that can be characterized by complexity, uncertainty, and conflicts. Advertisements were launched on YouTube sites in March 2006, in a short-lived collaboration with Google AdSense. The simplicity of the service was legendary, providing millions of videos in several categories to the market. In October 2006, Google, Inc., announced that it had reached an agreement to acquire the company for $1.65 billion in Google stock. The acquisition was completed at the end of 2006.

Google's basic idea is based on making information in the Web-world easier to find. It is a search engine that in mathematical terms can be defined by the logic of algorithms. Mathematical smartness, organizing skills, success, and the powerful technology that characterized Google also created many conflicts, some of which came about because of the acquisition of YouTube. The conflicts often start when Google attempts to expand beyond its core business. In March 2007, for example, Viacom filed a $1 billion copyright lawsuit against Google and its YouTube subsidiary. Google was strongly attracted to businesses beyond its core business activity, which created turmoil in the market and drew protests from its competitors (New York, April 9, 2007). These two competing companies are characterized by different corporate cultures based on win/lose propositions designed to win audiences at each other's expense. The legal action signifies the cultural clash between traditional media companies and internet start-ups, whose major attraction was to allow users to download music and video clips. When all is said and done, the legal conflict over copyright and customers creates attention and can be beneficial to both Google and Viacom by maintaining an air of uncertainty over the legality of YouTube.

High-tech entrepreneurs can be differentiated from other entrepreneurs by their interests, education, and previous experience. The decision by high-tech entrepreneurs to start a company may stem from a perceived opportunity, but negative circumstances from their former work lives may also prove important. Such negative circumstances led to the formation of Napster, which represented the first generation of online file-sharing virtual services. Online file sharing activities exposed the company to the accusation of piracy and to copyright infringement and resultant legal actions. The legal problems surrounding the copyright issue forced Napster from the virtual Web environment. The two founding entrepreneurs of YouTube learned hard lessons from observing the crash of Napster, which was based on the early dotcom spirit. This experience sensitized them to the uncertainties, risks, and high pressures in the Web environment. YouTube's perceptions about the fall of Napster and its intuitions about the future were crucial factors based upon which the risks were avoided and effective decisions were made.

The success of YouTube is the story of two high-tech professionals who were successful in identifying and exploiting the opportunities for growth and unmet potential for innovation. One characteristic of the process of creating YouTube was having fun while developing the company. Both founders had fun with the creation, growth and current sale of the company to Google, personal characteristics which Sotto (1997) called "playfulness." Their playfulness was an important element of the business development and was vital during the creation of the company products. In their success, the YouTube entrepreneurs assert that "they have not been driven by lofty ideals." Indeed, they have been contrasted with the high-minded ideology of Google, the potential new owner. Google's ideology and rhetoric are illustrative of their expressed vision of "making the world a better place and not being evil."

Another characteristic of the founders of the company, in addition to having fun, is their style of coping with costs and challenges of high technology. The entrepreneurs were distinguished by their down-to-earth manner and unassuming informality of coping in keeping with the times.

The YouTube founders had other personal characteristics. In addition to having fun, they were oriented toward modest income, informal lifestyles, and being autonomous and flexible about business and work life directions. They also faced and overcame the difficulties stemming from the emerging and fast-expanding internet-related business that was facing legal uncertainties and thriving on rapid changes. They approached such conditions with an adventurous spirit and eager to pursue opportunities in the virtual business world. The two founders generated new products from new technologies and were proactive, innovative, and relentless in their approach to business. They were part of the new wave of entrepreneurs and they developed their skills through high education, self-learning, and experience with the online world. One of the two entrepreneurs has been the engineering brain and possesses high mathematical skills, having been educated at the Illinois Mathematics and Science Academy for gifted children. He was a competent programmer, able to understand hard core engineering and, at the same time, remain playful in his private life, to have an addiction to poker games and habit of reading literature. The other partner had different personal characteristics. With his educational background in fine arts and design, he was responsible for the creative functions of the company. His personality could be characterized as playful and laid-back.

Working as a team of two, they developed the most innovative service for internet and Web users. Both entrepreneurs exhibit characteristics distinguishing them as the new-wave entrepreneurs with similarity in cognitive style and entrepreneurial mindsets that enabled their venture collaboration. They had effective "personal chemistry." They said of themselves: "We had a really good mix of skills. It takes the softer side of getting the community elements right, as well as the technology." They continued: "We got to experience a lot. We got to go through the bubble bursting to going public, to being acquired by Ebay, which eventually bought PayPal for $1.4 billion. We had the security that meant we were not afraid to fail, but we were motivated. We're gonna keep moving forward with our ad models, develop new concepts and expand our ideas around empowering users with content from professionally produced sources that they've embraced" (interviews with YouTube, *Financial Times*, October 13, 2006).

Another distinguishing characteristic that they exhibited was envisioning future possibilities that others failed to recognize, which created a new market niche for products. They not only envisioned but also were motivated by and believed in their own invention. It appears that their intuitive approach and alertness to the market needs, information and communication technology of YouTube, involved relying on their own impressions, explorations and synthesis when they made their decisions. Their mindsets illustrated broader professional and business skills that helped to capitalize on untapped business niches. For example, one innovation was to allow a Yale student to submit a video résumé, titled "Impossible Is Nothing," to investment bank UBS in 2006. It became a YouTube classic and spawned many imitations. By allowing user-based inventions, YouTube not only enlarged its own supply-side but also provided a creative example for its users who, having made their own films, videos or other film-related inventions, could use YouTube for their dissemination to other users. Another niche example that has been successful is YouTube Video Awards, a product within "the YouTube Community" that honours bygone achievements, in the spirit of the Oscars. Among the 2006 winners were "Hotness Prevails" (best comedy) and "Kiwi!" (most inspirational cartoon). It seems that the jury or authority that gives these awards had a preference for animated cartoons. The most crucial is that the idea of animation balances the aesthetic and ideological attitudes of the company. (*New York Times*, March 27, 2007).

YouTube entrepreneurs tend to have personal connections within the industry, communication skills, and persistence in bringing to the market novel and untested products that could be used at unconventional times and places.

The founders demonstrated personal skills for negotiation and communication, which have been a significant resource for securing contracts with big competitors such as Warner Music, Universal and Sony BMG. In these arrangements the founders faced problems associated with operating at the high level of 'legal uncertainty' and issues related to copyright practices. In appraising problematic situations they were likely to rely on their own subjective impressions.

Generally, the results of the analysis of the case of YouTube provide empirical support for the view that company success is complex and requires entrepreneurs to have several key characteristics: (1) exceptionally evocative products that attract not only buyers or users but also competitors, investors and other business actors, (2) perceptive thinking and intuitive mindsets, (3) high competency in communication and discussion skills, (4) high educational background, (5) good team-working and good interpersonal relations with partners and co-workers.

In the contemporary high-technology, global and competitive business environment, the ability to create novelty is complex and involves a Web of values, power relations, and personalities with entrepreneurial mindsets, knowledge and knowing, and strong motivation to succeed.

CONCLUSION

The illustration of the entrepreneurial constitution of ventures in a high-tech environment is encouraging in a number of respects. It illustrates that successful entrepreneuring in a high-tech environment benefits from the high technology, virtuality and cyber-culture of modern organizations. This chapter reflects on how contemporary entrepreneurs, by using intangible and knowledge-based resources, created a sophisticated technology. The findings draw attention to the technology-driven opportunism within virtual media and the role of entrepreneur within human–computer interaction.

Its point of departure was to challenge the new emerging understanding of the virtual communication mandates according to which new empowered workplaces are being created. This chapter also provides empirical evidence about the life cycle of YouTube business development and illuminated the virtuality-related process both descriptively and operationally. The development of knowledge about entrepreneuring in high-tech environment concerns understanding of both boundaryless flexible and highly competitive business context, characteristics of entrepreneurs and communication processes.

The global view of information and communication technology opens the way to evolution of the different forms of virtual organizations and entrepreneurship. Suffice it perhaps to recall that in our time computers are unable to perform cognitive tasks related to new venture creation that a person can. The YouTube example illustrated a combination of high-tech successful entrepreneuring process and virtual organization. These findings also indicate the distinguishing characteristics that high-tech entrepreneurs represent. In the case of YouTube, net-broker and entrepreneur share some of the characteristics. New-wave and net-brokering entrepreneurs can be depicted as well educated, receptive to newness, creative, playful, communicative and using their entrepreneurial mindset to look for new technical and business opportunities. The scholarly literature provides theoretical and empirical support for the view that entrepreneurs can be regarded as net-brokers identifying and leveraging business opportunities and also combining resources and assets to create new virtual businesses.

This chapter has presented some aspects of the impact of high-tech work life upon entrepreneurial action, drawing on the example of a currently active Google subsidiary. The story of YouTube is a distinctive and highly reliable example of successful high-tech entrepreneurs, which are in constant dialog with the worlds of others. They play an important role within human–computer interaction. A closer look at the virtual media activities and entrepreneurship may shed new light on these challenging issues, providing useful frameworks that differentiate high-tech entrepreneurs as cyborgs from others and predicting high technology-driven environments. Further research should include the subsequent and more systematic following of their business performance.

REFERENCES

Aggestam, M. (2003). Developing intellectual entrepreneurs of the universities: The example of Venture Lab. In S. Kwiatkowski & J. Sadlak (Eds.), *Intellectual entrepreneurship through higher education.* Leon Kozminski Academy of Entrepreneurship and Management. Warsaw: Unieco-Cepes.

Allinson, C., Chell, E., & Hayes, J. (2000). Intuition and entrepreneurial behaviour. *European Journal of Work Organizational Psychology, 9,* 31-43.

Barnatt, C. (1995). Office space, cyberspace, and virtual organization. *Journal of General Management, 20,* 78-92.

Birely, S., Cromie, S., & Myers, A. (1991). Entrepreneurial networks: Their emergence in Ireland and overseas. *International Small Business Journal, 9,* 56-74.

Burns, T., & Stalker, G. M. (1961). *The management of innovation.* London: Tavistock.

Burt, R. (1992). The social structure of competition. In N. Nohria & R. Eccles (Eds.), *Network and organizations* (pp. 57-91). Boston: Harvard Business School Press.

Byrne, J. (1993). The virtual corporation. *Business Week* (8 February), 98-102.

Chell, E. (2000). Towards researching the "opportunistic entrepreneur": A social constructionist approach and research agenda. *European Journal of Work and Organizational Psychology, 9,* 63-80.

Clark, A. (2003). *Natural-born-cyborgs: Minds, technologies, and the future of human intelligence.* Oxford: Oxford University Press.

Dick, P. (1968). *Do androids dream of electric sheep?* New York: Doubleday.

Featherstone, M. (1995). Post-bodies, ageing and virtual reality. In M. Featherstone & A. Wernick (Eds.), *Images of ageing.* London: Routlage.

Featherstone, M., & Burrows, R. (1995). *Cyberspace/cyberbodies/cyberpunk. Cultures of technological embodiment.* London: Sage Publications.

Ferguson, C. (1988). From the people who brought you voodoo economics. *Harvard Business Review, 66,* 55-62.

Florida, R., & Kenney, M. (1990). *The breakthrough illusion: Corporate America's failure to move from innovation to mass production*. New York: Basic Books.

Franke, U. (1999). The virtual Web as a new entrepreneurial approach to network organizations. *Entrepreneurship and Regional Development, 11,* 203-229.

Gates, B., Myhrvold, N., & Rhinearson, P. (1995). *The road ahead.* New York: Viking.

Gerlach, N., & Hamilton, S. (2000). Telling the future, managing the present: Business restructuring literature as SF. *Science Fiction Studies, 27,* 461-77.

Gibson, W. (1984). *Neuromancer.* New York: Ace Books.

Goldman, S., & Nagel, R. (1993). Management technology and agility: The emergence of a new era in manufacturing. *International Journal of Technology Management, 8,* 18-38.

Goldman, S., Nagel, R., & Preiss, K (1995). *Agile competitors and virtual organizations: Strategies for enriching the customer.* New York: Van Nostrand Reinhold.

Göranzon, B. (1991). *The practical intellect: Computers and skills.* London: Springer.

Hatch, C. (1995). *The network brokers handbook.* US Department of Commerce, National Institute of Standards and Technology, manufacturing Extension Partnership, Gaithersburg, MD 20899, USA.

Haraway, D. (1991). *Simians, cyborgs, and women: The reinvention of nature*. London: Free Association Books.

Hardwick, M., Spooner, D., Rando, T., & Morris, K. (1996). Sharing manufacturing information in virtual enterprises. *Communication of the ACM, 39,* 46-54.

Johannisson, B. (1990). Community entrepreneurship—Cases and conceptualisation. *Entrepreneurship and Regional Development, 2,* 71-88.

Jordan, T. (1999). *Cyberpower. The culture and politics of cyberspace and the internet.* London and New York: Routledge.

Kanet, J., & Faisst, W. (1997). *The role of information technology in supporting the entrepreneur for the virtual enterprise: a life-cycle-oriented description.* Working paper, Clemson University, Clemson, SC.

Keeley, R., & Tabrizi, B. (1995). High-tech entrepreneurs: Serious competitors or troublemakers? *Journal of High Technology Management Research, 4,* 127-143.

Kirzner, I (1997). Entrepreneurial discovery and the competitive market process: An Austrian approach, *Journal of Economic Literature, 35,* 60-85.

Miles, R., & Snow, C. (1986). Network organisations: New concepts and new forms. *California Management Reviews, 28,* 62-73.

Minzberg, H. (1976). *The structuring of organizations.* Englewood Cliffs, NJ: Prentice-Hall.

Moshowitz, A. (1986). Social dimension of office automation. *Advances in Computers, 25,* 335-404

Paul, S., Whittam, G., & Wyper, J. (2007). Towards a model of the business angel investment process. *Venture Capital, 9,* 107-125.

Pihkala, T., Varamaki, E., & Vesalainen, J. (1999). Virtual organisation and the SMEs: A review and model development. *Entrepreneurship and Regional Development, 11,* 335-349.

Pruitt, S., & Barett, T. (1991). The virtual workplace. In M. Benedikt (Ed.), *Cyberspace: first steps*. Cambridge, MA: MIT Press.

Reich, R. (1987). *Tales of a new America*. New York: Times Books.

Roberts, E. (1989). The personality and motivation of technological entrepreneur. *Journal of Engineering and Technology*. *6*, 5-23.

Rushkoff, D. (1994). *Cyberia. Life in the trenches of hyperspace*. San Fransisco: Harper.

Scholz, C. (1997). Das virtuelle Unternehmen—Schlagwort oder echte Vision? *Manager Bilanz*, *1*, 12-19.

Sherman, B., & Judkin, P. (1992). *Glimpses of heaven, visions of hell: Virtuality and its implications*. London: Hodder & Stoughton.

Sotto, R. (1997). The virtual organization. *Accounting, Management, and Information Technology*, *7*, 37-51.

Starr, J., & MacMillan, I. (1990). Resource cooption via social contracting: resource acquisition strategies for new ventures. *Strategic Management Journal*, *11*, 79-92.

Tomas, D. (1989). The technophilic body: On technicity in William Gibson's cyborg culture. *New Formation*, *8*, 113-29.

Tomas, D. (1991). Old rituals for new space. In M. Benedikt (Ed.), *Cyberspace: First steps*. London. MIT Press.

Upton, D., & McAfee, A. (1996). The real virtual factory. *Harvard Business Review*, *74*, 123-133.

Wiener, N. (1948). *Cybernetics*. New York: Wiley.

ENDNOTES

[a] See: Dick's (1968) novel, *Do androids dream of electric sheep?*, on which movies such as *Blade Runner* (1983), *Stepford Wives* (1975; 2004), *Neuromancer* (1984), *Idoru* (1996) and the *Matrix* trilogy were based.

[b] For discussion of both plastic surgery and biochip implants in the cyberpunk literature, see Tomas (1989; 1991) and Featherstone & Burrows (1995).

[c] Data about YouTube was collected from the *Financial Times*, *New York Times*, *Time* and Wikipedia. See: www.ft.com/google, http://www.time.com/time/business/article/0,8599,1595049,00.html, and http://www.en.wikipedia.org/wiki/YouTube.

Chapter XIX
Identifying Flexibilities

Marja-Liisa Trux
Helsinki School of Economics, Finland

ABSTRACT

This chapter takes you to a data security workplace in Finland. It presents reflections on the tensions of managing selves and others, as experienced by the employees and the managers. It argues that a generally critical approach to normative management may be overlooking the actual complexity and ambiguous nature of the late modern cultural environment. Both self-authoring and manipulative moves are made difficult by the amalgamating hegemonic and countercultural currents. The author points at chances for resistance through new forms of literacy. Instead of dropping "culture" as a conservative or managerial pursuit, we must learn to navigate successfully in the broken cultural landscape of today's workplaces. The very same images that can be used for manipulation are open to more solidary configurations by the cultural and social imagination of organisational members.

Knowledge economy is the dominant call today, as much in official national and multinational strategies as in the rhetoric of corporations. Organizations must meet demands for innovativeness and quick learning. As the human imagination is a hard thing to control, the managerial elites have tended to turn from Taylor-like physical control to more psychological forms that lead discreetly to the desired outcomes. No direct monitoring is needed to guarantee maximal input from the workforce. Manipulated workers themselves define high objectives and devote themselves to their tasks, sometimes at the risk of their health. The limits of this approach are attained in cases where increased pressure to produce ever new innovative solutions on time and respecting consumer demands has led *de facto* to decreased creativity and burnout symptoms. To the frustration of the fashionable faith in managerial omnipotence, it seems that the sought-after revolutionary innovation just cannot be merely called into being, but may instead appear as a byproduct of leisurely playful activity. Indeed, market pressures and innovations seem to mix like oil and water. How can any realist management strike a balance between these two?

Alongside rather primitive retreats to tight control, the past decade has also witnessed fairly "democratic" and participatory moves, partly as an inheritance from the high-tech industry's most typical professional culture, referred to alternatively as "nerd culture," "hacker ethics," or "techies culture." The field of high-tech management is by no means dominated by manipulative forms. Very different approaches coexist, and managers as much as workers rely upon one or another of them in the underdefined cultural conditions of late capitalism. Local societal and institutional conditions bring further variation to the arenas of industrial relations, regulatory practices, and work cultures.

This chapter is based on ethnography in a middle-sized Finnish software company F-Secure[a], mainly in its Helsinki headquarters and by comparison in a Silicon Valley subsidiary[b]. I went to the organization originally to inquire into ethnicity and its management, since the company had a relatively high proportion of foreign workers. I soon learned that there was no management related to ethnicity. The first round of my inquiry in 2000 suggested that foreigners in Helsinki were nevertheless exceedingly happy about what they called "Finnish management." Why? Could it be explained in terms of the workers being so lucky—middle-class with stable incomes, during a boom and with a common professional identity? Largely, of course. They were, however, happiest about the more sustainable traits in management style: investment in education and respect for personal autonomy etc. The foreigners summarized it as "air of democracy."

After the downturn, F-Secure's "democracy" came under attack, but survived and was even institutionalized in some organizational practices. Yet its acting out in the social exchange of daily work is continuously contested and defended in the face of the postmodern pressures of investor relations, financial insecurity, consumer demands, market strategies, and technological development – all of which may and do cause sudden turbulence that needs to be checked, often curtailing the freedom of the workers. In this day-to-day balancing action, cultural forms like the discourses around "flexibility" take the role of tools for self-management and managing others. Both workers and managers attempt to construct situated identity and agency, but their attempts are confused by the ambiguity of connotative links and the absence of clear addresses in dialog with the world.

"FLEXIBLE" WORK AND ITS CULTURAL CONTEXT

Many writers have drawn attention in recent years to the widespread use of the image of "flexibility" in dominant postmodern discourse, especially concerning work and production. For Richard Sennett, this is one of the cornerstones in the talk of the managerial elites or "Davos men" (Sennett, 1998). Liberated from the iron cage of rigid organizations, workers must now survive the fierce competition of postmodern, unpredictable markets, selling their capabilities anew each day and renewing their skills as best they can (Sennett, 2006). I have myself listened to a representative of the Confederation of Finnish Industries reciting the necessary qualities of today's workers as "creativity, curiosity, interaction, development, renewal, faith in one's own opportunities, target setting, interest in the customer, skill to tackle in the network of ever more differences, social skills, learning from other cultures, and transformability." He also quoted Charles Darwin as having said that "only the most transformable survive." (Pokela, 2005)

As the anthropologist Emily Martin has shown, the image of the flexible citizen is not confined to the realm of work alone. She has followed its occurrences in U.S. society from discourses concerning immunology to health and fitness and then to workplaces. Borrowed from medical jargon, the public discourse started during the

1960s and 1970s to apply the idea of resistance to various microbes as a key element in the popular image of the good citizen, who took care of her organism (by means of a healthy diet, exercise and a regular life style) so that it could react to each illness attack fittingly, nimbly and intelligently—beating off the intruding germs. This soon led to the corresponding image of the citizen as a flexible, trained body, adapting quickly to each new environmental challenge, and then to the present image of the ideal postmodern worker, seizing each new opportunity as it comes (See Martin, 1994). Like individuals, organizations must also be elastic. Even dinosaurs must dance. The latest episode can be discerned in repetitive accounts of the competitiveness of national economies against each other. Which country is going to get this year's flexibility prize?

Manipulative Hooks

Manipulative control in IT has played upon several hooks, caching the computer experts by elements in their heterogeneous cultural heritage. Since the beginnings of this professional subculture, mutual solidarity among fellow programmers was sustained through help and sharing of information. Reputations among colleagues were achieved by (sometimes extreme) devotion of time and energy. Countercultural elements nurtured suspicion towards authorities and underlined individuality in the professional identity. "Hackers" were the cowboys of the digital frontiers in a genuinely American way.[c] Liberation from the "iron cage" of hierarchical organizations was their watchword. While the boom continued, manipulative employers easily made workers "devote" themselves to the company vision, "help" the employer and feel free to work day and night like the good entrepreneurs they all were.

When the bubble burst, things got complicated. In the disillusionmentment that ended or at least problematized the flirtation between countercultural and neo-liberal elements, some employees at F-Secure seem to take sides with the employer, tightening their belt in the hope of a reward for their loyalty. Others seem to be questioning the motives of the employer more deeply. Judging by its decisions, the management oscillated between radical democratic experiments, cost cutting in panic and neo-liberal faith in flexible, globally distributed production. In this context it is not necessary to refer to "management" as a term of firmly organized action following a script or manipulative plot, and backed by massive power—as in fashionable accounts of "strategic management." Instead, I look at the management sideways, as a more threatened form of activity, managing (or not managing) to do something: to steer the boat, to prevent it from sinking, to negotiate with the crew, to guess where one might be amid the turbulent waters, and to choose among plausible options for the next move. In my case company, the managers on the whole share the professional subculture of their workers. The cognitive and moral confusion that followed the downturn has concerned them as well as the workers, though of course with different positioning. Members of the managerial class have significantly more power than their subordinates, but they may have no better understanding of what to do with that power; how should they understand the present situation, what does it mean—and, thus, who they are themselves. Despite the fact that these uncertainties may be poorly recognized and reflected upon by the people themselves, they point at the need to tackle the question of *agency* in a heterogeneous cultural environment under strong pressure to change.

Uncertain Agency

The American anthropologist Sherry Ortner has argued since the 1980s for taking up agency as the pivotal point between cultural (semiotic) and social (power relations) forces or structures (Ortner 1984, 2005). Building on the works of Bourdieu, Giddens, Sahlins, Geertz, and Sewell, who each in their distinct ways have tried to reintroduce agency into the heart of social theory, Ortner offers a picture of the individual's (or a group's) unique, historical existence, struggle for survival and a meaningful life. Her aim is to avoid both the danger of conceiving human subjects as blind carriers of tradition, and the opposite danger of seeing them as nameless marionettes of power structures. In her approach they have partial freedom, limited by fears, passions, hopes, restrictions and bonds of all kinds. Ortner is seeking a way to look at life from the subordinates' position: as a struggle for a foothold, however small. Thus her ideas lend theoretical support to many current ethnographies describing the problematic or narrow, even marginal life-spaces of various powerless groups in late modern societies—including immigrants and other inhabitants of transnational social places. As an extension, it may be adapted to accounts of the stratified and unstable life-spaces of employees in the high tech companies. Substantially wealthier than marginal immigrants as they are, their lives are still profoundly marked by the power differentials present in their own organizations and the encompassing business ecology.

Ortner presents two readings of *postmodern consciousness*: Fredric Jameson (Ortner, 2005, p. 41-42) and Sennett (ibid. p. 43-44). Jameson was much criticized in the 1980s. His account appeared then as somewhat fancy, over dramatized and socially ungrounded. But the 2000s have shown us how postmodernism can strike. In Sennett's description of work under the "new capitalism," futurological visions have become a reality, albeit a rather ugly one. According to Ortner, the two are talking about same things (Table 1).

Both are pointing to the need to restore meaning and orientation in the postmodern world, which has become uninterpretable and illegible. We must learn to read our world and show others how to read it, says Ortner (ibid. p. 44-46). Countercurrents of subjectivity exist just as countercurrents of culture do. According to Ortner, subjectivities are complex and reflexive: while few people fully embody the dominant culture, and some are totally subjected by it, most are, nevertheless, partial misfits. They partially internalize, partially reflect upon, and react against it (ibid. 45).

In Ortner's account, the multilayered and reflexive cultural consciousness provides the grounds for questioning the dominant culture. I would compare this to the historian William Sewell's theorizing on misfits between cultural model and reality that provide the grounds for cultural change and transformative/alternative movements (See Sewell, 1999). Both are dynamic, non-totalizing accounts of culture.

But how exactly is culture woven into social life? According to Holland, Lachicotte, Jr., Skinner, and Cain (1998), cultural forms are the tools of self-management (or self-authoring) as much as managing others. They are the currency of all social fields, including workplaces. These authors have combined the ideas of Bakhtin and Vygotsky and rewritten the borders of that combination to the theory of *fields* by Bourdieu. The result is a theoretical edifice that takes on the duty of explaining the intricacies of learning to live one's

Table 1. Comparison of Jameson's and Sennett's analysis of postmodernism according to Ortner

Jameson:	**Sennett:**
waning of affect, depthlessness	indifferent work ("just a job"), masks of cooperativeness by the bosses
spatial disorientation	temporal disorientation ("no long term")

identity—or identities. Their attempt can be seen as an answer to the call made by Ortner for bridging the cleft between semiotic structures and power structures. Their account of *heteroglossia* is useful for understanding the present conditions where people attempt to form their "voice" out of a multitude of different and differently positioned "voices" of others (ibid. p. 181-183). As *spaces of authoring,* workplaces like the one I describe are difficult to outline, marked by uneven power positions, insecure and changing. But they are not beyond human bricolage and not immune to countercultural offensives and dialogization. Accounting for the process of authoring enables us to understand the struggle involved and the time and effort it takes.

F-SECURE CORPORATION

F-Secure Corporation protects individuals and businesses against computer viruses and other threats spreading through the Internet and mobile networks. Its products include antivirus, network encryption, desktop firewall with intrusion prevention, anti-spam, and parental control. A constant vigil is kept at the company headquarters in Helsinki against any new viruses. Founded in 1988, F-Secure was listed on the Helsinki Exchanges in 1999. In addition to Finland, the company has offices in the USA, Sweden, Norway, France, Germany, Italy, Poland, The United Kingdom, India, Singapore, and Japan.

F-Secure's retailers and distributors have expanded to more than 60 countries. As its products are to an important degree distributed through Internet service providers and mobile equipment manufacturers, direct marketing to individual consumers is less important. In the Finnish context, F-Secure is a middle-sized organization. Its personnel doubled from 200 to 400 during the boom (1999-2000), and toward the end of this research (2004) had come down to an intermediate level of approximately 300. In Finland it is well known as one of the flagships of turn-of-millennia technology. Its enlistment created long lines of popular investors on the streets of Helsinki. Today it has a more consolidated reputation, at times a quasi-official position as the favorite source for journalists wanting to ask anything related to Internet threats. According to the management, F-Secure is well known in the Nordic countries, somewhat in Europe, but scarcely at all in the United States. Indeed, as I talked to the American employees, many of them started with reflections on working for a *small* company, overshadowed by large competitors.

At my first visit to F-Secure, I was given a list of the company's values (see Table 2). The HR manager took care to explain them to me, as they were—according to him—the basis of the

Table 2. F-Secure's company values, as given in 1999

Value	Explanation
1. People	Both "fellows" (refers to all organizational members) and customers are important.
2. Innovation	Mistakes are allowed. "There can be no learning if we fear mistakes."
3. Integrity	Legality, justice, openness. "We go beyond that stipulated by law in order to treat people well, and to protect the environment."
4. Building clocks	"When somebody makes an innovation at work, it must be put on the wall like a clock, so that others can use it too."
5. Fun and joy	"Working must also be fun."

attractiveness of F-Secure as an employer, both nationally and internationally.

While interviewing the workers, I asked their opinion of the values. Most agreed that they were both the right values for the business and principles that were observed in the everyday life. Some had reservations concerning the best ways to realize the values, and their relative weight. One employee (from the U.S.) said that "integrity" was a very Nordic value, a "social-democratic" feature of the organizational life.

F-Secure was no longer a start up when the bubble burst and the IT-sector began its downturn. Crisis management apparently took over, and management of creativity was set aside. In addition to cutting all extravagant benefits (and some rather modest ones), people were laid off. Those with temporary contracts were the first to go in Helsinki. The remaining employees were still recovering from this experience when I returned in 2002.[d] According to the Americas manager, the company had intended to enlist on NASDAQ, but this was dropped when the downturn hit. Its founder and long-term CEO is still the principal owner.

"FINNISH MANAGEMENT IS WONDERFUL"

Since the company had no experience in diversity management, but an ethnically mixed workforce[e], I was curious to see how they were doing. The first round of my inquiry suggested they were doing fine in Helsinki. To my unbelieving ears the foreign interviewees *all* insisted that they were exceedingly happy about what they called "Finnish management." In spite of differences in job titles, gender, and country of origin, all related spontaneously that they greatly appreciated the company culture and operating methods.

Management is really good here. Thwarting innovation wouldn't work in this business anyway. This is a good model, no matter where you are from. It is wonderful to see that you are getting respect.[f]

If there was any difference, westerners in particular admired the flexibility and efficiency of "Finnish management methods." People from Russia and Asia emphasized freedom and respect for individuals. But even these differences were slight. The foreigners gave a consistent picture of what the company was like. Table 3 presents the recurrent topics in this talk, as much in their words as possible.

Could the foreign employees' happiness be explained in terms of the workers being so lucky at the height of boom, soon after arrival in a new country? No doubt that is a major reason. It does, however, merit attention that what they were happiest about were not the stock options or the company sponsored holiday trips abroad, but the more sustainable traits in management style: investment in education, respect for personal autonomy, listening to the workers etc.

Finns were somewhat less excited, mostly noting that the workplace had "filled expectations." Some even criticized the company for overdoing it:

Sometimes informality and spoken agreements lead to ambiguity about who is going to do what.[g]

PROFESSIONAL AND MORAL DIMENSIONS OF WORK

As far as I tell from the glimpses offered in the interviews, work at F-Secure presents a fairly typical array of duties in the product-centered IT-sector. The pivotal function seems to be the anti-virus laboratory with its researchers, and the large number of people in "development teams" producing the programs destined to customers. Other works consist of supporting scientific expertise (a mathematician), localizing, pre- and

Table 3. Foreign employees' perception of management methods at boom time

Flow of information	Openness, no secrets (but one has to ask for information)
Operations	Efficiency; the organization is able to react quickly, action is taken following a decision The culture is sensible and looks for practical solutions The culture is flexible, not bound by formalities Oral agreements
Treatment of employees	Individual employees and their time are respected, and employees are listened to and trusted Mutual support among colleagues, supervisors offer help
Organization	Organization is flat and feels democratic

post sales support, sales, marketing, managerial work, secretarial work, legal assistance, HR, communications, etc. According to the HR manager, the requirements for recruiting anti-virus analysts include for instance familiarity with "old fashioned" programming languages, nowadays less common amongst western IT-professionals. Recruitment is extremely focused, sometimes causing long delays in filling vacant positions, as fitting applicants are rare even among the global workforce. Michael, a Spanish expert, had been recruited in 2002 in the midst of the toughest downturn when unemployment offices received a wave of IT-professionals as client. There had been a search of six months for a fitting applicant.

Despite his young age and only two years at F-Secure, Michael gives a typical expert professional's account of his work. To him, the work essence is the centre of identity and what most matters to him. He divides the tasks into two categories as "interesting cases," when there are some particularly evil viruses or something new technically, such as,

...cases where you really need to put a lot of time but that are interesting. And I even continue doing it at home, because it's like I have to find out what this does.[h]

and as "routine cases" when,

...there is a lot of very simple viruses like created by the teenager that need réclame from the internet, that put it together and put some insulting stuff inside it. ... That is never going to make it anywhere, maybe to something like a hundred computers and that's all. ... So here I see a lot of those. Sometimes I spend like four weeks analysing these like shit. And some of these are not even proper viruses, some of them don't even work well. ... We still have to check what they do, because we do good work. And we don't if we don't check them. ... So it's kind of a monotonous kind of work sometimes, looking at another one and another one, and another one. (Same source)

I have heard a popular opinion about anti-virus researchers, asserting their proximity to the "hackers" disseminating the viruses. While this can be understood from the perspective of the security industry (policemen and criminals; inspectors and stock speculators), as a wry sort of appreciation for the virtuoso criminal, I would not forget the more general tendency of any expert professionals to identify with their work, require a level of challenge matching or extending their capacities, to learn new things and to expect due appreciation. Michael also likes the media appearance, being interviewed by newspapers for instance, concerning special cases that attract public attention. Is he any different from the university researcher dreaming of public attention and a solid reputation among fellow scientists?

The anti-virus research is also based on constant vigil, and workers like Michael are bound during their free time to come to work within two hours in case of need. This is a feature he has come to "hate," something that casts a stressful shadow over all free time activities.

I never got woken up actually in the middle of the night. I've had to stay up to three or four at night, and I have woken up at seven. But nothing like four o'clock at night I haven't had to come to work at that time. ... But still having to stay in the office and work... Once it was my birthday and a Sunday, and I stayed the whole day in the office. There are those kinds of things. (Same source)

Michael doesn't have children, but some colleagues do, and their tightrope walk between the urgencies of home and work make Michael think twice about getting children.

Most of the employees are middle-class people, sons, and daughters of "good families" and between their twenties and forties. Hence I felt deceivingly at home with them, as they talked about their studies and travels and experiences at settling and forming a family. Among the foreigners interviewed were people from Australia, the United States, India, Switzerland, Norway, France, Portugal, Spain, Denmark, and Russia. Most of them could not be described as *visible minorities*. The Russians, however, do suffer a collective stigma in Finland, related to national history. There is a special derogatory reference to Russians, one that Wierzbowsky—one of the virus researchers—said was used on him by a customer, but never by a fellow colleague. It was as if they had invented a firewall to keep out all the discrimination, competition, neo-nationalism and ethnicizing evils of our time.

They were also pretty well off financially. The first round took place at the height of the hype. The young men (there were some women, but the male "voice" dominated, which is hardly a surprise) not only had stable incomes, but they all benefitted from stock options, a new phenomenon invested with many personal life prospects and hopes of a bright future. Perhaps even more importantly, the boom time atmosphere seemed to offer boundless opportunities for personal development, social and geographical mobility, and even social heroism for the avant-garde of high technology. At that time most were adherents of the "fresh page" doctrine (in Richard Sennett's terms) stating that human creativity would be liberated from the "iron cage" of military-like bureaucracies, work would become play, and new inventions, such as the Internet, would liberate the rest of human kind to a new dawn. I welcome any doubtful readers to review some of the boom time books and financial magazines for a reminder of the social context of my informants' comments that may look naive in retrospect.[i]

One more factor overcoming sociodemographic differences is professional identity. Most of the employees are IT-professionals, but many of them vigorously reject the term "nerd," since it has indeed been used as a stereotypic and degrading label. In my reading, "nerds" are not those clinically introvert Aspergerians the general prejudice would imagine, spending days literally glued to their keyboards with their fingers sticky from the grease of chips and the sugar of soft drinks. The offices of these professionals are not littered with scattered papers, pizza boxes, and empty bottles. In my limited knowledge, they are decent, if somewhat arid offices where variably social persons work alone, in pairs and in teams. Teamwork, sequentially organized into projects, is their bread and butter, and social skills a topic frequently coming up in their talk. Some have a room of their own, others share rooms. To my knowledge, there are no open-plan offices, but there are lobbies for coffee breaks. According to Michael, "we hang around in the lobby and talk and brain-storm, and things like that." Others, like Mark, the scientific advisor, have a more lonely sort of position. Some of the foreigners suggest that they may at times refrain from joining a social gathering because, as Bharat, the Indian localizer said, "just because of me they have to talk in English."

Some of them are close friends, and they participate in outdoor days and company parties like any Finnish workers. The most striking thing I observed happened in the first of the two Christmas parties that I attended. Contrary to the custom in Finland, where Christmas parties are a routine organizational event, spouses were invited. This had almost the same effect as bringing parents to an adolescents' party; there were no drunkards under the tables. The party was lively; people danced, but remained almost sober. In the other party, I found myself in the midst of a yelling and singing audience to a rap group, formed among the employees and cutting some dash even beyond the organization, I was told. In both cases, foreigners were present and mixing with Finns.

While the general prejudice has concentrated on personality issues, I find the moral dimensions of professional identity more interesting. It is difficult to find a reference word fitting and acceptable to all professionals in the digital sector. Some prefer this and others that, and any inquiry is likely to lead to lengthy discussions with multiple and controversial recommendations. Perhaps I developed a slight preference to "nerd" over "hacker" because the later uncomfortably seemed also to refer to my informants' professional antagonists in this particular sub-industry: the producers and disseminators of computer viruses. I understand a different preference, and apologize for possibly hurting somebody's professional or subcultural identity.

Computer experts are to a varying degree participants in a truly global network of fellow professionals. They have their own role models and their own Mecca, Silicon Valley. While most programmers never make a pilgrimage to California, and not everybody shares the ideas of the most radical bearded gurus, their existence lends to the whole professional current some aura of expertise, innovativeness and potential for social consequences—hence, moral importance. And the most obvious quality of this identity is its transnational nature. Through the 1960s on to the very recent past, the "nerds" formed one of the natural audiences, and gave birth for their part, to the still innocent and emancipatory "fresh page" discourse, in the variation that has been called "hacker ethics" (Gere, 2006; Himanen, 2001; Kaplan, 1999). *They* had broken the power of central computers and distributed computing to each one's desk, *they* were teaching IT-skills to all, *they* were providing everyman with CIA-proof encrypting programs, *they* had stunned the business elites with open source code—the fruit of *their* transnational communities.[j] Asking about the role of ethnicity at work, I heard a dozen variations on the idea: borders are absurd, distances can be overcome.[k] If these people would lower themselves to the petty game of ethnic distinctions, would they not betray, among other things, their professional identity?

MANAGEMENT PRACTICES AT HYPE AND DOWNTURN

Returning to the workplace in 2002 after the downturn in IT, I found that the work ethos had suffered a blow. Disillusionment and fear for jobs were not given an ethnic attribution, however. Rather, the whole issue of ethnicity and cultural differences was wiped away, neutralized. The dominant culture among most Finnish workers strongly advocated for "practical solutions" against "multicultural bureaucracy." Apart from the passing comments of the foreigners, according to which F-Secure was "a Finnish-like company," nobody gave me even half a reason to suspect that they might be discriminated against on the basis of ethnicity. Suddenly expelled from their formerly sheltered position, the workers had to face the same economic insecurities and corrosive pressures against their professional morale as any other group of workers under the postmodern conditions. Yet these tensions were not ethnicized. This is remarkable in the context of the steep economic cycle they had gone through, and the fact that no special attention had been directed to diversity by the management. Instead of looking for scapegoats from among their colleagues, many of the workers openly and passionately criticized the management in terms of market and technology choices, enlistment and other financial decisions, and the way costs were cut.

The workers' perception of "democracy" does not seem ungrounded in the light of the fact that like many high tech companies, F-Secure is a young organization founded by a handful of friends. The founder-CEO used to sit on the floor when they ran out of chairs at monthly meetings. That no longer happens, but more established forms of "democracy" flourish; in addition to the monthly meetings, an advisory workers' council with elected representatives and top management offers a regular forum for a two-way flow of information. This is in fact a mandatory body in all Finnish organizations of a certain size. The general opinion has recently much regretted the fact that these "cooperative councils" (*yt-neuvottelukunta*) only meet to hande the strictly stipulated matters involved in layoffs. The letter of law is thus observed, while its spirit is neglected. In this matter F-Secure follows its own path.

Burnout counter-offensives were taken already at boom time to discredit the self-imposed image of the tireless programmer. At the downturn, the bulk of the workforce attained a family founding age, and family-values now rule, making it legitimate to avoid traveling, for instance. Extra hours are counted. In spite of this, the distance between the workers and the management has grown most noticeably since the layoffs, and more subtly by the growth of the personnel, and by postmodern demands for consumer service and stakeholder impression management. This is not an exceptional course of development among Finnish IT-companies. According to a review of management practices (Ruohonen, 2004, p. 22), business has come to dominate technology, product-focused organizations have turned to service and all employees must now learn to deal with customers.

Layoffs, fusions, and re-engineering have created insecurity, which in turn has led the workers to unionize and join unemployment funds. Unionization is at a high level in Finland, like in the other Nordic countries, and extensive (post WW2) legislation and political culture has tied the triangle of unions, employers, and state authorities together, though not without tension. The IT-sector however, has been a notable exception, for reasons probably linked to its short and wealthy history and its (American) ideational roots in extreme individuality—including neo-liberal tenets of economic individuality (Gere 2006, 138). The downturn profoundly shook this constellation, causing a rush to more "Nordic" forms of industrial relations. At F-Secure, unionization also reached a level that, according to Finnish legislation, required appointment of a company-level representative of the technology workers' union from among the workers.

In Finnish ICT-companies, work pace and quality demands have increased (Ruohonen, 2004). Global outsourcing means that software development is no longer the best source of income[1]; companies earn better by continuous service contracts. Big software development projects are risky because margins have narrowed, and only a few Finnish companies have succeeded in them. Most companies look forward to the roles of service provider and partner in a knowledge network, though not all. Because of its products, F-Secure for instance is continuously oriented towards program development, and its service provision is tied to its products.

Business development has had its counterpart in management culture. The unofficial, spoken agreement type of management "by the kitchen table" with generous economic benefits was commonly used at the hype time to attract skilled workforce—a rather mobile workforce (ibid. 27). Now the companies appreciate somewhat more formal methods such as hierarchy, departments, guidelines and job descriptions. Workers pay attention to "sensible" things like attractive and meaningful jobs, a healthy workplace atmosphere, organizational culture, management methods and an interesting substance (ibid. 28). In sum, fancy has turned into the normal, solid and sensible. This picture applies fairly well to managerial culture at F-Secure, with the exception that solid and sensible things were always at the top of the list, even at hype.

As an organization, F-Secure seems to be relying on what the Finnish sociologists Juha Antila and Pekka Ylöstalo (2002) have termed the *proactive mode*, allowing the workers to influence both the products and the working conditions in exchange for the expectation that they bear responsibility and take initiative in the flow of events. Product design and marketing are coupled, and the mode also includes cooperation with various stakeholders. In the case of F-Secure, sustainability as a related idea is visible in efforts to shrink the "ecological foot print" by using green energy and by joining a corporate responsibility network.

"Democratic" management in a "sustainable" organization? I admit that the description may seem naive. Have I simply overlooked the clever workings of managerial exploitation? To the extent that the workers themselves may be doing so, this may be true. But the workers were not uncritical; they openly criticized the industry, the customers and the management. It was a very different discourse from those outlined by Gideon Kunda in "Engineering Culture" (1992). One of my informants with experience of organizations in Russia and the U.S. highly appreciated the fact that problems can be acknowledged and conflicting interests negotiated instead of pretending that there is always a win-win situation. Of course the management is not bound to its subordinates' views; There is only a kind of enlightened autocracy, not real democracy. Does the management after all hear the bottom-up flow of wisdom? The company does not escape the basic tension of any centrally led organization, and it is part of late modern capitalism. Something still seems to alleviate the worst pain that might be caused by these pressures. I have, however, only the workers' understandings of what that something might be. My material points at what they consider the sources of their feeling of an "air of democracy" and the varying ways in which they engage themselves in its defence.

I have myself worked in many Finnish organizations, and some have been almost like this case, but most have been much more authoritarian and controlling. It seems to me now that the firm is really exceptional, but in *general management culture*, and not in the absence of diversity management. DM practices are only now beginning to take foot in Finnish organizations, and mostly through the initiative of public authorities. At the turn of the millennium, they were virtually nonexistent in all Finland. In this desert of ethnicity-celebration, I had found an oasis full of content foreigners. So I started wondering: does generally "democratic" and participatory management take the place of diversity management—is it perhaps even better, considering all the criticism regarding hidden control and "divide-and-conquer" management that the later has received?[m] The other way round, discrimination is often just another name for exploitation, junk-job reality being the true source of culture-attributed or ethnicized problems (Trux, 2000). Does this mean that when there is no exploitation there are no problems?

Not all is well at F-Secure, however. In my present understanding they simply do not cope with diversity. They do not understand what is going on in the existing transcultural reality (not that many of us would). The HR manager was afraid that open discussion of the issue would spark conflict, and the Finns unfortunately believed that they can go on working in the "normal" way without bothering themselves about whatever understandings of normality other people might have. They want no self-reflection, and are not well prepared for it either. Between indolence and discrimination, there seems to be only a thin line.

THREATENED PROFESSIONAL IDENTITY

If "nerds" were culturally dominant during the boom, now the pace is set by economists. Research programs with a large investment of work and human commitment are abandoned because of managerial strategies based on the market situation. "The boys must stop polishing their products for ever and adding extra features," the HR manager said, "because consumers just want the basic, inexpensive, and easy-to-use." It is difficult to imagine a worse blow to the professional heirs of the Promised Land once called the *knowledge society*. Wierzbowsky, as one of the employees who must keep watch over the Internet, described his job after the downfall: "Something between janitor and medical." His irony hardly covered the disillusionmentment.

From labor market perspective, the employees have come down from expectations of social mobility. Instead of making the companies compete for their workforce, they must now be content if they even have workplace with some continuity. Below the higher ranks of management, the value of stock options has melted away. Fearing loss of economic control over the company, management cut off all boom era luxury benefits and some more common ones as well. Some workers strongly criticized this, as further proof of the management's inaccurate understanding of the situation, exaggeration of danger and undervaluation of the comfortable conditions necessary for innovative work. One of them was Noam. I met Noam in the autumn of 2003. When I had asked about "democracy" and the bottom-up flow of information in the company, the HR manager had mentioned him. According to the manager, the employees do use the chance to debate in the monthly meetings. "When Noam brings out his checked notebook, we can be sure that tough questions will follow."

Noam is an engineer, calling himself "a nerd," and a project manager. Though he is himself officially part of management, he works directly with products and near his team members when the teams are small and at F-Secure they often are. For him, the project manager is "the guy that does everything that the others don't do" and "the 911." He reflected on his stressful position and my comment on his light tone by stressing commitment to quality and learning and respect for deadlines. He took a retrospective look at his company to tell me how he has come to where he is now, professionally.

So, last project for instance we had a 12 per cent delay, in a project which was the biggest so far in this company. So, things are going more or less ok. ... I think one of the things that is important for this... Not just for me but the whole R&D organization where I work, was we had a manager that had a vision ... for the department, and he was able to implement that. So we were able to improve all the time and become I would say much more professional now than we were last time when you were here in 2000.[n]

Among all the casual, easy-to-talk-to F-Securians, Noam was one of the most open. Or did I simply feel that we had a common moral stand, which made me sympathize with him? I only know that we ended our conversation in "joint imagination" on the future of capitalism and on Brazilian culture.

Aside from passionate ideas about his profession and the organization in which he works, Noam also showed a markedly balanced approach, bringing up various perspectives and considering the limitations of his own information. He had come to F-Secure in typical boom-era fashion, by publishing one of his university projects on the Internet while he was finishing his studies in his native country in southern Europe. Somebody at F-Secure saw it and he soon had an invitation to a job interview.

And so basically I finished the project that I published on the internet. Then I was contacted to come here. Then I came here and talked to the people and what convinced me was talking to the actual people who were doing the actual work, not really the management ... But it was really seeing the people and how happy they were and the relationship they had with each other which was very close and friendly. And so that convinced me basically to try it, try it out. I always wanted to work abroad anyway, so it was a good first step. (Same source)

Noam lamented the disappearance of "vision" from the company. He felt himself very much at odds with the downturn atmosphere of disillusionment. In his view, the "nerds" do not live by salary alone. In addition they want to change the world and serve the larger society.

So my point or my... The perfect workplace is where there is always this vision. Of course for that you have a how should I say, a non-materialistic approach, or not just a materialistic approach. ... To the work. ... So you have to have a notion of what are then things that keep the people together. ... This was one of the things that I said once to [the CEO's first name] when we had this... Had this traditional talking to people as they came into the company, when the company was still small enough. It was in –98. ... And I told him that why are we going public, because there is more important things to do. A company has its responsibility towards its environment and especially to its people, because the company only exists... the capitalist way of looking at this is that companies exist to make profit... My way of looking at this is that the company exists to make profit so that it can invest in the society where it is. So that the society can grow and it can grow with the society. And I think that when we get to the point where the company is to make profit, that's where the pleasure ends. ... Because [then] profit is what drives you. And profit is not a vision. It's a number. (Same source)

Noam draws a line from this thought to the way the workers are treated. According to him, investing in people is no longer important for the company. Everything is being sacrificed to economize, which is also ridiculous, since the resulting cost cuts are small, while the entire productivity of the company rests on human input. The latest on the list for cuts was removing all plants from the offices, a decision that very much unnerved him.

While many of his colleagues had left the organization, Noam was still there. The reason he gives was not solidarity towards the company, but solidarity towards old friendships. He would not like to leave his colleagues.

Describing an ideal job or ideal work he would like to do, Noam vigorously defends his view of professional pride and craftsmanship against what he perceives as a financial shift in values.

So obviously knowledge was more important than money there. And the fact that I was finding a company in the capitalist world that was trying to do the things that I thought were important was something special. And I said ok that it's clear that I want to be in this company. I don't want to be in an open office where everybody wears a suit and everything we do is to work the day for making the money day after day and that's it ... Then the work is over. That's not what I was looking for. That's one of the reasons why I liked F-Secure, at that time: Data Fellows. Maybe the name change is also something telling about the company, because we changed the name because of the marketing value of "F-Secure." ... So the perfect work, again to go back to the initial question is where you can fulfill yourself, not by feeling happy for being at work but being proud of what you do. ... And not necessarily being proud, you know, you're doing a big piece of money, nobody loves that. You don't need to do that to be proud. You can do a chair and be proud. ... Because you can see that you have done it perfectly. ... And it works and it fulfills its cause. ... And you feel that you have done it better than the previous chair. (Same source)

The situation might be described from Richard Sennett's point of view as a workforce struggling beneath an imposed order of "new capitalism" and that would not be far from truth. The recent changes had indeed hollowed out the self-confidence of the people I had met three years earlier, and their efforts to explain and get a hold on the situation ran in many directions, incoherent. But that is only one part of the story. In 2000 many praised flexibility, innovativeness, risk, and irreversible change. At that time, these values were still framed within the "hacker ethics" and the workers believing in them were supported by a sense of economic security and social status. Work was "exciting." It was the ethos Sennett calls *fresh page thesis*, the predecessor of neo-liberal thought, as we would now call it. Those who do not admit that these values can do harm still tend to perceive them in that innocent light (including both managers and workers). Following Sherry Ortner, I distinguish between varying degrees of reflexivity toward the dominant forms, from total incorporation to becoming aware and turning against them. This process may be especially difficult for high-tech professionals because they had previously anchored their professional identity in good part in the same values now imposed upon them in a new merciless tone.[o] Flexibility is not what it used to be.

Concerning flexibility, a further irony can be observed; anti-virus work is yet another *mise en scène* of the flexible citizen, pointing to the older image of the human body equipped with an intelligent, adaptable defense system. Anti-virus products for the pc can thus be compared with anti-germ (typically lactobacillus, vitamins or vaccinations) products destined to the consumer as a biological organism. In this setting, the virus researchers are given the role of "flexibility champions" to help us all improve our flexibility.

PRAGMATISM

Despite the fact that it is real and disturbing, the tension between utopian and postmodern images does not in itself adequately describe the situation at F-Secure. For the Finnish members of the organization (and possibly some others as well), there is a third way to understand flexibility. In all my attempts to inspire reflection upon culturally constituted practices at work—and indeed anything to do with culture or ethnicity—I was discouraged by their firm belief in the primacy of practical and material matters over the cultural.

Niilo, a Finnish worker, made a very interesting observation; he said that the "company way" arises from avoiding "multicultural bureaucracy"—by simply doing the job and keeping the customer happy. His organizational ideal concentrates on the main goal and how to achieve it by concrete action. On the one hand, it coincides with the observation that joint action brings people together despite all divisions. On the other hand, it speaks for a pragmatic approach elevated to the degree of a value in itself. *Away with all bureaucracy, the main thing is that ordinary work gets done.* This is a tradition that has greatly shaped work and administration in Finland. Ethnology has drawn attention to it as both a peasant tradition with a history of self-reliant, marginal subsistence agriculture (Apo, 1996a, 1996b)—peasant pragmatism—and as a part of national self-understanding, an ideal image of "the pragmatic Finn" as a naive but faithful underclass (Lehtonen, Löytty, & Ruuska, 2004).

At F-Secure, pragmatist reasoning was a ubiquitous, unquestioned value that people like Niilo fully embodied without any awareness or reflection on its peculiarity or national symbolism. Other material also supports the view that pragmatist values prevail in the workplace. "Bureaucracy" was the belittling term applied to anything that would come between the people and their practical tasks. This is what the foreign workers praised as "Finnish management" or an "efficient way to work": cutting out ceremonies and time-consuming formalities, being brief, taking the initiative. This is also a common self-complacent discourse among Finnish business elites. Nevertheless it has some truth in it. It can be a way to a very flexible order, giving priority to "the real work." Its downside is that it sometimes becomes a straightforwardly advancing bulldozer, ignoring and pushing aside all questions and alternatives.

There is great power in pragmatism; it doubtless supports organizational cohesion. Although I present it here as a local culture, variations of pragmatism are numerous elsewhere as well. The protestant work culture described by Max Weber is the classic (compare with Prasad 1997)[p], but the "fresh page" endorsed by the boom-era utopians was another (*finally we are free to do "the real work"*), and so is the present postmodern ethos of flexibility (*you shall be paid only for "the real work"*). These forms are not completely reducible to each other. They lead instead in different directions and imply varying frames.

The peasant work ethic used to impose extreme persistence—to a degree of work cult, with delayed gratification. Unlike in most forms of capitalism, the delay is not understood to be life-long, however. Instead, a proper rhythm of work and leisure is expected. Neither is compensation expected to reach beyond the median level of one's reference group, but continuity and respect at work are crucial. The logic is of a bonding type, very different from the promiscuous indifference characteristic of the "new capitalism" (cf. Sennett 2006). What may remain unclear to the workers, and indeed to their managers as well, is that while according to the peasant pragmatism belt tightening promises a better fate in the next phase of the productive cycle, or by the next harvest, the same abstinent behavior under the efficiency demands of late capitalism will not bring a reward, only greater demands, based on the logic of constantly intensifying competition. In the workplace context this is both convenient and confusing. For instance, the workers may come to take sides with the management, in the way Niilo was suggesting, as far as peasant pragmatism is glossed over with capitalist ideas of efficiency, and this would be consistent with a more entrepreneurial strand of the professional culture. On the other hand, they may also come to question the company's goals (what is it that it is practical for), like Noam. This approach would find support in the professional culture's countercultural strands.

Why would any workers take Niilo's stand? In contrast to the recent demands for continuously growing productivity, the description by Antila et al. (2002) of the *proactive mode* of business organization can be read as a description of an ideal form of management under Finnish-like pragmatism (the peasant form or some derivative of it). The idea that the workers are allowed to influence products and working conditions in exchange for the expectation that they bear responsibility and take initiative is proof of the assumptions that (1) they are believed to have the capacity to do so and (2) that the employer is compelled or propelled to surrender her power to some extent. The first assumption obviously connotes the view that all wisdom is not located in the board room, but workers are an intellectual resource concerning business organization. The second assumption has often been linked to the moderative tension introduced by the former socialist bloc as a potential threat to Western market economy. I would suggest that a longer perspective be used here.

While some companies today certainly dispose of their workers easily and relocate production to sites where the workforce is replaceable, faceless and docile, others like F-Secure employ "empowering" measures to keep the people they need. Employees have not always been plentiful and exchangeable, and rarely were at Finnish latitudes during pre-industrial times. Whenever people have been exploited (and indeed they have been) and disposed of readily, the consequences have been hard for the rulers, who themselves could not escape the physical environment. Going through the annual cycles in the world's northernmost agricultural regions for centuries introduced an imperative as great or greater than the two generations of socialist neighbors. Workers were simply too scarce and precious to waste—and they knew how to survive better than their elite rulers. Taking into consideration that the present rulers—business managers—are in large part from modest, often rural families with their history galvanized by only two or three generations age, even a simple carrying over of the peasant current would not be surprising, but we need not assume it. Things can also be reinvented. Moreover, in the face of growing ecological and other global imperatives for drastic changes in the industrial-capitalist process, is a preoccupation with survival a matter of the past or of the future?

Pragmatism need not remain linked to rural forms. Urban forms of pragmatism embrace cultural items from impermeable use-and-rinse children's trousers for playing outdoors and economical combined generation of heat and electricity, to handy Nordic kitchens with hang-up-to-dry cupboards liberating spouses from drying the dishes, not to mention electronic equipment destined for modern consumers. In effect it says: *Save time and energy, find new intriguing technologies to liberate yourself!* In the proactive mode, the tension between employers and employees does not disappear, but is joined by the powerful quest to survive. This quest is seldom presented as such, but cut down to countless local innovations, nice and handy ways to cope with all the (working) life's small to middle-sized puzzles. There is a latent, tacit expectation of a certain kind of ideal attitude by the managers from the workers, and vice versa. Pleasurable work "proceeds smoothly," as the Finnish engineer Matti put it.[q]

There have been some inquiries into the way pragmatist attitudes mark the professional subculture of economists in Finland (the second important professional group at F-Secure). According to Leppälä and Päiviö (2001), main stream business students believe in the primacy of the working life and do not involve themselves too deeply in theoretical elaborations. They prepare themselves for business activity in which one must advance continuously and keep ahead of competitors, without hesitation or reflection. The relative roles of variations of pragmatism in this discourse fall beyond the scope of the present study. If its development has been anything akin to the vicissitudes of other professional cultures in Finland, however, it seems likely that at least peasant pragmatism alongside more generally Western (American?) capitalistic pragmatisms would be amalgamating in such attitudes; once again, the consequences are both convenient and confusing.

HETEROGLOSSIA

Three different approaches to flexibility (and related values) clash at F-Secure: the dominant postmodern form, the professional idealistic form, and the mostly unreflected, constitutive form of peasant pragmatism. My aim is not to confirm that the number of noticeable cultural forms at F-Secure is three. Instead, I use this limited example to draw attention to the multitudinous and contested nature of cultural environment and the chances for counterculture and alternative reactions offered by it.

In the present situation, while the late capitalist economy is demolishing pre-existing economic and social structures, insecurity and underdefined conditions have become every-day reality even at F-Secure. People struggle to gain a foothold by drawing from available cultural forms, but this has proved to be difficult. Surface similarity and deep differences in culturally constituted models make navigation a confusing experience. Description of the cultural forms (or currents) present at work is also a challenge for research. The challenge is worth meeting, however, since a light scratch in this direction reveals that (contrary to the claims of many speakers in hegemonic positions) even the management in all organizations is not totally converted to the new dominant forms; the same confusion is present in many organizational members' understanding. The degree of reflexivity varies, as do the particular forms applied.

A cultural critique can clarify underlying assumptions of older or "traditional" structures – we can expect several partially overlapping layers of them. In the same way, a cultural critique must shed light on underlying assumptions of the new cultural forms introduced by hegemonic distributors. Critiques of this kind would not be written for academic audiences only, but also for the people themselves, to meet their needs for meaning and orientation. It is their agency that research would, in the final analysis, try to serve.

Only after we see what is on the table, can we start to craft our answer. There is always more to be put on the table than the latest fashions of late capitalist "world culture." Locality is not dead. Even in organizations erasing all spontaneous production of it, in the centers and halls of the global economy, people still carry shreds of localities with them, under the skin so to say, in their minds, in the form of structures of feeling. "People" includes here all humans present, from customers and partners, secretaries, legal assistants, engineering and marketing professionals to all kinds of managers. Knowing that the struggle is not over—that whatever the hegemonic forms presently introduced, alternatives and ingredients for potential future alternatives are at hand—is vital.

Equally important is the observation previously made, that not even the management totally incorporate the new hegemonic forms, but rather hovers much like the workers (and the researcher), between different forms. As previously noted, the Finnish managers share (at least) with their Finnish workers the tendency to confuse peasant pragmatism with late capitalist endeavours. This may occur without any reflection, or it may cause varying degrees of mixed feelings. Some of the managers (including the founders of the organization) also share the professional subculture with their engineering workforce, again with strong urges for solidary conduct and countercultural forms blended with the entrepreneurial spirit and individualism that can in turn be combined with neo-liberal forms. The steps taken are therefore hesitant and controversial. Management tries to protect the social capital of the organization from the most predatory effects of the late capitalist environment, while still accepting (or half accepting) its dominant neo-liberal ideology. We all engage in fostering or resisting images like "flexibility" without the full awareness of their resonances, contradictions, and entanglements with various discourses. The identity of the ultimate speaker in any given utterance remains ambiguous.

Since identification of the "voices" in the heteroglossia is difficult and time-consuming, so is the crafting of one's own identity. To what address should one direct a response to claims of flexibility, for instance? Is it your fellow "nerd," a pragmatic Finnish manager appealing to you for doing a project flexibly, or is it a voice of the "Davos men" speaking with his mouth?[r] Managers have it no easier: how to dissipate lingering suspicions that the later is the case at moments when it is not? Are they sure they know themselves with whose words they are speaking at any given moment? Who are the persons, social forces and power positions in this kind of field that has become ghost wrestling? We need maps, no matter how broken and changing the cultural landscape is. Perhaps in that case even more desperately.

ACKNOWLEDGMENT

Thanks to Teppo Sintonen and Minna Ruckenstein for useful comments on the first draft.

ORAL REFERENCES

Pokela, P. (2005). Presentation at a conference on Diversity in Helsinki at 9.-10.12.2005. (Originally held in Finnish: Moninaisuuspäivän konferenssi. Kuntatalo, Helsinki. Organizer: SEIS - Suomi eteenpäin ilman syrjintää. Notes and translation by the present author.)

REFERENCES

Antila, J., & Ylöstalo, P. (2002). *Proaktiivinen toimintatapa. Yritysten ja palkansaajien yhteinen etu?* [Proactive mode. Common asset for companies and employees?] Työpoliittinen tutkimus 239. Helsinki: Työministeriö.

Apo, S. (1996a). Agraarinen suomalaisuus – rasite vai resurssi? [The agrarian Finnishness – burden or resource?] In P. Laaksonen & S. L. Mettomäki (Eds.), *Olkaamme siis suomalaisia. Kalevalaseuran vuosikirja* (pp. 75-76). Helsinki: SKS.

Apo, S. (1996b). Itserasismista positiivisiin suomalaisuuksiin. [From self-imposed racism to positive Finnishnesses.] In S. Apo & J. Ehrnrooth (Eds.), *Millaisia olemme. Puheenvuoroja suomalaisista mentaliteeteista.* Helsinki: Kunnallisalan kehittämissäätiö.

Gere, C. (2006). *Digitaalinen kulttuuri.* Turku: Faros. [Orig.: Digital Culture. London: Reaction Books, 2002.]

Himanen, P. (2001). *The hacker ethic. A radical approach to the philosophy of business.* New York: Random House.

Holland, D., Lachicotte, W. Jr., Skinner, D., & Cain, C. (1998). *Identity and agency in cultural worlds.* Cambridge: Harvard University Press.

Kaplan, D. A. (2000). *The Silicon boys and their valley of dreams.* New York: Harper Collins.

Kunda, G. (1992). *Engineering culture: Control and commitment in a high-tech corporation.* Philadelphia: Temple Univ. Press.

Lehtonen, M., Löytty, O., & Ruuska, P. (2004). *Suomi toisin sanoen.* [Finland in other words.] Tampere: Vastapaino.

Leppälä, K., & Päiviö, H. (2001). *Kauppatieteiden opiskelijoiden moraalijärjestys: Narratiivinen tutkimus kolmen eri pääaineen opiskelusta Helsingin kauppakorkeakoulussa.* [Moral order among students of economics: A narrative inquiry into studying three major subjects at Helsinki School of Economics.] Helsinki: Publications of the Helsinki School of Economics and Business Administration, B-34.

Litvin, D. R. (1997). The discourse of diversity: From biology to management. *Organisation, 4*(2), 187-209.

Lorbiecki, A., & Jack, G. (2000). Critical turns in the evolution of diversity management. *British Journal of Management,* Sep. 2000, Special Issue, *11*(3).

Martin, E. (1994). *Flexible bodies: The role of immunity in American culture from the days of polio to the age of AIDS.* Boston: Beacon Press.

Ortner, S. B. (1984). Theory in anthropology since the sixties. *Comparative Study of Society and History, 126*(1), 126-166.

Ortner, S. B. (2005). Subjectivity and cultural critique. *Anthropological Theory 5*(1), 31-52.

Prasad, P. (1997). The protestant ethic and the myths of the frontier: Cultural imprints, organizational structuring, and workplace diversity. In P. Prasad et al. (Eds.), *Managing the organisational melting pot. Dilemmas of workplace diversity.* Thousand Oaks, CA: Sage.

Roberts, F. M. (1989). The Finnish coffee ceremony and notions of self. *Arctic Anthropology 26,* 20-33.

Ruohonen, M. (2004). Johtamiskulttuurien muutos ICT-yrityksessä – hypestä todellisuuteen. [Changing management culture in an ICT-company – from hype to reality.] In M. Ruohonen et al. (Eds.), *Tietoyritysten muuttuvat työkulttuurit.* Tampere: Tampere University Press.

Sennett, R. (1998). *The corrosion of character. The personal consequences of work under the new capitalism.* New York: Norton.

Sennett, R. (2006). *The culture of the new capitalism.* New Haven, CT: Yale Univ. Press.

Sewell, W. (1999). Geertz, cultural systems, and history: From synchrony to transformation. In S. Ortner (Ed.), *The fate of "culture." Geertz and beyond.* Berkeley: Univ. of California Press.

Trux, M. (2000). Monimuotoinen työyhteisö. In M. Trux (Ed.), *Aukeavat ovet. Kulttuurien moninaisuus Suomen elinkeinoelämässä.* Helsinki: WSOY. [Published in English as Diversity under the Northern Star. In A. Forsander (Ed.), 2002. *Immigration and Economy in the Globalization Process. The Case of Finland.*] Electronic publication, ISBN: 951-563-416-4. http://www.sitra.fi/en/Publications/web/web.htm

Trux, M. (2005). "Ei sitä meillä kukaan kato" – kansainvälisyys ja monietnisyys helsinkiläisellä IT-työpaikalla. ["Nobody looks at that here" – internationality and multiethnicity at an IT-workplace in Helsinki.] *Työpoliittinen aikakauskirja 2/2005,* 49–69.

Trux, M. (2006). *Almost-equals: Transnationality in a Finnish ICT-company.* Paper presented at EURODIV Conference Qualitative diversity research: Looking ahead. Katholieke Universiteit Leuven, Belgium, September 19th-20th, 2006.

ENDNOTES

a According to a mutual agreement with the company, I use its true name in this text. All arguments and conclusions are my own, however, and the picture I give refers to my experiences and material from the period of 1999 to 2004. The research on which this chapter is based was supported in part by grants from the The Finnish Work Environment Fund, The Foundation for Economic Education and the Finnish Ministry of Labor. Sections of this chapter have appeared earlier in Trux (2000, 2005, 2006).

b The fieldwork had three "episodes": first I made a round of interviews and some participant observation at the headquarters in 1999-2000 as part of a larger project, reported in Trux 2000. I returned to the "site" (though they had moved within the Helsinki region) and conducted more interviews with some more participant observation during 2002-2004, the present tense referring to 2004. Although I had the opportunity to follow the organization and the workers through the ups and downs of those years, the central weakness of the ethnography is that I had no access to the actual working offices, beyond a few times I managed to slip in at the responsibility of individual informants, and there were a few parties and meetings I attended. In the manner of high-tech companies, F-Secure frustratingly limited my presence at their premises to the visitors' zone. Taking into consideration that the company is engaged in the security business, this was perhaps to be expected. As a pay-off, at the independence-access-dimension I ended up near the independent end. Still, one has to read my account against the fact that is it heavily biased towards the discursive level. Of course no ethnography has anything like a direct access to reality. Moreover, the process is completed only when the report is written and read. I have attempted to compose my account so that the "voice" of each informant appears like it was, a fully political doorkeeper to his or her experience, with a due reaction to my presence and many references to absent others. Within this longer period in and about the Helsinki headquarters, I made a two weeks trip to a Silicon Valley unit. The findings of that brief but intense visit are discussed in Trux 2006.

c See Gere 2006 for the history of the multiple roots of this hybrid form, ranging from engineers' and radio-amateurs' ideals, to Californian entrepreneurship to the hippy movement, anti-government dispositions and all kinds of liberalisms. Utopian and dystopian visions have gone hand in hand through this story that is revealingly compatible with the story of the digitalized (Western) society.

d I found no widespread use of temps as a regular means of "enhancing" productivity. It seems that people were first taken on as temps. Later on, their contracts were made permanent. At the downturn, those who "had not regularized their contracts" faced the greatest risk of being laid off. It remains unclear whether this tendency to avoid hosts of half-outsiders within the company is related to a communitarian management style or to the demands of risk management in the digital security industry (many workers told me that their backgrounds had been studied by the Finnish Security Police at the request of the company). Risk management has in any case not precluded the growing use of subcontractors, for instance the (cheaper) programming workforce in St. Petersburg.

e At the beginning of my inquiry, the proportion of foreigners was estimated to be 13% of the workforce. When I last asked about this, the HR secretary estimated that people who had come as immigrants accounted for as much as 20% of the personnel in Helsinki. The law in Finland, as in many other continental European countries, bans records based on ethnic identity.

f Wierzbowsky, a virus researcher, interview on 29.2.2000. All names of informants appearing in this text are pseudonyms that they invented for themselves.

g Matti, software engineer, interview on 19.5.2000. Translated from Finnish by the author.

h Michael, virus researcher, interview on 16.3.2004.

i The following is another example of the atmosphere of avant-gardism: The HR manager was visibly pleased with himself, as he explained to me how the company's new logo was reminiscent of Superman's badge. It was only half a joke, since in fact their job is to protect not only businesses, but also public organizations such as hospitals, municipalities and schools - and in the end individual consumers - against various digital threats.

j See Gere (2006) for an account of the role of countercultural movements in democratizing and demilitarizing the digital technologies after the Second World War.

k These were in fact words of Maria Cecilia Duffau Echevarren, an Uruguayan ex-prisoner of conscience, in her letter of thanks to Amnesty International: "The most important thing is that … between us, human beings, it has been proved that borders are absurd, languages are surmountable, that distances can be overcome, because the heart is big…and people like you keep the hope of a new dawn alight."

l Nokia was not among the surveyed companies. The conclusions are based on findings in small and middle-sized IT-service providers and multimedia companies.

m See, for example, Litvin (1997), Lorbiecki and Jack (2000).

n Noam, software engineer, interview on 23.9.2003.

o Though I concentrate here on the image of "flexibility," individualism is another hook between the "hacker ethic" and neo-liberalism, as described before.

p Following Weber, the religious roots of capitalism are most often located somewhere among the "protestants," though more specifically among Calvinists. Finns made no protests, but became protestants by their Swedish king's decision, and the form to which they were converted was Lutheran. Being no historian of religions, I content myself with noting that the idea of "providence" as legitimating overt celebration of one's riches in front of others, is still fundamentally against the grain here; witness the regrets of high ranking business observers that "it is very difficult to instill on entrepreneurial spirit in Finland, because people are envious of each other's success". The reference in such comments is usually North American. Whether the egalitarian strands were introduced by Lutheran ethics and/or had some older source, ultimately in pre-Christian values, is beyond the topic here. See however Roberts (1989) for the delicate balancing of egalitarian and individualist values in Finnish rural life of the 1970s.

q This is not to say that this kind of pragmatism is a cultural phenomenon found only in Finland. The precise combination of present-day features and frames is probably unique, but similar strands are most certainly to be found elsewhere, with a historical connection to this form or without. Neither is the description an argument for the absolute prevalence of the form within the territory of the Finnish nation-state (there are competing ideals) or of the certainty of its being carried over to the future. Judging by its wide circulation in present-day society, and its speedy conversion into ever new interpretations in new fields of activity, I am tempted to predict that it will not slow down very soon or in the face of only moderate contesting forms. The fact that the idea of survival has also repeatedly offered an easy 'hook' by which to rally all social classes to the rescue of the nation-state – including the elites' privileges – is no disproof of its existence or a warrant for any simple reduction to *upper structure ideology*.

r As a matter of fact there was a woman at the head of the anti-virus team at F-Secure. The Bulgarian immigrant expert was unfortunately too busy to participate in my study. In spite of recent appearance of women in technical professions, however, the technical-managerial positions in the Finnish IT-sector are still overwhelmingly male dominated. So, in waiting for the day to say, factually, "with her mouth," we must be content with the masculine pronoun.

Chapter XX
Disciplining Innovation?
Mobile Information Artefacts in a Telco Innovation Center

Chris Russell
University of Wales Institute, Cardiff, UK

ABSTRACT

This chapter introduces the pattern of bargains with the devil as a means of examining the adoption and use of mobile information artefacts. It argues that, in contrast to other attempts to impose discipline on innovation work practices, the introduction of mobile information artefacts is requested by knowledge workers. Vendor advertising plays a tempting role: proclaiming that the artefact is a consumer good for the knowledge-worker and a productivity tool for the employer. This enlightenment--and deception--results in knowledge workers persuading their employer by appropriating the productivity discourse of the vendor. There follows a honeymoon phase of play and pleasure for these "techies." But this is the prelude to destruction, as the knowledge-worker faces demands for their promised productivity. The artefact disciplines their innovation work, even erodes it; thus the situation of the employment relationship within relations of consumption results in the knowledge-worker (and their employer) being exploited by the vendor.

INTRODUCTION

Mobile information artefacts are an increasing presence in the pockets and palms of knowledge workers. A fact recognized by their nomenclature: popular types are the Pocket Personal Computer (PC) and the Palm (Pilot). Accompanying the person from home to the office and back, such artefacts disregard and disrupt the work life boundary. One study considers the consequences for employees can be "very positive" (Breu, Hemingway, & Ashurst, 2005); another "a dual-edged sword" (Towers, Duxbury, & Thomas, 2005). Both focus on work flowing into life: a flow unhindered, for no resistance beyond switching off the artefact is to be found. Neither considers the possibility of an inverse flow of life into work. This chapter contributes conceptually and empirically to such findings by positing the possibility of an inverse flow of life into work and exploring the significance of mobile information artefacts for line-, peer-, and self-management. Empirically this chapter draws upon nine months experience of work in the innovation centre of a European mobile network operator in the United Kingdom.

Such experience is appropriate for seeking to uncover practices of resistance, misbehavior, and dissent concerning mobile information artefacts whilst looking through the factory gate to the extra-organizational influence of the vendor. As practices of resistance, misbehavior and dissent are "covert and subterranean [they] are inevitably...difficult to identify and research" (Collinson & Ackroyd, 2005, p. 306). They are even more difficult to research when depending upon "large data sets and increasingly available quantitative survey materials, rather than in-depth or longitudinal fieldwork" (Thompson & Ackroyd, 1995, p. 619): it would be difficult to construct a survey that might penetrate through the false or faulty consciousness of employees on issues such as the impact of vendor discourse on their actions. Even in an interview, employees may not be forthcoming about practices of resistance, misbehavior, and dissent. The experience of work in the innovation centre is viewed through a critical conceptual lens, shaped by theorists such as Adorno, Baudrillard and Milbank.

During this time in the innovation centre a new service was developed:

It was just a bit of fun—not yet a business project [i.e. not something that had been officially approved and funded]—but it could be a useful service, for parents to track their children, for instance, not just, or specifically, for employers to track employees. It was just unfortunate that we should use Mike [a fellow Graduate Trainee] in the demonstration to Colin [a Senior Consultant and Manager of one stream of research in the innovation centre]. (Dave, Graduate Trainee)

The service ran on a smartphone—an artefact that blends the information power of a Pocket PC and the telecommunication power of a mobile phone—and enabled one to see the location of fellow subscribers to that network. The innovation lay not so much in the technology, but in the fusion of data: the information of a subscriber's strength of signal vis-à-vis three or more masts, used by any network to determine which mast your call will be directed through, and a standard digital mapping service. The phone numbers used in the trial were those of Graduate Trainees (in some cases with, in some cases without, their consent). During the demonstration of the service to a manager, the unexpected location of one such trainee was exposed: instead of being in a neighbouring town—for a medical appointment—they were several hundred miles away.

One might consider that this event of the disciplining of a member of staff through the use of a technological innovation, a mobile information artefact, is a relatively unique event. However, the more standard services of mobile information artefacts—such as a task tool for "to-do" lists and a calendar for time management—may be used to discipline—for good or ill—the innovation of knowledge workers. Such artefacts can be seen as one of many attempts to reduce knowledge worker discretion and autonomy, other attempts in the innovation centre being: the roll out of a company-wide innovation inducement scheme; the global integration (i.e., reduction) of research and development resources; the re-layering of the management hierarchy; the recruitment of cheap labor (i.e., centrally-funded Graduate Trainees on placement) and the abolition of "seed money" to develop concept demonstrators. However, whilst these attempts were opposed by the knowledge workers, the introduction of mobile information artefacts was welcomed, indeed, requested, suggesting a more complex picture.

SOME PERTINENT POINTS FROM MANAGEMENT LITERATURE

This section will explore some pertinent points from critical literature and how they may impact upon—and be impacted by—the study of the adoption of mobile information artefacts.

Resistance, Misbehavior, and Dissent

Since the late 1990s, the leftist camp within university business schools where the mainstream orthodoxy is "associated with some version of rightist politics" (Grey, 2005, p. 175) has been divided. On the one hand, critical management studies (CMS) researchers; on the other more materialist labor process theory (LPT) researchers. A recent expression of LPT *contra* CMS may be found in Thompson (2005a); a recent expression of CMS *contra* LPT in Grey (2005).

Foucauldian CMS researchers have often adopted the conception of a panopticon. Such "panopticons are now electronic and informational" providing for 'the increased and successful use of monitoring and surveillance' (Thompson et al., 1995, p. 623). One study has "found that, unsurprisingly, supervisors prefer electronic monitoring more than non-supervisors, and that the great majority of the latter group considered that electronic monitoring might cause heightened tension between managers and workers" (Bradley, Erickson, Stephenson, & Williams, 2000, p. 106).

Such a conception can be "particularly dangerous in closing the space to 'see' resistance" (Thompson et al., 1995, p. 624). This is, in part, because CMS can conflate the intent of managerial practice—as embodied in discourse or texts—and its outcome. However, "Foucauldian theory and research…is not the only mechanism for taking labor out of the process' (Thompson et al., 1995, p. 627). There has been 'the virtual removal of labor as an active agency of resistance in a considerable portion of theory and research' (Thompson et al., 1995, p. 615) as management has been seen as the prime mover in bargaining since the changed context of the 1980s. This constituted a departure from post-war research which revealed how employees adapted wage-effort bargains in their favor including "the (re)appropriation of time, in which workers use time for themselves instead of productive work" and "the reappropriation of products, in which the employee uses materials for some other purpose than the productive process" (Thompson et al., 1995, p. 616). In such research employees were "active and innovative in attempts to survive in employment, recurrently breaking rules, and actively re-negotiating them on a continuous basis, with a management frequently tolerant of, accommodating to, or conniving in, such practices" (Thompson et al., 1995, p. 616).

However, this one concept is not sufficient to capture employee practices. Resistance suggests merely a response to managerial initiatives. It may be that employees can express discontent towards management through the pro-active adoption of an information artefact. Thompson et al. developed the broader concept of misbehavior: misbehaving employees "may anticipate management, and may lead rather than follow" (1995, p. 617). However, this concept, too, is limited for it ignores forms of action "that don't involve overt conflict with management" (Jaros, 2005, p. 17). In practice, there is a "continuum of overlapping worker responses" (Jaros, 2005, p. 22).

> *Whether it is the establishment of Taylorism, bureaucracy, human relations, or new technology, extravagant claims as to the rationality and effectiveness were made by managerial advocates and too often believed by academics. We now know that workers learned to bend the bars in these particular iron cages. Why should the current crop of new management practices be any different?* (Thompson et al., 1995, p. 629)

One might therefore ask how and where are employees bending the bars of mobile information artefacts?

Production and Consumption: Work and Life

A key text relating religion to management—or at least capitalism more generally—is Weber's *The Protestant Ethic and the Spirit of Capitalism* (1930). Buchholz summarizes Weber's conception of the Protestant Ethic as the "maximization of production and minimization of consumption" (2000, p. 77). In the Protestant countries of Northern Europe—such as the United Kingdom, the location of this study—and their colonies, the emphasis was on hard work for the glory of God, whether such work was inherently secular or religious. Such sentiment may be found in the work of the Anglican clergyman and poet George Herbert:

A servant with this clause
Makes drudgery divine;
Who sweeps a room as for Thy laws
Makes that and th'action fine.
(Herbert, 1982)

In the post-war period the work element of the Protestant Ethic persisted but "consumption activities became separated from whatever moral limits and justifications the Protestant Ethic provided"; it was "no longer relevant to a consumer culture that emphasized instant gratification and increased consumption" (Buchholz, 2000, p. 80). Mainstream management literature—and practice—"represents a return to or even a revival of the Protestant ethic" (Bell & Taylor, 2003, p. 344) "by emphasizing work as having the potential to fulfil all human needs, including physical, mental, emotional and existential needs" (Bell et al., 2003, p. 337). The implication of this is that "the maintenance of a work life balance ceases to be a problem" (Bell et al., 2003, p. 337). There is little recognition that the interests of labor and capital are—or at least should be acted upon as if they are (Milbank, 1990)—antagonistic.

A more (Anglo-)Catholic theological approach by Milbank retains a scepticism towards consumption, recognising surplus desire, value generated through the creation and manipulation of consumer preferences, whilst, in drawing upon Marx, recognising surplus value generated in production. This "dual generation" (1990, p. 192) of value means that both consumption and production require critical attention. Such an approach engages with Marx's labor theory of value (LTV), in contrast with "Scholars in a labor process tradition, both Marxist and non-Marxist, [who] have long managed to debate the key issues without recourse to it" (Thompson, 2005b). It also recognizes "that the extraction of surplus value from labor is no longer an imperative, because profits can be achieved in the exchange process by leveraging product market power" (Jaros, 2005, p. 8).

How might mobile information artefacts provide for employee's life needs? Do they alter the work life balance? Does this matter? How are they implicated in the generation of surplus desire and surplus value?

Line-, Peer- and Self-Management of Knowledge-Work

The assumption of mainstream knowledge-worker management literature—and practice—that "complementary personal and organizational transformation is...achievable" (Bell et al., 2003, p. 345) has enabled a supposed change in the approach of management from "imposing control" to "eliciting commitment" (Walton, 1985). Knowledge workers, in particular, are granted supposed privileges, such as the right to work from home or to dress casually. However, such privileged flexibility is: "as long as they fulfil commitments and do not arouse attention... If people overtly come and go as they please, they may also be covertly watched by supervisors ("If he comes late, I kinda take notice if he stays late; a private arithmetic—you never know when you have to use it.")" (Kunda, 1992, p. 48). Such attention is understandable for "to balance the forces of technology and profit, managers must be able to ship products on time" (Kunda, 1992, p. 44). In striving to ship on time "managers often check up on the engineer's work… both to provide resources and to allay their own fears that their engineers are in danger of failing to meet deadlines" (Perlow, 1997, p. 80). As a consequence:

creativity gets killed much more often than it gets supported. For the most part, this isn't because managers have a vendetta against creativity. On the contrary, most believe in the value of new and useful ideas. However, creativity is undermined unintentionally every day in work environments that were established—for entirely good reasons—to maximize business imperatives such as coordination, productivity, and control. (Amabile, 1998, p. 77)

An alternative approach suggested for enhancing creativity is "information sharing and collaboration" (Amabile, 1998, p. 84). This can build expertise and excitement: "the more often people exchange ideas and data by working together, the more knowledge they will have"(Amabile, 1998, p. 84); "with the exception of hardened misanthropes, information sharing and collaboration heighten people's enjoyment of work and thus their intrinsic motivation" (Amabile, 1998, p. 84). However, "engineers also frequently inquire about one another's progress" (Perlow, 1997, p. 80) and "such disruptions can impede attainment of the organization's goals" (Perlow, 1997, p. 84). Such peer-management "may be seen as an additional overlay of control that complements other means of eliciting compliance" (Kunda, 1992, p. 220), not just commitment. "Some of the more overt trappings of bureaucratic control systems and managerial power are relaxed, disguised, or reinterpreted" but "the enhanced flexibility, structural ambiguity, looser behavioral controls, and less rigid supervision of work behavior…are balanced by management's relentless efforts to define and impose the member role" (Kunda, 1992, p. 220) which:

is "supervised" by a broader group of control agents—practically anyone who is a member of the organization. Management, as it were, sets the stage…and reserves the right to the final word…but most of the work is done by the members, in whose interest it is to continuously reinforce in each other and in themselves overt adherence to the member role. This type of control is usually manifested in very subtle forms of group pressure; occasionally there are more dramatic outbursts. (Kunda, 1992, p. 219)

The team may "attack its members who fail to deliver" (Casey, 1999, p. 168); the introduction of a team-oriented, flatter hierarchy "covertly revives interpersonal suspicion, sibling-like rivalry and nepotism at the same time as it overtly, officially, promotes egalitarian team mate cooperation, familial warmth, and over-riding commitment to the product" (Casey, 1999, p. 172).

How may knowledge workers exploit mobile information artefacts now that they "have been 'empowered'…which boils down to bearing responsibility for making themselves relevant to the company" (Bauman, 2002, p. 34)?

BARGAINS OLD AND NEW

From Falstaff to Faust, individuals have struck bargains with the devil. Such bargains are rather different in nature to those traditionally struck for the adoption of new technologies. Pacts with the devil tend to balance—or not—the pleasure of consumption now with the pain of labor later; pacts with employers tend to balance—or not—the pain of labor now for the pleasure of consumption later. These latter pacts, whether formal or informal, explicit or implicit, have tended to be collective, and have tended to commence with the employer proposing a new technological artefact, initiating the bargaining process. The employee would then oppose the artefact, but after a period of negotiation would be appeased by a concession such as "technology pay" (Noon 1989). From thence on, the employee would be resigned to its adoption. Such a pattern is prevalent as an assumption behind much of the literature whether of a managerial or more critical orientation. Such a pattern may still be relevant to most new technologies. However, in the case of mobile information artefacts such a pattern is being supplemented, if not supplanted. This section explores the adoption of mobile information artefacts by applying the three characteristic roles of Satan: tempter, deceiver, and destroyer (Ling, 1961, p. 61).

Temptation

How are employees tempted to adopt mobile information artefacts? Certainly not with time or money: technology pay is near forgotten and mobile information artefacts capture downtime, intensifying work. Employers present the artefact itself as the concession, a benefit for the employee.

You'd be amazed at some of the things we came up with: free phone and reduced call charges. (Orange, 2006)

*A dazzling financial package like this certainly helps... A laptop and a mobile phone *...*
**Graduates joining the Finance Professional Programme are not eligible.* (Lloyds TSB, 2006)

A benefit that Lloyd TSB's finance graduates give up for that other employee "benefit": professional development.

Some employers, rather than paying you to adopt the artefact, let you pay to adopt the artefact, giving you the benefit of saving the tax.

Been thinking of splashing out on a handheld computer for personal use? Well, now you can... Paid over a two year period all handheld PDAs [personal digital assistants, which include Pocket PCs and Palms] are with a warranty and benefit from tax free status. (AstraZeneca, 2006)

Such schemes may shortly end (Collinson, 2006) but their creation points to a (con)fusion of consumption with production. If it is for personal use, why is it tax free, unlike other products for personal use? If it is for organizational use why should the employee pay at all? It may be that employees have internalised their employer's objectives and seek "biographic solutions to structural contradictions" (Ulrich Beck cited in Bunting, 2004, p. xxv). However, it may be that mobile information artefacts, to a greater extent than the new technologies of old, permit a variety of uses of varying benefit for employer and employee.

Since the first software product in 1966 (Campbell-Kelly, 2003) information artefacts have been put to different uses than those intended at the time of purchase. The magnitude of difference between these uses has, however, increased, especially since the advent of networked computers in the 1990s (Greenbaum, 2004). An information artefact purchased for maintaining accounts could then be used to send e-mail. As well as communicating with colleagues email can be used to communicate with friends and family. Later on, such an information artefact could be used to shop online or download music and videos. The use of a production-provided information artefact for consumption can be a form of resistance, misbehavior or dissent as it is an example of appropriating "materials for some other purpose than the productive process" (Thompson et al., 1995, p. 616). As such it can circumscribe the totalising rationalization of the labor process. Mobile information artefacts extend this potential for consumption, for they can be used outside the office, away from the gaze of colleagues and managers. The smaller screen may even enable use for consumption within the office: a backflow of life into work to counter the intended flow of work into life. In the innovation center, knowledge workers used their devices to: follow football scores; check their bank accounts; download music; view (often pornographic) pictures and videos. Employers may have little control for the artefacts—where traces of such activities may be left—are rarely, if ever, inspected or serviced by their technicians and the data traffic across a mobile phone network—in contrast to the office network—is beyond their surveillance. Some foresee or have experienced problems (Sørensen & Pica, 2004); others see that such use for consumption may not preclude—and may be the prelude to—use as management intended.

It may be that the employer acts as something of an "*agent provocateur*" (Caird, 1956, p. 36) encouraging the adoption of the artefact, the eating of the apple. For as Ling explains: "while the ultimate source of temptation is Satan, much, and, possibly, all temptation, in the event, is experienced as coming *not* directly from a personal spiritual adversary, but rather in ways which merit the description, "socially mediated" temptation" (1961, p. 62). Such mediated temptation may take the form of an apple in the Garden of Eden (Caird, 1956, p. 38); it may also take the form of a mobile information artefact. However, if the employer experiences problems perhaps they are not themselves identifiable with the serpent. Much of the advertising for mobile information artefacts plays on the fusion of consumption and production capabilities, for instance, for the PalmOne: "For work. For play. For life." (Shaw, 2004). Vendor advertising, rather than the systems developer (Howcroft & Wilson, 2003), plays the Janus role: one head proclaiming that the artefact is a productivity tool for the employer; the other that it is a consumer good. Consumption now drives acceptance directly, for instead of financial inducement the inducement is the consumption potential of the mobile information artefact—can the artefact let me store music, play games, communicate with friends and family?—and the mobile information artefact's own status as a consumer good.

In an era when company cars or suits are no longer a sign of professional status community health visitors and psychiatric nurses—in wishing to keep up with the Joneses—have demanded mobile information artefacts because "as always, GPs seem to be better equipped" (Amicus, 2002, p. 5). Bargains for mobile information artefacts may initially be individual and idiosyncratic: "i-deals" (Rousseau, 2005). However, they may become collective through emulation, as employees break the tenth commandment, coveting that which belongs to their neighbour (Exodus 20.17). Further evidence of the mobile information artefact as consumer good—and of employee agency in appropriating them—is found in the sales of Motorola pagers in the early 1990s. Motorola changed the colour of their pager from basic black to bright green, a change that "actually cost…nothing could get…fifteen bucks extra per unit" (Former Head of Motorola's Pager Division cited in Postrel, 2003, p. 66).

Deception

Shortly after the apple has been eaten or the mobile information artefact has been bought the deception is exposed. Whatever value—aesthetic or otherwise—ascribed to the mobile information artefact, the engagement phase of anticipation and the honeymoon phase of play and exploration are the prelude to resentment, when the word "blackberry" brings pain rather than pleasure (Hodgkinson, 2005) as the employee says farewell to downtime and time off (Jacques, 2006). They run up against the "unmeetable payment-due dates" (Baudrillard, 1996, p. 161) for the productivity promised to the employer by the other head of Janus. Knowledge workers in the innovation centre appropriated such vendor discourse—via copy and paste—in the business justification section of the asset request forms though which they ordered their mobile information artefacts. They later suffered post-consumption blues:

> *We are tempted to believe that certain achievements and possessions will guarantee us an enduring satisfaction. We are led to imagine ourselves scaling the steep sides of the cliff face of happiness to reach a wide, high plateau on which to continue our lives; we are not reminded that soon after reaching the summit we will be called down again into fresh lowlands of anxiety and desire.* (de Botton, 2004, p. 207)

Sensitivity to the creation of surplus desire requires attention to that forgotten (f)actor in the adoption of mobile information artefacts: the vendor. Adopting Adorno (Adorno, 1991; Horkheimer & Adorno, 2002), one can picture the vendor manipulating the consumer's idea of a product's worth to them—its use value—through the medium of advertising which deceives under the guise of enlightening. Such manipulation helps to create the surplus desire that enables the product to sell at its exchange value: use value is the alibi for exchange value (Baudrillard, 1996). The organizational use value, the consideration—or prediction—of what is best for the employer, and the personal use value, what is best for the employee, whether in their work or non-work life, are influenced by vendor discourse. Exchange value, traditionally conceived as the product of such manipulation of use value, was of little significance for the employees of the innovation centre for that was the concern of the person who pays, the cost-centre manager. However, it is possible that a dissatisfied employee would wish to maximise the exchange value so as to have the most—detrimental—impact upon the organization's success; also they may wish to minimise the exchange value as there may be a trade-off, for instance, a laptop comes out of the same training/resources budget as attendance at conferences.

Vendor discourse is not the only influence upon conceptions of use value. Bourdieu's capitals have been adopted as broad categories to encompass what actors bring from their habitus into the field of play. Such an approach has worked well for exploring another case of (non)adoption: the digital divide (Kvasny, 2002). It is considered that these capitals influence the adoption process in addition to the employee's conception of the artefact's use values. Various capitals exist such as educational capital and literacy capital. The set in *Distinction* (Bourdieu, 1984) are considered most relevant to the adoption of a cultural artefact: economic capital, cultural capital, social capital and symbolic capital. Such an application of the thought of Adorno and Bourdieu assumes that mobile information artefacts are cultural artefacts; an assumption which is justified as the software and computer industry is considered to form part of the creative economy according to the Department of Culture Media and Sport (cited in Warhurst & Thompson, 2005). It is also evidenced by the "interpretive flexibility" (Pinch & Bijker, 1987) that such artefacts enjoy.

Destruction

As with the introduction of desktop visual terminals in the early 1980s (Storey, 1986) variation continues to exist in the extent to which companies adopt new technology and the degree to which they realise its control possibilities. An employer may provide a PDA; used without a mobile phone or card and independently of an office computer this is essentially an automated Filofax. The Task function of a Pocket PC could be said to automate a "to-do" list or jottings on the back of an envelope. It has the advantage of keeping such lists together in one place and reminding one of when tasks fall due. As a tool that is personal, not accessible to other team members, it does not present a threat to knowledge workers' autonomy or workload. Due to the lack of central control over mobile information artefact adoption (Sørensen et al., 2004)—whether by the technology, human resource or finance functions—the artefact may well end up hidden at the back of a drawer, never to be used again, as happened in the innovation centre. In this case, there will be little disruption to the work life boundary, minimal advance of the frontier of control between employer and employee and no erection of—or extension to—a Foucauldian panopticon.

However, their use may be more destructive. The use of mobile information artefacts may be monitored—just as employers have subjected email communication to surveillance and restriction (Duane & Finnegan, 2005)—and this can incorporate into production one's outside life. Employees may be provided with a Blackberry and be expected to check their emails at home and on holiday, synchronise their device with their desktop in the office—thereby enabling surveillance by support staff—and make available their calendar—thereby enabling monitoring of their work arrangements by colleagues. Peer-management—and one might even postulate self-management—is potentially as arbitrary and destructive—if not more so—than that of line management; take, for instance, Lord of the Flies (Golding, 1997). Innovators routinely checked the calendars of those not in the office to ascertain the validity of their absence. Invalid or insufficient explanations would result in—at best—subtle or—at worst—sly remarks to the person concerned, their peers or more senior members of the centre.

The Lutheran theologian Tillich defines the demonic as "the perversion of the creative" (1936, p. 77). The creativity of the consumers of mobile information artefacts is eroded as their loss of downtime means less time for reflection, for those quieter moments—such as a train journey (Silf, 1998)—when a divine spark ignites a new idea. Because knowledge workers "do not know when they will be interrupted, even long periods without interruption are not recognized as opportunities for deep concentration" (Perlow, 1997, p. 78); this may be done only in retrospect. Satan "deceives men [and women] into following courses of action which are contrary to their own true welfare, that is to say contrary to the development of true personality...which thus ultimately brings men [and women] to destruction" (Ling, 1961, p. 64). If the creative potential of the employee is not realised, if their attention to non-work life is diminished, then the mobile information artefact (and its vendor) may be said to bring the employee to destruction.

IMPLICATIONS FOR (UNDERSTANDING) THE MANAGEMENT OF KNOWLEDGE WORKERS

The new bargains struck for the adoption of mobile information artefacts reveal shortcomings in the models used to predict—and plan for—information artefact adoption and the lifecycles or phases used to structure the production and consumption of such artefacts. As such a new conceptual framework is required.

Models of Adoption

Most researchers investigating the adoption of new technologies select a model from the eight available "off-the-shelf": the theory of reasoned action, the technology acceptance model (TAM), the motivational model, the theory of planned behavior, the model of personal computer utilization, the innovation diffusion theory, the social cognitive theory and the unified theory of acceptance and use of technology (UTAUT). TAM, the most commonly selected until very recently, forms the basis of several studies of mobile information artefact adoption (Breu et al., 2005; Yu, Yu, Liu, & Yao, 2003); UTAUT, which seeks to supersede the other models (and more) forms the basis of several others (Anderson & Schwager, 2004; Garfield, 2005; Ristola & Kesti, 2005).

Whichever of these models the researcher chooses, they are choosing a managerialist frame. The authors of the UTAUT, for instance, declare that it "provides a useful tool for managers to assess the likelihood of success for new technology introductions and helps them understand the drivers of acceptance in order to proactively design interventions (including training, marketing, etc.) targeted at populations of users that may be less inclined to adopt and use new systems" (Venkatesh, Morris, Davis, & Davis, 2003, p. 425). Managerialism is explicit in the intention for it to be used by—and serve the purposes of—management. It is also implicit in that it assumes that the manager will be initiating the adoption of a new technology by others. This does not permit the possibility of employee-initiated adoption, as is proposed here for mobile information artefacts. It may well be that often the employees who have the power to initiate the adoption of mobile information artefacts will be managers, however, it is their own adoption that they are concerned with. This could be considered a form of managerial misbehavior, something that "few researchers have concerned themselves with" (Collinson et al., 2005, p. 306). So "that it is possible to "see" resistance and misbehavior" (Thompson et al., 1995, p. 629) a different frame is required.

These models are a packaging of selected elements of reference disciplines such as "psychology, and sociology" (Venkatesh et al., 2003, p. 426). The elements selected may, within the reference discipline, be far from universally accepted, for instance, TAM is strongly behavioral. A common emphasis in all of the models is that adoption is planned and reasoned. In new bargains it may well be that emotionality as well as rationality (Avgerou & McGrath, 2005) has a part to play. In constructing a conceptual framework it may therefore be necessary to return to the reference disciplines and engage with alternative perspectives.

These models were primarily created for quantitative research. When they have been appropriated for qualitative research (e.g., TAM in Breu et al., 2005; UTAUT in Garfield, 2005) their findings have been prematurely prescribed. A different frame, if designed with qualitative research in mind, could move away from narrow and shallow variables, such as age and gender in UTAUT, to broader and deeper concepts.

Phases of Information Artefact Production and Consumption

The pattern of bargaining with the devil reveals the shortcomings of the stages of adoption conceived in the fields of information systems (IS) and industrial relations (IR). The systems development life cycle (SDLC) is the standard expression of the IS development process. It usually consists of ten phases which may be structured sequentially (e.g., structured systems analysis and design methodology [SSADM]) or iteratively (after adaptation) (e.g., rapid applications development [RAD]):

1. *Initiation Phase*
2. *System Concept Development Phase*
3. *Planning Phase*
4. *Requirements Analysis Phase*
5. *Design Phase*
6. *Development Phase*
7. *Integration and Test Phase*
8. *Implementation Phase*
9. *Operations and Maintenance Phase*
10. *Disposition Phase*

(Department of Justice, 2003)

The existence of such phases may be argued for on logical grounds (Weaver, 1998) (i.e., one must plan [phase 3] before one implements [phase 8]). However, the specificity of these phases also reflects the "pernicious form" (Braverman, 1998, p. 227) of the division of labor that arose in the systems development industry. Such a division of labor no longer exists today when most large-scale systems, such as those for enterprise resource planning (ERP), are customised rather than custom-written. Customization provides for little control, enabling, for instance, the roles of systems analyst and programmer to be largely re-united in the role of systems developer (responsible for, as a minimum, phases 4 to 6). Without the logic of the division of labor behind it, these software developers pay "lip service" to the SDLC and the methodologies based upon it. If this is the relevance of the SDLC to the customization of large-scale systems it is unlikely to prove relevant in its current form for the adoption of mobile information artefacts.

An alternative conceptualization of technological change may be found in the field of IR, consisting of three stages:

1. *Initial planning of technological change*
2. *Selection of equipment/system*
3. *Implementation of change*

(Williams & Steward, 1985, p. 64)

Produced to assist in the analysis of the "application of new technology" (Williams & Steward, 1985, p. 58) in general, rather than the development of custom-written software, this conceptualization may be better-suited to the definition of information systems accepted in the IS discipline: "any hardware/software capability employed within organizations to do some task" (Orlikowski, 1988, p. 23). However, this conception does not look through the factory gate to the extra-organizational influence of the vendor nor does it display concern for what might happen after the implementation of change, for instance, when the devilish deception is exposed and the employee is destroyed.

Refining a Conceptual Framework

This research adopted a qualitative method, exploring patterns of adoption and their driving or influencing factors within a specific organizational context. By drilling to some depth it is hoped that ecological validity can be claimed. Recognising that the particular arrangement of patterns and factors will differ between spatio-temporal contexts, the research will proceed to explore further organizations—two such sites are envisaged—each time revising the analysis. An analysis applicable to a number of cases will therefore be built up, an analysis transferable, at least in part, to other contexts.

Figure 1. A conceptual framework for exploring the adoption of mobile information artefacts

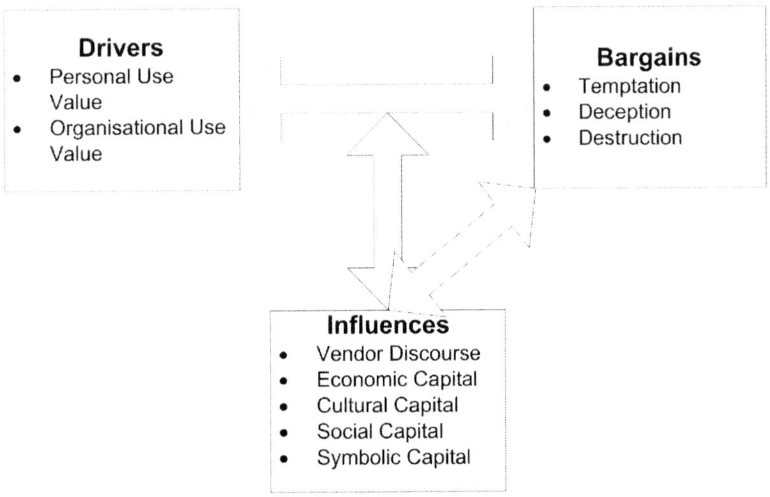

Utilising the resulting conceptual framework (see above) in critical fieldwork is not without potential pitfalls. Two aims of critical research are emancipation and non-performative intent (Howcroft & Trauth, 2004). If this research results in employers or vendors restricting or exploiting the personal use value of mobile information artefacts then these will not be fulfilled. Employers may also strive to set the boundary and scope of the investigation (Wilson & Greenhill, 2004) and employees in such organizations may not be especially keen to speak of resistance, misbehavior and dissent. Due to the lack of central control over mobile information artefact adoption (Sørensen et al., 2004) access will need to be of sufficient breadth, depth and duration. Furthermore, when exploring the role of vendor discourse it may be challenging to reconcile the possible false consciousness of the employee with the possible false presumption of the researcher regarding the effect of such discourse, the latter a leap of which Adorno has been accused (Crook, 1994).

CONCLUSION

It is hoped that this chapter succeeds in providing for the possibility of identifying new bargains, bargains that might follow a pattern such as this:

1. **Temptation:** The play discourse of the vendor appeals to the employee who may then propose the artefact, the employer being appeased by the productivity discourse of the vendor.
2. **Deception:** The honeymoon phase of play and exploration is a period of particularly intense pleasure for these "techies" but the employee soon finds that just as Satan can masquerade as an angel of light, this artefact is not (just) a tool of consumption, it is a tool of production as they run up against the "unmeetable payment-due dates" (Baudrillard, 1996, p. 161) for their productivity.
3. **Destruction:** The word "blackberry" brings pain rather than pleasure (Hodgkinson, 2005) as in return for the temporary satisfaction of their surplus desire—manipulated by marketing—they have consented to disruption of their work life boundary; they may even be addicted to their "crackberry."

Such a pattern may be a little too linear, especially as these knowledge workers were themselves responsible for some elements of the artefacts that destroyed them; but the pattern seems to reflect their experience that the fusion of consumption with production can be more of a con than a bargain. Despite their intention, the result, due to their being deceived, may be a new form of the protestant ethic, the ethic which shaped production and consumption in the United Kingdom. As their work life encroaches further upon their non-work life production is maximised and consumption minimised. The empowerment of self-managing one's work life boundary means little in practice for the knowledge-worker: the structural pressures and demands make it impossible not to let work permeate into life; the permeation of life into work is still controlled by the employer. The difference with this Protestant Ethic is that it is not the gratification that is deferred but the dissatisfaction. The artefact may also result in further disciplining of their innovation work; even its erosion as their loss of downtime means less time for reflection, for those quieter moments when a divine spark ignites a new idea. The situation of the employment relationship within relations of consumption may result in the knowledge-worker (and with them their employer) entering into a new relationship of exploitation where the (con)fusion of consumption and production results in a bargain only for the vendor.

However, if the bargains for mobile information artefacts are similar to those with the devil there is some hope: just as Jesus resists temptation, perhaps we can also, and if we can, the devil will have no hold over us (Ling, 1961, p. 61). The path of "radical discipleship," "unmasking and resisting" (Myers, 1988, p. 452) such deceptions, is rocky and far from straightforward:

Even if the ruses involved in the extraction of surplus labor and surplus desire were to become "fully transparent," it is still perfectly possible that the majority could be persuaded to accept... this mechanism of seduction, for it is true that it alone guarantees the kind of society which capitalism delivers, with its further worship of all forms of empty, sublime equivalence. (Milbank, 1990, p. 193)

REFERENCES

Adorno, T. W. (1991). *The culture industry*. London: Routledge.

Amabile, T. M. (1998). How to kill creativity. *Harvard Business Review, September-October*.

Amicus. (2002). *Make it happen. A CPHVA and CPNA campaign*. Amicus.

Anderson, J. E., & Schwager, P. H. (2004). SME adoption of wireless LAN technology: Applying the UTAUT model. In *Proceedings of the 7th Annual Conference of the Southern Association for Information Systems* (pp. 39-43).

AstraZeneca. (2006). *Rewarding you*. Retrieved February 21, 2006, from www.astrazeneca.co.uk/ideas/rewardingyou/lifestyle_shopping.asp

Avgerou, C., & McGrath, K. (2005). Rationalities and emotions in is innovation. In D. Howcroft & E. M. Trauth (Eds.), *Handbook of critical information systems research* (pp. 299-324). Cheltenham: Edward Elgar.

Baudrillard, J. (1996). *The system of objects* (J. Benedict, Trans.). London: Verso.

Bauman, Z. (2002). *Society under siege*. Cambridge: Polity Press.

Bell, E., & Taylor, S. (2003). The elevation of work: Pastoral power and the new age work ethic. *Organization, 10*(2), 329-349.

Bourdieu, P. (1984). *Distinction: A social critique of the judgement of taste*. London: Routledge.

Bradley, H., Erickson, M., Stephenson, C., & Williams, S. (2000). *Myths at work*. Cambridge: Polity.

Braverman, H. (1998). *Labor and monopoly capital* (25th Anniversary ed.). New York: Monthly Review Press.

Breu, K., Hemingway, C., & Ashurst, C. (2005). The impact of mobile and wireless technology on knowledge workers: An exploratory study. In D. Bartmann, F. Rajola, J. Kallinikos, D. E. Avison, R. Winter, P. Eindor, J. Becker, F. Bodendorf, & C. Weinhardt (Eds.), *Proceedings of the 13th European Conference on Information Systems*. Regensburg, Germany.

Buchholz, R. A. (2000). Toward a new ethic of production and consumption: A response to sagoff. *Ruffin Series in Business Ethics*, p.75.

Bunting, M. (2004). *Willing slaves: How the overwork culture is ruling our lives*. London: HarperCollins.

Caird, G. B. (1956). *Principalities and powers: A study in pauline theology*. Oxford: Clarendon.

Campbell-Kelly, M. (2003). *From airline reservations to sonic the hedgehog: A history of the software industry*. Cambridge, MA: MIT Press.

Casey, C. (1999). "Come, join our family": Discipline and integration in corporate organizational culture. *Human Relations, 52*(2), 155-176.

Collinson, D., & Ackroyd, S. (2005). Resistance, misbehavior, dissent. In S. Ackroyd, R. Batt, P. Thompson, & P. S. Tolbert (Eds.), *The Oxford handbook of work and organization*. Oxford: Oxford University Press.

Collinson, P. (2006). Computer scheme logged off. *The Guardian*.

Crook, S. (1994). Introduction: Adorno and authoritarian irrationalism. In *Adorno: The stars down to earth and other essays on the irrational in culture* (pp. 1-33). London: Routledge.

de Botton, A. (2004). *Status anxiety*. London: Hamish Hamilton.

Department of Justice. (2003). The department of justice systems development life cycle guidance document: Information Resources Management, Justice Management Division, Department of Justice.

Duane, A., & Finnegan, P. (2005). Reinforcing control of email systems through electronic monitoring and control. The *4th International Critical Management Studies Conference*. Cambridge, England.

Garfield, M. J. (2005). Acceptance of ubiquitous computing. *Information Systems Management, Fall* (pp. 24-31).

Golding, W. (1997). *Lord of the flies*. London: Faber and Faber.

Greenbaum, J. (2004). *Windows on the workplace*. New York: Monthly Review Press.

Grey, C. (2005). Critical management studies: Towards a more mature politics. In D. Howcroft & E. M. Trauth (Eds.), *Handbook of critical information systems research* (pp. pp.174-194). Cheltenham: Edward Elgar.

Herbert, G. (1982). To thee my god and king. In *Hymns ancient and modern*. London: Hymns Ancient and Modern.

Hodgkinson, T. (2005, Oct 15). First it was apple, then orange and now, blackberry. *The Guardian*.

Horkheimer, M., & Adorno, T. W. (2002). *Dialectic of enlightenment: Philosophical fragments* (E. Jephcott, Trans.). Stanford: Stanford University Press.

Howcroft, D., & Trauth, E. M. (2004). The choice of critical information systems research. In B. Kaplan, D. Truex, D. Wastell, T. Wood-Harper, & J. I. DeGross (Eds.), *Information systems research: Relevant theory and informed practice* (pp. 195-211). London: Kluwer.

Howcroft, D., & Wilson, M. (2003). Paradoxes of participatory practices: The janus role of the systems developer. *Information and Organization, 13*, 1-24.

Jacques, A. (2006). Farewell to time off. *The Independent on Sunday, 01/01/06.*

Jaros, S. J. (2005). Marxian critiques of Thompson's (1990) "core" labor process theory: An evaluation and extension. *Ephemera, 5*(1), 5-25.

Kunda, G. (1992). *Engineering culture: Control and commitment in a high-tech corporation*. Philadelphia: Temple University Press.

Kvasny, L. (2002). *Problematizing the digital divide: Cultural and social reproduction in a community technology initiative*. Atlanta: Georgia State University.

Ling, T. (1961). *The significance of Satan: New Testament demonology and its contemporary relevance*. London: SPCK.

Lloyds TSB. (2006). *Rewards and benefits*. Retrieved February 21, 2006, from www.adarkhorse.com/rewardsBenefits.php

Milbank, J. (1990). *Theology and social theory: Beyond secular reason*. Oxford: Blackwell.

Myers, C. (1988). *Binding the strong man: A political reading of Mark's story of Jesus*. Maryknoll, NY: Orbis.

Orange. (2006). *Benefits*. Retrieved 21 February 2006, from www.orange.co.uk/jobs/working_for_orange/benefits.php

Orlikowski, W. J. (1988). Computer technology in organizations: Some critical notes. In D. Knights & H. Willmott (Eds.), *New technology and the labor process*. Basingstoke: Macmillan.

Perlow, L. A. (1997). *Finding time: How corporations, individuals, and families can benefit from new work practices*. Ithaca: Cornell University Press.

Pinch, T. J., & Bijker, W. E. (1987). The social construction of facts and artifacts or how the sociology of science and the sociology of technology might benefit one another. In W. E. Bijker, T. P. Hughes, & T. J. Pinch (Eds.), *The social construction of technological systems*. Cambridge, MA: MIT Press.

Postrel, V. (2003). *The substance of style: How the rise of aesthetic value is remaking commerce, culture, and consciousness*. New York: HarperCollins.

Ristola, A., & Kesti, M. (2005). The effect on familiar mobile device and usage time on creating perceptions towards mobile services. *The IEEE 4th International Conference on Mobile Business*. Sydney, Australia.

Rousseau, D. M. (2005). *I-deals: Idiosyncratic deals employees bargain for themselves*. London: M.E. Sharpe.

Shaw, K. (2004). Cooltools: Quick takes on high-tech toys. *NetworkWorld, 05/10/04*.

Silf, M. (1998). *Landmarks: Exploration of ignatian spirituality*. London: Dartman, Longman and Todd.

Sørensen, C., & Pica, D. (2004). *Out-of-sight shouldn't mean out-of-mind: Why corporates need to get control of their wireless assets*. London: London School of Economics and Political Science.

Storey, J. (1986). The phoney war? New office technology. In D. Knights & H. Willmott (Eds.), *Managing the labor process* (pp. 44-66). Aldershot: Gower.

Thompson, P. (2005a). Brands, boundaries, and bandwagons: A critical reflection on critical management studies. In C. Grey & H. Willmott (Eds.), *Critical management studies: A reader* (pp. pp.364-382). Oxford: Oxford University Press.

Thompson, P. (2005b). *Labor and the labor process conference*. Critical Management Workshop. Retrieved October 25, 2005, from www.jiscmail.ac.uk

Thompson, P., & Ackroyd, S. (1995). All quiet on the workplace front? A critique of recent trends in British industrial sociology. *Sociology, 29*(4), 615-633.

Tillich, P. (1936). *The interpretation of history*. London: Scribner.

Towers, I., Duxbury, L., & Thomas, J. (2005). Shifting boundaries: Technology, time, place, and work. The 23rd *Annual International Labor Process Conference*. Glasgow, Scotland.

Venkatesh, V., Morris, M. G., Davis, G. B., & Davis, F. D. (2003). User acceptance of information technology: Toward a unified view. *Management Information Systems Quarterly, 27*(3), 425-478.

Walton, R. E. (1985). From control to commitment in the workplace. *Harvard Business Review, March-April*.

Warhurst, C., & Thompson, P. (2005). From conception to consumption: Myopic analysis of the creative industries. The 23rd *Annual International Labor Process Conference*. Glasgow, Scotland.

Weaver, P. (1998). *Practical SSADM version 4+*. London: Pitman.

Weber, M. (1930). *The protestant ethic and the spirit of capitalism*. London: Allen and Unwin.

Williams, R., & Steward, F. (1985). Technology agreements in Great Britain: A survey 1977-83. *Industrial Relations Journal, 16*(3), 58-73.

Wilson, M., & Greenhill, A. (2004). Theory and action for emancipation. In B. Kaplan, D. Truex, D. Wastell, T. Wood-Harper, & J. I. DeGross (Eds.), *Information systems research: Relevant theory and informed practice* (pp. 667-674). London: Kluwer.

Yu, J., Yu, C. S., Liu, C., & Yao, J. E. (2003). Technology acceptance model for wireless internet. *Internet Research, 13*(3), 206-222.

Compilation of References

Abbott, A. (1988). *The system of professions. An essay on the division of expert labor.* Berkeley, CA: The University of Chicago Press.

Adler, P. (2001). Market, hierarchy, and trust: The knowledge economy and the future of capitalism. *Organization Science, 12*(2), 215-234.

Adorno, T. W. (1991). *The culture industry.* London: Routledge.

Aggestam, M. (2003). Developing intellectual entrepreneurs of the universities: The example of Venture Lab. In S. Kwiatkowski & J. Sadlak (Eds.), *Intellectual entrepreneurship through higher education.* Leon Kozminski Academy of Entrepreneurship and Management. Warsaw: Unieco-Cepes.

Aggestam, M., & Keenan, J. (2000). *Communication capital in organisation: Managerial mindsets and activities during dramatic change.* Paper presented at International Communication Association, 1-5 June 2000, Acapulco, Mexico.

Aggestam, M., & Keenan, J. (2001). *A constitutive meta-analysis of communication capital in organisations undergoing dramatic change: A case study,* Paper presented at International Communication Association, 24-28 May, 2001, Washington, DC.

Aggestam, M., & Keenan, J. (2001). Corporate governance and organizational knowledge capital: Some conceptualisations. *Corporate Governance Journal, 9*(4), (October), 259-275.

Aggestam, M., & Keenan, J. (2005). *Narrating organizational knowledge: A communicationist turn.* Paper delivered at the Irish Academy of Management meetings, Galway, Ireland, September 2005.

Ailon-Souday, G., & Kunda, G. (2003). The local selves of global workers: The social construction of national identity in the face of organizational globalization. *Organization Studies, 24*(7), 1073-1096.

Akrich, M. (1992). The description of technical objects. In W. E. Bijker & J. Law (Eds.), *Shaping technology/Building society: Studies in sociotechnical change* (pp. 205-224). Cambridge, MA and London, England: MIT Press.

Akrich, M., Callon, M., & Latour, B. (2002). The key to success in innovation—Part 2: The art of choosing good spokespersons. *International Journal of Innovation Management, 6,* 207-225.

Aldrich, H., & Herker, D. (1977). Boundary-spanning roles and organization structure. *Academy of Management Review, 2*(2), 217-230.

Allen, R. (1990). *Thinkers of our time.* Polanyi. London: Claridge Press.

Allinson, C., Chell, E., & Hayes, J. (2000). Intuition and entrepreneurial behaviour. *European Journal of Work Organizational Psychology, 9,* 31-43.

Allport, F. H. (1962). A structuronomic conception of behavior: Individual and collective. *Journal of Abnormal and Social Psychology, 64,* 3-30.

Alstom Power s.r.o. (2002). *100 years of steam turbines in Brno.* Brno, Czech Republic: Alstom Power.

Altheide, D. L., & Johnson, J. M. (1994). Criteria for assessing interpretative validity in qualitive research. In N. K. Denzin, & Y. S. Lincoln (Eds.), *Hanbook of qualitative research* (pp. 485-499). Thousand Oaks: Sage.

Alvesson, M. (2000). Social identity and the problem of loyalty in knowledge-intensive companies. *Journal of Management Studies, 37*(8), 1101-1123.

Alvesson, M. (2001). Knowledge work: Ambiguity, image, and identity. *Human relations, 54*(7), 863-886.

Alvesson, M. (2002). *Understanding organizational culture*. London: Sage.

Amabile, T. M. (1998). How to kill creativity. *Harvard Business Review, September-October.*

Amicus. (2002). *Make it happen. A CPHVA and CPNA campaign*. Amicus.

Amin, A. (1994). *Post-Fordism. A reader.* Oxford: Blackwell.

Anastasi, A. (1983). Evolving trait concepts. *American Psychology, 38*, 172-183.

Ancona, D. G., Okhuysen, G. A., & Perlow, L. A. (2001). Taking time to integrate temporal research. *The Academy of Management Review, 26*(4), 512-529.

Anderson, J. E., & Schwager, P. H. (2004). SME adoption of wireless LAN technology: Applying the UTAUT model. In *Proceedings of the 7th Annual Conference of the Southern Association for Information Systems* (pp. 39-43).

Andersson, J. (2005). "Mellan tillväxt och (o)trygghet: idéer om tillväxt och socialpolitik i socialdemokratisk ideology." In J. Andersson, J. Björkman, & I. Humlesjö, (Eds.), *Välfärdsstatens skräpvind: historiska spår i dagens välfärdspolitik*, (pp. 97-119). Lund: Studentlitteratur.

Ang, S., & Straub, D.W. (1998). Production and transaction economies and IS outsourcing: A study of the U.S. banking industry. *MIS Quarterly, 22*(2), 535-552.

Angell, I. (2000). *The new barbarian manifesto: How to survive in the information age.* London: Kogan Page.

Antila, J., & Ylöstalo, P. (2002). *Proaktiivinen toimintatapa. Yritysten ja palkansaajien yhteinen etu?* [Proactive mode. Common asset for companies and employees?] Työpoliittinen tutkimus 239. Helsinki: Työministeriö.

Apo, S. (1996). Agraarinen suomalaisuus – rasite vai resurssi? [The agrarian Finnishness – burden or resource?] In P. Laaksonen & S. L. Mettomäki (Eds.), *Olkaamme siis suomalaisia. Kalevalaseuran vuosikirja* (pp. 75-76). Helsinki: SKS.

Apo, S. (1996). Itserasismista positiivisiin suomalaisuuksiin. [From self-imposed racism to positive Finnishnesses.] In S. Apo & J. Ehrnrooth (Eds.), *Millaisia olemme. Puheenvuoroja suomalaisista mentaliteeteista.* Helsinki: Kunnallisalan kehittämissäätiö.

Appadurai, A. (1996). *Modernity at large: Cultural dimensions of globalization.* Minneapolis: University of Minnesota Press.

Apple, M. W. (1995). *Education and power* (2nd ed.). New York: Routledge.

Apte, U. M., Sobol, M. G., Hanaoka, S., Shimada, T., Saarinen, T., Salmela, T., & Vepsalainen, A. P. J. (1997). IS outsourcing practices in the USA, Japan, and Finland: A comparative study. *Journal of Information Technology, 12*(4), 289-304.

Argyris, C. (1990). *Overcoming organizational defenses: Facilitating organizational learning.* Englewood Cliffs, NJ: Prentice Hall.

Argyris, C., & Schön, D. (1978). *Organizational learning: A theory of action perspective.* Reading, MA: Addison-Wesley.

Arora, A., & Gambardella, A. (2005). Bridging the gap: Conclusions. In A. Arora & A. Gambardella (Eds.), *From underdogs to tigers: The rise and growth of the software industry in Brazil, China, India, Ireland, and Israel* (pp. 275-302): Oxford, UK: Oxford University Press.

Arthur, M. B., & Rousseau, D. M. (1996). Introduction: The boundaryless career as a new employment principle. In M. B. Arthur & D. M. Rousseau (Eds.), *The boundaryless career: A new employment principle for a new organizational era* (pp. 3-20). Oxford: Oxford University Press.

Asch, S. E. (1952). *Social psychology.* Englewood Cliffs, NJ: Prentice-Hall.

Ashforth, B. E., & Mael, F. (1989). Social identity theory and the organization. *Academy of Management Review, 14*(1), 20-39.

AstraZeneca. (2006). *Rewarding you.* Retrieved February 21, 2006, from www.astrazeneca.co.uk/ideas/rewardingyou/lifestyle_shopping.asp

Atkinson, D. (1998). The production of the pupil as a subject within the art curriculum. *Journal of Curriculum Studies, 30*(1), 27-42.

Atkinson, R. D., & Correa, D. K. (2007). *The 2007 State New Economy Index—Benchmarking economic transformation.* The Ewing Marion Kauffman Foundation and the Information Technology and Innovation Foundation.

Audebert-Lasrochas, P. (2000). *Les équipes intelligentes.* Paris : Éd. d'organisation.

Austin, J. L. (1975). *How to do things with words.* In J. O. Urmson & M. Sbisà (Eds.), *The Williams James Lectures delivered at Harvard University in 1955.* Oxford: Oxford University Press.

Avgerou, C., & McGrath, K. (2005). Rationalities and emotions in is innovation. In D. Howcroft & E. M. Trauth (Eds.), *Handbook of critical information systems research* (pp. 299-324). Cheltenham: Edward Elgar.

Avison, D. E., & Myers, M. D. (1995). Information systems and anthropology: An anthropological perspective on it and organizational culture. *Information Technology & People, 8*(3), 43-56.

Avolio, B. J. (1999). *Full leadership development: Building the vital forces in organizations.* Thousand Oaks, CA: Sage.

Baier, A. (1986). Trust and antitrust. *Ethics, 96*(2), 231-260.

Bakshi, M. (2005, March 28). *Heard you, loud and clear.* The Hindu Business Line. Retrieved March 9, 2007, from http://www.thehindubusinessline.com/ew/2005/03/28/stories/2005032800200300.htm

Ball, K., & Wilson, D. (2000). Power, control, and computer-based performance monitoring: A subjectivist approach to repertoires and resistance. *Organization Studies, 21*(3), 539-565.

Ballantine, J., Levy, M., Munro, I., & Powell, P. (2003). An ethical perspective on information systems evaluation. *International Journal of Agile Management Systems, 2*(3), 233-241.

Ballard, M. (2006). *NHS IT probe useless.* Retrieved August 20, 2006, from http://www.theregister.co.uk/2006/03/24/nao_npfit_too_late/

Ballay, J. F. (1997). *Capitaliser et transmettre les savoir-faire de l'entreprise.* Paris: Eyrolles.

Ballay, J. F. (2002). *Tous managers du savoir! La seule ressource qui prend de la valeur en la partageant.* Paris : Éd. d'organisation.

Barber, B. (1983). *The logic and limits of trust.* New Brunswick, NJ: Rutgers University Press.

Bariff, M., & Galbraith, J. R. (1978). Intraorganizational power considerations for designing information systems. *Accounting, Organizations, and Society, 3*(1), 15-27.

Barinaga, E. (forthcoming). *The power effects of discourse—inclusion and exclusion in the information society* (preliminary title). Stockholm: EFI.

Barley, S. (1988). On technology, time, and social order: technically induced change in the temporal organization of radiological work. In F. A. Dubinskas, (Ed.). *Making time: Ethnographies of high-technology organizations.* Philadelphia: Temple University Press.

Barley, S. R., & Kunda, G. (2001). Bringing work back in. *Organization Science, 12*(1), 76-95.

Barley, S. R., & Kunda, G. (2004). *Gurus, hired guns, and warm bodies: Itinerant experts in a knowledge economy.* Princeton, NJ: Princeton University Press.

Barley, S. R., & Kunda, G. (2006). Contracting: A new form of professional practice. *Academy of Management Perspectives, 20*(1), 45-66.

Barnatt, C. (1995). Office space, cyberspace, and virtual organization. *Journal of General Management, 20*, 78-92.

Barret, R. (2005). *Management, labor process, and software development*. London: Routledge.

Barringer, B. R., & Harrison, J., S. (2000). Walking a tightrope: Creating value through interorganizational relationships. *Journal of Management, 26*(3), 367-403.

Bass, B. M. (1990). *Bass and Stogdill's handbook of leadership*. New York: Free Press.

Batt, R., Christopherson, S., Rightor, N., & Van Jaarsveld, D. (2001). *Networking. Work patterns and workforce policies for the new media industries*. Washington, DC: Economic Policy Institute.

Batt, R., Doellgast, V., & Kwon, H. (2006). Service management and employment systems in U.S. and Indian call centers. In L. Brainard & S. M. Collins (Ed.), *Offshoring white-collar work: The issues and implications*. The Brookings Trade Forum, 2005. (pp. 335-361). Washington, DC: The Brookings Institution.

Batt, R., Doellgast, V., Kwon, H., Nopany, M., Nopany, P., & da Costa, A. (2005). *The Indian call centre industry: National benchmarking report strategy, HR practices, & performance* (CAHRS Working Paper #05-07). Ithaca, NY: Cornell University, School of Industrial and Labor Relations, Center for Advanced Human Resource Studies. http://digitalcommons.ilr.cornell.edu/cahrswp/7

Baudrillard, J. (1996). *The system of objects* (J. Benedict, Trans.). London: Verso.

Bauman, Z. (2002). *Society under siege*. Cambridge: Polity Press.

Bauman, Z. (2004). *Wasted lives: Modernity and its outcasts*. Cambridge: Polity Press

Beck, U. (2000). *The brave new world of work* (P. Camiller, Trans.). Cambridge: Polity.

Beck, U. (2000). *What is globalization? Translated by Patrick Camiller*. Cambridge: Polity.

Beck, U. (2002). A life of one's own in a runaway world. In U. Beck & E. Beck-Gernsheim (Eds.), *Individualization: Institutionalized individualism and its social and political consequences* (pp. 22-29). London: Sage.

Beck, U., & Beck-Gernsheim, E. (2002). Authors' preface. In U. Beck & E. Beck-Gernsheim (Eds.), *Individualization: Institutionalized individualism and its social and political consequences* (pp. xx-xxv). London: Sage.

Beck, U., & Beck-Gernsheim, E. (2002). Losing the traditional. In U. Beck & E. Beck-Gernsheim (Eds.), *Individualization: Institutionalized individualism and its social and political consequences* (pp. 1-21). London: Sage.

Beckert, J. (2005). *Trust and the performative construction of markets* (MPIfG Discussion Paper 05/8). Cologne: Max Planck Institute for Studies of Societies.

Beirne, M., Ramsay, H., & Panteli, A. (1998). Developments in computing work: Control and contradiction in the software labor process. In P. Thompson & C. Warhurst (Eds.), *Workplaces of the future* (pp. 142-62). London: Macmillan.

Bélanger, J. (1991). La production sociale des normes de production: monographies d'entreprises en Grande-Bretagne et au Québec. *Sociologie et sociétés, 23*(2), 183-197.

Bell, D. (1973). *The coming of post-industrial society*. New York: Basic Books.

Bell, E., & Taylor, S. (2003). The elevation of work: Pastoral power and the new age work ethic. *Organization, 10*(2), 329-349.

Benoit, C. (1989). The professional socialisation of midwives: Balancing art and science. *Sociology of Health and Illness, 11*(2), 160-180.

Berger, P. L., & Luckmann, T. (1966). *The social construction of reality*. Garden City, NY: Doubleday.

Berger, P. L., & Luckmann, T. (1966/1983). *Społeczne tworzenie rzeczywistości*. Warszawa: PIW.

Berger, P., & Luckman, P. (1966). *The social construction of reality: A treatise in the sociology of knowledge*. Garden City, NY: Doubleday.

Berniker E. (1987). *Understanding technical systems*. Paper presented at Symposium on Management Training Programs: Implications of New Technologies, Geneva, Switzerland.

Bettenhausen, K., & Murnighan, J. K. (1985). The emergence of norms in competitive decision-making groups. *Administrative Science Quarterly, 30*, 350-372.

Beyer, H., & Holzblatt, K. (1998). *Contextual design: Defining customer-centred systems*. San Francisco: Morgan Kaufmann.

Bhatnagar, S. C., & Madon, S. (1997). The Indian software industry: Moving towards maturity. *Journal of Information Technology, 12*(4), 277-288.

Bijker, W. (1995). *Of bicycles, bakelite, and bulbs: Toward a theory of sociotechnical change*. Cambridge, MA: MIT Press.

Billig, M., Condor, S., Edwards, D., Gane, M., Middleton, D., & Radley, A. (1988). *Ideological dilemmas: A social psychology of everyday thinking*. London: Sage.

Birely, S., Cromie, S., & Myers, A. (1991). Entrepreneurial networks: Their emergence in Ireland and overseas. *International Small Business Journal, 9*, 56-74.

Blair, H. (2001). You're only as good as your last job: The labor process and the market in the British film industry. *Work, Employment and Society, 15*(1), 149-169.

Blomquist, Å., & Arvola, M. (2002). Personas in action: Ethnography in an interaction design team. In *Proceedings of the 2nd Nordic Conference on Human-Computer Interaction* (pp. 197-200). Aarhus, Denmark: ACM Press.

Bloomfield, B., & Coombs, R. (1992). Information technology, control, and power: The centralization and decentralization debate revisited. *Journal of Management Studies, 29*(4), 459-484.

Bloomfield, J., & Bianchini, F. (2004). *Planning for the intercultural city*. Bournes Green: Comedia.

Boland, R. J., & Tenkasi, R. V. (1995). Perspective making and perspective taking in communities of knowing. *Organization science, 6*(4), 350-372.

Borja, J., & Castells, M. (1997). *Local and global: Management of cities in the information age*. London: Earthscan.

Bormann, E. G. (1983). Symbolic convergence: Organizational communication and culture. In L. L. Putnam & M. E. Pacanowsky (Eds.), *Communication and organizations*. Beverly Hills: Sage.

Bormann, E. G. (1985). Symbolic convergence theory: A communication formulation. *Journal of Communication, 35*, 128-138.

Borthakur, M. (2006, Nov. 9). *BPO: Only business*. The Times of India. Retrieved March 9, 2007, from http://timesofindia.indiatimes.com/articleshow/ 374855.cms

Borys, B., & Jemison, D. B. (1989). Hybrid arrangements as strategic alliances: theoretical issues in organizational combinations. *Academy of Management Review, 14*(2), 234-248.

Bourdieu, P. (1984). *Distinction: A social critique of the judgement of taste*. London: Routledge.

Bradley, H., Erickson, M., Stephenson, C., & Williams, S. (2000). *Myths at work*. Cambridge: Polity.

Braithwaite, J., & Drahos, P. (2000). *Global business regulation*. Cambridge: Cambridge University Press.

Brattberg, L. (1977). Järvastaden—en kommunal satsning. In *Sankt Erik årsbok*. Stockholm: Samfundet S: t Erik.

Braverman, H. (1974). *Labour and monopoly capital*. London: Monthly Review Press.

Braverman, H. (1998). *Labor and monopoly capital* (25th Anniversary ed.). New York: Monthly Review Press.

Brecknock, R. (2006). *More than just a bridge: Planning & designing culturally*. Bournes Green: Comedia.

Breu, K., Hemingway, C., & Ashurst, C. (2005). The impact of mobile and wireless technology on knowledge workers: An exploratory study. In D. Bartmann, F. Rajola, J. Kallinikos, D. E. Avison, R. Winter, P. Eindor, J. Becker, F. Bodendorf, & C. Weinhardt (Eds.), *Proceedings of the 13th European Conference on Information Systems.* Regensburg, Germany.

Briers, M., & Chua, W. F. (2001). The role of actor-networks and boundary objects in management accounting change: A field study of an implementation of activity-based costing. *Accounting, Organizations and Society, 26*(3), 237-269.

Brockett, C. (1975). Toward a clarification of the need hierarchy theory: Some extensions of Maslow's conceptualization. *Interpersonal Development, 6*, 77-90

Brocklesby, J., & Cummings, S. (1996). Foucault plays Habermas: An alternative philosophical underpinning for critical systems thinking. *Journal of the Operational Research Society, 47*(6), 741-754.

Brohm, R. (2005). *Polycentric order in organizations. A dialogue between Michael Polanyi and IT-consultants on knowledge, morality, and organization.* Erasmus Research Institute of Management (ERIM).

Brown, D. (2005). *The black book of outsourcing.* Chichester: Wiley.

Brown, J. S., & Duguid, P. (1991). Organizational learning and communities of practice: Towards an unified view of working, learning, and innovation. *Organization Science, 2*(1), 40-57.

Brown, M. E., & Gioia, D. A. (2002). Making things click: Distributive leadership in an online division of an offline organization. *Leadership Quarterly, 13*(4), 397-419.

Bruckman, A. (1992). Identity workshop: Emergent social and psychological phenomena in text-based virtual reality. Retrieved November 27, 2006, from ftp://ftp.cc.gatech.edu/pub/people/asb/papers/identity-workshop.rtf

Bucciarelli, L. (1994). *Designing engineers.* Cambridge, MA and London, England: MIT Press.

Buchholz, R. A. (2000). Toward a new ethic of production and consumption: A response to sagoff. *Ruffin Series in Business Ethics*, p.75.

Buechler, H., & Buechler, J. M. (2003). *Contesting agriculture: Cooperativism and privatization in the New Eastern Germany.* SUNY Series in the Anthropology of Work. New York: State University of New York Press.

Bukowitz, W. R., & Williams, R. L. (1999). *The knowledge-management fieldbook.* London: Prentice Hall.

Bunting, M. (2004). *Willing slaves: How the overwork culture is ruling our lives.* London: HarperCollins.

Burbach, R., Nuñez, O., & Kagarlitsky, B. (1997). *Globalization and its discontents.* London: Pluto Press.

Burgelman, R. A., & McKinney, W. (2006). Managing the strategic dynamics of acquisition integration: Lessons from HP and Compaq. *California Management Review, 48*(3), 5-27.

Burke, K. A. (1969). *A rhetoric of motives.* London: University of California Press.

Burns, T., & Stalker, G. M. (1961). *The management of innovation.* London: Tavistock.

Burt, R. (1992). The social structure of competition. In N. Nohria & R. Eccles (Eds.), *Network and organizations* (pp. 57-91). Boston: Harvard Business School Press.

BusinessWeek Online. (2003). *The new global shift.* Retrieved December 20, 2006, from http://www.businessweek.com/magazine/content/03_05/b3818001.htm

Byrne, J. (1993). The virtual corporation. *Business Week* (8 February), 98-102.

Caird, G. B. (1956). *Principalities and powers: A study in pauline theology.* Oxford: Clarendon.

Caldwell, D. H., & O'Reilly III, C. A. (1982). Boundary-spanning and Individual Performance: The Impact of Self-Monitoring. *Journal of Applied Psychology, 67*(1), 124-127.

Callon, M. (1986). Some elements of sociology of translation. In J. Law (Ed.) *Power, action, and belief. A new sociology of knowledge.* (pp. 196-233). London: Routledge & Kegan Paul.

Callon, M. (1999). Actor-network theory: The market test. In J. Law, & J. Hassard (Ed.), *Actor network theory and after* (pp. 181-196). Oxford: Blackwell Publishing.

Campbell-Kelly, M. (2003). *From airline reservations to sonic the hedgehog: A history of the software industry.* Cambridge, MA: MIT Press.

Carlile, P. R. (2002). A pragmatic view of knowledge and boundaries: Boundary objects in new product development. *Organization Science, 13*(4), 442-455.

Carlile, P. R. (2004). Transferring, translating, and transforming: An integrative framework for managing knowledge across boundaries. *Organization Science, 15*(5), 555-568.

Carte, T. A., Chidambaram, L., & Becker, A. (2006). Emergent leadership in self-managed-virtual teams. *Group Decision and Negotiation, 15*, 324-343.

Case, P., & Pineiro, E. (2006). Aesthetics, performativity, and resistance in the narratives of a computer programming community. *Human Relations, 59*(6), 753-782.

Casey, C. (1999). "Come, join our family": Discipline and integration in corporate organizational culture. *Human Relations, 52*(2), 155-176.

Castells, M. (1996). *The rise of the network society.* Cambridge, MA: Blackwell.

Castells, M. (2000). Materials for an exploratory theory of the network society. *The British Journal of Sociology, 51*(1), 5-24.

Chasserio, Stéphanie, & Marie-Josée Legault. (2008, in press). Strategic human resources management is irrelevant when it comes to highly skilled professionals in the Canadian new economy! *International Journal of Human Resource Management.*

Chaston, I. (1998). Self-managed teams: Assessing the benefits for small service-sector firms. *British Journal of Management, 9*, 1-12.

Chatman, J. A., & Cha, S. A. (2002). Leading though organizational culture. In S. Chowdhury (Ed.), *Next generation business series: Leadership.* Financial Times. Prentice Hall Publishers.

Chatterjee, M. B., & Ramnarayan, A. (2006, April 10). *For you and your family.* The Hindu Business Line. Retrieved March 10, 2007, from http://www.hinduonnet.com/thehindu/thscrip/print.pl?file=2006041000010100.htm&date=2006/04/10/&prd=ew&

Chatterjee, S. R., & Cecil, A. P. (2000). Indian managers in transition: Orientation, work goals, values and ethics. *Management International Review, 40*, 81-95.

Checkland, P. (1981). *Systems thinking, systems practice.* London: John Wiley and Sons.

Checkland, P., & Scholes, P. (1990). *Soft systems methodology in action.* Chichester: John Wiley and Sons.

Chell, E. (2000). Towards researching the "opportunistic entrepreneur": A social constructionist approach and research agenda. *European Journal of Work and Organizational Psychology, 9*, 63-80.

Chiapello, E., & Fairclough, N. L. (2002). Understanding the new management ideology: A transdisciplinary contribution from critical discourse analysis and new sociology of capitalism. *Discourse and Society, 13*(2), 185-208.

Churchland, P. M. (1992). A deeper unity: Some Feyerabendian themes in neurocomputational form. In S. Davis (Ed.), *Connectionism: Theory and practice* (pp. 30-50). New York: Oxford University Press.

Churchman, C. W. (1968). *The systems approach.* New York: Dell Publishing.

CiteHR. (2004). *BPOs encourage parents day out.* Retrieved March 10, 2007, from http://www.citehr.com/bpo-india=bpo-indusrty-vt7396.html

Clark, A. (2003). *Natural-born-cyborgs: Minds, technologies, and the future of human intelligence.* Oxford: Oxford University Press.

Clark, E., & Soulsby, A. (1999). *Organizational change in Post-Communist Europe: Management and transformation in the Czech Republic.* London: Routledge

Clawson, D. (1980). *Bureaucracy and the labor process: The transformation of U.S. industry, 1850-1920.* New York: Monthly Review Press.

Clegg, S. (1981). Organization and control. *Administrative Science Quarterly, 26*, 545-562.

Coleman, J. C. (1990). *Foundations of social theory.* Cambridge, MA: Harvard University Press.

Collinson, D. L. (1994). Strategies of resistance, power, knowledge, and subjectivity in the workplace. In J. M. Jermier, D. Knights, & W. N. Nord (Eds.). *Resistance and power in organizations.* London, UK: Routledge.

Collinson, D. L. (2003). Selves at work. *Organization, 10*(3), 527-547.

Collinson, D., & Ackroyd, S. (2005). Resistance, misbehavior, dissent. In S. Ackroyd, R. Batt, P. Thompson, & P. S. Tolbert (Eds.), *The Oxford handbook of work and organization.* Oxford: Oxford University Press.

Collinson, P. (2006). Computer scheme logged off. *The Guardian.*

Cook, K., Hardin, R., & Levi M. (2005). *Cooperation without trust.* New York: Russell Sage Foundation.

Cook, S. D. N., & Yanow, D. (1996). Culture and organizational learning. In M. D. Cohen & L. S. Sproull (Eds.), *Organizational learning* (pp. 430-59), Thousand Oaks, CA: Sage.

Cooper, A. (1999). The inmates are running the asylum: Why high-tech products drive us crazy and how to restore the sanity. Indianapolis, IN: SAMS Publishing.

Cooper, R. G. (2001). *Winning at new products: Accelerating the process from idea to launch* (3rd ed.). NY: Perseus Publishing.

Corbett, M. F. (2004). *The outsourcing revolution: Why it makes sense and how to do it right.* Chicago: Dearborn.

Córdoba, J. R., & Midgley, G. (2006). Broadening the boundaries: An application of critical systems thinking to IS planning in Colombia. *Journal of the Operational Research Society, 57*(9), 1064-1080.

Costlow, T. (2003). Globalization drives changes in software careers. *IEEE Software, 20*(6), 12-16.

Cotte, D. (1999). *Les enjeux du management de l'information dans les organisations: Usages, outils, techniques.* Lille, France: ADBS.

Courpasson, D. (2000). *L'action contrainte. Organisations libérales et domination.* Paris: PUF.

Coutu, A. & Willmott, H. (2003). Re-embedding situatedness: The importance of power relation in learning theory. *Organization Science, 14*(3), 283-296.

Crabtree, A. (2003). *Designing collaborative systems: A practical guide to ethnography.* London: Springer-Verlag.

Creighton, S., & Hodson, R. (1997).Whose side are they on? Technical workers and management ideology. In S. R. Barley & J. E. Orr (Eds.), *Between craft and science: Technical work in U.S. settings.* Ithaca, NY: ILR Press, Cornell University.

Critchley, S. (1999). *The ethics of deconstruction: Derrida and Levinas* (2nd ed.). Edinburgh University Press.

Crook, S. (1994). Introduction: Adorno and authoritarian irrationalism. In *Adorno: The stars down to earth and other essays on the irrational in culture* (pp. 1-33). London: Routledge.

Cross, N. (2006). *Designerly ways of knowing.* London: Springer-Verlag.

Cross, R. L., & Parker, A. (2004). *The hidden power of social networks: Understanding how work gets done in organizations.* Boston: Harvard Business School Press.

Crossen, M., Lane, H. W., & White, R. E. (1999). An organization learning framework: From intuition to institution. *Academy of Management Review, 24*(3), 522-537.

Czarniawska, B (1997). *Narrating the organisation: Dramas of institutional identity.* Chicago: University of Chicago Press.

Czarniawska, B., & Sévon, G. (1996). *Translating organizational change.* Berlin: Walter de Gruyter.

Czarniawska-Joerges, B. (1992). *Exploring complex organizations: A cultural perspective*. Newbury Park, CA: Sage.

D'Cruz, P., & Noronha, E. (2006). Being professional: Organizational control in Indian call centers. *Social Sciences Computer Review, 24*(3), 342-361.

D'Cruz, P., & Noronha, E. (2006, May 27). *Organising call centre agents: Emerging issues*. Economic and Political Weekly. Retrieved February 27, 2007, from http://www.epw.org.in

D'Mello, M. (2005). Thinking local, acting global": issues of identity and related tensions in global software organizations in India. *Electronic Journal for Information Systems in Developing Countries, 22*(2), 1-20.

Damsgaard, J., & Scheepers, R. (1999). Power, influence, and intranet implementation: A safari of South African organizations. *Information Technology and People, 12*(4), 333.

Dash, D. P. (Date not available). *Strategic assumption surfacing and testing (SAST)*. Retrieved on March 16, 2007, from http://www1.ximb.ac.in/users/fac/dpdash/dpdash.nsf/pages/CP_SAST

Davenport, T. H., & Prusak, L. (1998). *Working knowledge: How organizations manage what they know*. Boston: Harvard Business School Press.

Davies, B., & Harré, R. (1990). Positioning: The discursive production of Selves. *Journal of Theory of Social Behaviour, 20*, 43-65.

de Botton, A. (2004). *Status anxiety*. London: Hamish Hamilton.

Deal, T. E., & Kennedy, A. A. (1982). *Corporate cultures: The rites and rituals of corporate life*. Reading, MA: Addison-Wesley.

Deery, S., & Kinnie, N. (2004). *Call centres and human resource management*. Hampshire: Palgrave Macmillan.

Denison, D. R., & Mishra, A. K. (1995). Toward a theory of organizational culture and effectiveness. *Organization Science, 6*, 204-223.

Denison, D. R., Hooijberg, R., & Quinn, R. E. (1995). Paradox and performance: Toward a theory of behavioral complexity in managerial leadership. *Organization Science, 6*, 524-540.

Department of Education Science and Training, & Howard Partners. (2003). *Evaluation of the cooperative research centres programme*. Canberra, Australia: Howard Partners Pty Ltd.

Department of Education Science and Training. (2006). *CRC Directory 2006*. Retrieved October 1, 2006, from https://www.crc.gov.au/HTMLDocuments/Documents/PDF/CRC_Directory_2006.pdf

Department of Health. (2005). *Hewitt announces action to turnaround NHS finances*. Retrieved July 21, 2006, from http://www.dh.gov.uk/PublicationsAndStatistics/PressReleases/PressReleasesNotices/fs/en?CONTENT_ID=4127292&chk=HDOR9C

Department of Justice. (2003). The department of justice systems development life cycle guidance document: Information Resources Management, Justice Management Division, Department of Justice.

Derber, C. (1982). *Professionals as workers: Mental labor in advanced capitalism*. Boston: GK Hall & Co.

Derber, C. (1983). Managing professionals: Ideological proletarianisation and post-industrial labor. *Theory and Society, 12*(5), 561-603.

Derber, C., & Schwartz, W. A. (1988). Les hiérarchies à l'intérieur des hiérarchies. Le pouvoir professionnel à l'œuvre. *Sociologie et Sociétés, 20*(2), 55-76.

Derrida, J. (1991). Letter to a Japanese friend [original 1987]. In P. Kamuf (Ed.), *A Derrida reader: Between the blinds* (pp. 270-276). New York: Columbia University Press.

Derrida, J. (1995). *Points: Interviews, 1974-1994*. Stanford, CA: Stanford University Press.

Derrida, J. (2001). Deconstructions: The impossible. In S. Lotringer & S. Cohen (Eds.), *French theory in America* (pp. 12-32). New York and London: Routledge.

Deutsch, M. (1973). *The resolution of conflict: Constructive and destructive processes*. New Haven, CT: Yale University Press.

Dev, S. (2004, Jan. 26). ITES Players offer long-term careers. *IT People, 8*(18), 1-2.

Dhillon, G. (2004). Dimensions of power and IS implementation. *Information & Management, 41*, 635-644.

Dick, P. (1968). *Do androids dream of electric sheep?* New York: Doubleday.

DiMaggio, P. W., & Powell, W. W. (1983). The iron cage revisited: Institutional isomorphism and collective rationality in organizational fields. *American Sociological Review, 48*, 147-160.

Dixon, J. C., & Rosenbaum, M. S. (2004). Nice to know you? Testing contact, cultural, and group threat theories of anti-black and anti-Hispanic stereotypes. *Social Science Quarterly, 85*(2), 257-280.

Dobers, P., & Schroeder, J. (2003). *The image economy—organizational processes and representational practices in a Swedish context*. Unpublished paper; presented at a seminar in the Stockholm School of Entrepreneurship.

Doherty, N., & King, M. (2001). An investigation of the factors affecting the successful treatment of organizational issues in systems development. *European Journal of Information Systems, 10*, 147-160.

Doke, D. D. (2003, April 29). What makes Indian call centers tick. *Personnel Today*, 6-8.

Dolinina, I. B., & Cecchetto, V. (1998). Facework and rhetorical strategies in intercultural argumentative discourse. *Argumentation, 12*, 167-181.

Doolin, B. (2004). Power and resistance in the implementation of a medical management information system. *Information Systems Journal, 14*(4), 343-362.

Dossani, R., & Kenney, M. (2003). Went for cost, stayed for quality? Moving the back office to India. *Berkeley Roundtable on the International Economy*. Berkeley, CA: University of California.

Doukas, D. (2003). *Worked over: The corporate sabotage of an American Community*. Ithaca, NY: Cornell University Press.

Doussard, M. & Mastracci, S. (2003). *Uncertain futures: The real impact of the high-tech boom and bust on Seattle's IT workers*. Center for Urban Economic Development: University of Illinois at Chicago.

Drazin, R. (1990). Professionals and innovation: Structural-functional vs. radical-structural perspectives. *Journal of Management Studies, 27*(3), 245-263.

DTA Consultants. (2004). *The Indian IT and IT-enabled services Industry*. Retrieved January 20, 2007, from http://www.dtainternational.com/ITES_in_India11.pdf

Du Gay, P. (1996). *Consumption and identity at work*. London: Sage.

Duane, A., & Finnegan, P. (2005). Reinforcing control of email systems through electronic monitoring and control. The *4th International Critical Management Studies Conference*. Cambridge, England.

Dubinskas, F. A. (1988). *Making time: Ethnographies of high-technology organizations*. Philadelphia: Temple University Press.

Dunn, E. C. (2004). *Privatizing Poland: Baby food, big business, and the remaking of labor*. Ithaca, NY: Cornell University Press.

Easterby-Smith, M., Thorpe, R., & Lowe, A. (2002). *Management research: An introduction* (2nd ed.). London: Sage.

Eccles, R. G. (1981). Bureaucratic vs. craft administration: The relationship of market structure to the construction firm. *Administrative Science Quarterly, 26*(3), 449-469.

Economist Intelligence Unit. (2003). Finding staff and keeping them. *Business India Intelligence, 10*(1), 5-8.

Edley, N. (2001). Analysing masculinity: Interpretative repertoires, ideological dilemmas and subject positions. In M. Wetherell, S. Taylor, & S. Yates (Eds.), *Discourse as data: A guide for analysis* (pp. 189-228). London: Sage.

Elden, M. (1983). Democritization and participative research in developing local theory. *Journal of Occupational Behaviour, 4*, 21-33.

Elden, M., & Leven, M. (1991). Cogenerative learning: Bringing participation into action research. In W. F. White (Ed.), *Participatory action research* (pp.127-142). Newbury Park, NJ: Sage Publications.

Elgin, B. (2005). Revenge of the nerds: Again. *Business Week, July 28, 2005.*

Eraly, A. (1988). *La structuration de l'entreprise. La rationalité en action.* Bruxelles: Éditions de l'Université de Bruxelles.

Erduran, S., Simon, S., & Osbourne, J. (2004). TAPing into argumentation: Developments in the application of Toulmin's argument pattern for studying science discourse. *Science Education, 88*(6), 915-933.

Erez, A., Lepine, J., & Elms, H. (2002). Effects of rotated leadership and peer evaluation on the functioning and effectiveness of self-managed teams: A quasi experiment. *Personnel Psychology, 55*, 932-947.

Ericson, R., & Doyle, A. (2003). *Risk and morality.* Toronto: University of Toronto Press

Etkowitz, H., & Leydesdorff, L. (2000). The dynamics of innovation: From national systems and "mode 2" to a triple helix of university-industry-government relations. *Research Policy, 29*, 109-123.

Evetts, J. (2001). New directions in state and international professional occupations: Discretionary decision-making and acquired regulation. *Proceedings of the SASE 13th Annual Meeting on Socio-Economics Knowledge: The New Wealth of Nations*, University of Amsterdam, Pays-Bas, 28 juin - 1er juillet, retrieved from http://www.sase.org/conf2001/papers/papers.html

Eyal, G. (2003). *The origins of post-communist elites.* Minneapolis: University of Minnesota Press.

Eyal, G., Szelenyi, I., & Townsley, E. (1998). *Making capitalism without capitalists: The new ruling elites in Eastern Europe*. London: Verso.

Featherstone, M. (1995). Post-bodies, ageing and virtual reality. In M. Featherstone & A. Wernick (Eds.), *Images of ageing.* London: Routlage.

Featherstone, M., & Burrows, R. (1995). *Cyberspace/cyberbodies/cyberpunk. Cultures of technological embodiment.* London: Sage Publications.

Ferguson, C. (1988). From the people who brought you voodoo economics. *Harvard Business Review, 66,* 55-62.

Ferguson, J. (1999). *Expectations of modernity: Myths and meanings of urban life on the Zambian Copperbelt.* Berkeley, CA: University of California Press.

Fleming, P., & Spicer, A. (2003). Working at a cynical distance: Implications for power, subjectivity, and resistance. *Organization, 10*(1), 157-179.

Florida, R., & Kenney, M. (1990). *The breakthrough illusion: Corporate America's failure to move from innovation to mass production.* New York: Basic Books.

Foucault, M. (1977). *Discipline and punish: The birth of the prison.* London: Allen Lane.

Foucault, M. (1977). *The history of sexuality volume one: The will to knowledge* (Vol. 1). London: Penguin.

Foucault, M. (1978/1990). *The history of sexuality: An introduction* (Vol. I). New York: Vintage.

Foucault, M. (1980). *Power/knowledge: Selected interviews & other writings 1972-1977.* New York: Pantheon Books.

Foucault, M. (1980). Truth and power. In P. Rabinow (Ed.), *The Foucault reader: An introduction to Foucault's thought* (pp. 51-75). London: Penguin.

Foucault, M. (1980). Two lectures. In C. Gordon (Ed.), *Power/knowledge: Selected interviews and other writings Michel Foucault* (pp. 78-108). New York: Harvester Wheatsheaf.

Foucault, M. (1982). Afterword: The subject and power. In H. Dreyfus & P. Rabinow (Eds.), *Michel Foucault: Beyond structuralism and hermeneutics* (pp. 208-226). Brighton: The Harvester Press.

Foucault, M. (1982). On the genealogy of ethics: An overview of work in progress. In P. Rabinow (Ed.), *The Foucault reader: An introduction to Foucault's thought* (pp. 340-372). London: Penguin.

Foucault, M. (1984). The ethics of the concern of the self as a practice of freedom (R. e. a. Hurley, Trans.). In P. Rabinow (Ed.), *Michel Foucault: Ethics subjectivity and truth: Essential works of Foucault 1954-1984* (pp. 281-301). London: Penguin.

Foucault, M. (1984). *The history of sexuality volume two: The use of pleasure*. London: Penguin.

Foucault, M. (1984). What is enlightenment? (C. Porter, Trans.). In P. Rabinow (Ed.), *The Foucault reader: An introduction to Foucault's thought* (pp. 32-50). London: Penguin.

Foucault, M. (1985). *The history of sexuality volume three: The care of the self*. London: Penguin.

Foucault, M. (1989). *The archaeology of knowledge*. London: New York: Routledge.

Foucault, M. (1994). *La Hermeneutica del Sujeto (Hermeneutics of the Subject)* (F. Alvarez-Uría, Trans.). Madrid: Ediciones de la Piqueta.

Foucault, M. (1994). La práctica del cuidado de uno mismo como práctica de la libertad. In F. Alvarez-Uría (Ed.), *La Hermeneutica del Sujeto (Hermeneutics of the Subject)* (pp. 106-142). Madrid: Ediciones de la Piqueta.

Foucault, M. (1994). Technologies of the self. In P. Rabinow (Ed.), *Michel Foucault: Ethics, subjectivity and truth*. (pp. 223-251). London: Penguin.

Frank, T. (2001). *One market under god—Extreme capitalism, market populism, and the end of economic democracy*. New York: Anchor Books

Franke, U. (1999). The virtual Web as a new entrepreneurial approach to network organizations. *Entrepreneurship and Regional Development, 11*, 203-229.

Freeman, C. (2000). *High tech and high heels in the global economy: Women, work, and pink collar identities in the Caribbean*. Durham, NC: Duke University Press.

Freidson, E. (1994). *Professionalism reborn: Theory, prophecy, and policy*. Chicago: University of Chicago Press.

Friedberg, K., & Friedberg, J. (1996). *Nuts! Southwest Airlines' crazy recipe for business and personal success*. Austin, TX: Bard Press.

Friedman, A. L. (1977). *Industry and labour: Class struggle at work and monopoly capitalism*. London, UK: Macmillan.

Friedman, R. A., & Podolny, J. (1982). Differentiation of boundary-spanning roles: Labor negotiations and implications for role conflict. *Administrative Science Quarterly, 37*(1), 28-47.

Friedman, T. L. (2005). *The world is flat: A brief history of the twenty-first century*. Waterville: Thorndike Press.

Friedmann, G. (1964). *Le travail en miettes*. Paris: Gallimard.

Frissen, V. (1995). Gender is calling: Some reflections on past, present and future uses of the telephone. In R. Gill, & D. Gring, (Eds.), *The gender-technology relation*. London: Taylor & Francis.

Gabriel, Y. (2005). Glass cages and glass palaces: Images of organization in image-conscious times. *Organization, 12*(1), 9-27.

Galbraith, J. R. (1973). *Designing complex organizations*. MA: Addison-Wesley.

Gambetta, D. (1988). Foreword. In D. Gambetta (Ed.). *Trust: Making and breaking cooperative relations*. Oxford: Basic Blackwell.

Gambetta, D. G. (1988). Can we trust trust? In D. G. Gambetta (Ed.), *Trust: Making and breaking cooperative relations* (pp. 213-237). New York: Basil Blackwell.

Ganesh, V. (2006, December 1). *Want to retain staff? Keep it in the family, say BPOs*. Hindustan Times. Retrieved from http://www.hindustantimes.in/news/181_1857532,00020021.htm

Garfield, M. J. (2005). Acceptance of ubiquitous computing. *Information Systems Management, Fall* (pp. 24-31).

Gates, B., Myhrvold, N., & Rhinearson, P. (1995). *The road ahead*. New York: Viking.

Gelwich, R. (1977). *The way of discovery: An introduction to the thought of Michael Polanyi*. New York: Oxford University Press.

Gere, C. (2006). *Digitaalinen kulttuuri*. Turku: Faros. [Orig.: Digital Culture. London: Reaction Books, 2002.]

Gergen, K. J. (1999). *An invitation to social construction*. London: Sage.

Gerlach, N., & Hamilton, S. (2000). Telling the future, managing the present: Business restructuring literature as SF. *Science Fiction Studies, 27*, 461-77.

Gibbons, M., Limoges, C., Nowotny, H., Schwartzman, S., Scott, P., & Trow, M. (1994). *The new production of knowledge: The dynamics of science in contemporary societies*. London: Sage Publications.

Gibson, J. W., & Tesone, D. V. (2001). Management fads: Emergence, evolution, and implications for managers. *Academy of Management Executive, 15*(4), 122-133.

Gibson, W. (1984). *Neuromancer*. New York: Ace Books.

Giddens, A. (1990). *The consequences of modernity*. Cambridge: Polity Press.

Giddens, A. (1991). *Modernity and self-identity: Self and society in the late modern age*. Cambridge: Polity Press.

Gill, J. H. (2000). *The tacit mode: Michael Polanyi's postmodern philosophy*. Albany, NY: State University of New York Press.

Gill, R. (2000). Discourse analysis. In M. Bauer & G. Gaskell (Eds.), *Qualitative researching with text, image and sound: A practical handbook for social research* (pp. 172-190). London: Sage.

Gill, R. (2002). Cool, creative, and egalitarian? Exploring Gender in Project-based New Media Work in Europe. *Information, Communication, and Society, 5*(1), 70-89.

Gillies, A. (2000). *Information and IT for primary care: Everything you need to know but are afraid to ask*. Oxon: Radcliffe Medical Press.

Gioia, D. A., & Sims, H. P. (1986). Introduction: Social cognition in organizations. In H. P. Sims, Jr., D. A. Sims, et al. (Eds.), *The thinking organization*. San Francisco: Jossey-Bass Publishers.

Glaser, B., & Strauss, A. (1957). *Discovery of grounded theory: Strategies for qualitative research*. Chicago: Aldine.

Goffman, E. (1959). *The presentation of self in everyday life*. London: Penguin.

Golding, W. (1997). *Lord of the flies*. London: Faber and Faber.

Goldman, S., & Nagel, R. (1993). Management technology and agility: The emergence of a new era in manufacturing. *International Journal of Technology Management, 8*, 18-38.

Goldman, S., Nagel, R., & Preiss, K (1995). *Agile competitors and virtual organizations: Strategies for enriching the customer*. New York: Van Nostrand Reinhold.

Goman, C. K. (2002). *Five reasons people don't tell what they know*. DestinationKM-Viewpoint. Retrieved June 22, from DestinationKM.com

Göranzon, B. (1991). *The practical intellect: Computers and skills*. London: Springer.

Gotterbarn, D. (1997). Software engineering: A new professionalism. In C. Myers, T. Hall, & D. Pitt (Eds.), *The responsible software engineer: Selected readings in IT professionalism* (pp. 21-31). London: Springer.

Grabher, G., & Stark, D. (1997). Organizing diversity: Evolutionary theory, network analysis, and post-socialism. In G. Grabher & D. Stark (Eds.), *Restructuring networks in post-socialism: Legacies, linkages, and localities* (pp. 1-32). Oxford: Oxford University Press.

Gray, D. O., Lindblad, M., & Rudolph, J. (2001). Industry-university research centers: A multivariate analysis of member retention. *Journal of Technology Transfer, 26*, 247-254.

Greenbaum, J. (2004). *Windows on the workplace*. New York: Monthly Review Press.

Gregory, A. (2000). Problematizing participation: A critical review of approaches to participation in evaluation theory. *Evaluation, 6*(2), 179-199.

Grey, C. (2005). Critical management studies: Towards a more mature politics. In D. Howcroft & E. M. Trauth (Eds.), *Handbook of critical information systems research* (pp. pp.174-194). Cheltenham: Edward Elgar.

Griffeth, R. W., Hom, P. W., & Gaertner, S. (1998). A meta-analysis of antecedents and correlates of employee turnover: Update, moderate testes, and research implications for the millennium. *Journal of Management, 26*, 463-488.

Griffith, T. L., Devadas, R., & Neale, M. A. (2003). Virtualness and knowledge in teams: Managing the love triangle of organizations, individuals, and information technology. *MIS Quarterly, 27*, 265-287.

Gupta, S., & Chaudhari, N. S. (2006). Information technology offshore outsourcing: A perspective of advanced countries. In H. S. Kehal & V. P. Singh (Eds.), *Outsourcing and offshoring in the 21st century: A socio-economic perspective* (pp. 122-139). Hershey, PA: Idea Group Publishing.

Haddon, L., & Kommonen, K. H. (2003). *Interdisciplinary explorations: A dialogue between a sociologist and a design group*. Helsinki: University of Art and Design.

Hakken, D. (2000). Resocialing work: The future of the labor process. *Anthropology of Work Review, XXI*(1), 8-10.

Halaby, C. N. (1986). Worker attachment and workplace authority. *American Sociological Review, 51*, 634-649.

Hale, N. M. (1998). Effects of age and interpersonal contact on stereotyping of the elderly. *Current Psychology, 17*(1), 28-48.

Hall, E. T. (1981). *Beyond culture*. New York: Doubleday.

Hall, R. H. (1986). *Dimensions of work*. California: Sage Publications Inc.

Hammond, D. (2002). Exploring the genealogy of systems thinking. *Systems Research and Behavioural Science, 5*(19), 429-439.

Hampden-Turner, C., & Trompenars, A. (1993). *The seven cultures of capitalism*. New York: Currency-Doubleday.

Handy, C. (1976). *Understanding organizations*. Aylesbury: Penguin.

Handy, C. B. (1994). *The empty raincoat: Making sense of the future*. London: Hutchinson/Arrow.

Hannerz, U. (1987). The world in creolization. *Africa, LVII*, 546-549.

Hannerz, U. (1996). *Transnational connections: Culture, people, places*. London: Routledge.

Haraway, D. (1991). *Simians, cyborgs, and women: The reinvention of nature*. London: Free Association Books.

Hardin, R. (2004). *Distrust*. New York: Russell Sage Foundation.

Hardwick, M., Spooner, D., Rando, T., & Morris, K. (1996). Sharing manufacturing information in virtual enterprises. *Communication of the ACM, 39*, 46-54.

Hardy, C, Phillips, N., & Lawrence T. (1998). Distinguishing trust and power in interorganizational relations: forms and facades of trust. In C. Lane & R. Backman (Ed), *Trust within and between organizations: Conceptual issues and empirical applications* (pp. 64-87). Oxford: Oxford University Press.

Hargreaves, A., & Goodson, I. (1996). Chapter 1: Teachers' professional lives: Aspirations and actualities. In I. F. Goodson & A. Hargreaves (Eds.), *Teachers' professional lives* (pp. 1-27). London: Falmer Press.

Harris, T. E., & Sherblom, J. C. (2005). *Small group and team communication*. Boston: Pearson.

Harvey, D. (1989). *The condition of postmodernity: An enquiry into the origins of cultural change*. Cambridge, MA: Blackwell.

Harvey, D. (2000). *Spaces of hope*. Berkeley, CA: University of California Press.

Hatch, C. (1995). *The network brokers handbook*. US Department of Commerce, National Institute of Standards and Technology, manufacturing Extension Partnership, Gaithersburg, MD 20899, USA.

Hatch, M. J., Kostera, M. I. & Koźmiński, A. K. (2005). *The three faces of leadership:*

Hatchuel, A., & Weil, B. (1992). *L'expert et le système. Gestion des savoirs et métamorphose des acteurs dans l'entreprise industrielle*. Paris: Economica.

Hayes, K. (2007, Sept 6-7, 2007). *Triple helix organisations, knowledge-stewarding communities of practice and perceptions of time: The hunters and gatherers of commercialisation*. Paper presented at the 8th European Conference on Knowledge Management, Consorci Escola Industrial de Barcelona, Spain.

Hayes, K., & Fitzgerald, A. (2005, Sept 4-7). *Preliminary findings of an investigation into interactions between commercial and scientific occupational cultures in hybrid research organisations*. Paper presented at the Proceedings of the 6th International CINet Conference, Brighton, UK.

Hayri, A., & McDermott, G. A. (1998). The network properties of corporate governance and industrial restructuring: A post-socialist lesson. *Industrial and Corporate Change, 7*(1), 153-193.

Health Informatics Community. (2004). NPfIT says vast unforeseen expenditure claims are "complete nonsense." Retrieved July 15, 2006, from http://www.informatics.nhs.uk/cgibin/item.cgi?id=922&d=11&h=24&f=46&dateformat=%25o-%25b-%25Y

Heeks, R. (1996). *India's software industry: State policy, liberalization, and industrial development*. New Delhi: Sage.

Heery, E., & Salmon, J. (2000). *The insecure workforce*. London, UK: Routledge.

Hekkert, P., & Van Dijk, M. (2001). Designing from context. In P. Lloyd & H. Christiaans (Eds.), *Designing in context* (pp. 383-394). Delft: Delft University Press Science.

Helsby, G., & McCulloch, G. (1996). Chapter 3: Teacher professionalism and curriculum control. In I. F. Goodson & Hargreaves, A. (Eds.), *Teachers' professional lives*, (pp. 56-71). London: Falmer Press.

Herbert, G. (1982). To thee my god and king. In *Hymns ancient and modern*. London: Hymns Ancient and Modern.

Heritage, J. (1984). *Garfinkel and ethnomethodology*. Cambridge: Polity Press.

Herman, E. S., & Chomsky, N. (2002). In Dan Frank (Ed.), *Manufacturing consent. The political economy of the Mass Media*.

Himanen, P. (2001). *The hacker ethic. A radical approach to the philosophy of business*. New York: Random House.

Hirst, P. H. (1982). Professional authority: Its foundation and limits. *British Journal of Educational Studies, 30*(2), 172-182.

Hjorth, D. (2004). Creating space for play/invention—concepts of space and organizational entrepreneurship. *Entrepreneurship & Regional Development, 16*, 413-432.

Hochschild, A. R. (1983). *The managed heart: Commercialization of human feeling*. Berkeley, CA: University of California Press.

Hochschild, A. R. (1997). *The time bind: When work becomes home and home becomes work*. New York: Metropolitan Books.

Hodgkinson, T. (2005, Oct 15). First it was apple, then orange and now, blackberry. *The Guardian*.

Hodgson, D. (2000). *Discourse, discipline and the subject: A Foucauldian analysis of the UK financial services industry*. Aldershot: Ashgate.

Hoepfl, H. (1995). Performance and customer service: The cultivation of contempt. *Studies in Cultures, Organizations and Societies, 1*(1), 47-62.

Hofstede, G. (1980). *Culture's consequences*. Beverly Hills, CA: Sage.

Holland, D., Lachicotte, W. Jr., Skinner, D., & Cain, C. (1998). *Identity and agency in cultural worlds*. Cambridge: Harvard University Press.

Holy, L. (1996). *The little Czech and the great Czech Nation: National identity and post communist social transformation*. Cambridge: Cambridge University Press.

Horkheimer, M., & Adorno, T. W. (2002). *Dialectic of enlightenment: Philosophical fragments* (E. Jephcott, Trans.). Stanford: Stanford University Press.

Horton, K. S. (2000). *The exercise of power and information systems strategy: The need for a new perspective*. Paper presented at the 8th European Conference on Information Systems (ECIS), Vienna.

Howcroft, D., & Trauth, E. M. (2004). The choice of critical information systems research. In B. Kaplan, D. Truex, D. Wastell, T. Wood-Harper, & J. I. DeGross (Eds.), *Information systems research: Relevant theory and informed practice* (pp. 195-211). London: Kluwer.

Howcroft, D., & Wilson, M. (2003). Paradoxes of participatory practices: The janus role of the systems developer. *Information and Organization, 13*, 1-24.

Hoyle, E., & John, P. D. (1995). *Professional knowledge and professional practice*. London: Cassell.

Huemer, L. (1998). *Trust in business relations: Economic logic or social interaction?* Umea: Borea bokvorlag.

Hulin, C. L., & Roznowski, M. (1985). Organizational technologies: Effects on organizations' characteristics and individual's responses" In L. L. Cummins & B. M. Staw (Eds.), *Research in organizational behavior*. Greenwich, CT: JAI Press.

IEEE-USA. (2006). *About us*. Retrieved November 27, 2006 from http://www.ieeeusa.org/about/default.asp

IndiatimesInfotech. (2006). *BPOs outsourcing work to smaller cos*. Retrieved January 5, 2007, from http://infotech.indiatimes.com/BPO__ITES/Outsourcing/BPOs_outsourcing_work_to_smaller_cos_/articleshow/2011023.cms

Infosys. (2001). *Management of human-assets at Infosys*. Retrieved March 10, 2007, from http://www.infosys.com/about/cases/INFOSYS6%20case%20withchanges.pdf

InfotechOnline. (2004). *BPO workers more satisfied now*. Retrieved November 10, 2006, from http://infotech.indiatimes.com/articleshow/msid-919912.cms

InfotechOnline. (2004). *Invited! Your family for the BPO job*. Retrieved January 5, 2007, from http://infotech.indiatimes.com/articleshow/msod-940751.cms

InfotechOnline. (2006). *BPOs take care of your career too*. Retrieved September 5, 2006, from http://infotech.indiatimes.com/articleshow/msid-1456697.cms

Inkeles, A., & Levinson, D. (1954). National character: The study of modal personality and sociocultural systems. In G. Lindzey (Ed.), *Handbook of social psychology* (pp. 977-1020). Cambridge: Addison-Wesley.

Introna, L. D. (1997). *Management, information and power: A narrative of the involved manager*. Basingstoke: Macmillan.

ISO. (1999). ISO 13407: *Human-centred design processes for interactive systems*. Geneva, Switzerland: International Organization for Standardization.

Jackson, M. (2000). *Systems approaches to management*. Wiley: Chichester.

Jackson, M. (2003). *Systems thinking: Creative holism for managers*. Wiley: Chichester.

Jackson, M. C. (1992). An integrated programme for critical thinking in information systems research. *Information Systems Journal, 2*, 83-95.

Jacob, R. L. & Pariat, L. (2000). *Gérer les connaissances: un défi de la nouvelle compétitivité du 21e siècle. Information, interaction, innovation*, CEFRIO. Retrieved from http://www.cefrio.qc.ca/rapports/gererconnaissance.pdf

Jacques, A. (2006). Farewell to time off. *The Independent on Sunday, 01/01/06.*

Janssens, M., & Steyaert, C. (1999). The inhuman space of HRM: Sssssssssssssssensing the subject. *Organization, 6*(2), 371-383.

Jaros, S. J. (2005). Marxian critiques of Thompson's (1990) "core" labor process theory: An evaluation and extension. *Ephemera, 5*(1), 5-25.

Jasperson, J. S., Carte, T., Saunders, C. S., Butler, B. S., Croes, H. J. P., & Zheng, W. (2002). Power and information technology research: A metatriangulation review. *MIS Quarterly, 26*(4), 397-459.

Jemielniak, D. (2004). Kultura: Zawody i profesje. *Prace i Materiały Instytutu Studiów Międzynarodowych SGH*, 32.

Johannisson, B. (1990). Community entrepreneurship—Cases and conceptualisation. *Entrepreneurship and Regional Development, 2,* 71-88.

Johansson, I. (1991). *Stor-Stockholms bebyggelsehistoria - Markpolitik, planering och byggande under sju sekler.* Stockholm: Gidlunds.

Jones, C. (1996). Career in project networks: The case of the film industry. In M. B. Arthur & D. M. Rousseau (Eds.), *The boundaryless career: A new employment principle for a new organizational era* (pp. 58-75). Oxford: Oxford University Press.

Jordan, T. (1999). *Cyberpower. The culture and politics of cyberspace and the internet.* London and New York: Routledge.

Kakumanu, P., & Portanova, A. (2006). Outsourcing: Its benefits, drawbacks, and other related issues. *Journal of American Academy* of *Business, 9*(2), 1-7.

Kalakota, R., & Robinson, M. (2004). *Offshore outsourcing: Business models, ROI, and best practices.* Alpharetta, GA: Mivar Press.

Kallinikos, J. (2003). Work, human agency, and organizational forms: An anatomy of fragmentation. *Organization Studies, 24*(4), 595-618.

Kanet, J., & Faisst, W. (1997). *The role of information technology in supporting the entrepreneur for the virtual enterprise: a life-cycle-oriented description.* Working paper, Clemson University, Clemson, SC.

Kanter, R. M. (1995). *World class: Thriving locally in the global economy.* London: Simon & Schuster.

Kanungo, R. N. (1995). Culture and work alienation: Western models and eastern realities. In H. S. Kao, D. Sinha, & N. Sek-Hong (Eds.), *Social values and development: Asian perspectives* (pp. 233-250). New Delhi: Sage.

Kaplan, D. A. (2000). *The Silicon boys and their valley of dreams.* New York: Harper Collins.

Katzenbach, J., & Smith, D. (1993). *The wisdom of teams.* Boston: Harvard Business School Press.

Keeley, R., & Tabrizi, B. (1995). High-tech entrepreneurs: Serious competitors or troublemakers? *Journal of High Technology Management Research, 4,* 127-143.

Keenan, J. (2000). Organizational knowledge capital perspectives for informatic systems. In G. Bradley (Ed.), *Humans on the net: Information and communication technology, work organization, and human beings.* Stockholm, Sweden: Prevent.

Kehal, H. S., & Singh, V. P. (Eds.). (2006). *Outsourcing and offshoring in the 21st Century: A socio-economic perspective.* Hershey, PA: Idea Group Publishing.

Kendall, G., & Wickham, G. (1999). *Using Foucault's methods.* London: Sage.

Kessels, J. (1996). Knowledge productivity and the corporate curriculum. In J. F. Schreinemakers (Ed.), *Knowledge management: Organization, competence, and methodology* (Vol. 1, pp. 168-174). Würzburg: Ergon Verlag.

Khan, N. (2006). *The road to interculturalism: Tracking the arts in a changing world.* Bournes Green: Comedia.

Kirkman, B. L., & Rosen, B. (1999). Beyond self-management: Antecedents and consequences of team empowerment. *Academy of Management Journal, 42*(1), 58-74.

Kirkman, B. L., & Shapiro, D. L. (2001). The impact of cultural values on job satisfaction and organizational commitment in self-managing work teams: The mediating role of employee resistance. *Academy of Management Journal, 44*(3), 557-569.

Kirkman, B. L., Gibson, C. B., & Shapiro, D. L. (2001). "Exporting" teams: Enhancing the implementation and effectiveness of work teams in global affiliates. *Organizational Dynamics, 30*(1), 12-29.

Kirzner, I (1997). Entrepreneurial discovery and the competitive market process: An Austrian approach, *Journal of Economic Literature, 35,* 60-85.

Knell, J. (2000). *Most wanted: The quiet birth of the free worker*. London, UK: The Industrial Society.

Knights, D., & Sturdy, A. (1990). New technology and the self-disciplined worker in the insurance industry. In I. Varcoe, M. McNeil, & S. Yearley (Eds.), *Deciphering science and technology* (pp. 126-154). London: Macmillan.

Knights, D., & Willmott, H. (1989). Power and subjectivity at work. *Sociology, 23*(4), 535-558.

Knowledge@Wharton. (2004). *A view from the developing world.* Retrieved January 25, 2007, from http://knowledge.wharton.upenn.edu/article.cfm?articleid=860

Knowledge@Wharton. (2004) *Why corporations pursue BPO.* Retrieved February 25, 2007, from http://knowledge.wharton.upenn.edu/article.cfm?articleid=860

Koch, T., & Thomas, M. (1997). The social and cultural embeddedness of entrepreneurs in Eastern Germany. In G. Grabher & D. Stark (Eds.), *Restructuring networks in post-socialism: Legacies, linkages, and localities* (pp. 242-262). Oxford: Oxford University Press.

Koen, P., Ajamian, G., Boyce, S., Clamen, A., Fisher, E., Fountoulakis, S. et al. (2002). Fuzzy front end: Effective methods, tools, and techniques. In *PDMA toolbook for new product development* (pp. 2-35). John Wiley & Sons.

Kolankiewicz, G. (1996). Social capital and social change. *British Journal of Sociology, 47*(3), 427-441.

Korten, D. C. (2001). *When corporations rule the world.* San Francisco: Berrett-Koehler.

Koskinen, I., Battarbee, K., & Mattelmäki, T. (2003). *Empathic design: User experience in product design.* Helsinki: Edita Publishing.

Kostera, M. (1996). *Postmodernizm w zarządzaniu.* Warszawa: PWE.

Kotamraju, N. P. (2002). Keeping up: Web design skill and the reinvented worker. *Information, Communication, & Society, 5*(1), 1-26.

Kraft, P. (1977). *Programmers and managers: The routinisation of computer programming in the United States.* New York: Springer Verlag.

Kraft, P. (1987). Computers and the automation of work. In R. Kraut (Ed.), *Technology and the transformation of white collar work* (pp. 91-111). NJ: Lawrence Erlbaum.

Kraft, P., & Dubnoff, S. (1986). Job content, fragmentation, and control in computer software work. *Industrial Relations, 25*(2), 184-196.

Kramer, R. M., & Cook, K. S. (2004). *Trust and distrust in organizations. Frontiers of theory and research.* Thousand Oaks, CA: Sage Publications.

Krippendorf, K. (1971). Communication and the genesis of structure. *General Systems, 20,* 15-35.

Kuczi, T., & Mako, C. (1997). Toward industrial districts? Small-firm networking in Hungary. In G. Grabher & D. Stark (Eds.), *Restructuring networks in post-socialism: Legacies, linkages, and localities* (pp. 176-189). Oxford: Oxford University Press.

Kujala, S. (2003). User involvement: A review of the benefits and challenges. *Behaviour and Information Technology, 22,* 1-17.

Kunda, G. (1992). *Engineering culture: Control and commitment in a high-tech corporation.* Philadelphia: Temple University Press.

Kvasny, L. (2002). *Problematizing the digital divide: Cultural and social reproduction in a community technology initiative.* Atlanta: Georgia State University.

Laflamme, R. (1998). Mobilisation ou manipulation? In R. Laflamme (Ed.), *Mobilisation et efficacité au travail*, Actes du 9e congrès de *l'Association internationale de psychologie du travail de langue française* (AIPTLF), Tome 6 (p. 87-96). *Proceedings of Gestion des paradoxes dans les organisation*, Cap Rouge, Presses Interuniversitaires.

Lane, C. (1998). Introduction: Theories and issues in the study of trust. In C. Lane, & R. Bachmann (Eds.), *Trust within and between Organizations – conceptual issues and empirical applications* (pp. 1-30). Oxford: Oxford University Press.

Lane, C., & Bachmann, R. (1998). *Trust within and between organizations – conceptual issues and empirical applications*. Oxford: Oxford University Press.

LaPalombara, J. (2001). The underestimated contributions of political science to organizational learning. In M. Dierkes, A. B. Antal, J. Child, & I. Nonaka (Eds.), *Handbook of organizational learning and knowledge*, (pp. 56-71). Oxford: Oxford University Press.

Larson, M. S. (1977). *The rise of professionalism: A sociological analysis*. Berkeley, CA: University of California Press.

Lash, S., & Urry, J. (1987). *The end of organised capitalism*. Oxford: Polity

Latour, B. (1987). *Science in action. How to follow scientists and engineers through society* Cambridge, MA: Harvard University Press.

Latour, B. (1991). Technology is society made durable. In J. Law (Ed.). *A sociology of monsters. essays on power, technology, and domination* (pp. 103-131). London: Routledge.

Latour, B., & Woolgar, S. (1986). *Laboratory life: The construction of scientific facts* (2nd ed.). Princeton, NJ: Princeton University Press.

Lawler, E. E., Mohrmann, S. A., & Ledford, G. (1995). *Creating high performance organizations: Practices and results of employee involvement and total quality management, practice, and results in Fortune 1000 companies*. San Francisco: Jossey-Bass.

Lawson, B. (1997). *How designers think: The design process demystified*. Architectural Press (an imprint of Butterworth-Heineman, a division of Reed Educational and Professional Publishing).

Leadbeater, C., & Oakley, K. (1999). *The independents: Britain's new cultural entrepreneurs*. London: Demos.

Lee, H., & Liebenau, J. (1999).Time in organizational studies: Towards a new research direction. *Organization Studies, 20*(6), 1035-1058.

Legault, M. (1988). Le métier de chercheur scientifique en sciences sociales et la sociologie des professions. *Sociologie et sociétés, 20*(2), 163-176.

Legault, Marie-Josée. (2005). Differential gender effects of project management and management by project on skilled professionals. In Kay S. Devine, & Jean-Noël Grenier (dir.), *Reformuler les relations industrielles dans une économie de marché libérale, Actes du 41e colloque de l'Association canadienne des relations industrielles (ACRI) - Canadian Industrial Relations Association (CIRA)*, (p. 105-124). Concord, Ontario: ACRI-CIRA et Captus Press.

Lehrer, M., & Asakawa, K. (2004). Pushing scientists into the marketplace: Promoting science entrepreneurship. *California Management Review, 46*(3), 55-76.

Lehtonen, M., Löytty, O., & Ruuska, P. (2004). *Suomi toisin sanoen*. [Finland in other words.] Tampere: Vastapaino.

Leppälä, K., & Päiviö, H. (2001). *Kauppatieteiden opiskelijoiden moraalijärjestys: Narratiivinen tutkimus kolmen eri pääaineen opiskelusta Helsingin kauppakorkeakoulussa*. [Moral order among students of economics: A narrative inquiry into studying three major subjects at Helsinki School of Economics.] Helsinki: Publications of the Helsinki School of Economics and Business Administration, B-34.

Lesser, E. L., & Storck, J. (2001). Communities of practice and organizational performance. *IBM Systems Journal, 40*(4), 831-841.

Letiche, H. (1998). Business ethics: (In-)justice and (anti-)law. In M. Parker (Ed.), *Ethics and organizations* (pp. 122-149). London: Sage.

Levina, N., & Vaast, E. (2005). The emergence of boundary-spanning competence in practice: Implications for implementation and use of information systems. *MIS Quarterly, 29*(2), 335-363.

Levinas, E. (1996). Transcendence and height [original 1962]. In A. Peperzak, S. Critchley, & R. Bernasconi (Eds.), *Emmanuel Levinas: Basic philosophical writings* (pp. 11-32). Bloomington and Indianapolis, IN: Indiana University Press.

Levinas, E. (1996). Transcendence and intelligibility [original 1984]. In A. Peperzak, S. Critchley, & R. Bernasconi (Eds.), *Emmanuel Levinas: Basic philosophical writings* (pp. 149-159). Bloomington and Indianapolis, IN: Indiana University Press.

Levine, R. V., Norenzayan, A., & Philbrick, K. (2001). Cross-cutural differences in helping strangers. *Journal of Cross-Cultural Psychology, 32*, 543-560.

Levine, S. (2006). High performance organizations: Creating a culture of agreement. *Handbook of Business Strategy, 7*(1), 375-380.

Lewicki, R. and Bunker, B. (1996) Developing and maintaining trust in work relationships. In Kramer, R., & Tyler, T. (Eds.). *Trust in organizations – frontiers of theory and research* (pp. 114-139). Thousand Oaks, CA: Sage.

Lewicki, R., McAlliser, D., & Bies, R. (1998). Trust and distrust: New relationships and realities. *Academy of Management Review, 23*(3), 438-458.

Lincoln, J. R., & Kalleberg, A. L. (1990). *Culture, control, and commitment. Study of work organization and work attitudes in the United States and Japan.* Cambridge: Cambridge University Press.

Lincoln, Y. S., & Guba, E.G. (1985). *Naturalistic inquiry.* Beverly Hills, CA: Sage

Ling, T. (1961). *The significance of Satan: New Testament demonology and its contemporary relevance.* London: SPCK.

Litvin, D. R. (1997). The discourse of diversity: From biology to management. *Organisation, 4*(2), 187-209.

Liu, Y. (1999). Justifying *my* position in *your* terms: Cross-cultural argumentation in a globalized world. *Argumentation, 13*, 297-315.

Liyanage, S., & Mitchell, H. (1993). Organizational management in Australian cooperative research centres. *Technology Analysis and Strategic Management, 5*(1), 3-14.

Lloyds TSB. (2006). *Rewards and benefits.* Retrieved February 21, 2006, from www.adarkhorse.com/rewards-Benefits.php

Locke, T. (2001). Questions of professionalism: Erosion and reclamation. *CHANGE. Transformations in Education, 4*(2), 30-50.

Lorbiecki, A., & Jack, G. (2000). Critical turns in the evolution of diversity management. *British Journal of Management,* Sep. 2000, Special Issue, *11*(3).

Lowestedt, J. (1993). Organizing frameworks in emerging organizations: A cognitive approach to the analysis of change. *Human Relations, 46*(4), 501-526.

Lukes, S. (1974). *Power: A radical view.* London: Macmillan.

MacDonald, K. M. (1995). *The sociology of the professions.* London, Thousand Oaks, CA, New Delhi: Sage Publications.

Machacek, D. (2003). Stale uznavame 'Zlaté české ručičký'. *Mlada Fronta Dnes*, April 17, 2003.

MacKenzie, D., & Wajcman, J. (1999). *The social shaping of technology.* Buckingham: Open University Press.

Manager, artist, priest. Oxford: Blackwell Publishing.

Mangham, I. L., & Overington, M. A. (1987). *Organizations as theater: Social psychology and dramatic performance.* Chichester: Wiley.

Mankekar, P. (2002). "India shopping": Indian grocery stores and transnational configurations of belonging. *Ethnos, 67*(1), 75-97.

Manske, A. (2003). Arbeits- und Lebensarrangements in der Multimediabranche unter Vermarktlichungsdruck - Rationalisierungspotenzial für den Markterfolg. In E. Kuhlmann & S. Betzelt (Eds.), *Geschlechterverhältnisse im Dienstleistungssektor - Dynamiken, Differenzierungen und neue Horizonte* (pp. 133-146). Baden-Baden: Nomos.

March, J. G., & Simon, H. A. (1958). *Organizations*. New York: Wiley.

Mareš, P. (1995). Legitimacy. In I. Možny, L. Musil, P. Mareš, I Řezniček, & T. Sirovatka (Eds.), *Social consequences of change in ownership: Two case studies in industrial enterprises in the Czech Republic– Spring 1993* (pp. 53-70). Brno, Czech Republic: Masarykova Univerzita.

Marglin, S. (1973). What do bosses do? The origins and functions of hierarchy in capitalist production. *Review of Radical Political Economics*, *6*(2), 60-112.

Markus, M. L. (2002). Power, politics, and MIS implementation. In M. Myers & D. Avison (Eds.), *Qualitative research in information systems*. London: Sage.

Marquard, O. (1991). *In defense of the accidental: Philosophical studies*. New York: Oxford University Press.

Martin, E. (1994). *Flexible bodies: The role of immunity in American culture from the days of polio to the age of AIDS*. Boston: Beacon Press.

Martin, J. (2002). *Organizational culture: Mapping the terrain*. Thousand Oaks, CA: Sage Publications Inc.

Marx, K., & Engels, F. (1996/1848). *The communist manifesto*. London: Junius.

Mason, R. O., & Mitroff, I. I. (1981). *Challenging strategic planning assumptions*. Chichester: John Wiley & Sons.

Massey, D. (1984). *Spatial divisions of labour: Social structures and the geography of production*. London: Macmillan.

McAulay, L., Keval, N., & Doherty, N. (2002). The stakeholder dimension in information systems evaluation. *Journal of Information Technology, 17*(4), 241-255.

McCue, A. (2005, January 21). *High staff turnover hits Indian offshore firms*. Offshoring Special Report. Retrieved December 15, 2006, from http://www.silicon.com/ research/specialreports/offshoring/0,3800003026,39127243,00.htm

McEvily, B. (2003). Trust as an organizing principle. *Organization Science, 14*(1), 91-103.

McGovern, P. (1998). *HRM, technical workers and the multinational corporation*. London, UK: Routledge.

McKerrow, R. E. (1980). *Argument communities*. Paper presented at the 1st Summer Conference on Argumentation, Annandale, VA.

McLaughlin, J., Rosen, P., Skinner, D., & Webster, A. (1999). *Valuing technology: Organizations, culture, and change*. London: Routledge.

Mead, G. H. (1934). *Mind, self, and society*. Chicago: University of Chicago Press.

Mead, M. (1951). *Soviet attitudes towards authority*. New York: W. Morrow.

Menon, S., Sinha, S., & Urs, A. (2004, August 1). Tug of talent. *The Economic Times*, p. 7. Kolkata, India: The Times Group.

Merali, Y. (2000). Individual and collective congruence in the knowledge management process. *Journal of Strategic Information Systems*, *9*(2-3), 213-234.

Merton, R. K. (1957). Priorities in scientific discovery: A chapter in the sociology of science. *American Sociological Review, 22*(6), 635-660.

Meyerson, D., & Martin, J. (1987). Cultural change: An integration of three different views. *Journal of Management Studies, 24*(6), 623-647.

Michaelson, K. L. (2000). Laboring in cyberspace: Internet realities and the future of work. *Anthropology of Work Review, XXI*(1), 11-14.

Midgley, G., Munlo, I., & Brown, M. (1998). The theory and practice of boundary critique: Developing housing services for older people. *Journal of the Operational Research Society*, *49*(5), 467-478.

Milbank, J. (1990). *Theology and social theory: Beyond secular reason.* Oxford: Blackwell.

Miles, R., & Snow, C. (1986). Network organisations: New concepts and new forms. *California Management Reviews, 28,* 62-73.

Minzberg, H. (1976). *The structuring of organizations.* Englewood Cliffs, NJ: Prentice-Hall.

Mitchell, T. R., Holtom, B. C., Lee, T. W., Sablynski, C. J., & Erez, M. (2001). Why people stay: Using job embeddedness to predict voluntary turnover. *Academy of Management Journal, 44,* 1102-1121.

Möllering, G. (2006). *Trust: Reason, routine, reflexivity.* Oxford: Elsevier.

Montgomery, D. (1979). *Worker's control in America.* Cambridge: Cambridge University Press.

Moravec, M., Johannessen, O. J., & Hjelmas, T. A. (1997). Thumbs up for self-managed teams. *Management Review, 86*(7), 42-47.

Morgan, G. (1997). *Images of organization.* Thousand Oaks, CA: Sage.

Moshowitz, A. (1986). Social dimension of office automation. *Advances in Computers, 25,* 335-404

Možny, I., Musil, L., Mareš, P., Řezniček, I., & Sirovatka T. (1995). *Social consequences of change in ownership: Two case studies in industrial enterprises in the Czech Republic– Spring 1993.* Brno, Czech Republic: Masarykova Univerzita.

Mukherjee, W. (2004, June 17). *Most BPO workers change careers.* The Economic Times. Retrieved October 6, 2006, from http://www1.economictimes. indiatimes. com/articleshow/msid-742807.cms

Murphy, R. (1990). Proletarianization or bureaucratization: The fall of the professional? In R. Torstendahl, & M. Burrage (Eds.), *The formation of professions: Knowledge, state, and strategy* (pp. 71-96). London: Sage.

Murray, P., & Moses, M. (2005). The centrality of teams in the organizational learning process. *Management Decision, 43*(9), 1186-1202.

Musil, L. (1995). Participation. In I. Možny, L. Musil, P. Mareš, I Řezniček, & T. Sirovatka (Eds.), *Social consequences of change in ownership: Two case studies in industrial enterprises in the Czech Republic– Spring 1993* (pp. 26-52). Brno, Czech Republic: Masarykova Univerzita.

Myers, C. (1988). *Binding the strong man: A political reading of Mark's story of Jesus.* Maryknoll, NY: Orbis.

Nardi, B, A., & O'Day, V. L. (1999). *Information ecologies: Using technology with heart.* Boston: MIT Press.

Nash, J. C. (1989). *From tank town to high tech: The clash of community and industrial cycles.* Albany, NY: SUNY Press.

Nash, J. C. (2001). *Mayan visions: The quest for autonomy in an age of globalization.* New York: Routledge.

NASSCOM. (2004). *NASSCOM Strategic Review 2004.* New Delhi: NASSCOM.

NASSCOM. (2005). *NASSCOM Strategic Review 2005.* New Delhi: NASSCOM.

NASSCOM. (2005). *Strategic review: The IT industry in India.* New Delhi: NASSCOM.

NASSCOM. (2007). *NASSCOM Strategic Review 2007.* New Delhi: NASSCOM.

Nemeth, C. J. (1997). Managing innovation: When less is more. *California Management Review, 40*(1), 59-74.

Nichols, T. (1996). Russian Democracy and Social Capital. *Social Science Information, 35*(4), 629-642.

Nicholson, B., & Sahay, S. (2004). Embedded knowledge and offshore software development. *Information and Organization, 14*(4), 329-365.

Noble, D. F. (1979). *America by design: Science, technology, and the rise of corporate capitalism.* Oxford, UK: Oxford University Press.

Nonaka, I., & Takeuchi, H. (1995). *The knowledge creating company: How Japanese companies create the dynamics of innovation.* New York: Oxford University Press.

Nowotny, H., Scott, P., & Gibbons, M. (2001). *Re-thinking science: Knowledge and the public in an age of uncertainty.* Malden, MA: Blackwell Publishers Inc.

NZZ. (2004). Auftrieb für Zürich - Positive Reaktionen auf den Standortentscheid von Google. *Neue Zuericher Zeitung, 29 January 2004.*

O'Connell, M. S., Doverspike, D., & Cober, A. B. (2002). Leadership and semi-autonomous work team performance: A field study. *Group and Organization Management, 27*(1), 50-65.

O'Driscoll, S. (2005, November 8). *BPOs bring fun back to the workplace.* The Times of India. Retrieved March 10, 2007, from http://www.timesofindia.indiatimes.com/articleshow/msid-1288762,prtpage-1.cms

O'Reilly, C. A., & Pfeffer, J. (2000) Cisco systems: Acquiring and retaining talent in hypercompetitive markets. *Human Resource Planning, 23*, 38-52.

Ó'Riain, S. (2000). Net-working for a living: Irish software developers in the global workplace. In M. Burawoy, J. A. Blum, S. George, Z. Gille, T. Gowan, L. Haney, M. Klawitter, S. H. Lopez, S. Ó'Riain, & M. Thayer (Eds.), *Global ethnography: Forces, connections, and imaginations in a postmodern world.* Berkeley: University of California Press.

Ó'Riain, S. (2002). High-tech communities: Better work or just more work? *Contexts, Fall/Winter 2002.*

OffshoringTimes. (2005). BPOs-Those who look for a serious career. Retrieved January 9, 2007, from http://www.offshoringtimes.com/Pages/2005/BPO_news330.html

Ohmae, K. (1990). *The borderless world.* New York: Harper Business.

Oliga, J. (1996). *Power, ideology, and control.* New York: Plenum.

Olufowote, J. O. (2006). Rousing and redirecting a sleeping giant. *Management Communication Quarterly, 19*, 431-492.

Ong, A. (1999). *Flexible citizenship: The cultural logics of transnationality.* Durham, NC: Duke University Press.

Orange. (2006). *Benefits.* Retrieved 21 February 2006, from www.orange.co.uk/jobs/working_for_orange/benefits.php

Orlikowski, W. (1988). The data processing occupation: Professionalisation or proletarianisation. *Research in the Sociology of Work, 4*, 95-124.

Orlikowski, W. (1989). Software work. *Industrial Relations, 25*(2), 184-96.

Orlikowski, W. J. (1988). Computer technology in organizations: Some critical notes. In D. Knights & H. Willmott (Eds.), *New technology and the labor process.* Basingstoke: Macmillan.

Orlikowski, W. J. (2002). Knowing in practice: Enacting a collective capability in distributed organizing. *Organization Science, 13*(3), 249-273.

Orr, J. (1990). *Talking about machines: An ethnography of a Modern Job.* Ithaca, NY: Cornell University.

Orr, J. E. (1996). *Talking about machines.* Ithaca, NY, London: Cornell University Press.

Ortner, S. B. (1984). Theory in anthropology since the sixties. *Comparative Study of Society and History, 126*(1), 126-166.

Ortner, S. B. (2005). Subjectivity and cultural critique. *Anthropological Theory 5*(1), 31-52.

Osterman, P., Kochan, T. A., Locke, R., & Piore, M. J. (2001).*Working in America.* Cambridge, MA: MIT Press.

Oudshoorn, N., & Pinch, T. (2003). Introduction: How users and non-users matter. In N. Oudshoorn, & T. Pinch (Eds.). *How users matter: The co-construction of users and technology* (pp. 1-25). Cambridge, MA and London, England: MIT Press.

Pandey, A., & Singh, P. (2005). *Women in call centres.* Economic and Political Weekly. Retrieved February 10, 2006, from http://www.epw.org.in

Parker, M. (2002). *Against management.* Cambridge: Polity Press

Paul, S., Whittam, G., & Wyper, J. (2007). Towards a model of the business angel investment process. *Venture Capital, 9*, 107-125.

Pawlowski, S. D., & Robey, D. (2004). Bridging user organizations: Knowledge brokering and the work of information technology professionals. *MIS Quarterly, 28*(4), 645-672.

Pech, R. J. (2001). Termites, group behaviour, and the loss of innovation: Conformity rules! *Journal of Managerial Psychology, 16*(7), 559-574.

Perlow, L. (1999).The time famine: Toward a sociology of work time. *Administrative Science Quarterly, 44*(1), 57-81.

Perlow, L. A. (1997). *Finding time: How corporations, individuals, and families can benefit from new work practices*. Ithaca: Cornell University Press.

Peter-Martin, H., & Schumann, H. (1997). *The global trap: Globalization and the assault on democracy and prosperity*. London: ZED Books.

Peters, T. (2001). *The brand you 50*. New York: Alfred A. Knopf.

Pettigrew, A. (1979). On studying organizational cultures. *Administrative Science Quarterly, 27*, 641-652.

Pihkala, T., Varamaki, E., & Vesalainen, J. (1999). Virtual organisation and the SMEs: A review and model development. *Entrepreneurship and Regional Development, 11*, 335-349.

Pinch, T. J., & Bijker, W. E. (1987). The social construction of facts and artifacts or how the sociology of science and the sociology of technology might benefit one another. In W. E. Bijker, T. P. Hughes, & T. J. Pinch (Eds.), *The social construction of technological systems*. Cambridge, MA: MIT Press.

Pinchot, G. (1985). *Intrapreneuring: Why you don't have to leave the corporation to become an entrepreneur*. New York: Harper Collins.

Pink, D. H. (2001). *Free agent nation: How America's new independent workers are transforming the way we live*. New York: Warner Books.

Piore, M. (1973). Notes for a theory of labor market stratification. In M. Reich, R. Edwards, & D. Gordon (Eds.), *Labor market segmentation* (pp. 125-150). Lexington, MA: Heath and Company.

Poel, I. v. d. & Verbeek, P. P. (2006). Ethics and engineering design. *Science Technology & Human Values, 31*, 223-236.

Polanyi, M. (1958). *Personal knowledge: Towards a post-critical philosophy*. Chicago, IL: University of Chicago Press.

Polanyi, M. (1966). *The tacit dimension*. London: Routledge & Kegan Paul.

Pongratz, H. J., & Voß, G. G. (2003). From employee to "entreployee": Towards a "self-entrepreneurial" work force? *Concepts and Transformation, 8*(3), 239-254.

Portes, A., & Sensenbrenner, J. (1993). Embeddedness and immigration: Notes on the social determinants of economic action. *American Journal of Sociology, 98*(6), 1320-50.

Postrel, V. (2003). *The substance of style: How the rise of aesthetic value is remaking commerce, culture, and consciousness*. New York: HarperCollins.

Postuła, A. (2007). *Rola informatyków w organizacjach*. Doctoral thesis. Warszawa: Uniwersytet Warszawski.

Potter, J. (1996). Discourse analysis and constructionist approaches: Theoretical background. In J. T. E. Richardson (Ed.), *Handbook of qualitative research methods for psychology and the social sciences* (pp. 125-140). Leicester: BPS.

Potter, J., & Wetherell, M. (1987). *Discourse and social psychology: Beyond attitudes and behaviour*. London: Sage.

Pottruck, D. S., & Pearce, T. (2001). *Clicks and mortar: Passion-driven growth in an Internet-driven world*. San Francisco: Jossey-Bass.

Powell, W. W. (1990). Neither market nor hierarchy: Network forms of organization. *Research in Organizational Behavior, 12*, 295-336.

Power, M., Desouza, K. C., & Bonifazi, C. (2006). *The outsourcing handbook: How to implement a successful outsourcing process.* London: Kogan Page.

Pradhan, J. P., & Abraham, V. (2005). Social and cultural impact of outsourcing: Emerging issues from Indian call centres. *Harvard Asia Quarterly, Summer.* Cambridge, MA: Harvard Asia Center.

Prahalad, C. K., & Bettis, R. A. (1986). The dominant logic: A new linkage between diversity and performance. *Strategic Management Journal, 7*, 485-501.

Prahalad, C. K., & Hamel, G. (1994). *Competing for the future*. Boston: Harvard Business School Press.

Prasad, P. (1997). The protestant ethic and the myths of the frontier: Cultural imprints, organizational structuring, and workplace diversity. In P. Prasad et al. (Eds.), *Managing the organisational melting pot. Dilemmas of workplace diversity.* Thousand Oaks, CA: Sage.

Pratt, A. C. (2000). New media, the new economy, and new spaces. *Geoforum, 31*, 425-436.

Prax, J. Y. (2003). *Le manuel du knowledge-management.* Paris: Dunod.

Premus, R. (2002). Moving technology from labs to market: A policy perspective. *International Journal of Technology Transfer and Commercialisation, 1*(1/2), pp. 22-39.

Pruitt, S., & Barett, T. (1991). The virtual workplace. In M. Benedikt (Ed.), *Cyberspace: first steps*. Cambridge, MA: MIT Press.

Quan, M. (2003). *IEEE USA presses Congress on visa curbs*. CommsDesign. Retrieved November 27, 2006, from http://www.commsdesign.com/news/market_news/OEG20030326S0029

Quinlan, P. (1991). Connectionism and psychology. Chicago: University of Chicago Press.

Quinn, R. E. (1984). Applying the competing values approach to leadership: Toward an integrative model. In J. G. Hunt, R. Stewart, C. Schriesham, & D. Hosking (Eds.), *Managers and leaders: An international perspective*. Pergamon, NY.

Quinn, R., Faerman, S., Thompson, M., & McGrath, M. (1996*). Becoming a master manager.*New York: John Wiley & Sons.

Rabinow, P. (1984). Introduction. In P. Rabinow (Ed.), *The Foucault reader: An introduction to Foucault's thought* (pp. 3-29). London: Penguin.

Ralph, W. & Wyndrum, J. (2006). *Welcome to the new year.* IEEE-USA, President's Column. Retrieved November 26, 2006, from http://www.ieeeusa.org/communications/presidentscolumn/Wyndrum/jan06.html

Rathi, N. (2003). *Human resource challenges in the Indian software industry: An empirical study of employee turnover.* Unpublished doctoral dissertation, Indian Institute of Technology, Mumbai Mumbai.

RediffNews. (2005). Salary hike, a threat to Indian BPO. Retrieved January 25, 2007, from http://www.rediff.com/money/2005/apr/20bpo2.htm

Reich, R. (1987). *Tales of a new America.* New York: Times Books.

Remenyi, D., & Sherwood-Smith, M. (1999). Maximise information systems value by continuous participative evaluation. *Logistics Information Management, 12*(1/2), 145-156.

Remesh, B. P. (2004). Cyber Coolies' in BPO: Insecurities and vulnerabilities of non-standard work. *Economic and Political Weekly.* Retrieved June 20, 2006, from http://www.epw.org.in

Revathy, L. N. (2004, July 19). *Live it up in office.* The Hindu Business Life. Retrieved March 6, 2007, from http://www.hinduonnet.com/the hindu/thscrip/ptint.pl?file-2004071900040100.htm&date=2004/07/19/&prd=we&

Reynaud, J. D. (1991). La régulation sociale. *Revue internationale d'action communautaire, 25*(65), 121-126.

Ristola, A., & Kesti, M. (2005). The effect on familiar mobile device and usage time on creating perceptions towards mobile services. *The IEEE 4th International Conference on Mobile Business.* Sydney, Australia.

Ritzer, G. (1998). The McDonaldization thesis: Is expansion inevitable? *International Sociology, 13*, 291-308.

Roberto, M. A. (2005). *Why great leaders don't take yes for an answer: Managing for conflict and consensus.* Upper Saddle River, NJ: Wharton School Publishing.

Roberts, E. (1989). The personality and motivation of technological entrepreneur. *Journal of Engineering and Technology. 6*, 5-23.

Roberts, F. M. (1989). The Finnish coffee ceremony and notions of self. *Arctic Anthropology 26*, 20-33.

Robertson, M., & O'Malley Hammersley, G. (2000). Knowledge management practices within a knowledge-intensive firm: The significance of the people management dimension. *Journal of European Industrial Training, 24*(2-4), 241-253.

Robertson, M., & Swan, J. (2003). Control—What control? Culture and ambiguïty within a knowledge intensive firm. *Journal of Management Studies, 40*(4), 831-858.

Robertson, M., Scarbrough, H., & Swan, J. (2003). Knowledge creation in professional service firms: Institutional effects. *Organization Studies, 24*(6), 831-857.

Robertson, R. (1992). *Globalization: Social theory and global culture*. London: Sage.

Robertson, S. L. (1996). Teachers work, restructuring, and postfordism: Constructing the new professionalism. In I. F. Goodson, & A. Hargreaves (Eds.), *Teachers' professional lives* (pp. 28-55). London: Falmer Press.

Robey, D., & Boudreau, M. (1999). Accounting for the contradictory organizational consequences of information technology: Theoretical directions and methodological implications. *Information Systems Research, 10*(2), 167-185.

Rochet, C. (2003). *Conduire l'action publique des objectifs aux résultats.* Paris: Éd. Village mondial.

Rohracher, H. (2005). The diverse roles of users in innovation processes. In H. Rohracher (Ed.), *User involvement in innovation processes: Strategies and limitations from a socio-technical perspective* (pp. 9-35). Munchen/Wien: Profil Verlag.

Roozenburg, N. F. M., & Eekels, J. (1995). *Product design, fundamentals and methods.* Chichester, UK: John Wiley and Sons.

Rosen, M. (1991) Coming to terms with the field: Understanding and doing organizational ethnography. *Journal of Management Studies, 28*(1), 1-24.

Rosenhead, J. (1987). From management science to workers' science. In M. C. Jackson, & P. Keys (Eds.), *New directions in management science* (pp. 109-131). Aldershot: Gower.

Rousseau, D. M. (2005). *I-deals: Idiosyncratic deals employees bargain for themselves.* London: M.E. Sharpe.

Rumelhart, D. E. (1992). Towards a microstructural account of human reason. In S. Davis (Ed.), *Connectionism: Theory and practice* (pp. 69-83). New York: Oxford University Press.

Ruohonen, M. (2004). Johtamiskulttuurien muutos ICT-yrityksessä – hypestä todellisuuteen. [Changing management culture in an ICT-company – from hype to reality.] In M. Ruohonen et al. (Eds.), *Tietoyritysten muuttuvat työkulttuurit.* Tampere: Tampere University Press.

Rushkoff, D. (1994). *Cyberia. Life in the trenches of hyperspace.* San Fransisco: Harper.

Ryle, G. (1949). *The concept of mind.* Chicago: University of Chicago Press.

Saffold, G. S. (1988). Culture traits, strength, and organizational performance: Moving beyond "strong" culture. *Academy of Management Review, 13*, 546-558.

Sahay, S., & Walsham, G. (1997). Social structure and managerial agency in India. *Organization Studies, 18*(3), 415-444.

Sahay, S., Nicholson, B., & Krishna, S. (2003). *Global IT outsourcing: Software development across borders.* Cambridge, UK: Cambridge University Press.

Sahlin-Anderson, K. (1996). Imitating by editing success. In B. Czarniawska & G. Sèvon (Eds.), *Translating organizational change.* Berlin: Walter de Gruyter.

Sako, M. (1992). *Prices, quality, and trust: Inter-firm relations in Britain and Japan*. Cambridge: Cambridge University Press.

Sako, M. (1998). Does trust improve business performance? In C. Lane, & R. Bachmann (Eds.), *Trust within and between organizations: Conceptual issues and empirical applications* (pp. 88-117). Oxford: Oxford University Press.

Sandelands, L. E., & Stablein, R. E. (1987). The concept of organization mind. In S. Bacharach, & DiTomaso (Eds.), *Research in the sociology of organizations* (Vol. 5, pp. 131-161). Greenwich, CT: JAI Press.

Sanders, A. F. (1988). *Michael Polanyi's post critical epistemology. A reconstruction of some aspects of "tacit knowing."* Amsterdam: Rodopi.

Sanders, E. B. N. (2000). Generative tools for codesigning. In Scrivener, Ball, & Woodcock (Eds.), *Collaborative design*. London: Springer-Verlag.

Scarbrough, H. (1999). Knowledge as work: Conflicts in the management of knowledge workers. *Technology Analysis and Strategic Management, 11*(1), 5-16.

SCB (Statitiska centralbyrån) (2005). *Statistisk årsbok 2004*. City of Stockholm.

Schein, E. H. (1984). Coming to a new awareness of organizational culture. *Sloan Management Review, 25*(2), 3-16.

Schein, E. H. (1985). *Organizational culture and leadership*. San Francisco: Jossey-Bass.

Schein, E. H. (1995). *Organizational and managerial culture as a facilitator or inhibitor of organizational transformation* (MIT Sloan School of Management Working Paper). Cambridge, MA, USA.

Scholz, C. (1997). Das virtuelle Unternehmen—Schlagwort oder echte Vision? *Manager Bilanz, 1*, 12-19.

Schuler, D., & Namioka, A. (1993). *Participatory design: Principles and practices*. NJ: Lawrence Erlbaum Associates.

Senker, J., & Sharpe, M. (1997). Organizational learning in cooperative alliances: some case studies in biotechnology. *Technology Analysis and Strategic Management, 9*(1), 35-51.

Sennett, R. (1998). *The corrosion of character: The personal consequences of work in the new capitalism*. New York: Norton.

Sennett, R. (2000). The new political economy and its culture. *The Hedgehog Review, Spring* (2000), (Online).

Sennett, R. (2006). *The culture of the new capitalism*. New Haven, CT: Yale Univ. Press.

Serafeimidis, V., & Smithson, S. (1999). Rethinking the approaches to information systems evaluation. *Logistics Information Management, 12*(1-2), 94-107.

Serafeimidis, V., & Smithson, S. (2003). Information systems evaluation as an organizational institution experience from a case study. *Information Systems Journal, 13*, 251-274.

Sèvon, G. (1996). Organizational imitation in identity transformation. In B. Czarniawska, & G. Sèvon (Eds.), *Translating organizational change*. Berlin: Walter de Gruyter.

Sewell, W. (1999). Geertz, cultural systems, and history: From synchrony to transformation. In S. Ortner (Ed.), *The fate of "culture." Geertz and beyond*. Berkeley: Univ. of California Press.

Shahane, P. S. (2005, August 11). *Wanna have fun sans sex? Come, join our BPO*. Infotech Online. Retrieved March 10, 2007, from http://infotech.indiatimes.com/articleshow/msid-1197310,prtpage-1.cms

Shahane, P. S., & Savitha, V. (2007, January 6). *It's family first and tech cos and BPOs*. Infotech Online. Retrieved March 11, 2007, from http://infotech.indiatimes.com/articlesehow/msid-1067797,prtpage-1.cms

Shaw, K. (2004). Cooltools: Quick takes on high-tech toys. *NetworkWorld, 05/10/04*.

Sheikh, Z. (2004, Nov. 1). *Want a free international trip? Work for a BPO*. Rediff News. Retrieved from http://in.rediff.com/getahead/2004/nov/01ga-zia.htm

Shenhav, Y. (1999). *Manufacturing rationality. The engineering foundations of the managerial revolution*. Oxford University Press.

Sheppard, B. H. (1995). Negotiating in long-term mutually interdependent relationships among relative equals. In R. J. Bies, R. J. Lewicki, & B. H. Sheppard (Eds.), *Research on negotiation in organizations* (Vol. 5, pp. 3-44), Greenwich: JAI Press.

Sherman, B., & Judkin, P. (1992). *Glimpses of heaven, visions of hell: Virtuality and its implications*. London: Hodder & Stoughton.

Sherriff, L. (2004). *What's the point of NHS IT? No one knows*. Retrieved January 15, 2006, from http://www.theregister.co.uk/2004/07/28/nhs_it_bens/

Shih, J. (2004). Project time in Silicon Valley. *Qualitative Sociology, 27*(2), 223-245,

Shrinate, S. (2004, June 6). Truly, a learning experience. *Business Today,* 156-157.

Siegel, D. S., Waldman, D. A., Atwater, L. E., & Link, A. N. (2004). Toward a model of the effective transfer of scientific knowledge from academicians to practitioners: Qualitative evidence from the commercialization of university technologies. *Journal of Engineering and Technology Management, 21*, 115-142.

Silf, M. (1998). *Landmarks: Exploration of ignatian spirituality*. London: Dartman, Longman and Todd.

Sillince, J. A. A. (2002). A model of the strength and appropriateness of argumentation in organizational contexts. *Journal of Management Studies, 39*(5), 585-618.

Silverman, D. (1970). *The theory of organizations*. London: Heinemann.

Silverman, D. (2001). *Interpreting qualitative data*. Thousand Oaks, CA: Sage.

Simon, H. A. (1991). Bounded rationality and organizational learning. *Organization science, 2*, 125-134.

Sinha, D. (1988). Basic Indian values and behavior dispositions in the context of national development. In D. Sinha & H. S. Kao (Eds.), *Social values and development: Asian perspectives* (pp. 31-55). New Delhi: Sage.

Sinha, J. B., & Kumar, R. (2004). Methodology for understanding Indian culture. *The Copenhagen Journal of Asian Studies, 19*, 89-104.

Sinha, J. B., Vohra, N., Singhal, S., Sinha, R. B., & Ushashree, S. (2002). Normative predictions of collectivist-individualist intentions and behaviour of Indians. *International Journal of Psychology, 37*(5), 309-319.

Sinha, J. P. B., Sinha, T. N., Verma, J., & Sinha, R. B. N. (2001). Collectivism coexisting with individualism: An Indian scenario. *Asian Journal of Social Psychology, 4*, 133-145.

Six, F. (2005). *The trouble with trust – the dynamics of interpersonal trust building*. Cheltenham: Edward Elgar.

Smircich, L. (1983). Concepts of culture and organizational analysis. *Administrative Science Quarterly, 28*(3), 339-358.

Smircich, L., & Stubbart, C. (1985). Strategic management in an enacted world. *Academy of Management Review, 10*(4), 724-736.

Smith, B., & Thomas, J. (1972). Cross-cultural attitudes among managers: A case study. *Sloan Management Review, 13*, 34-51.

Smith, K., Carroll, S., & Ashford, S. (1995) Intra- and interorganizational cooperation: Toward a research agenda. *Academy of Management Journal, 38*, 7-23.

Smithson, S., & Tsiavos, P. (2004). Re-constructing information systems evaluation. In C. Avgerou, C. Ciborra, & F. Land (Eds.), *The social study of information and communication technology: Innovation, actors and contexts* (pp. 207-230). Oxford: Oxford University Press.

Sobol, M., & Apte, U. (1995). Domestic and global outsourcing practices of America's most effective IS Users. *Journal of Information Technology, 10*(4), 269-81.

Sørensen, C., & Pica, D. (2004). *Out-of-sight shouldn't mean out-of-mind: Why corporates need to get control of their wireless assets*. London: London School of Economics and Political Science.

Sotto, R. (1997). The virtual organization. *Accounting, Management, and Information Technology, 7,* 37-51.

Soulsby, A. (2002). The impact of societal transformation on Czech managers: a study of post-communist careers. In M. Kelemen & M. Kostera (Eds.), *Critical management research in Eastern Europe* (pp. 164-186). London: Palgrave.

Srinivas, M. N. (1997). Practising social anthropology in India. *Annual Review of Anthropology, 26*, 1-24.

Srinivas, S. (2005, March 28). Stress Kills 6 IT Geeks (India). *The Asian Age*.

Srinivasan, T. N. (2006). Information technology and India's growth prospects. In L. Brainard & S. M. Collins ((Ed.), *Offshoring white-collar work- the issues and implications, The Brookings Trade Forum, 2005.* (pp. 203-240). Washington, DC: The Brookings Institution.

Stangor, C., Jonas, K., Stroebe, W., & Hewstone, M. (1996). Influence of student exchange on national stereotypes, attitudes, and perceived group variability. *European Journal of Social Psychology, 26*, 663-675.

Star, S. L., & Griesemer, J. R. (1989). Institutional ecology, "translations," and boundary objects: Amateurs and professionals in Berkeley's Museum of Vertebrate Zoology 1907-1939. *Social Studies of Science 19*, 387-420.

Stark, D. (1997). Recombinant property in East European Capitalism. In G. Grabher & D. Stark (Eds.), *Restructuring networks in post-socialism: Legacies, linkages, and localities* (pp. 35-69). Oxford: Oxford University Press.

Starr, J., & MacMillan, I. (1990). Resource cooption via social contracting: resource acquisition strategies for new ventures. *Strategic Management Journal, 11*, 79-92.

Steadman, J. (2004). *Career trends*. IEEE-USA. Retrieved November 27, 2006, from http://www.ieeeusa.org/communications/Steadman/steadmannov04.html

Steen, M. (2006). *Our need is to do something about that problem—Studying how researchers and developers interact with informal carers during an innovation project.* Presented at EASST 2006 Conference (European Association for Studies of Science and Technology), August 23-26, Lausanne, Switzerland.

Steen, M. (2006). We don't want woollen trousers—Studying how researchers and developers interact with police officers during an innovation project. In R. ten Bos & R. Kaulingfreks (Eds.), *Proceedings of SCOS 2006 (Standing Conference on Organisational Symbolism)* (pp. 644-666), July 13-15, Nijmegen, The Netherlands.

Steiner, C. (2000). Teaching scientists to be incompetent: Educating for industry Work. *Bulletin of Science, Technology, and Society, 20*(2), 123-132.

Stevenson, W. B., Pearce, J. L., & Porter, L. W. (1995). The concept of "coalition" in organizational theory and research. *Academy of Management Review, 10*, 256-258.

Stewart, G. L., & Manz, C. C. (1995). Leadership for self-managing work teams: A typology and integrative model. *Human Relations, 48*(7), 747-770.

Stewart, J., & Williams, R. (2005). The wrong trousers? Beyond the design fallacy: Social learning and the user. In H. Rohracher (Ed.), *User involvement in innovation processes: Strategies and limitations from a socio-technical perspective* (pp. 39-71). Munchen/Wien: Profil Verlag.

Stinchcombe, A. L. (1959). Bureaucratic and craft administration of production: A comparative study. *Administrative Science Quarterly*, (4), 168-187.

Stone, K. (1974). The origins of job structures in the steel industry. *The Review of Radical Political Economy, 2*(6), 113-173.

Storey, J. (1986). The phoney war? New office technology. In D. Knights & H. Willmott (Eds.), *Managing the labor process* (pp. 44-66). Aldershot: Gower.

Strannegård, L., & Friberg, M. (2001). *Already elsewhere. Play, identity, and speed in the business world.* Raster Forlag, Sweden.

Strašil, K. (1994). *180 Let Prvni Brnenska Strojirna (180 Years of PBS)*. Brno, Czech Republic: Moravska Typografie

Sutton, R. I., & Rafaeli, A. (1992). How we untangled the relationship between displayed emotion and organizational sales: A tale of bickering and optimism. In P. J. Frost & R. E. Stablein (Eds.), *Doing exemplary research* (pp. 115-128). Newbury Park, CA: Sage Publications Inc.

Swan, J., Robertson, M., & Bresnen, M. (2001). Knowledge management and the colonization of knowledge. *Special Issue of the Best papers from the Critical Management Studies Conference, The Electronic Journal of Radical Organization Theory, 7*(2), (online).

Swan, J., Scarbrough, H., & Robertson, M. (2002). The construction of "communities of practice" in the management of innovation. *Management Learning, 33*(4), 477-496.

Symons, V., & Walsham, G. (1988). The evaluation of information systems: A critique. *Journal of Applied Systems Analysis, 15*, 119-132.

Taket, A. (1994). Undercover agency? Ethics, responsibility, and the practice of OR. *Journal of the Operational Research Society, 45*(2), 123-132.

Tarondeau, J. C. (1998). *Le management des savoirs*. Paris: PUF, coll. Que sais-je?

Tayeb, M. (2001). Conducting Research across cultures: Overcoming drawbacks and obstacles. *International Journal of Cross Cultural Management, 1*, 91-108.

The Economic Times. (2004). BPOs make staffers boogie woogie. Retrieved March 5, 2007, from http://www.247customer.com/newsandevents/Article-56.php

The Economic Times. (2005). Why BPO execs are given training. Retrieved October 5, 2006, from http://www.bpo.nasscom.org/artdisplay.aspx?art_id=4491&cat_id=608

The Economic Times. (2006). BPOs' no-poaching pacts fall apart. Retrieved November 26, 2006, from http://economictimes.indiatimes.com/BPOs_nopoaching_pacts_fall_apart/articleshow/498749.cms

The Economic Times. (2006). BPOs get serious, axe "fun job" stage. Retrieved March 9, 2007, from http://economictimes.indiatimes.com/Hot_Links/BPO/ BPOs_get_serious_axe_fun_job_tag/articleshow/1486707.cms

The Economic Times. (2007). How to bell the BPO cat. Retrieved March 10, 2007, from http://economictimes.indiatimes.com/articleshow/msid-1736650,prtpage-1.cms

The Times of India. (2006). Two Knights @ The Call Centre. Retrieved March 9, 2006, from http://www-935.ibm.com/services/us/gbs/bus/html/dakshnews_21apr06.html

The Tribune. (2006). Future on the line. Retrieved March10, 2007, from http://www.tribuneindia.com/2006.20060111/jobs/main1.htm

Thomas, R. (1995). Access and inequality. In N. Heap, R. Thomas, G. Einon, R. Mason, & H. Mackay (Eds.), *Information technology and society*. London: Sage in association with Open University.

Thomas, R., & Linstead, A. (2002). Losing the plot? Middle managers and identity. *Organization, 9*(1), 71-93.

Thompson, P. (2005). Brands, boundaries, and bandwagons: A critical reflection on critical management studies. In C. Grey & H. Willmott (Eds.), *Critical management studies: A reader* (pp. pp.364-382). Oxford: Oxford University Press.

Thompson, P. (2005). *Labor and the labor process conference*. Critical Management Workshop. Retrieved October 25, 2005, from www.jiscmail.ac.uk

Thompson, P., & Ackroyd, S. (1995). All quiet on the workplace front? A critique of recent trends in British industrial sociology. *Sociology, 29*(4), 615-633.

Thrupp, S. (1963). The guilds. *Cambridge Economical History*, (3), 230-280.

Tichy, N. (1982). Managing change strategically: The technical, political, and cultural keys. *Organizational Dynamics* (Autumn), 59-80.

Tidd, J., Bessant, J., & Pavitt, K. (1997). *Managing innovation: Integrating technological, market and organizational change*. Chichester, NY: Wiley.

Tillich, P. (1936). *The interpretation of history*. London: Scribner.

Tillmar, M. (2002). *Swedish Tribalism and Tanzanian Agency: Preconditions for trust and cooperation in a small-business context*. Linköping: Linköpings universitet.

Tilly, C. (1999). *Durable inequality*. Berkeley, CA: University of California Press.

Tomas, D. (1989). The technophilic body: On technicity in William Gibson's cyborg culture. *New Formation, 8*, 113-29.

Tomas, D. (1991). Old rituals for new space. In M. Benedikt (Ed.), *Cyberspace: First steps*. London. MIT Press.

Tonkiss, F. (1998). Analysing discourse. In C. Seale (Ed.), *Researching society and culture* (pp. 245-260). London: Sage.

Topsfield, J. (2006). Migrants blamed for IT jobs cut. *The Age*. Retrieved November 27, 2006, from http://www.theage.com.au/articles/2006/01/09/1136771500496.html?from=top5

Torsello, D. (2002). Managing instability: Trust, social relations, and the strategic use of ideas and practices in a southern Slovakian Village. In D. Torsello & M. Pappova (Eds.), *Social networks in movement: Time, interaction, and interethnic spaces in Central Eastern Europe*. Nostra Tempora 8, Forum Minority Research Institute. Samorin, Slovakia.

Torstendahl, R., & Burrage, M. (1990). *The formation of professions: Knowledge, state, and strategy*. London, Sage.

Towers, I., Duxbury, L., & Thomas, J. (2005). Shifting boundaries: Technology, time, place, and work. The 23rd *Annual International Labor Process Conference*. Glasgow, Scotland.

Triandis, H. (1995). *Individualism and collectivism*. Boulder, CO: Westview.

Triandis, H. (2000). Dialectics between cultural and cross-cultural psychology. *Asian Journal of Social Psychology, 3*, 185-196.

Trice, H. M. (1993). *Occupational subcultures in the workplace*. Ithaca, NY: ILR Press, Cornell University.

Trux, M. (2000). Monimuotoinen työyhteisö. In M. Trux (Ed.), *Aukeavat ovet. Kulttuurien moninaisuus Suomen elinkeinoelämässä*. Helsinki: WSOY. [Published in English as Diversity under the Northern Star. In A. Forsander (Ed.), 2002. *Immigration and Economy in the Globalization Process. The Case of Finland*.] Electronic publication, ISBN: 951-563-416-4. http://www.sitra.fi/en/Publications/web/web.htm

Trux, M. (2005). "Ei sitä meillä kukaan kato" – kansainvälisyys ja monietnisyys helsinkiläisellä IT-työpaikalla. ["Nobody looks at that here" – internationality and multiethnicity at an IT-workplace in Helsinki.] *Työpoliittinen aikakauskirja 2/2005*, 49–69.

Trux, M. (2006). *Almost-equals: Transnationality in a Finnish ICT-company*. Paper presented at EURODIV Conference Qualitative diversity research: Looking ahead. Katholieke Universiteit Leuven, Belgium, September 19th-20th, 2006.

Tuan, Y. F. (1977). *Space and place: The perspectives of experience*. MN: University of Minnesota Press.

Tuckman, B., & Jensen, M. (1977). Stages of small group development. *Group and Organizational Studies, 2*, 419-427.

Tushman, M., & O'Reilly, C. (1997). *Winning through innovation: A practical guide to leading organizational change and renewal*. Boston: Harvard Business School Press.

Twigger-Ross, C. L., & Uzzell, D. L. (1996). Place and identity process. *Journal of Environmental Psychology, 16*, 205-220.

Upadhya, C. (2004). A new transnational capitalist class? Capital flows, business Nnetworks, and entrepreneurs in the Indian software industry. *Economic and Political Weekly*, 5141-5151.

Upton, D., & McAfee, A. (1996). The real virtual factory. *Harvard Business Review, 74,* 123-133.

Urry, J. (2001). *Sociology beyond societies: Mobilities for the twenty-first century.* London: Routledge.

Ursell, G. (2000). Television production: Issues of exploitation, commodification, and subjectivity in UK television labor markets. *Media, Culture, & Society 2000, 22*(6), 805-825.

USK (Utrednings- och statistikkontor) (2004). *Befolkningsstatistik 2003.* City of Stockholm.

USK. (2003). *Arbetandes och boendes uppfattning om trafiksituationen i Kista.* City of Stockholm.

v. Werder, A. (1999). Argumentation rationality of management decisions. *Organization Science, 10*(5), 672-690.

van Kleef, E., van Trijp, H. C. M., & Luning, P. (2005). Consumer research in the early stages of new product development: A critical review of methods and techniques. *Food Quality and Preference, 16,* 181-201.

Van Maanen, J., & Barley, S. R. (1984). Occupational communities: Culture and control in organizations. In L. L. Cummings, & B. M. Staw (Eds.), *Research in organizational behaviour* (Vol. 6, pp. 287-365). Greenwich, CT: JAI Press.

Veblen, T. ([1921] 1964). *The engineers and the price system.* New York: A.M. Kelley.

Venkatesh, V., Morris, M. G., Davis, G. B., & Davis, F. D. (2003). User acceptance of information technology: Toward a unified view. *Management Information Systems Quarterly, 27*(3), 425-478.

Vinton, D. E. (1992). A new look at time, speed, and the manager. *Academy of Management Executive, 6*(4), 7-16.

Von Hippel, E. (2005). *Democratizing innovation.* Cambridge, MA: MIT Press.

Von Krogh, G., Ichijo, K., & Nonaka, I. (2000). *Enabling knowledge creation. How to unlock the mystery of tacit knowledge and release the power of innovation.* Oxford: Oxford University Press.

Voß, G. G. (1999). Menschen als Unternehmer Ihrer Selbst. Der Arbeitskraftunternehmer - Ein neuer Typus von Arbeitskraft und seine Konsequenzen. *Leben im Transit. Neue Arbeitsgesellschaft und die Suche nach dem Beruf, Evangelische Akademie Tutzing, 15th-17th January, 1999.*

Voß, G. G., & Pongratz, H. J. (1998). Der Arbeitskraftunternehmer: Eine neue Grundform der Ware Arbeit? *Kölner Zeitschrift für Soziologie und Sozialpsychologie, 30*(1), 131-158.

Voß, G. G., & Pongratz, H. J. (2000). Vom Arbeitnehmer zum Arbeitskraftunternehmer - Zur Entgrenzung der Ware Arbeit. In H. Minssen (Ed.), *Begrenzte Entgrenzungen: Wandlungen von Organization und Arbeit* (pp. 225-247). Berlin: Edition Sigma.

Walsham, G. (1993). *Interpreting information systems in organizations.* Chichester: John Wiley and Sons.

Walsham, G. (1999). Interpretive evaluation design for information systems. In L. Willcocks & S. Lester (Eds.), *Beyond the IT productivity paradox* (pp. 363-380). Chichester: John Wiley and Sons.

Walsham, G., & Waema, T. (1994). Information systems strategy and implementation: A case study of a building society. *ACM Transactions on Information Systems, 12*(2), 150-173.

Walton, R. E. (1985). From control to commitment in the workplace. *Harvard Business Review, March-April.*

Warhurst, C., & Thompson, P. (2005). From conception to consumption: Myopic analysis of the creative industries. The 23rd *Annual International Labor Process Conference.* Glasgow, Scotland.

Waterson, P., Clegg, C., & Axtell, C. (1997). The dynamics of work organization, knowledge, and technology during software development. *International Journal of Human-Computer Studies, 46*, 79-101.

Weaver, P. (1998). *Practical SSADM version 4+*. London: Pitman.

Weber, M. (1930). *The protestant ethic and the spirit of capitalism*. London: Allen and Unwin.

Webster, F. (2002). *Theories of the information society* (2nd ed.). Routledge.

Wegner, D. M., Erber, R., & Raymond, E. (1991). Transactive memory in close relationships. *Journal of Personality and Social Psychology, 61*, 923-929.

Weick, K. (1979). *The social psychology of organizing* (2nd ed.). Reading, MA: Addison-Wesley.

Weick, K. (1979). *The social psychology of organizing*. New York: McGraw-Hill

Weick, K. (1995). *Sensemaking in organization*. Thousand Oaks, CA: Sage.

Weick, K. (2001). *Making sense of the organization*. Malden, MA: Blackwell Publishing.

Weick, K. E., & Roberts, K. H. (1993). Collective mind in organizations: Heedful interrelating on flight desks. *Administrative Science Quarterly, 38*(3), 357-381.

Weinreich, P., Bacova, V., & Rougier, N. (2003). Basic primordialism in ethnic and national identity. In P. Weinreich & W. Saunderson (Eds.), *Analyzing identity: Cross-cultural, societal and clinical contexts* (pp. 113-169). London: Routledge.

Wenger, E. (1999). *Communities of practice: Learning, meaning, and identity*. Cambridge University.

Wenger, E. (2002). *Cultivating communities of practice*. Boston: Harvard Business School.

Westrup, C., Al Jaghoub, S., El Sayed, & Liu, W. (2002). *Taking culture seriously: ICTs, cultures, and development*. Paper presented at the Proceedings of the 7th International Working Conference of IFIP WG 9.4 on Information and Communication Technologies and Development: New Opportunities, Perspectives and Challenges. May 29-31., Indian Institute of Management, Bangalore, India. pp. 17-28.

Wetherell, M., & Potter, J. (1988). Discourse analysis and the identification of interpretative repertoires. In C. Antaki (Ed.), *Analysing everyday explanation: A case book of methods* (pp. 168-183). London: Sage.

Whalley, P. (1986). Markets, managers, and technical autonomy. *Theory and Society, 15*(1-2), 223-247.

Wiener, N. (1948). *Cybernetics*. New York: Wiley.

Williams, R., & Steward, F. (1985). Technology agreements in Great Britain: A survey 1977-83. *Industrial Relations Journal, 16*(3), 58-73.

Williamson, O. E. (1993). Calculativeness, trust, and economic organization, *Journal of Law and Economics*, (36), 453-502.

Willmott, H. (1995). The odd couple? Re-engineering business processes; managing human relations. *Technology, Work, and Employment, 10*(2), 89-98.

Wilson, M., & Greenhill, A. (2004). Theory and action for emancipation. In B. Kaplan, D. Truex, D. Wastell, T. Wood-Harper, & J. I. DeGross (Eds.), *Information systems research: Relevant theory and informed practice* (pp. 667-674). London: Kluwer.

Wilson, T. D. (2002). The nonsense of knowledge-management. *Information research, 8*(1), paper no 144.

Winner, L. (1993). Upon opening the black box and finding it empty: Social constructivism and the philosophy of technology. *Science Technology & Human Values, 18*, 362-378.

Wittel, A. (2001). Toward a network sociality. *Theory, Culture, & Society, 18*(6), 51-76.

Wolf, E. R. (1972). American anthropologists and American society. In H. Dell, & H. Hymes (Ed.), *Reinventing anthropology* (pp. 251-263). New York: Pantheon Books.

Wood, P. (2004). *The intercultural city reader*. Bournes Green: Comedia.

Woolgar, S. (1991). Configuring the user: The case of usability trials. In J. Law (Ed.), *A sociology of monsters: Essays on power, technology, and dominations* (pp. 57-102). London and New York: Routledge.

Xu, L. D. (2000). The contribution of systems science to information science research. *Systems Research and Behavioural Science*, *2*(17), 105-116.

Young, E. (2003). Joined-up thinking. *New Scientist, 179*(2408), 46-52.

Yu, J., Yu, C. S., Liu, C., & Yao, J. E. (2003). Technology acceptance model for wireless internet. *Internet Research, 13*(3), 206-222.

Zack, M. H. (1999). Managing codified knowledge. *Sloan Management Review*, 45-58.

Zaheer, A., McEvily, B., & Perrone, V. (1998). Does trust matter? Exploring the effects of inter-organizational and interpersonal trust on performance. *Organization Science, 9*(2), 141-159.

Zara, O. (2004). *Le management de l'intelligence collective.* Retrieved from http:www.axiopole.com/intelligence_collective.php

Zarraga, C., & Bonache, J. (2005). The impact of team atmosphere on knowledge outcomes in self-managed teams. *Organization Studies*, *26*(5), 661-681.

Zarubavel, E. (1979). *Patterns of time in hospital life*. Chicago: Chicago University Press.

Zarubavel, E. (1996). Social memories: Steps to a sociology of the past. *Qualitative Sociology, 19*(3), 283-299.

Ziman, J. (2000). Are debatable scientific questions debatable? *Social Epistemology, 14*(2/3), 187-199.

Zuboff, S. (1988). *In the age of the smart machine: The future of work and power.* Oxford: Heinemann Professional.

Zucker, L. G. (1986). Production of trust: Institutional sources of economic structure. In S. Barry & L. Cummings (Eds.), *Research in organizational behavior* (Vol. 8, pp. 53-111). Greenwich: JAI Press.

About the Contributors

Maria Aggestam is a researcher at School of Business, University of Lund, in Sweden. Her research revolves around such themes as entreneurship, environmental jolts, corporate governance, and former Central and Eastern Europe organizational issues. Aggestam is the author of several research articles on management, entrepreneurship, and corporate governance during environmental jolts. In addition, Aggestam is studying and writing about entrepreneuring in various industries such as in art-related industries, high-tech industry, and entrepreneurs within the biotechnology. Her research is cross disciplinary, combining perspectives from among the social sciences. Her most recent interests, for example, involve the relations between language and entrepreneurial achievement. She has served on committees and chaired several academic conferences in North America and Europe. Most recently, Aggestam has been teaching graduate course in France and the U.S.

Ester Barinaga (BA, MBA, PhD) is currently a visiting scholar at SCANCOR, Stanford University, and Wallander Fellow at the department of industrial economics and management at the Royal Institute of Technology in Stockholm. She obtained her PhD from the Stockholm School of Economics. She has published in human relations and is at the moment finalizing a book on inclusion and exclusion in the information society. Her research interests include cultural diversity, the information society, and the intersection between technology and culture.

Peter Case is a professor of organization studies, Bristol Business School, University of the West of England, UK. He serves as chairperson of the Standing Conference on Organizational Symbolism and is a member of the editorial boards of *Leadership*, *Culture and Organization*, and the *Leadership and Organizational Development Journal*. He has held visiting scholarships at Helsinki School of Economics and the Royal Institute of Technology of Stockholm. Peter is interested in the social and organizational impact of information and communication technologies. Recent publications include *The Speed of Organization* (with S. Lilley and T. Owens, 2006: CBS & Liber) and *John Adair: the Fundamentals of Leadership* (with J. Gosling and M. Witzel, 2007: Palgrave).

Rajneesh Chowdhury is senior consultant with CHR Global Consulting Services, New Delhi, India. His areas of interest include change management, management systems consulting, executive competency assessments, and process-based organizations. He has a first degree in sociology (India), and holds double masters in management systems (UK) and sociology (India). Rajneesh has a background in high profile research and consulting projects, both in the public and private sectors. Amongst several prestigious accolades, Rajneesh was nominated to the British Ministry of Trade and Industry in 2005 for the award of Business Leader of Tomorrow for making outstanding contribution to healthcare information systems.

José-Rodrigo Córdoba-Pachón (www.hull.ac.uk/php/sbswdr) is a lecturer in management systems at the Business School, University of Hull (UK). His research interests are on the use of systems methodologies, ethics and power in information systems practice. Before coming to the UK, he worked for several years as a project manager in information technology. Córdoba-Pachón obtained his MA (with distinction) and PhD in Management at Hull, followed by an Economic and Social Research Council of the UK (ESRC) fellowship and other nominations from the Academy of Management and the British Academy. Since 2004, he researches on the development of e-government systems and has established the ELIS group (e-government and local integration with sustainability. He has organized several international workshops on information systems and critical research Currently, he is a member of the British Computer Society, the Operational Research Society, and the Academy of Management. He has published several papers in internationally recognised journals including the Journal of the Operational Research Society, Electronic Journal of Information Systems Evaluation, Journal of Information, Communication, and Ethics in Society and the Information Systems Journal (forthcoming 2009).

Marisa D'Mello has a doctorate from the Centre for Technology, Innovation and Culture (TIK), University of Oslo, Norway. Her doctoral project, completed in 2006, examined identities of Indian information technology (IT) workers and their relationship with globalization processes,mobilities, culture, gender and organizational practices.Marisa's postgraduate degrees are in psychology from India and the USA. She was a lecturer in psychology for several years in Mumbai before working in the IT industry in India as a Human Resources professional in 1993. Currently, she is an independent consultant and researcher with IT firms in Mumbai. She is also affiliated with the TIK Centre, University of Oslo, Norway.

Anneke Fitzgerald is the research studies program coordinator in the College of Business, University of Western Sydney, Australia. Since completing her PhD in 2002, Anneke's research interests include employing mixed methodology in organizational behavior and organizational psychology research. She recently received a College Excellence award for her research into professional identity in health care organizations. Fitzgerald has published her work in peer-reviewed journals, at academic and professional conferences and in edited books. Her current research focuses on interactions between professional and occupational groups.

Jasmine Folz, a part time anthropology instructor at Seattle Central Community College, holds a master's degree in social anthropology from Dalhousie University. The data for this chapter was taken from original research she conducted for her master's thesis "*Reengineering Work and Reworking Engineers: High-Tech Workers in the 'Post-Industrial' Context*," which, in turn, was inspired by her interest in studying up, emerging labour issues, and her experience as an on-site vendor manager at Microsoft in the late 1990s.

Pauline Gleadle is lecturer in management at the Open University in the UK. Her research interests include financialization, debates around enterprise, and the role of knowledge workers. She has been lead guest editor for a special issue of the *Journal of Organizational Change Management* on change, identity, and employment and is currently (lead) guest editing a special issue for "*Organization*" on enterprising selves. She has published in the British Journal of Management and in Critical Perspectives on Accounting.

Kate Hayes is a doctoral candidate at the University of Western Sydney. She holds an MBA and spent 18 years working for the IBM Corporation in Sydney, Australia, and in White Plains and Somers, New York. Her research interests include occupational subcultures, occupational identity, organizational culture, innovation and hybrid research organizations. Hayes is the recipient of a Commonwealth of Australia APA Award and holds the 2004 scholarship from the UWS College of Business. Her current research program deals with interactions between occupational subcultures in commercialization processes.

Vidar Hepsø is principal researcher and project manager at Statoil Research and Technology in Trondheim, Norway. Hepsø has a PhD in anthropology, but has worked with new ways of collaboration enabled by information and communication technology since he started to work for Statoil in 1991. Most of this work has its basis in action research and aims at helping oil industry professionals to improve collaboration within and across teams. These diverse professionals and teams range from the community of crane operators, operations and maintenance technicians, subsea engineers, production engineers and subsurface specialists. He also holds an adjunct professor position at the Norwegian School of Management in Oslo in the field of leadership and organizational management.

James Keenan is a professor of Communication at Fairfield University in Fairfield Connecticut, USA. His doctorate is from Columbia University. Keenan's teaching and research interests are broadly in organizational communication with particular focus on human communication, teamwork, and creative performances in high-technology environments. His theoretic-practitioner experiences have included designing and evaluating human-technology interfaces in spacecraft, aircraft, and information systems. Keenan is currently directing the Graduate Program in Organizational Communication at University College at Fairfield University.

Elisabeth K. Kelan is a research fellow in the Lehman Brothers Centre for Women in Business at London Business School. She completed her PhD at the London School of Economics and Political Science. Her main research interests are identities in organizations, technology at work, management education, performative gender theories, and qualitative methods, particularly discourse analysis and ethnography.

Sarosh Kuruvilla is currently professor of collective bargaining, comparative industrial relations, and Asian studies. He joined Cornell's faculty in 1990 after obtaining a doctorate in business administration from the University of Iowa in 1989, and after a career as a labor relations manager in the industry. Kuruvilla's research interests focus broadly in the area of comparative industrial relations and specifically on the linkages between industrial relations policies and practices, national human resource policies and practices and economic development policies.

Dominika Latusek is an assistant professor of management and organization theory at the Kozminski Business School (Poland). Before that, she was studying and conducting research in Poland, Germany and the U.S. She is a Junior Fulbright grant recipient for research at Stanford University. Her research is generally in the area of inter-organization relationships; in particular, her interests focus on dynamics of trust and distrust, as well as their role in cooperation within and between organizations. Her previous research includes field studies of culture and management practices in international corporations.

Marie-Josée Legault teaches labor relations and labor studies at Université du Québec à Montréal; her fields of research include highly qualified professional workers in the knowledge economy and their working conditions, project management as a model for the organization of work, its consequences and gender effects, and the theoretical consequences for the traditional labor relations models.

Alan Nobbs is currently working as Manager of the North and East Yorkshire and Northern Lincolnshire (NEYNL) Cardiac Network, a post held since July 2004, with responsibility for coordinating and delivering strategic service improvement work across 11 National Health Service (NHS) healthcare organizations in the UK. He has significant experience in managing the complexity of change across organizations. Alan Nobbs commenced his clinical career in the NHS in 1984. He joined NHS management in 2000 within the acute care sector, holding a post with organizational strategic responsibility for the delivery of the National Service Framework for Coronary Heart Disease.

Ben Passmore is a cultural anthropologist specializing in work, education, and organizational culture. He holds a PhD from Southern Methodist University in Dallas Texas. His current work focuses on labor markets, symbolic capital, and the creation of policy frameworks within bureaucracies. He is currently the Director of Policy Research at the University System of Maryland.

Erik Pineiro is an associate professor of industrial economics and management at the Royal Institute of Technology, Stockholm. He's currently a visiting scholar at the Scandinavian Consortium for Organizational Research at Stanford University. Pineiro's research focuses on the effects that the particular characteristics of software code have on the organization of software development. Recent publications include *Aesthetics at the heart of logic: On the role of beauty in computing innovation* In C. Gustafsson, P. Guillet de Monthoux & S. E. Sjöstrand (Eds.) (2007) "Aesthetic leadership: Managing fields of flow in art and business," Palgrave McMillan.

Agnieszka Postula received her master's degree from the School of Management, Warsaw University, Poland in 2000. In 2007, she received PhD ("IT professionals' role in organizations") at the same faculty. From 2000, she works at Management Systems Department as an assistant. Her main research interests lie in the area of organizational anthropology, communication and socio-professional roles in Polish organizations. Other interests and experience of Agnieszka Postula include team building trainings, group work, interpersonal communication, and negotiations which she uses in coaching. She teaches negotiation, international negotiation, organization and management theory, organizational anthropology and cross-cultural management at several faculties.

Aruna Ranganathan is a 2nd year PhD student at the School of Industrial and Labor Relations at Cornell University and has explored the outsourcing industry in India along with her advisor, Professor Sarosh Kuruvilla. She is interested in new work arrangements made possible by the emergence of global value chains.

Wendy Robson lectures in information systems at the Business School, University of Hull (UK). She has worked in the information systems field for many years, first working commercially on IS developments and strategies for a number of multi-nationals in the oil and oil service sector before moving to an academic career. Her interests are in interpreting, evaluating, and informing the strategy development process in information systems. Taking, in this work, an information systems perspective in which e-business is the contemporary manifestation of a concern with effective exploitation of information systems. Because the organizational implications of IS/e-business play out so significantly in the SME and public services sectors, Robson is often working in those contexts. She has been the fund-holder for a major ESF research project into the role of e-mentoring for effecting labour market improvements and is now involved with its successor (empathy-networks) concerned with e-mentoring systems in logistics and supply chain management. Wendy is currently a member of the British Computer Society and the Operational Research Society. She has published books, articles, and book chapters in the IS strategy and evaluation field.

Chris Russell is a senior lecturer in information systems at the Cardiff School of Management, part of the University of Wales Institute, Cardiff. Prior to returning to academia in 2003, he worked as an analyst, applications architect, and project manager in the telecommunications industry. He has degrees from the Universities of Oxford and Aston.

Marc Steen works at TNO ICT as a consultant and project leader for user research, product development and marketing strategy. Before that, he worked at Delft University of Technology, at Philips Consumer Electronics and at KPN Research. He studied Industrial Design and Marketing Management. He feels attracted to *human-centred design* and attempts to cooperate with so-called "end-users" during research and design. He is currently (2007) writing a PhD/DBA thesis on this topic (www.marcsteen.nl).

Marja-Liisa Trux (Lic.Psy.) has made her interdisciplinary career trough psychology, cultural anthropology, and organizational studies. She has worked among immigrants struggling with learning problems and initiated the field of inquiry into multiethnic workplaces in Finland, in the sectors of cleaning industry and high-tech industry. Her academic interests include cross-cultural contact zones, questions of power and agency, Finnishness in the present world, late modern ideologies such as diversity management, and possible dialogising moves. Trux prepares her doctoral dissertation on the meanings of ethnicity in a Finnish based high tech company for Helsinki School of Economics /Organizations and Management.

Index

I

K

L

M

N

O

P

R

S

T

U

V

W

Y